SELECTED PAPERS ON PRECALCULUS

THE
RAYMOND W. BRINK SELECTED MATHEMATICAL PAPERS

Published by
THE MATHEMATICAL ASSOCIATION OF AMERICA

―――

Committee on Publications
EDWIN F. BECKENBACH, Chairman

*The Raymond W. Brink
Selected Mathematical Papers*

VOLUME ONE

SELECTED PAPERS ON PRECALCULUS

Reprinted from the

AMERICAN MATHEMATICAL MONTHLY
(Volumes 1–81)

and from the

MATHEMATICS MAGAZINE
(Volumes 1–49)

Selected and arranged by an editorial committee consisting of

TOM M. APOSTOL, Chairman
California Institute of Technology

GULBANK D. CHAKERIAN
University of California, Davis

GERALDINE C. DARDEN
Hampton Institute

JOHN D. NEFF
Georgia Institute of Technology

Published and distributed by
THE MATHEMATICAL ASSOCIATION OF AMERICA

© 1977 by
The Mathematical Association of America (Incorporated)
Library of Congress Catalog Card Number 77-792-79

Complete Set   ISBN 0-88385-200-4
Vol. 1   ISBN 0-88385-202-0

Printed in the United States of America

Current printing (last digit):

10 9 8 7 6 5 4 3 2 1

FOREWORD

By a resolution unanimously and enthusiastically adopted at its meeting in Toronto on August 25, 1976, the Board of Governors of the Mathematical Association of America expressed its deepest appreciation to Mrs. Carol Ryrie Brink for a most generous gift to the Association in honor of her late husband, Professor Raymond W. Brink.

The Board was particularly pleased to note that Mrs. Brink made this gift to the Association at a time when our organization has become increasingly dependent on private sources for support. The Board therefore felt grateful to Mrs. Brink not only for her generosity but also for her wish that the gift will stimulate others to contribute to the projects of the Association.

The present gift has enabled the Association to establish a new regular series of publications to be called RAYMOND W. BRINK SELECTED MATHEMATICAL PAPERS. The series provides a fitting and lasting memorial to Professor Brink, who served the Association in many significant ways including terms as Governor (1934–39 and 1943–48), as Vice-President (1940–41), and as President (1941–42).

The Board pledged itself to use Mrs. Brink's gift to make the RAYMOND W. BRINK SELECTED MATHEMATICAL PAPERS an outstanding series of which Professor Brink would have been especially proud.

Articles for inclusion in the SELECTED PAPERS volumes are selected from past issues of the Association's journals.

SELECTED PAPERS IN PRECALCULUS, the first volume of the series, contains Professor Brink's *Retiring Presidential Address*. In this address, Professor Brink was concerned with the mathematical situation during a particular phase of our national development, but the philosophy he expressed applies to all periods.

The previously existing SELECTED PAPERS IN CALCULUS thus becomes the second numbered volume in the series. In mathematics, it has several times been the case that a theorem has been known for some time before an individual, not the discoverer of the theorem, points out its true significance. The theorem then flourishes and bears fine fruit. Just so, the RAYMOND W. BRINK SELECTED MATHEMATICAL PAPERS series, to which Mrs. Brink has called attention, will surely flourish and bear fruit.

EDWIN F. BECKENBACH
Chairman, Committee on Publications
Mathematical Association of America

PREFACE

This volume contains selected papers on precalculus reprinted from THE AMERICAN MATHEMATICAL MONTHLY and the MATHEMATICS MAGAZINE. The selection was made by an editorial committee appointed by E. F. Beckenbach during his tenure as Chairman of the Committee on Publications of the Mathematical Association of America. The charge to the editorial committee was to scan the back volumes of the MONTHLY and the MATHEMATICS MAGAZINE and to select articles that would be helpful to teachers and students in courses at the precalculus level.

The editorial committee read more than 1000 articles of which 180 were finally selected for reproduction. The papers have been arranged into ten categories, which are treated as chapters of the book. Some of the chapters are further divided into subcategories.

In each subcategory the selections from the MONTHLY appear first, followed by those from the MATHEMATICS MAGAZINE. Most subcategories are followed by a bibliography listing further related papers. Some of the bibliographic entries are accompanied by brief editorial comments describing in more detail the contents of the paper.

The editorial committee followed a selection procedure similar to that described in the Preface of the companion volume SELECTED PAPERS ON CALCULUS published by the Association in 1969. The selection process took a little more than a year to complete, and there was remarkably good agreement on all the papers that were finally selected.

The committee attempted to keep the volume to a reasonable size, while including a broad spectrum of ideas and maintaining some sort of balance among the various categories. Many of the papers contain material that can be used directly in the classroom. Others provide insights, background, or source material for special projects. Some papers discuss matters of controversy. The inclusion of any paper or bibliographic entry is not to be considered in any way as an endorsement by the editorial committee of the point of view expressed by its author.

It is the committee's hope that there will be much of interest in this volume, both for the beginning teacher and the more experienced veteran, as well as for students.

THE EDITORIAL COMMITTEE

CONTENTS

FOREWORD v

PREFACE vii

RETIRING PRESIDENTIAL ADDRESS: College Mathematics During Reconstruction   RAYMOND W. BRINK   1

1. PEDAGOGY

Marginal Notes	T. H. HILDEBRANDT	15
On the Teaching of Elementary Mathematics	M. RICHARDSON	20
Some Variations of the Multiple-Choice Question	H. F. S. JONAH AND M. W. KELLER	28
Geometry and Empirical Science	C. G. HEMPEL	34
A Theorem in Elementary Mathematics	A. D. WALLACE	44
Objectives in the Teaching of College Mathematics	F. S. NOWLAN	45
Mathematics and Intellectual Honesty	MOSES RICHARDSON	54
The Proof of the Law of Sines	ALFRED BRAUER	60
On the Definition of Functions	H. P. THIELMAN	61
A Kind of Problem That Effectively Tests Familiarity with Functional Relations	K. O. MAY	64
"If This Be Treason..."	R. P. BOAS, JR.	65
'Rye Whiskey' in Contrapositive	W. P. COOKE	67
The Professor's Song	TOM LEHRER	68
Club Topics	H. D. LARSEN (Editor)	69
Bibliographic Entries: Pedagogy		71

2. NUMBERS

(a) INTEGERS

Pascal's Triangle and Negative Exponents	L. V. ROBINSON	73

CONTENTS

On Some Mathematical Recreations	RICHARD BELLMAN	74
Problems about Problems	I. D. MACDONALD	76
Another Proof of the Infinite Primes Theorem	M. WUNDERLICH	80
More on the Infinite Primes Theorem	R. L. HEMMINGER	80
On the Geometry of Numbers in Elementary Number Theory	G. E. ANDREWS	81
Musimatics or the Nun's Fiddle	A. L. LEIGH SILVER	83
Galileo Sequences, a Good Dangling Problem	KENNETH O. MAY	89
Recurrent Sequences and Pascal's Triangle	THOMAS M. GREEN	91
Properties of a Game Based on Euclid's Algorithm	EDWARD L. SPITZNAGEL, JR.	100
Bibliographic Entries: Integers		105

(b) IRRATIONAL NUMBERS

The Limit for e	RICHARD LYON AND MORGAN WARD	106
Another Proof of an Estimate for e	R. F. JOHNSONBAUGH	107
The Irrationality of $\sqrt{2}$	EDWIN HALFAR	108
The Irrationality of $\sqrt{2}$	ROBERT GAUNTT	109
On the Irrationality of Certain Trigonometric Numbers	E. A. MAIER	109
A Simple Irrationality Proof for Quadratic Surds	M. V. SUBBARAO	110
Irrational Numbers	J. P. JONES AND S. TOPOROWSKI	110
A Method of Establishing Certain Irrationalities	I. NIVEN AND E. A. MAIER	111
Bibliographic Entries: Irrational Numbers		114

(c) COMPLEX NUMBERS

A Note on Geometrical Applications of Complex Numbers	S. A. SCHELKUNOFF	115
Complex Numbers and Vector Algebra	D. E. RICHMOND	117
Introduction of Complex Numbers as Vectors in the Plane	L. FUCHS AND T. SZELE	123
An Isometry of the Plane	EDWARD D. GAUGHAN	127
The Proof of the Triangle Inequality	M. F. SMILEY	128
Bibliographic Entries: Complex Numbers		129

(d) INFINITY

Infinite and Imaginary Elements in Algebra and Geometry	R. M. WINGER	130
Infinite and Imaginary Elements in Algebra and Geometry	TOMLINSON FORT	138

The Infinite and Imaginary in Algebra and Geometry: A Reply
                                                W. L. G. WILLIAMS   139

Infinite and Imaginary Elements in Algebra and Geometry: Reply to Criticisms   R. M. WINGER   146

3. INDUCTION, IDENTITIES, AND INEQUALITIES

(a) INDUCTION

On Mathematical Induction	J. W. A. YOUNG	151
On Proofs by Mathematical Induction	E. T. BELL	159
Discussions	W. A. HURWITZ	161
A Paradox Relating to Mathematical Induction	R. G. ALBERT	164
Mathematical Induction and Recursive Definitions	R. C. BUCK	165
A Remark on Mathematical Induction	V. L. KLEE, JR.	173
On the Danger of Induction	LEO MOSER	173
On the Number of Subsets of a Finite Set	DAVID S. GREENSTEIN	174
Bibliographic Entries: Induction		174

(b) IDENTITIES

Some Trigonometric Products	R. C. MULLIN	175
A Note on Sums of Powers of Integers	DAVID ALLISON	176
A Combinatorial Proof that $\Sigma k^3 = (\Sigma k)^2$	ROBERT G. STEIN	177
Bibliographic Entry: Identities		178

(c) INEQUALITIES

Arithmetic, Geometric Inequality	D. J. NEWMAN	178
A Proof of the Arithmetic-Geometric Mean Inequality	BENGT ÅKERBERG	179
On Two Famous Inequalities	L. H. LANGE	180
Bibliographic Entries: Inequalities		183

4. TRIGONOMETRY AND TRIGONOMETRIC FUNCTIONS

On the Representation of the Trigonometric Functions by Lines
                                                R. D. CARMICHAEL   184

A Curious Case of the Use of Mathematical Induction in Geometry   J. V. USPENSKY   185

The Trigonometric Functions of Half or Double an Angle	Roscoe Woods	189
Geometric Proofs of Multiple Angle Formulas	Wayne Dancer	190
The Addition Formulas for the sine and cosine	E. J. McShane	192
The Addition Formulas in Trigonometry	A. S. Householder	194
Values of the Trigonometric Ratios of $\pi/8$ and $\pi/12$	H. L. Dorwart	195
A Substitution for Solving Trigonometric Equations	R. W. Wagner	196
Derivation of the Tangent Half-Angle Formula	F. E. Wood	198
The Laws of sines and cosines	L. J. Burton	198
Angles with Rational Tangents	T. S. Chu	199
The General sine and cosine Curves	E. L. Eagle	200
A Derivation of the Formulas for $\sin(\alpha+\beta)$ and $\cos(\alpha+\beta)$	A. K. Bettinger	201
Terse Trigonometry	C. M. Fulton	203
Elementary Proofs for the Equivalence of Fermat's Principle and Snell's Law	Michael Golomb	204
A Geometric Proof of the Equivalence of Fermat's Principle and Snell's Law	Daniel Pedoe	206
$\sin(A+B)$	F. H. Young	208
Concerning $\sin(A+B)$ (Letters from G. Matthews and F. H. Young)	H. V. Craig, (Editor)	209
A Simple Proof of the Formula for $\sin(A+B)$	Norman Schaumberger	210
Matrices in Teaching Trigonometry	A. R. Amir-Moéz	211
The Law of sines and Law of cosines for Polygons	R. B. Kershner	214
Bibliographic Entries: Trigonometry and Trigonometric Functions		218

5. ELEMENTARY ALGEBRA

(a) POLYNOMIALS

On the Irreducibility of Certain Polynomials	J. Westlund	219
Construction of an Algebraic Equation with an Irrational Root Approximately Equal to a Given Value	L. S. Dederick	220
On Nonnegative Polynomials	Louis Brickman and Leon Steinberg	222
Bibliographic Entries: Polynomials		225

(b) VECTOR ALGEBRA

Fundamental Identity of Vector Algebra	C. J. COE AND G. Y. RAINICH	226
Coordinate Geometry from the Vector Point of View	SAMUEL BOURNE	227
A Vector Proof of Euler's Theorem on Rotations of E^3	M. K. FORT, JR.	231
On the Vector Triple Cross Product Identity	DANIEL T. DWYER	232
Vector Proofs in Solid Geometry	M. S. KLAMKIN	233

(c) CONSTRUCTIONS WITH STRAIGHTEDGE AND COMPASS

On the Trisection of an Angle and the Construction of Regular Polygons of 7 and 9 Sides	L. E. DICKSON	247
On Who First Proved the Impossibility of Constructing Certain Regular Polygons with Ruler and Compass Alone	N. D. KAZARINOFF	250
Bibliographic Entries: Constructions with Straightedge and Compass		251

(d) MISCELLANEOUS

Rationalizing Factors and the Method of Undetermined Coefficients	L. J. PARADISO	251
A Note on Partial Fractions	R. A. GARVER	253
A Note on Partial Fractions	L. S. JOHNSTON	255
A Note on Joint Variation	R. A. ROSENBAUM	256
New Proof of a Classic Combinatorial Theorem	S. W. GOLOMB	257
Another Generalization of the Birthday Problem	J. E. NYMANN	258
Bibliographic Entries: Miscellaneous		260

6. SOLUTIONS OF EQUATIONS

Relating to Solutions of Quadratic Equations	G. R. DEAN	261
Note on the Algebraic Solution of the Cubic	E. J. OGLESBY	263
A Graphical Method of Solving Simultaneous Linear Equations	J. P. BALLANTINE	265
Note on the Solution of a Set of Linear Equations	J. P. BALLANTINE	266
A Note on the Roots of a Cubic	E. C. KENNEDY	267
The Graphical Interpretation of the Complex Roots of Cubic Equations	G. HENRIQUEZ	269
Complex Roots of a Polynomial Equation	H. M. GEHMAN	270

Cows and cosines	L. R. Ford	273
A Rule for Computing the Inverse of a Matrix	A. A. Albert	275
A Modern Trick	Claire Adler	276
On an Elementary Derivation of Cramer's Rule	D. E. Whitford and M. S. Klamkin	277
A Unifying Technique for the Solution of the Quadratic, Cubic, and Quartic	Morton J. Hellman	278
The Insolvability of the Quintic Re-examined	Morton J. Hellman	281
Bibliographic Entries: Solutions of Equations		282

7. SYNTHETIC GEOMETRY

(a) TRIANGLES

Historical Note	F. Cajori	283
A Modification of a Proof by Steiner	Otto Dunkel	285
A Simple Proof of the Theorem of Morley	Jacob O. Engelhardt	289
Generalization, Specialization, Analogy	George Pólya	290
Angle Bisectors of an Isosceles Triangle	W. E. Bleick	293
The Steiner-Lehmus Theorem	G. Gilbert and D. MacDonnell	294
On (What Should Be) a Well-known Theorem in Geometry	Daniel Pedoe	295
Morley's Triangle Theorem	R. J. Webster	297
Bibliographic Entries: Triangles		298

(b) OTHER CONFIGURATIONS

A Note on Knots	F. V. Morley	299
On the Division of a Circumference into Five Equal Parts	H. C. Bradley	302
On Two Intersecting Spheres	N. A. Court	303
Two Isoperimetric Problems	M. S. Knebelman	307
A Generalization of the Median Theorem for Triangles	Roger Burr Kirchner	311
Squaring Rectangles and Squares	N. D. Kazarinoff and Roger Weitzenkamp	313
Packing Cylinders into Cylindrical Containers	Sidney Kravitz	325
Packing Equal Circles in a Square	Michael Goldberg	331

A Geometric Application of $f(n)=n/(n+1)$	Francine Abeles	339
Bibliographic Entries: Other Configurations		341

(c) MISCELLANEOUS

How to Trisect an Angle with a Carpenter's Square	Henry T. Scudder	342
Hilbert's Axioms of Plane Order	C. R. Wylie, Jr.	343
After the Deluge	D. A. Moran	349
Geometric Interpretations of Some Classical Inequalities	Joseph L. Ercolano	350
Two Mathematical Papers Without Words	Rufus Isaacs	352
Bibliographic Entries: Miscellaneous		353

8. CONIC SECTIONS

(a) EQUATIONS

A Method of Defining the Ellipse, Hyperbola and Parabola as Conic Sections	W. W. Landis	354
A Biquadratic Equation Connected with the Reduction of a Quadratic Locus	A. C. Lunn	356
An Elementary Analysis of the General Equation of Second Degree	E. S. Allen	357
Simplification of the Equations of Conics	H. B. Thornton	360
Conic Sections from whose Equations the xy-term May Be Eliminated by a Rotation of Axes Involving No Surd Numbers	D. C. Duncan	361
Simplification of Equations of Conics	L. S. Johnston	362
The Reduced Equation of the General Conic	A. E. Johns	364
Simplification of Equations of Conics	M. T. Bird	368
Derivation of the Equation of Conics	F. Hawthorne	370
On Integral Coordinates	Norman Anning	371
Bibliographic Entries: Equations of Conic Sections		371

(b) CLASSIFICATION AND CONSTRUCTION

Simple Constructions for the Conics	L. S. Johnston	372
Classification of the Conics	R. C. Yates	373
Folding the Conics	R. C. Yates	375

The Ellipse as a Circle with a Moving Center	F. H. Young	378
On Defining Conic Sections	G. B. Huff	380
On Defining the Conic Sections	Michael Pascual	381
A Historical Note on a Problem in this Monthly	Joy B. Easton	382
Bibliographic Entries: Classification and Construction of Conic Sections		385

(c) TANGENTS

On the Chord of Contact of Tangents to a Conic	W. D. Lambert	386
Tangent to a Circle from an Exterior Point	F. H. Young	387
Tangent Lines and Planes	F. H. Young and J. L. Ericksen	388
One Side Tangents	William R. Ransom	389

(d) APPLICATIONS

Mathematical Forms of Certain Eroded Mountain Sides	T. M. Putnam	391
The "Reflection Property" of the Conics	R. T. Coffman and C. S. Ogilvy	393
A Study of Conic Section Orbits by Elementary Mathematics	Raphael Coffman	395
Bibliographic Entry: Applications		404

9. ANALYTIC GEOMETRY

(a) LINES AND PLANES

A General Method of Deducing the Equation of a Tangent to a Curve	G. W. Greenwood	405
Derivation of the Normal Form of the Equation of the Straight Line	Kenneth May	408
On the Equation of a Line	Carol S. Scott	408
Distance from a Line, or Plane, to a Point	J. P. Ballantine and A. R. Jerbert	409
Determinants and Plane Equations	R. R. Stoll	410
On the Slopes of Perpendicular Lines	S. Leader	412
The Straight Line Treated by Translation and Rotation	Kenneth May	412
Point to Line Distance	William R. Ransom	413
The Distance Formula and Conventions for Sign	Thomas E. Mott	414
An Easy Way from a Point to a Line	R. L. Eisenman	417
Bibliographic Entries: Lines and Planes		418

(b) CURVES AND SURFACES

The Hyperboloid as a Ruled Surface	J. P. Wilson	418
Simple Hints on Plotting Graphs in Analytic Geometry	Aubrey Kempner	420
A New Curve Connected with Two Classic Problems	G. M. Juredini	424
An Unusual Spiral	L. S. Johnston	426
Plotting Curves in Polar Coordinates	P. C. Hammer	428
Equations and Loci in Polar Coordinates	R. W. Wagner	429
A Lesson in Graphing	F. Max Stein	432
The Equation of a Sphere	Murray S. Klamkin	435
Bibliographic Entries: Curves and Surfaces		437

10. AREA AND VOLUME

The Volume of the Sphere	H. L. Coar	438
The Expression of the Areas of Polygons in Determinant Form	R. P. Baker	441
Note on the Volume of a Tetrahedron in Terms of the Coordinates of the Vertices	L. E. Dickson	442
Some Limit Proofs in Solid Geometry	W. R. Longley	444
The Analytic Determination of the Area of a Triangle in Terms of its Sides	K. P. Williams	450
A Historically Interesting Formula for the Area of a Quadrilateral	J. L. Coolidge	452
Solid Angles	J. W. Cell	454
The Area of a Triangle as a Function of the Sides	Victor Thébault	456
Formula for the Area of a Triangle	M. K. Fort, Jr.	458
The Method of Archimedes	S. H. Gould	460
A Simpler Proof of Heron's Formula	Claude H. Raifaizen	464
Bibliographic Entry: Area and Volume		465

AUTHOR INDEX   467

COLLEGE MATHEMATICS DURING RECONSTRUCTION*†

RAYMOND W. BRINK, University of Minnesota

1. Introduction. If we pause for a little while amid the turmoil of the war-training courses in order to look forward to the period just after the war, it is not because we feel that the war is over nor even that its end is hidden from us by being just around the corner. The war-training programs will doubtless be with us for a long time to come. Doubtless they will be subject to more of the sudden and arbitrary changes that we have all found so baffling. Doubtless we shall continue to be faced with shortages of staff, irregular schedules, budgetary difficulties, and elusive academic standards, as in the recent past. Yet I feel that the general pattern of training in the colleges is fairly well established or at least that we are psychologically better prepared than we were twelve months ago to meet such wartime changes as may occur. On this plateau, then, rather than summit, that we have reached I should like to pause to take stock of our gains and losses, and to look forward toward the land ahead. Perhaps it is not too early to see something of this land of college mathematics after the war is won and of the problems we shall find there. Perhaps, by looking back at the way we have come, we can find help in solving those problems.

2. College enrollments after the war. In order to form some notion of the size of the task ahead of us, let us first consider the probable general university and collegiate enrollment, without special regard to its effect on mathematics. We shall almost certainly have a flood of former students returning to the colleges from their service with the armed forces, governmental agencies, and defense plants. To these will be added the recent high school graduates of perhaps several years, who will be entering college for the first time.

Another force, also tending in the direction of increased enrollments, and, to my mind, at least equally important for the long haul, is the probable broader base of education that will result from the war. American education, in constrast to that in most European countries, has long been characterized by the theory of the broad base. Not for us has been the system of higher education only for the intellectually or economically elite. Of course, this has resulted in the attempted education of many young persons never intended by their Maker to breathe the rare atmosphere of the intellectual life. Yet if, as educators would probably be the first to assert, education is a chief bulwark of de-

---

* From AMERICAN MATHEMATICAL MONTHLY, vol. 51 (1944), pp. 61–74.
† Retiring presidential address delivered to the Mathematical Association of America, Chicago, Ill., November 27, 1943.

mocracy, it would seem to be a sound theory which both permits and encourages each citizen to carry his education as far as his native gifts permit, regardless of his economic status and arbitrary academic standards. In the experiences that our young men are undergoing in the training camps and on the battlefields of the world, I can see a great levelling force at work. It will level upward, I hope, as well as downward. After observing that some eighty per cent of their officers have had some college background, after having taken part in Army Institute and other courses preparing them for their immediate jobs or more remotely for their places as citizens, after having worked and fought on even terms with comrades who expect to return to college, is it not likely that many thousands of our young soldiers will demand for themselves the privileges of further education?

When they knock at our college doors, I hope that we shall apply some intelligent guidance, not so much in the form of outright rejection as in advice as to the most suitable types of training. But somewhere in our educational system we shall accept them without too rigid an application of scholastic requirements. For one thing, during our recent experiences, we have learned not to be too choosey about our raw material. Many colleges that previously accepted only students in the upper twenty per cent of their high school classes have been delighted to take men selected for them on the doubtful basis of an Army General Classification Test. Good men are still good men, and stand out as such. But we have found that with the aid of proper refresher courses even men of inferior preparation can be given useful training at a level beyond what might at first be expected from their previous experience. I shall revert to this point later on.

Most of all we will give an ungrudging welcome to these young men and women because we think that it will be advantageous to the nation to do so. Much as we may fear future wars and foreign aggression, our greatest dangers after the war will come from within. They will come from the disunity of one special interest opposing others. To dispel such dissensions we must teach as large a part of your young population as possible to distinguish information from dishonest propaganda, to seek logical instead of emotional bases of action, to understand the essentially cooperative nature of society, to respect he other man's manner of thought, and to earn a livelihood in socially useful work.

Not all of the tendencies will be in the direction of increased enrollments. Many of the returning men will be impatient to secure permanent positions. Many will have new opportunities open to them as a result of their war training and experience. Others, who married on the strength of wartime salaries and find that they married on a shoestring, will be forced to seek work to support their families.

But, on the whole, the signs point to enrollments far exceeding anything that the colleges have known before. A committee at one of our large middle-western state universities* has estimated that during the first year after the

---

* Report of the Senate Committee on Education, University of Minnesota, Senate Minutes, October 21, 1943.

war its enrollment will be from eleven per cent to seventeen per cent, depending on the war's duration, greater than it ever was before, and that during the second year after the war it will be from twenty-eight per cent to thirty-four per cent greater. By 1950–1951, the committee predicts an enrollment forty per cent greater than the maximum hitherto. These estimates do not take into account the stimulus of any possible plans for federal subsidization of students after the war. If these plans develop, it is likely that the forty per cent increase may occur within a year or two after the war's end.

With reference to the share that returning service men will occupy in our load, Colonel John N. Andrews of the Reemployment Division of Selective Service Headquarters states that the general indication now is that of the eleven to twelve million men who will have been in service by the time the war ends, at least one million will wish to return to vocational schools or colleges and universities to pursue further study. Many of these entered the service directly from high schools. Large numbers will have completed from one to three years of college work, and will wish to continue in their original courses or in other directions. And the indications are that those who desire to return to school will be given financial support to enable them to do so.

3. Plans for federal subsidization of education. Without yet knowing the exact form that a scheme of federal subsidies will assume, we can take it as probable that some scheme will develop. The most definite suggestions from any authoritative source are contained in the recommendations of the Conference on Postwar Readjustment of Civilian and Military Personnel under the auspices of the National Resources Planning Board.* The general principles of these recommendations were contained in the report of a special committee on education appointed by the President of the United States. On October 27, 1943, the President embodied these principles in a message† to Congress with a request for the necessary legislation. In brief outline, the recommendations for subsidy of education call for two plans. Under the first plan of general education free tuition and reasonable allowances for maintenance would be provided for a period not exceeding one year to any honorably discharged ex-service man who might apply for it. Both general and vocational or professional education and training would be provided, but the latter would not be provided in those fields and for those occupations in which the supply of trained personnel is already sufficient to meet anticipated requirements. Training and education would begin at the level appropriate to the individual and should have such forms and methods as are suited to the needs of mature men regardless of the academic level. (Recommendations 50–58.) In addition to this general plan of education the Conference recommended a plan of supplementary education. This plan recognizes the special responsibility of the Nation to those who had

---

* Demobilization and Readjustment—Report of the Conference on Postwar Readjustment of Civilian and Military Personnel, National Resources Planning Board, June, 1943, Superintendent of Documents, Washington, D. C.

† Congressional Record, October 27, 1943, p. 8881.

entered upon an extended course of education which was interrupted by their military service and it also recognizes the need of the Nation for specially trained personnel. Under the program of supplementary education scholarships not to exceed four years in duration would be granted on a competitive basis in the fields of higher education and in such technical and professional fields as offer some likelihood of satisfactory and useful employment. (Recommendations 59–66.)

We cannot be certain that such a program of federal subsidization will be adopted. It seems very probable, however, that some very similar plan will take effect. For the purpose of discussion, I should like to assume that this is the case. And I should like to consider some aspects of the situation in college mathematics with such a program imposed on our other tasks.

4. The load in mathematics. What the Conference called the "general" plan of education provides for one year of either general education or training for vocations or professions in which there is a predictable opportunity for employment. Even before a comprehensive survey of the needs in the various trades and professions has been made, we can make some prediction of the impact of this program on the mathematical load. Certainly, if we recall the importance that has been attached by the armed forces and by industry to elementary training in mathematics for their mechanics and foremen, we cannot doubt that there will be a great deal to do in teaching what we think of as "shop mathematics." I hope that most of this teaching can be done in trade schools, some special junior colleges, and in special extension courses of the universities or in industry itself. For I am sure that collegiate staffs will be fully occupied at more nearly the collegiate level.

But what of the young men who in ordinary times would have gone directly from high school into jobs requiring no further academic training? Many thousands of them who will have received training and seen service as electricians' mates or airplane mechanics or radio technicians will be stirred to seek advancement in similar lines. Many of them will wish to spend the time of their subsidized training as students of engineering or science. To the extent to which they can meet or nearly meet the entrance requirements of our colleges, they can be absorbed into the regular engineering and scientific curricula. Since mathematics is a subject that cannot be bypassed nor easily replaced by experience in the field, mathematics staffs will be under special stress to carry this additional burden. Indeed it is not unlikely that the load of teaching elementary college mathematics will exceed our present one, including the present Army and Navy programs.

5. Advanced mathematics in engineering and scientific courses. I foresee also a considerable expansion at the upper end of the engineering and scientific curricula. Many young men will have completed the somewhat abbreviated or at least hurried courses called advanced engineering in the specialized service programs. These young men as well as many recent engineering graduates on leaving the army will see in the federal educational program an opportunity to continue their studies from an apprentice to a professional level. It would probably be idle here to try to fix the point at which preprofessional work in engineering ends and professional training begins. It is certainly true that in such professions as law, medicine, and the ministry, formal schooling does not end after four years beyond the high school. Perhaps we are willing to entrust our rights, our health, and our souls, to the schools but not our material possessions. No more than we should expect every physician to be using basic science in research should we expect every professional engineer to employ mathematics at the research level. Yet, because of the essentially quantitative character of engineering work, we might not be too far off the mark in accepting the skill with which an engineer uses applied mathematics as one measure of his professional attainment. It has been repeatedly pointed out of late, notably by Thornton C. Fry in his report on mathematics in industry,* and by R. G. D. Richardson in this MONTHLY,† that the study of applied mathematics in America has not progressed as it has in certain foreign countries nor as has the study of pure mathematics here. It is perhaps a natural consequence of this that in many of our four- and five-year engineering schools the curricula are less mathematical and to an extent less professional in character than they might well be. Not in all but in many of them it is the tradition to consider mathematics as a tool subject taught on a relatively formal and mechanical basis. In such cases the special power of applied mathematics is largely lost. Perhaps the ideal in engineering training would be to combine the power of generalization, which implies the discernment of the essential in a problem, and which is so characteristic of modern American mathematics, with an interest in and familiarity with the materials of some field of technological application. I am inclined to believe that such considerations will ultimately lead away from the formal toward the more theoretical, that is, the ultimately more practical type of mathematics course in engineering colleges. Immediately after the war this tendency may not exist and may even be reversed for the courses in the first two years of the curriculum. This reversal will probably result from the lack of qualified teachers, the weak preparation of the mass of entering students, and from the desire of men already advanced in years and maturity of experience to pursue their training in as rapid and concentrated a form as possible.

But the war will itself have produced such advances in engineering practice that there will be a great demand for men of really superior training for the mere manufacture, installation and operation, let alone the design, of equipment.

Some days ago I talked with a young man, a recent first-year graduate student of mathematics, who is employed in one of the large government laboratories. The work he is doing is not concerned with the discovery of basic

---

* Research—A National Resource—II, VI, 4, pp. 268–288—A House Document, 77th Congress.

† Applied Mathematics and the Present Crisis. Vol. 50, No. 7, Aug.-Sept., 1943, pp. 415–423.

scientific facts, but is exceedingly practical in character and involves the application of familiar principles to strictly engineering problems. He was able to solve some of these problems by means of somewhat complicated differential equations, the application of potential theory, contour integration, and the use of Fourier transforms. He feels that, though he has had no engineering training, he is at an advantage as compared with others in the laboratory because he finds it natural to formulate a problem in mathematical language and to solve it mathematically. Another first-year graduate student of my acquaintance was employed by a company that manufactures intricate apparatus now used in the war. In this case approximations by means of orthogonal polynomials furnished the key to important improvements in equipment. These are merely two examples that many of you could doubtless duplicate many times. The point that I wish to emphasize is not the applicability of mathematics to engineering, which needs no emphasis. It is rather the fact that these were first-year graduate students and the techniques that they employed would be accessible to many young engineers with only a moderate increase in their mathematical background.

From these considerations I should expect a heavy increase in registration in our courses of intermediate and moderately advanced mathematics in engineering and scientific schools. I also expect a strong demand for additional or more extensive courses at this level, such as more courses in statistics with special emphasis on manufacturing and quality control, potential theory, advanced differential equations, theory of elasticity, group theory and fluid dynamics. How far and how promptly our depleted and overburdened staffs will be able to meet this demand is another story.

I should like to quote from a letter that I have just recently received from Colonel Andrews of the Reemployment Division of Selective Service, whom I have already mentioned. It seems to indicate that my own opinions are shared by those studying the problem from the point of view of employment. He writes: "Because of the unusual technical advances which are certain to occur following the war, all kinds of mathematicians, scientists, and research specialists will be in great demand. Such opportunities are already being discussed in the large industries and, eventually, the men in the services will be informed of the need for such specialists. It is believed that thousands of the young men will seek further training in these technical fields as soon as they are no longer needed in the armed forces. This program will call for a tremendous mathematics and science emphasis at the freshman and sophomore levels, especially, and also a continuation of these subjects in the junior and senior years. Perhaps the graduate work in these subjects will have the greatest challenge of their history. In order to meet these expanding demands, colleges and universities which have the proper staffs and laboratories should begin now the development of curricula to meet the needs."

6. Need of interdepartmental collaboration; liaison courses. What I have said applies equally well to the training of engineers and of other scientists. In both cases there are certain lessons to be learned from the military courses that we have taught. Some of these have required the closest collaboration between mathematicians and physicists. The pre-meteorological courses in mathematics, vectorial mechanics, and physics well illustrate the advantage of such collaboration in the organization of material. Especially in their earlier parts, a careful ordering of topics supplied in one course the materials necessary in another in time for their utilization there, and avoided unnecessary duplications of material. I hope that with the return of peace these lessons of cooperation will not be forgotten. In our standard curricula there is usually sufficient time to allow a course in one subject to follow logically certain prerequisite courses in other fields. Thus a man-sized course in physics ordinarily follows at least a year of college mathematics. But probably we shall have many mature students seeking rapid advancement in rather narrow scientific or technical training. It may then be necessary and, in their cases, desirable to teach them their mathematics and physics concurrently instead of in sequence. Our experience with the war training courses has shown that in case of necessity this can be done with considerable success. Such procedures depend, however, on the closest and most sympathetic interdepartmental collaboration. I shall refer later to the other necessity of planning for courses that are less narrow and concentrated in design.

With the increased interest in applied mathematics after the war I expect a tendency toward what I may call "liaison" courses. By these I mean courses that do not lie wholly in one conventional field, but furnish contact between two or more departments. The pre-meteorological course in Vectors and Mechanics is an example. It formed a most useful link between mathematics and its applications. It was most interesting also in another respect. It showed the feasibility of introducing relatively advanced material early in the college course. In the "C" program it successfully presented to freshmen concurrently with the first course in calculus and after only twelve weeks of college mathematics, material of vector analysis that had usually been reserved for juniors and seniors. The early presentation of this material undoubtedly greatly strengthened the calculus course by giving greater reality to such notions as curvilinear velocity and acceleration. It may well be that other entire courses or special topics require less maturity than we have hitherto assumed and that they can be introduced advantageously earlier in the curriculum than we supposed, especially for students seeking rapid advancement in a special field.

7. Statistics. Another interesting type of liaison course is one in elementary statistics. An immense amount of statistical work will be required as a result of the war. To realize this one has to think only of such sample problems as the rehabilitation of veterans, job surveys, reconversion of industry, quality control in the manufacture of new products, distribution of food and other commodities to our own and foreign populations, taxation and debt refunding, social security, and the redistribution of populations here and abroad. And I believe that mathematics departments should give careful study to their responsibilities in training for this work.

We have probably all considered this problem; some but not all of us have reached successful solutions. I know of one university where beginning courses in statistics are taught in departments of sociology, psychology, educational psychology, economics, preventive medicine, agriculture, mathematics and probably some others. It is as if, in a college of engineering, the basic mathematics were taught in separate departments such as physics, chemistry, electrical engineering, and mechanical engineering since each mistrusted the applicability of general mathematics to its own field. Yet, in statistics as elsewhere, it is the special power of mathematics to recognize the essence of a problem regardless of its immediate origin, to formulate quantitative definitions that really correspond to intuitive notions, to supply a tractable notation, and to analyze the relations existing between the elements. The inauguration of liaison courses in statistics involves not only the presentation of the theory common to all fields but also such interdepartmental cooperation that the results can really be utilized in the different departments. The problem is difficult, as we all know, but is likely to grow in importance. In relatively few institutions, I believe, has it been solved in a satisfactory way.

Not the least of the difficulties in connection with these liaison courses is that of securing instructors who have the ideal training already suggested for engineers, a thorough knowledge of fundamental mathematics combined with an interest in and real familiarity with some field of statistical or other application. In our graduate schools we should probably encourage or even require our students to become familiar with at least one field of applied mathematics as well as with their own line of specialization.

8. The need for general rather than limited courses. Before passing on to other matters I should like to make an observation suggested by these remarks on general or fundamental courses of statistics in contrast to courses limited to the applications in special fields. It has been remarkable and gratifying that in their training programs, the Army and Navy have sought mathematics. Not army mathematics nor navy mathematics, but fundamental mathematics. In their outlines of courses and in their lists of suggested materials of instruction, the services have very evidently appreciated that a general understanding of principles can more easily be applied to special situations than techniques learned only in connection with special problems can be transferred to unfamiliar areas. Of course, this does not mean that our teaching should deal only with general theorems and principles unrelated to any special problems that are familiar or exciting to the student. On the contrary, our courses should be as rich in varied and timely applications as we can possibly make them. Such applications provide the keenest motivation and the most satisfying means of clarification. Some of the books published during the emergency have made a genuine contribution just because they supplied a wealth of timely illustration. There have been relatively few, and most of them have been short-lived, of the meretricious publications designed to catch the reader with the bait of being, as they put it, "completely streamlined for the emergency," but written without regard to general principles or to the soundness that is the very strength of mathematics. Tendencies in this direction appeared early in the emergency, but received little encouragement from the armed services themselves. The most mathematically minded of navigators is not inclined to solve long problems in spherical trigonometry while navigating an airplane. But the Army and Navy showed great appreciation of the value of a general mathematical background even for men in training for very specific tasks. They were glad to enlist the aid of the colleges in securing this background for their trainees, and were willing themselves to provide the special applications required in service. I believe that this is the case even in the Pre-Flight course. It is true that the content of this course is extremely elementary. That is a question of the level of instruction and the quality of the men selected, not of generality. In fact it is shockingly disturbing to find that at the college age it is necessary for the men to receive this type of instruction, which is so generally necessary that all should have received it as children.

How fortunate it is that the Army did not build its curricula too narrowly around a single limited objective, but preferred to supply fundamental training in the basic sciences is well illustrated in the recent developments in the pre-meteorological programs. What a tragic waste it would have been had all the mathematical and physical training been centered in a tight circle about immediate meteorological applications! As it was, once the rude shock of the suspension of the "A" course had been absorbed, the morale of the men came back again, for we could assure them in all good faith that their effort was not wasted and that what they had learned could be transferred with no loss whatever to many other interesting and useful tasks.

These thoughts also have their lessons for the period after the war. If the Army can go wrong in predicting the requirements for meteorological officers (and few of us would quarrel with it for overestimating rather than underestimating them), is it likely that errors will not be made in surveys of the requirements in various trades and professions? And can any counselling service be so perfect that it can foretell with certainty the best outlets for all of the individual men? To a great extent we must resist the pressure to make our courses too limited and specific. We must so imbue them with the content and spirit of real mathematics that their results can be transferred to any field of application or of thought.

9. The necessary and the unnecessary in mathematics. I should like to make another remark closely related to this one. Some of the persons engaged in planning for the education of returning veterans are insistent, and, in a measure, rightly so, that this education be practical and concentrated in form. They point out that a man in his twenties will be unable to afford a long time for continuing his studies and that he will demand a maximum of content in a minimum of time. I agree with such statements and I believe that for many of these returning soldiers we shall have to give highly concentrated courses, as I have said before. But I cannot follow these gentlemen quite so far as they would have us go. One of them, in referring to some of the wartime mathematics courses says, "Mathematics has had much of the unnecessary subject matter 'squeezed out,' and it is very likely that this condition in the future will maintain." If by this is meant that in the future we should scrutinize our subject matter to see that it contributes to some really useful purpose, I agree. But I reserve the right to judge what is and what is not "unnecessary." And I should not consider subject matter unnecessary because it contributes to logical understanding and independence of thought instead of merely to formal technique. Nor should I consider a course too dilute merely because it covers the lapse of time necessary for some contemplation. By too much "squeezing" we may lose the life-giving juices and obtain indeed a desiccated product.

10. Placement problems. When our soldiers return, there will be great difficulty in placing them at suitable levels in our educational program. Our colleges should take steps at once to establish counselling systems adequate for the immense responsibility. It is to be hoped that these systems will not be entrusted to narrow groups of experts without a broad knowledge of the possibilities in the various fields, and that the basic sciences will be duly represented. Obviously, because of their maturity, the returning soldiers cannot be sent to high schools. But many of them will not have completed the prerequisites for the college courses they will wish to follow. We must share the responsibility of fitting these men into the educational scheme. In some wartime programs we have experienced the possibility of giving courses at elementary levels to prepare men for their collegiate work. Distasteful as it may be in some respects, I foresee the necessity of giving such courses for the returning veterans. I am not entirely sure that we should be wrong in making such courses available in peacetime to civilians. In foreign languages and in the natural sciences it is possible for college students to obtain beginning courses. In such courses it is possible for the students to proceed more rapidly than in high school and soon to press on the heels of those entering with one or two years of high school preparation. The chief danger in a similar procedure in mathematics would be in the fact that some high school advisers and administrators, knowing that these courses are available in college, would encourage pupils to omit mathematics from their high school programs. They would overlook the highly sequential nature of mathematics and not realize the necessity of beginning that subject as early in one's experience as possible. But the courses I have suggested may nevertheless be desirable solely as a means of rescue for the returning veterans and for the young civilians who, through poor advice or change of plans, may have omitted the high school mathematics necessary for their collegiate work.

11. Mathematics in general education. Up to this point I have spoken of mathematics chiefly in connection with its applications. Perhaps this is because these aspects have been stressed in the specialized training programs of the Army and Navy. There is a strong danger that these aspects are so obvious now that many educators will come to think of mathematics as useful only in its most crassly practical connections. There are some who have that opinion now.

In some quarters there will be a strong reaction away from the technological toward general education. They ask, as people are asking in so many realms of life, of what avail will it be to win the war unless a better world emerges from it. To help make this better world, we must educate our young people in the social sciences, in literature, in the fine arts, in philosophy. We must give them a general education. To all of this I agree, of course. The danger lies in the very fact that mathematics has become so conspicuous for its immediate practical necessity that its utility in general education will be forgotten.

In one instance, at least, the Army gave a sort of left-handed recognition to mathematics as a tool of general education. When asked why mathematics had to be included in a certain one of its training programs, the officer in charge replied, "Well, mathematics matures the men." Sometimes, I am sure, we have felt that it not merely matured the men but made them grow old before their time. What the officer meant, of course, was that mathematics is especially useful in the development of the general qualities that it values in its specialized personnel—qualities, perhaps, of clear, disciplined, hard-headed thought and clear expression.

The proposed federal program of subsidization of education that I have already outlined provides for general education as well as for vocational or professional training. If we are to do our share in carrying out the program, we must keep vividly in our own minds how mathematics can contribute to the objectives of a general education. And we must keep it in the minds of administrators in high schools and colleges and of the public as well. Of all the topics that I have mentioned this evening this one of the place of mathematics in general education lies as close as any to my heart. I had originally intended to make it the central theme of this address. It was my thought to recall the objectives of a college education and to examine the manner and the extent to which mathematics can definitely and in some respects uniquely contribute to attaining those objectives. The subject is too important and too large for the cursory treatment that I could give it now. With the hope that it will be discussed at length in some more suitable time and place, I shall assume that, as mathematicians, you share my feelings in this matter, and pass on to some of the immediate implications.

If it is true that we shall have many students returning from the service or coming to us from the high schools for a program of general education to which mathematics can make valuable contributions, then we must see to it that those contributions are made. Besides our extensive mathematical curricula, which secure many of the advantages of general education while training the student in his specialty, we must provide elementary terminal courses for this purpose. These courses, if less complete in the techniques necessary for continuation courses, should be rich in essential ideas. They should insist on precision and clarity of language. They should reveal mathematics as a logical structure arising from simple and natural assumptions, explicitly recognized, and leading to important and interesting results. They should indicate the necessity in any system of thought of exact definition of terms and provide examples of definitions that clarify and correspond to one's intuitive notions. We must not lead the student to believe that mathematics is vague or inconclusive by attempting to include too great a range of material or any material too remote from his experience. The time was when it was believed that the world could be saved by education—any education. But we have learned that education built on authority alone can work as efficiently for evil as for good. If education is to contribute to our national welfare, it must conduce to independence of thought, by which the individual is able to examine for himself the assumptions and conclusions behind dogma. I am convinced that the courses I have in mind can aid materially in establishing desirable habits of scientific thinking. As you know, such courses already exist in many institutions. I believe that we should make still further provision for them.

12. Shortages of staff. As I have surveyed these tasks that lie ahead of us, I have been haunted by a certain fear. It is the fear that we shall not have the personnel properly trained to carry the great load. As I have indicated, there is every prospect that at the freshman and sophomore levels our burden of teaching, far from decreasing, will actually increase after the war. Added to this, we should have more students than ever before at the higher levels, where the war-time load is most conspicuous by its absence. Just who is to do this teaching? Some of our staff, away on war duty, will return. But many who have tasted the flesh-pots of industry will not wish to go back to the genteel poverty of college teaching. Many teachers lent to us from other fields will be called back to duty in their own departments. Many scientists have been sacrificing their own interests to the common good by teaching instead of devoting their energy to research. It would dry up the very sources of progress to ask them to continue this sacrifice indefinitely.

And what of the supply of replacements? Almost we can say there isn't any. The normal crop of young students has been mostly absorbed into the Army or Navy or other war service. Selective Service, it is true, has recognized mathematics as a field in which there exists a critical shortage. But so far as students are concerned this has resulted chiefly in the deferment of undergraduates. Most graduate students, under the pressure of war-training courses and the attractions of high salaries, have been teaching instead of studying. A full-time graduate student or a part-time assistant is a *rara avis*.

13. The training of mathematicians. Of all our future tasks, and I have referred to a number of them this evening, perhaps the most pressing, most important, and most difficult is this one of the training of mathematicians for our college and university staffs. The plant must be erected before munitions can be produced. We can begin even now to do a little work on the problem. Just the other day a former high school teacher whose previous maximum salary for a nine-month year was $1620 was offered a college position at $3600 for twelve months. Such opportunities are tempting. But I believe that our younger staff should be informed of the opportunities that will exist after the war, and should be encouraged to take every present chance to carry on their studies and their research. The present load of teaching must be carried on. However, young people of special promise should be encouraged to do their teaching under conditions that offer some opportunity for study. So far as the slightest break in the wartime schedules permits, we should provide graduate courses and encourage our young teachers and students to take them. If we realize the prime necessity of this part of our program, we can do something to relieve the situation, but only in a piecemeal and inadequate way.

Under the present war training contracts, institutions are able to offer attractive salaries. It is not likely that the scholarships offered under any program of federal subsidization after the war will be large enough wholly to pay the costs of instruction. Colleges and universities will be hard put to it to offer salaries commensurate with the demand. It is not too early to begin to bring the prospective needs of education before the public with the hope of winning financial support. Even if such a campaign of public education were successful, it would not solve our problem. Money can attract mathematicians, but it cannot make them.

14. A possible source of teachers. There is only one large source of partly trained manpower. That is in the armed services themselves and especially in their present educational programs. The Army and Navy should recognize their responsibility to the servicemen after demobilization. If it is true, as I believe it is, that the scarcity of teachers will be as critical for these men as the scarcity of physicians, it would seem reasonable that the Army should immediately undertake a long-range program of training scientists similar to its program of medical training. If, for example, the best fifteen or twenty per cent of graduating pre-meteorology students were to be offered a year or more of concentrated but not too narrow training in mathematics and its applications, wonders could be done in creating a supply of men capable of giving at least elementary college instruction. This particular plan is fantastically improbable. But it might not at all be a waste of effort for the officers of the mathematical organizations to present to the A.S.T.D. the desperate necessity of some such scheme of long-range advanced instruction in mathematics.

15. The unity of the mathematical program at all levels. I should like to mention just one other thing before I close. I refer to the essential unity of the mathematical program at all levels from the junior high school, through the stages of high school, junior college, and college, to the graduate school. Research and graduate teaching vitalize college work. The colleges stimulate the teaching and set the standards for the high schools; in turn they are dependent on the high schools for both the number and quality of their students. Because of the extremely sequential nature of our subject this unity is even more important in mathematics than it is in most fields. Students who have been badly advised or badly taught in high school have special difficulty in repairing the damage to their mathematical training. I do not need to recall here the trends in attitude toward high school mathematics. It is enough to say that many educators expect and some of them desire that recent trends will be reversed and those of a few years ago will be renewed. We must furnish what help we can to the high school program. Quite apart from what we are tempted to feel are malicious attitudes toward mathematics, we must educate ourselves to the very genuine difficulties in the high school mathematical curriculum. Some of these difficulties result from the individual differences in ability of students; others from their differences in purpose. Others arise from an ignorance of the number of paths barred to them by an early neglect of mathematics.

The Mathematical Association of America has always had close contacts with the other mathematical organizations. We have held joint meetings with both the American Mathematical Society and with the National Council of Teachers of Mathematics. I can only urge that in the highly critical period just after the war, as an organization and as individuals we participate even more actively in the problems of our high school colleagues. Only by considering the entire mathematical program as a unit can we achieve the strength that comes from unity.

In this survey of what might be called "War's Math and Aftermath," I have attempted to consider the general magnitude of the task ahead, with some of its opportunities and problems. At moments I have stated my own opinions while quite aware that many of you will not agree with them. If I have spoken dogmatically at certain points, I did so only for purposes of economy in presentation and not because I felt that I had direct inspiration in the matter. I do believe that it is time for the subject to be brought before the house, and I hope that my own speculations may give rise to profitable discussion.

1

PEDAGOGY

MARGINAL NOTES*†

T. H. HILDEBRANDT, University of Michigan

1. This is not a discourse on the desirability of making marginal notes in one's mathematical reading or teaching. It is simply a collection of notes of the type that every teacher who is alive to the subjects which he is teaching makes in the margins of his textbook. As a consequence there is no particular claim to originality or priority. In presenting these remarks, it is hoped that perhaps some other teacher of the first two years of collegiate mathematics may find something which will be helpful in his own teaching.

My first note refers to the area of a triangle, the coordinates of the vertices being $P_1:(x_1,y_1)$; $P_2:(x_2,y_2)$; $P_3:(x_3,y_3)$. The usual formula for this area is either a third order determinant, or the same expanded and arranged as follows:

$$\tfrac{1}{2}\left[x_1(y_2-y_3)+x_2(y_3-y_1)+x_3(y_1-y_2)\right],$$

I do not understand why (if third order determinants are to be avoided) this latter is preferred to the equivalent expansion

$$\tfrac{1}{2}\left[(x_1y_2-x_2y_1)+(x_2y_3-x_3y_2)+(x_3y_1-x_1y_3)\right],$$

which has the advantage of an easy geometric interpretation and proof and which is easily remembered and easily extended. For $\tfrac{1}{2}(x_1y_2-x_2y_1)$ is the directed area of the triangle OP_1P_2O, a fact which is easy to prove. Then one finds that the directed area $P_1P_2P_3P_1$ is equal to the sum of the directed areas $OP_1P_2O, OP_2P_3O, OP_3P_1O$. Ease in remembering is of course increased by using second order determinants, which gives in reality a method for expanding the third order determinant form.‡ By way of extension, we note that it is an easy matter to write

---

* From AMERICAN MATHEMATICAL MONTHLY, vol. 36 (1929), pp. 216–221.

† Read before the Michigan Section of the Mathematical Association of America, March 31, 1927, at Ann Arbor, Mich.

‡ As a side remark, I should like to voice the opinion that it would be advantageous to drop from our teaching of determinants the so-called diagonal method of expansion of third order determinants and to replace it by the expansion by minors. The diagonal method is confusing because it is not applicable to determinants of order higher than the third, as one might have a right to expect. Expansion by minors is not only applicable to determinants of any order, and hence need not be unlearned, but even for determinants of the third order is in most cases simpler than the diagonal method. Moreover, it can be made (as any one can convince himself very easily) a logical foundation for the definition of the value of a determinant of any order and can be used as a basis for proof (by induction) of the properties of determinants.

down the area of a closed polygon. For instance the directed area of the polygon $P_1P_2P_3P_4P_5P_1$ is

$$\frac{1}{2}\left\{\begin{vmatrix} x_1 & y_1 \\ x_2 & y_2 \end{vmatrix}+\begin{vmatrix} x_2 & y_2 \\ x_3 & y_3 \end{vmatrix}+\begin{vmatrix} x_3 & y_3 \\ x_4 & y_4 \end{vmatrix}+\begin{vmatrix} x_4 & y_4 \\ x_5 & y_5 \end{vmatrix}+\begin{vmatrix} x_5 & y_5 \\ x_1 & y_1 \end{vmatrix}\right\}.$$

The geometric interpretation of the answer would in general require a figure, especially where the sides intersect between the vertices (as for instance in a five pointed star). If we write

$$x_1y_2 - x_2y_1 = x_1(y_2 - y_1) - y_1(x_2 - x_1) = x_1\Delta y - y_1\Delta x$$

it is easy to see that we are not far here from the rectangular equivalent of the polar element of area, which has to a large extent disappeared from our textbooks in calculus.

The extension of these considerations to three dimensions is not so simple. Obviously a cyclic performance such as

$$\tfrac{1}{6}\left[|x_1y_2z_3|+|x_2y_3z_4|+|x_3y_4z_1|+|x_4y_1z_2|\right],$$

in which $|x_1y_2z_3|$ expresses the third order determinant on the elements $x_1,y_1,z_1;x_2,y_2,z_2;x_3,y_3,z_3$, is not the volume of the tetrahedron $P_1P_2P_3P_4$, though $|x_1y_2z_3|$ is the volume of $OP_1P_2P_3$. The difficulty lies in the fact that a directed volume is slightly more complicated than a directed area. It can be shown that the result desired is:

$$\tfrac{1}{6}\left[|x_1y_2z_3|+|x_1y_3z_4|+|x_1y_4z_2|+|x_4y_3z_2|\right],$$

which shows a certain amount of symmetry especially if placed in juxtaposition with the fundamental equality which we used in connection with the area of a triangle, viz.,

$$OP_1P_2 + OP_2P_3 + OP_3P_1 + P_3P_2P_1 = 0.$$

The addition of a fifth point P_5 to the melee brings up the question what volume is defined by the points $P_1P_2P_3P_4P_5$. Obviously higher dimensions increase the difficulties. We are touching here upon the notion of directed volumes in higher dimensions.

2. In the margin of the section on curve tracing in polar coordinates I find the adjoining well known (?) figure from trigonometry. It is being omitted from trigonometries to some extent, mainly because it is seldom put to use later on. If used in connection with polar coordinate paper and a pair of dividers, it yields a simple and elegant method for plotting without laborious computation enough points of curves with polar equations of the following types: $r = a\cos\theta + b, r = a\sec\theta + b, r = a\tan\theta, r = a\tan\theta + b\sec\theta$, to enable even freshmen to produce fine sketches. Methods may be developed for using this figure even for such curves as $r = a\cos 2\theta$, $r\cos 2\theta = a$, and others containing trigonometric functions.*

---

* More extensive remarks on the same subject are contained in a note by H. A. Bender in this Monthly, vol. 34 (1927), pp. 481–484.

3. My next note occurs in connection with the remark that the tangent to a circle at a given point (x_1,y_1) on the circle can be obtained by writing down the equation of the point circle, center at (x_1,y_1) and finding the common chord of this circle and the given circle. The question arises whether this idea is extensible to conic sections and is to be answered in the affirmative. Thus the tangent to $a^{-2}x^2+b^{-2}y^2=1$ at (x_1,y_1) is the common chord of this conic and the degenerate conic $a^{-2}(x-x_1)^2+b^{-2}(y-y_1)^2=0$; the tangent to $a^{-2}x^2-b^{-2}y^2=1$ is the common chord with $a^{-2}(x-x_1)^2-b^{-2}(y-y_1)^2=0$; and the tangent to $y^2=4ax$ is the common chord with $(y-y_1)^2=0$. This leads to the general statement that the tangent to

$$Ax^2+Bxy+Cy^2+Dx+Ey+F=0$$

at (x_1,y_1) on the curve is the common chord of this conic and the similar degenerate conic:

$$A(x-x_1)^2+B(x-x_1)(y-y_1)+C(y-y_1)^2=0.$$

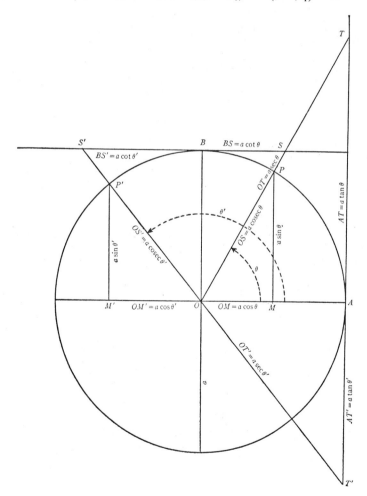

In addition to giving us the equation of a tangent to a conic without the necessity of finding the slope by a limits process, i.e., the use of derivatives, it has the additional advantage of making it possible to write down the equation of the tangent plane at a given point to a quadric in three-space. For example the equation of the tangent plane to the quadric $Ax^2 + By^2 + Cz^2 + D = 0$ at the point (x_1, y_1, z_1) on the quadric is the common plane of this quadric and the similar degenerate quadric: $A(x-x_1)^2 + B(y-y_1)^2 + C(z-z_1)^2 = 0$. I leave unanswered the question of what this process gives when the point is not on the conic or quadric.

4. At a point in the proof of the fact that the locus of points, the sum of whose distances from two fixed points is a constant, is an ellipse, I find the word "stop" in the margin. If for simplicity the fixed points are $F_1:(b,0)$, $F_2:(-b,0)$, and the constant is $2a$, then the first equation reads:

$$\left[(x-b)^2 + y^2\right]^{1/2} + \left[(x+b)^2 + y^2\right]^{1/2} = 2a.$$

Rationalizing* this equation and collecting terms ready for the second rationalization, one gets the expression to which the remark applies

$$a\left[(x-b)^2 + y^2\right]^{1/2} = bx - a^2.$$

Interpreting this equation we find that it says that the distance of (x,y) from $(b,0)$ is b/a times its distance from the line $x = a^2/b$; i.e., we have arrived at the focus-directrix property of the ellipse, which is usually the basis of discussion in analytic geometry. Obviously this observation will give us the eccentricity and equations of directrices in any case where the foci and major axis are given.

5. In connection with the section on transformation of coordinates, I find the query, "why not use on curves other than just conic sections?" For instance, in calculus, one finds occasionally such equations as $x^3 - 3axy + y^3 = 0$, and $x^2y + xy^2 = a^3$, having $x = y$ as an obvious line of symmetry. If this is made the new x-axis, the equations are of relatively simple type, solvable for y^2, and in the first case, it makes it even possible to find the area of the loop of the curve without resorting to the trick of polar coordinate representation.

6. In my calculus, I find such notes as the following:

(a) $1 - \cos x$ can be rationalized by multiplying by $1 + \cos x$, a remark which helps to avoid the introduction of half angles in proving that $\lim_{x \to 0}(1 - \cos x)/x = 0$, and in the integration of expressions such as $\int (1-\cos x)^{1/2} dx \cdot \int dx/(1-\cos x)$.

(b) For some curves it is simpler to apply the fundamental definition of curvature: $d\alpha/ds$, instead of the expression in terms of y' and y''. Examples $y = \log \cos x$; the circle $x = a \cos t$, $y = a \sin t$; the four cusped hypocycloid $x = a\cos^3 t$, $y = a\sin^3 t$; the cycloid; and the involute of the circle.

---

* Professor Shohat remarked in discussing this paper that this process could be simplified by noting that $(PF_1)^2 - (PF_2)^2$ is a rational expression, so that by division by $PF_1 + PF_2$ we get $PF_1 - PF_2$ from which PF_1 and PF_2 can be determined at once, giving the same result as above. My point is that even the straightforward usual methods have interesting interpretations.

(c) The tangential and normal components of acceleration of a body moving in a plane curve are the components of velocity of a point on the corresponding hodograph, along and perpendicular to the radius vector to the origin. Since this vector on the hodograph is v, the polar angle α (the angle between the tangent and x-axis on original curve), the tangential acceleration is dv/dt, and the normal acceleration is $v\,d\alpha/dt = v(d\alpha/ds)(ds/dt) = v^2/\rho$, where ρ is the radius of curvature of the original curve.

(d) A rational function which is a proper fraction can be uniquely expressed as the sum of proper fractions having the distinct factors of the original fraction as denominators. For example

$$\frac{4x^2+x+5}{(x-2)^3(x^2+1)^2} = \frac{Ax^2+Bx+C}{(x-2)^3} + \frac{Dx^3+Ex^2+Fx+G}{(x^2+1)^2},$$

and the coefficients in the numerator are uniquely determinable. (They are no longer unique if the fractions on the right are not proper.) The emphasis being upon the *proper* fraction, it follows that the numerator is always assumed to be one degree less than that of the denominator, and we have a method covering all possible cases. Incidentally this idea does not really increase the difficulty of integration. A term like the first, with a power of a linear expression in the denominator always yields to the substitution of a new variable for the linear expression. For terms of the second type, note that for instance $x^2/(x^2+1)^2$ is easier to integrate than $1/(x^2+1)^2$. Integration by parts, or substitution of $x = a + b\tan\theta$, for a term whose denominator is of the form $[(x-a)^2+b^2]$ will always work.

(e) An expression of the form

$$e^{\alpha x}\left[(A_n x^n + \cdots + A_1 x + A_0)\cos\beta x + (B_n x^n + \cdots + B_1 x + B_0)\sin\beta x\right]$$

is a solution of a linear homogeneous differential equation with constant coefficients in which the associated algebraic equation (obtained by putting $y = e^{mx}$ and dividing by e^{mx}) has the roots $\alpha \pm i\beta$ repeated $n+1$ times. This remark enables us to write down the differential equations whose general solutions are similar to

$$y = (c_1 + c_2 x) + c_3 e^{-x} + e^{3x}(c_4 \cos 2x + c_5 \sin 2x).$$

It also suggests a method for writing down the form of the particular integral for a linear differential equation with constant coefficients, in which the terms which are functions of the independent variable only, are linear combinations of expressions like the above. A particular example will illustrate best. Consider the differential equation

$$y'' - 4y' - 5y = 3xe^{-x} + 6\cos 2x + 5\sin 2x + 4e^x \sin 3x.$$

Since the general solution of this differential equation will also be a solution of a linear *homogeneous* differential equation with constant coefficients but of higher order (at least eighth) the solution will be determined by an algebraic equation whose roots are 5 and -1 (from the left hand side of the equation), and $-1, -1, \pm 2i, 1 \pm 3i$ (from the right). Note that $3xe^{-x} = (3x+0)e^{-x}$ and $4e^x \sin 3x = e^x(0 \cos 3x + 4 \sin 3x)$. As a consequence the complementary function is $y = c_1 e^{5x} + c_2 e^{-x}$ and the particular integral has the form

$$y = (c_3 x + c_4 x^2)e^{-x} + c_5 \cos 2x + c_6 \sin 2x + e^x(c_7 \cos 3x + c_8 \sin 3x),$$

the constants $c_3, \ldots, c_8$ being determined by the usual method of undetermined coefficients. Note that -1 counts as a triple root, as it actually is.

ON THE TEACHING OF ELEMENTARY MATHEMATICS*†

MOSES RICHARDSON, Brooklyn College

The following remarks are addressed to those who are preparing themselves to be teachers of elementary mathematics. They begin with some horrible examples of what I think a good teacher should not do, and some general principles that I think a good teacher would do well to follow; they conclude with a serious thesis concerning the proper emphasis of various aspects of elementary mathematics.

Most of the examples come out of my own first-hand experience at various institutions (although some do not) and could be, but will not be, documented. If this be undiplomatic, it is at least not appeasement. I shall not discuss what should be taught to freshmen because this depends to some extent on what they are being prepared for, and because I have expressed my views fully on this subject elsewhere. Rather, I shall confine myself to remarks which I consider to be valid regardless of what topics the curriculum contains.

There are at least three different French languages: Parisian French, Gascon French, and a language spoken by American tourists called High School French. In our subject, we have Pure Mathematics, Applied Mathematics, and a peculiar subject sometimes called "Freshman Mathematics." Every instructor, in the normal exercise of his duties, accumulates his own collection of specimens of "Freshman Mathematics." An innocuous example is that of the student who divided a large number by 1 by the process of long division. I call this innocuous because at least he was using a general algorithm which could be applied, however inadvisedly, to his problem. A more common and harmful sample is the following:

(1) $$\frac{2+x}{2x} = \frac{\cancel{2}+\cancel{x}}{\cancel{2}\cancel{x}} = 1 + 1 = 2.$$

This illustrates the "law of universal cancellation" which says that whenever a

---

* From AMERICAN MATHEMATICAL MONTHLY, vol. 49 (1942), pp. 498–505.

† Based on an address delivered at the triennial convention of Pi Mu Epsilon held at Lehigh University, January 1, 1942.

symbol occurs twice on the same paper or blackboard it may be crossed out in both places. A well-known "proof" of it proceeds as follows:

$$\frac{1\cancel{6}}{\cancel{6}4} = \frac{1\cancel{66}}{\cancel{66}4} = \frac{1\cancel{666}}{\cancel{666}4} = \cdots = \frac{1}{4}.$$

Very common specimens of "Freshman Mathematics" are the following:

(2) $\qquad \sin(x+y) = \sin x + \sin y, \qquad \log(x+y) = \log x + \log y,$
$$\sqrt{x+y} = \sqrt{x} + \sqrt{y}.$$

Other common errors are "transposing" $2x = 1$ to obtain

(3) $$x = \frac{1}{-2}$$

(presumably because whenever a symbol is "brought" from one side of an equation to the other its sign must be changed); and adding fractions by adding numerators and denominators separately as in

(4) $$\frac{a}{b} + \frac{c}{d} = \frac{a+c}{b+d}.$$

A favorite specimen of mine is the following "proof" of the remainder theorem:

$$\begin{array}{r|l} x-a & f(x) \quad\big|\, f \\ \hline & f(x) - f(a) \\ & \qquad f(a) \end{array}$$

A young teacher may well feel at times as though he has somehow blundered into a psychopathic ward. But successful psychoanalytic technique requires sympathy, understanding of the difficulties, and probing into the past of the patient to find the events leading up to the tragedy. I can speak at first hand of this latter aspect since I have taught both high school and college freshman classes for many years.

It is essential for the teacher to understand both the student's preparation and difficulties, and, if I may be so bold as to say so, the subject he is teaching. I emphasize this because I once heard a man high in educational circles assert that one who knows too much can't be a good teacher. This is unquestionably true, but only because one who knows too much cannot exist. Of course, there are distinguished scholars who do not teach well, and excellent teachers who are not distinguished scholars. But if one has the patience and sympathy needed for good teaching and is willing to devote time and thought to the problems of teaching, then increased scholarship can only enrich one's teaching, not spoil it. On the other hand it is impossible for any one to teach well something he doesn't himself understand. There are, I am afraid, some people teaching the third reader who have never progressed beyond the fourth. By this I do not mean at all that it is necessary for a teacher of elementary subjects to be well versed in partial differential equations or the calculus of variations in the large, or similar advanced technical subjects; but I do think it essential that he have a broad, deep, and clear understanding of the fundamental concepts underlying elementary mathematics. That such understanding is not always present becomes obvious upon contact with some teachers and some textbooks.

For example, I was once asked to settle a bet between two ex-colleagues, one of whom contended that $0/0=0$ because 0 divided by anything is 0, while the other held that $0/0=1$ because anything divided by itself is 1. Neither was willing to listen to an explanation of why both were wrong; they wanted only a verdict as to which was right. This would make a better story if there were a third who claimed that $0/0=\infty$ because anything divided by 0 is ∞, but I shall not yield to the temptation to embellish the facts.

Then there is the teacher who seriously claimed that it was literally correct to write $1/0=\infty$ in a trigonometry class on the grounds that Veblen and Young write this equation in their Projective Geometry. Needless to say, Veblen and Young are considering a particular coordinate system on a line in the projective plane and division means to them not numerical division, but a certain quadrilateral construction.

Another instructor "derived" an equation for vertical lines as follows:

$$y - b = \infty \ (x - a),$$
$$x - a = \frac{1}{\infty} (y - b) = 0.$$

In fact, so many people and textbooks have the courage of their confusion on the subject of infinity that it is small wonder that some of our brightest students come away with the impression that the cardinal number of the set of all natural numbers is ∞ which is obtained by dividing 1 by 0 and represents the place where parallel lines meet. I have often wondered why no one has ever "proved" the statement $1/0=\infty$ by appealing to the fact that if one tries to divide 1 by 0 on a calculating machine it will run until worn out.

Then there are many books which carelessly ask the student to write the nth term of a sequence whose first few terms are given, as though this were a reasonable mathematical question. To answer it, as you know, requires not mathematics but clairvoyance. For example, the nth term of a sequence beginning with 2, 4, 6, 8, $\cdots$ may be $2n$ but it may also be $2n+(n-1)(n-2)(n-3)(n-4)f(n)$ where $f(n)$ is almost completely arbitrary.

Perhaps more serious than these troublesome confusions is the failure of some teachers and some textbooks to pay due attention to converse propositions. For example, one of my ex-colleagues once remarked to me that he liked trigonometric identities because nothing could be more satisfyingly certain than to come out with the same expression on both sides. I answered, "Yes, as long as all the steps are reversible." The man's jaw dropped. He had never heard of that, and it took me 45 minutes to persuade him that the trouble lay with the converse proposition. It turned out that, although he liked trigonometric identities and had taught them for years, he had never understood them. Similarly, many texts on analytic geometry leave the student without the slightest suspicion that a converse proposition is needed in discussing equations of loci.

I say that this is perhaps a more serious matter because it is symptomatic of a more or less widespread tendency that I regard as vicious: namely, the tendency to *say* that one of the main objectives of teaching mathematics is to train the student to think logically and then to *teach* the subject as though it had nothing to do with logical reasoning at all. The man just referred to paid careful attention to converses in plane geometry because that was traditional but had little suspicion that algebra or trigonometry had anything to do with reasoning. In fact, I recently had the distressing experience of hearing someone high in educational circles say that after all plane geometry is mostly a matter of memory. Is it any wonder that this opinion of mathematics is so widespread when it is all too often taught as a matter of memorized formulas, routine meaningless manipulations, and undigested proofs? It is my opinion that many of the ills of mathematics come from such excessive formalization and neglect of fundamental principles and common sense. If mathematics is presented as a collection of arbitrary rules of thumb for performing peculiar manipulations which somehow solve problems in which the student is not interested anyway, it is not surprising that many intelligent students are repelled from the subject for life. Such a collection of tricks is merely a branch of parlor magic, while the ideas of mathematics constitute an important current in the history of human thought. When these disgruntled individuals grow up some of them lead movements to abolish mathematics as a required subject. I am convinced that this often happens because they have never had contact with genuine mathematics at all, but only with formalized, memorized regurgitated techniques.

This souring of public opinion among non-technical people is by no means the only ill effect of excessive formalization. I am convinced that many purely technical difficulties arise from the same source. Thus, the errors given in (2) arise from thoughtless, superficial, formal analogy with the distributive law $a(x+y)=ax+ay$. The troubles with cancellation, transposition, and fractions ((1), (3), (4)) are due to thoughtless, unrationalized formalization. A memorized rule that has not been understood is difficult to recapture when the smallest part of it is forgotten.

I do not know whether one can attribute the tendency toward excessive formalization to a unique source. Perhaps it is partly due to the fact that teachers are sometimes judged by the results of their students on stereotyped examinations which can usually be passed with only formal techniques and without understanding. This practice may put pressure on the teacher to forget about the difficult (though worth while) task of teaching mathematics and cause him to give instead a coaching course for the examination. This makes the teacher's life easier, but harder to justify. He can then "cover" all the topics required for the examination easily, by the simple process of cramming memorized techniques into his students and ignoring the opportunity to give real insight into mathematics. He can teach "box" methods for solving problems since he is sure that the examination will not contain a strange "type" of problem that will not fit into his boxes. I have actually seen a secondary teacher who taught the logarithmic solution of triangles at the outset of the term before the very definitions of the trigonometric functions. He "justified" this procedure on the grounds that he found students lost most credits on those questions on the Regents' examinations and consequently should get as much drill as possible on them. Beginning on the first day of the term surely gives the maximum drill. I met another secondary teacher who disliked teaching plane geometry because he was unable to formalize it as he did algebra. He did, however, like to teach some "calculus" in his advanced algebra section. By "teaching calculus" he seems to have meant training his students to write nx^{n-1} whenever they saw x^n, much as a mouse is trained to run a maze when a gong rings.

By now, I think I have complained enough about some of the extreme cases that I have had the misfortune to encounter. I hope they are uncommon. Let me now say a few words about some much less rare situations. There is the teacher who does "take up" the fundamental ideas and the reasoning behind his processes. But he does it once and rapidly, hopes that the best students will get it, and races on. This practice arises from a point of view with which I am not in sympathy. I do not consider the fundamentals important only for the few best students and the techniques important for everybody. I think that the fundamentals are important for all and will justify, in the long run, the time and effort needed to get them over to most of the class. I think it is part of the teacher's job to disseminate the ideas of mathematics as widely as possible, not to keep them within an esoteric circle of witch-doctors. The "devil take the hindmost" theory of teaching, if pushed to extremes, leads the teacher to give lectures which only he understands.

Another practice of which I disapprove and which is prevalent among college teachers, is that of blaming the student's incompetence on his high school training and letting it go at that. In fact, the high school teachers proceed similarly to blame the elementary school teachers, who blame the parents, who presumably can blame nothing but heredity, so that the original fault seems to lie with Adam. I can see no sense whatever in grumbling because a freshman has not mastered the routine techniques of high school algebra and then rushing him on into further ill-understood memorized techniques. We must accept our students as incontrovertible data; if they never learned to add fractions sensibly let us not consider it beneath our dignity to teach it to them. Even where they have had prerequisite material, they often have it no longer. Unless one is willing to face the facts as they are and do the best job possible under the circumstances, one might as well lapse into one's anecdotage and go about mumbling about the good old days.

I have said a great deal about what I think a good teacher of freshman mathematics should not do. It is harder to say what he should do since that depends on his own point of view, his temperament, training, and interests. However I shall undertake to enunciate a few generalities.

I think it is important for a teacher of elementary mathematics to be thoroughly familiar with the foundations of the subjects he teaches and to have as broad a mathematical background as possible. A teacher who has only superficial knowledge himself, slips comfortably into the habit of regarding things as "easy" because he is "familiar" with them. On the other hand, genuine familiarity with the deep-lying difficulties underlying elementary mathematics will, if nothing else, prevent false emphases and tend to make one more tolerant towards the students' difficulties. This was brought home to me in the first high school class I ever taught when a student bisected an angle thus:

An angle had been defined as a figure consisting of two rays with a common vertex. But insufficient stress had been placed on a careful definition of what is meant by "bisecting" an angle. For example, one may be more inclined to sympathize with a student's troubles with minus signs if one realizes oneself that $-(-2)$ and $(-1)(-2)$ represent different situations. In any case, you will not teach them that "two minuses make a plus" because you can lay one minus sign vertically across the other thus: $+$. Many an advanced student, well trained in advanced techniques, remains unfamiliar with fundamental concepts because he has been expected to acquire familiarity with them by osmosis. This process does not always take place, perhaps because it proceeds only from the more dense to the less dense.

I think it is important that you try not to teach falsehoods that have to be corrected later, unless, as Forder remarks,* it be contended that an unsound proof has an educational value not possessed by a sound one. Needless to say, an unsound proof may well have educational value not possessed by a sound one provided it is used as an example of faulty reasoning. By this I do not mean at all that elementary teaching should be rigorous. I mean only that where gaps occur they should be pointed out, and that theorems which cannot be treated soundly at the student's level should be assumed or discussed informally rather than be given a bad proof.

Try to keep alive your interest in mathematics and its teaching. It is easier to give your students enthusiasm for mathematics if you have it yourself. Despite heavy schedules and other adverse conditions, it is always possible to come

---

* H. G. Forder, The Foundations of Euclidean Geometry, Cambridge Press, 1927, p. viii.

yet a little closer to attaining your ideal objectives, provided you know what you want them to be.† Try to teach in such a way that the students will acquire: (1) an appreciation of the natural origin and evolutionary growth of the basic mathematical ideas from antiquity to the present; (2) a critical logical attitude and a wholesome respect for correct reasoning, precise definitions, and clear grasp of underlying assumptions; (3) an understanding of the role of mathematics as one of the major branches of human endeavor and its relations with other branches. Try to emphasize the distinction between familiarity and understanding, between proof and routine manipulation, between a critical attitude of mind and habitual unquestioning belief, between scientifically organized knowledge and both encyclopedic collections of facts and mere opinion and conjecture. Try to give them not only formulas but a wholesome appreciation of the nature and importance of mathematics.

Let me try to summarize my point of view in another way. There are three main aspects that must be brought out to some extent in the teaching of elementary mathematics, namely:

A. The routine techniques which are basic for future work. This aspect will be designated by *"techniques."*

B. The concrete applications of mathematics, the concrete settings in which mathematics originated, the interrelations among the "branches" of mathematics, and it relevance to the real world in general. For lack of a better term, this aspect will be referred to as the *"relevance"* of mathematics.

C. The reasonable justification of the techniques, and the logical nature of mathematics in general. This aspect will be designated by *"reasonableness."*

There seems to be some fear in certain quarters that overemphasis of C leads to excessive abstraction which will repel the general public from mathematics. This fear seems to be founded on the undeniable fact that some abstractions which appear in research papers lack a rich concrete background. This is true, but it is also true that few mathematicians are seriously interested in such an abstraction per se except possibly for the purpose of acquiring a Ph.D. or a promotion. Nevertheless, it must be granted that even such an empty abstract science is pure mathematics even though it is not important pure mathematics. It does not seem to me that anything is to be gained by hiding this innocent fact from the general public as though it were an important trade secret, much as the Pythagoreans are said to have tried to hide the irrationality of the square root of two. The truth will gain more for our subject than suppression of allegedly harmful ideas, which are, in my opinion, not harmful at all as long as one emphasizes that mathematicians are almost always interested only in abstractions which have rich concrete interpretations and which serve to unify or clarify

---

† For an excellent discussion of the objectives of teaching elementary mathematics, see the Fifteenth Yearbook of the National Council of Teachers of Mathematics, which was drawn up by a joint committee of the Council and the Mathematical Association of America.

concrete subjects which are interesting in themselves. As for this tendency to abstraction seeping harmfully into the teaching of elementary mathematics, such fear strikes me as being not only groundless but actually pointed in precisely the wrong direction.

In the actual practice of elementary teaching, as I have observed it, "techniques," "relevance," and "reasonablesness" are usually given attention in precisely that order, with "techniques" far in the lead almost to the total exclusion of the other two, with "relevance" running a very bad second, and "reasonableness" a much worse third. It is my contention that this order should be reversed; in any case "reasonableness" and "relevance" deserve at least as much importance as "techniques." For students whose major interest is in the sciences, all three aspects should be virtually tied for first place. By this, I do not mean at all to belittle the importance of "techniques" for technical students, but I am convinced that when their reasonable justification is properly stressed, the techniques themselves require far less attention than is ordinarily given them. Understanding really does improve technical facility.

For students with little or no interest in the sciences, who really have little need for such facility, "techniques" should trail considerably behind. It is definitely my experience that these students can become very interested in the logical nature of mathematics, whereas they have no personal interest whatever in solving problems of a practical (to a technician) nature, no matter how interesting such problems may seem to a mathematician. It may seem surprising to many teachers, but it is nevertheless a fact that to many college freshmen the idea that algebra is a reasonable subject comes as a complete revelation. If there is danger to the status of mathematics, it does not arise from overemphasis of its "reasonableness." It comes from the deadly overemphasis on routine "techniques," and the unwholesome neglect of its "reasonableness" and of its "relevance" to the real world.

SOME VARIATIONS OF THE MULTIPLE-CHOICE QUESTION*

H. F. S. JONAH and M. W. KELLER, Purdue University

1. Introduction. In attempting to construct multiple-choice items for objective tests in mathematics which will test the students' ability to solve particular kinds of problems certain difficulties are encountered. Thus, for some types of problems, it is possible for the student to determine the correct answer without actually solving the problem when it is stated in the conventional manner. For example, in the problem

Example 1. The roots of the equation $x^2+5x+6=0$ are (1) $-2, -3$ (2) 2, 3 (3) $-6, 1$ (4) none of the proposed answers is correct.†

the correct response can be found by substitution. Consequently, such an item does not necessarily test the ability of the student to solve a quadratic equation. To obviate this possibility some constructors state the problem in such a manner that it is necessary to solve the problem first, and then perform some operation or operations on the answer. Thus, instead of stating the problem in the form given above, it would be given in a manner similar to the following:

Example 2. The sum of the squares of the roots of the equation $x^2+5x+6=0$ is (1) -13 (2) 13 (3) 37 (4) none of the proposed answers is correct.

For problems of this type this is a simple and satisfactory solution of the difficulty. There are other types of problems, however, where such a solution is not possible.

2. Object. It is the purpose of this paper to propose for other types of problems a different variation of the multiple-choice item which the authors have devised in an effort to minimize the possibility of the student determining the correct response without solving the problem. In addition some suggestions will be made on the use of multiple-choice items for testing the students' ability to graph and sketch curves. This is a type of item which has not been used very extensively by other constructors.

The items we wish to discuss formed part of a test which the authors prepared in cooperation with the Division of Educational Reference for the Army

---

* From AMERICAN MATHEMATICAL MONTHLY, vol. 52 (1945), pp. 1–6.

† This fourth response has been used uniformly on the various tests constructed by the authors. If the student is merely guessing this does not change the theoretical probability of guessing the correct response. Since it is often difficult, however, to construct a third plausible distractor it is believed that this procedure is a satisfactory solution in such cases because the student may have devised a better distractor. This scheme permits him to use it. When the fourth response is the correct one then such a scheme permits the student to solve the problem incorrectly yet mark the correct response. It is the opinion of the authors that this weakness is more than compensated for by its advantages.

Specialized Training Program. Hence the exact problems are restricted. Consequently, the examples which we shall use will not be the items included on this test but they will be similar. The comparison, therefore, cannot be exact. Since the problems will follow very closely the problems actually used it is believed that it will be possible to obtain a relatively clear picture of their effectiveness.

The test was given to all students regularly enrolled in AST Mathematics 407. The number taking the test was 385. The validity of each item, that is, how well each item discriminates between students making high grades on the test and those making low grades on the test, was determined. The method used for determining the validity index was that proposed by J. C. Flanagan.* This index is the estimate of the product moment coefficient of correlation between total test score and a given item based on the per cent obtaining the correct response to the item of those scores which were in the upper and lower 27% of the group. Since this is the product moment coefficient of correlation, an index of 0.19 is significant at the 5% level, and an index of 0.25 is significant at the 1% level. The reliability of the test was 0.90 when the short approximation form of the Richardson-Kuder formula was used. It should be pointed out that this formula tends to underestimate so that the actual value will be more than this.

3. A proposed solution. In analytical geometry when it is desired to determine if the student can find the equation of a line through two given points the objective form of this problem commonly used is

Example 3. The equation of the line passing through the points (2, 0) and (1, 2) is

(0.37; 78) (1) $2x-y=4$ (2) $2x+y=4$ (3) $2x-3y+4=0$ (4) none of the proposed answers is correct.

The first number, 0.37, in the parentheses is the validity index and the second number, 78, is the per cent of correct responses. The same notation will be used in the remainder of the examples which will be discussed.

As was mentioned previously the correct response to problems of this kind can be determined by substitution without actually solving the problem even though, in making up the distractors, the incorrect responses have one of the given points as a solution. It should be noted that the validity index of this item was definitely significant. Nevertheless one cannot be sure that the desired objective—to determine whether the student can find the equation of a line

---

* J. C. Flanagan, General considerations in the selection of test items and a short method of estimating the product moment coefficient from data at the tails of the distribution. Journal of Educational Psychology 30: 674–680, 1939.

through two given points—has been achieved. In trying to devise an item which would require the student to work the problem we included on the test some experimental items. These items directed the student to find the required equation and then identify the determined coefficients with the general form. Thus, in this form Example 3 would be given:

Example 4. The equation of the line through (2, 0) and (1, 2) is

$$ax + by + c = 0 \quad (a \text{ is a positive integer})$$

where
(0.61; 66) $a =$ (1)   2 (2) 3 (3) 1 (4) none of the proposed answers is correct,
(0.54; 63) $b =$ (1) -3 (2) 1 (3) 2 (4) none of the proposed answers is correct,
(0.42; 42) $c =$ (1) -4 (2) 4 (3) 6 (4) none of the proposed answers is correct.

For those topics which lend themselves to the use of this method of identification of coefficients it is believed that this type of objective item will encourage the student to work the problem. It certainly makes the trial and error method of substitution for determining the correct answer practically impossible. At the same time it has a tendency to discourage random guessing.

To machine score such a problem each part was scored separately. The correct identification of each coefficient was counted one on the total score.

The validity indexes for this problem indicates the possibility that the validity of the item is increased by this method over that of the same problem when it is given in the conventional form. Further experimentation is being conducted to determine the correctness of these tentative conclusions which were suggested by this pair of parallel problems. At least, for the ten experimental problems of this type included on the test, it can be stated that in every case each response—of which there were 37—was significant at well above the 1% level.

Because the method of determining or stating a correct answer by identification of coefficients is not generally used for solving problems in the classroom it might be thought that this would be confusing to the student, and consequently he would omit these problems. Apparently this was not the case in general since only about 1% omitted problems of this type while items of the conventional type were frequently omitted more often. There is some ambiguity in the statement of problems like Example 4 where the only restriction placed on a is that it be a positive integer. Although this might be confusing to an instructor of mathematics there was no indication during the administration of the test that it caused the student any difficulty. The statistical data also gives no such indication.

4. Some illustrative examples. In order to suggest some of the possible uses of this form of objective problem, and at the same time to give more information about the difficulty and general high validity of such items two additional examples will be given.

Example 5. The equation of the ellipse with semi-major axis of 5 on the *x*-axis, semi-minor axis of 3, and center at the origin is

$$ax^2 + bx + cy^2 + dy = e \quad (a \text{ is a positive integer})$$

where

(0.57; 66) $a =$ (1) 10 (2)   25 (3)   9 (4) none of the proposed answers is correct,
(0.51; 62) $b =$ (1) 0 (2) -10 (3) -6 (4) none of the proposed answers is correct,
(0.45; 57) $c =$ (1) 6 (2)   25 (3)   9 (4) none of the proposed answers is correct,
(0.72; 74) $d =$ (1) 0 (2) -10 (3) -6 (4) none of the proposed answers is correct,
(0.29; 53) $e =$ (1) 60 (2) 225 (3)   30 (4) none of the proposed answers is correct.

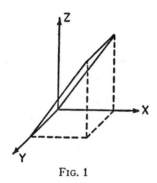

FIG. 1

Example 6. In Fig. 1 the plane *PT* passes through the point (3, 2, 4). The equation of the plane is

$$Ax + By + Cz + D = 0 \quad (A \text{ is a positive integer})$$

where

(0.45; 30) $A =$ (1) 4 (2)   3 (3)   1 (4) none of the proposed answers is correct,
(0.55; 33) $B =$ (1) 3 (2)   2 (3)   4 (4) none of the proposed answers is correct,
(0.34; 26) $C =$ (1) 0 (2) -3 (3) -2 (4) none of the proposed answers is correct,
(0.49; 54) $D =$ (1) 0 (2) -20 (3) -9 (4) none of the proposed answers is correct.

It should be noted that for any single problem there is considerable variation in the validity index and difficulty from response to response. This suggests the possibility of experimental investigations to determine the factors which are operating to cause these significant differences.

5. Curve sketching. Although no satisfactory method has thus far been proposed to test the ability of a student to sketch curves by the use of multiple-choice items various kinds of problems have been used to a limited extent to test certain parts of the process. Problems asking the student to determine the symmetry, extent, intercepts, *etc.*, of different equations have been used. These items are quite satisfactory to indicate whether the student understands that particular part of curve sketching. Since it is not possible to ask the student to sketch a curve when the test is an objective test using multiple-choice items it was believed that problems which require the recognition of the correct curve for a given equation and the correct equation for a given graph of a conic would help to give a more complete picture of the students' ability to sketch curves if used in conjunction with the kinds of problems previously suggested. Problems of this type were included on this test. The following examples will suggest some of the possibilities of this type of item. The statistical data included indicates the general effectiveness of such items in terms of their validity and difficulty.

Example 7. The graph of the equation $y = x^3 + x^2 - 2x$ has the general shape

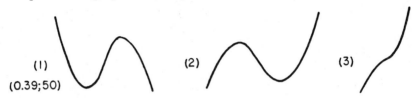

(4) none of the proposed answers is correct.

Example 8. The area bounded by the curves $y = -x^2$, and $4x - 2y = 1$ is sketched in

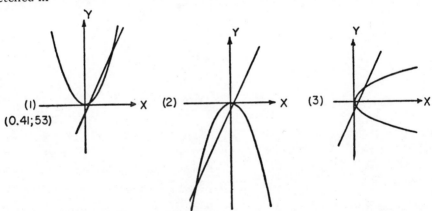

(4) none of the proposed answers is correct.

Example 9. The graph of $y = \sin x$ for values of x between 90° and 180° has the general shape

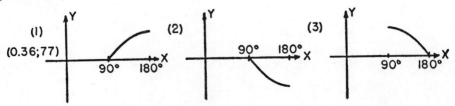

(4) none of the proposed answers is correct.

Example 10. The volume in the first octant bounded by the surfaces $x^2-y^2-z^2=0$, $x=4$, the xy-plane and the xz-plane has the general form

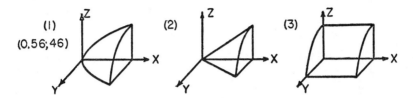

(4) none of the proposed answers is correct.

The reverse type of problem was also tried. One example is included.

Example 11. The equation of the conic in Fig. 2 is (0.42; 53) (1) $x^2+y^2=5$ (2) $2y^2-x=0$ (3) $x^2+4y=0$ (4) none of the proposed answers is correct.

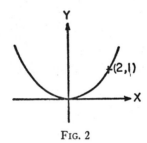

Fig. 2

The authors recognize that in several of these examples the problems are not stated with complete mathematical precision. The determining factor in each case was to make the statement as precise and exact as it seemed possible to do without confusing the student. This policy is consistent, it is believed, with that generally practiced by teachers and texts at this level of instruction.

6. Summary. In this paper we have suggested a new form of objective item for multiple-choice tests. For certain types of problems we believe it to be superior to the conventional item because it is so stated as to encourage the student to solve the problem, it does not permit the student to determine readily the answer by substitution, and at the same time it tends to minimize guessing. Since the items have such a universally high validity index for this sample their use is justified pending further investigation.

The latter part of the paper was devoted to indicating some of the possibilities for using multiple-choice questions for recognition of curves in helping to determine whether students can sketch curves. It is realized that if the student has the ability to recognize curves it does not necessarily follow that he can sketch a curve. However, we believe it does follow that a student who cannot recognize the correct curve also cannot sketch it. To that degree curve recognition does test the ability of a student to sketch curves.

The high validity, however, of these various types of items for this particular sample indicates that in so far as they do measure the ability of the student

GEOMETRY AND EMPIRICAL SCIENCE*

C. G. HEMPEL, Queens College

1. Introduction. The most distinctive characteristic which differentiates mathematics from the various branches of empirical science, and which accounts for its fame as the queen of the sciences, is no doubt the peculiar certainty and necessity of its results. No proposition in even the most advanced parts of empirical science can ever attain this status; a hypothesis concerning "matters of empirical fact" can at best acquire what is loosely called a high probability or a high degree of confirmation on the basis of the relevant evidence available; but however well it may have been confirmed by careful tests, the possibility can never be precluded that it will have to be discarded later in the light of new and disconfirming evidence. Thus, all the theories and hypotheses of empirical science share this provisional character of being established and accepted "until further notice," whereas a mathematical theorem, once proved, is established once and for all; it holds with that particular certainty which no subsequent empirical discoveries, however unexpected and extraordinary, can ever affect to the slightest extent. It is the purpose of this paper to examine the nature of that proverbial "mathematical certainty" with special reference to geometry, in an attempt to shed some light on the question as to the validity of geometrical theories, and their significance for our knowledge of the structure of physical space.

The nature of mathematical truth can be understood through an analysis of the method by means of which it is established. On this point I can be very brief: it is the method of mathematical demonstration, which consists in the logical deduction of the proposition to be proved from other propositions, previously established. Clearly, this procedure would involve an infinite regress unless some propositions were accepted without proof; such propositions are indeed found in every mathematical discipline which is rigorously developed; they are the *axioms* or *postulates* (we shall use these terms interchangeably) of the theory. Geometry provides the historically first example of the axiomatic presentation of a mathematical discipline. The classical set of postulates, however, on which Euclid based his system, has proved insufficient for the deduction of the well-known theorems of so-called euclidean geometry; it has therefore been revised and supplemented in modern times, and at present various adequate systems of postulates for euclidean geometry are available; the one most closely related to Euclid's system is probably that of Hilbert.

---

* From AMERICAN MATHEMATICAL MONTHLY, vol. 52 (1945), pp. 7–17.

2. The inadequacy of Euclid's postulates.

The inadequacy of Euclid's own set of postulates illustrates a point which is crucial for the axiomatic method in modern mathematics: Once the postulates for a theory have been laid down, every further proposition of the theory must be proved exclusively by logical deduction from the postulates; any appeal, explicit or implicit, to a feeling of self-evidence, or to the characteristics of geometrical figures, or to our experiences concerning the behavior of rigid bodies in physical space, or the like, is strictly prohibited; such devices may have a heuristic value in guiding our efforts to find a strict proof for a theorem, but the proof itself must contain absolutely no reference to such aids. This is particularly important in geometry, where our so-called intuition of geometrical relationships, supported by reference to figures or to previous physical experiences, may induce us tacitly to make use of assumptions which are neither formulated in our postulates nor provable by means of them. Consider, for example, the theorem that in a triangle the three medians bisecting the sides intersect in one point which divides each of them in the ratio of 1:2. To prove this theorem, one shows first that in any triangle ABC (see figure) the line segment MN which connects the centers of AB and AC is parallel to BC and therefore half as long as the latter side. Then the lines BN and CM are drawn, and an examination of the triangles MON and BOC leads to the proof of the theorem. In this procedure, it is usually taken for granted that BN and CM intersect in a point O which lies between B and N as well as between C

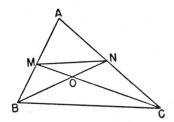

and M. This assumption is based on geometrical intuition, and indeed, it cannot be deduced from Euclid's postulates; to make it strictly demonstrable and independent of any reference to intuition, a special group of postulates has been added to those of Euclid; they are the postulates of order. One of these—to give an example—asserts that if A, B, C are points on a straight line l, and if B lies between A and C, then B also lies between C and A.—Not even as "trivial" an assumption as this may be taken for granted; the system of postulates has to be made so complete that all the required propositions can be deduced from it by purely logical means.

Another illustration of the point under consideration is provided by the proposition that triangles which agree in two sides and the enclosed angle, are congruent. In Euclid's Elements, this proposition is presented as a theorem; the alleged proof, however, makes use of the ideas of motion and superimposition of figures and thus involves tacit assumptions which are based on our geometric intuition and on experiences with rigid bodies, but which are definitely not warranted by—*i.e.* deducible from—Euclid's postulates. In Hilbert's system, therefore, this proposition (more precisely: part of it) is explicitly included among the postulates.

3. Mathematical certainty. It is this purely deductive character of mathematical proof which forms the basis of mathematical certainty: What the rigorous proof of a theorem—say the proposition about the sum of the angles in a triangle—establishes is not the truth of the proposition in question but rather a conditional insight to the effect that that proposition is certainly true *provided that* the postulates are true; in other words, the proof of a mathematical proposition establishes the fact that the latter is logically implied by the postulates of the theory in question. Thus, each mathematical theorem can be cast into the form

$$(P_1 \cdot P_2 \cdot P_3 \cdot \ldots \cdot P_N) \to T$$

where the expression on the left is the conjunction (joint assertion) of all the postulates, the symbol on the right represents the theorem in its customary formulation, and the arrow expresses the relation of logical implication or entailment. Precisely this character of mathematical theorems is the reason for their peculiar certainty and necessity, as I shall now attempt to show.

It is typical of any purely logical deduction that the conclusion to which it leads simply re-asserts (a proper or improper) part of what has already been stated in the premises. Thus, to illustrate this point by a very elementary example, from the premise, "This figure is a right triangle," we can deduce the conclusion, "This figure is a triangle"; but this conclusion clearly reiterates part of the information already contained in the premise. Again, from the premises, "All primes different from 2 are odd" and "n is a prime different from 2," we can infer logically that n is odd; but this consequence merely repeats part (indeed a relatively small part) of the information contained in the premises. The same situation prevails in all other cases of logical deduction; and we may, therefore, say that logical deduction—which is the one and only method of mathematical proof—is a technique of conceptual analysis: it discloses what assertions are concealed in a given set of premises, and it makes us realize to what we committed ourselves in accepting those premises; but none of the results obtained by this technique ever goes by one iota beyond the information already contained in the initial assumptions.

Since all mathematical proofs rest exclusively on logical deductions from certain postulates, it follows that a mathematical theorem, such as the Pythagorean theorem in geometry, asserts nothing that is *objectively* or *theoretically new* as compared with the postulates from which it is derived, although its content may well be *psychologically new* in the sense that we were not aware of its being implicitly contained in the postulates.

The nature of the peculiar certainty of mathematics is now clear: A mathematical theorem is certain *relatively* to the set of postulates from which it is derived; *i.e.* it is necessarily true *if* those postulates are true; and this is so because the theorem, if rigorously proved, simply re-asserts part of what has been stipulated in the postulates. A truth of this conditional type obviously implies no assertions about matters of empirical fact and can, therefore, never get into conflict with any empirical findings, even of the most unexpected kind; consequently, unlike the hypotheses and theories of empirical science, it can never suffer the fate of being disconfirmed by new evidence: A mathematical truth is irrefutably certain just because it is devoid of factual, or empirical content. Any theorem of geometry, therefore, when cast into the conditional form described earlier, is analytic in the technical sense of logic, and thus true *a priori*; *i.e.* its truth can be established by means of the formal machinery of logic alone, without any reference to empirical data.

4. Postulates and truth. Now it might be felt that our analysis of geometrical truth so far tells only half of the relevant story. For while a geometrical proof no doubt enables us to assert a proposition conditionally—namely on condition that the postulates are accepted—, is it not correct to add that geometry also unconditionally asserts the truth of its postulates and thus, by virtue of the deductive relationship between postulates and theorems, enables us unconditionally to assert the truth of its theorems? Is it not an unconditional assertion of geometry that two points determine one and only one straight line that connects them, or that in any triangle, the sum of the angles equals two right angles? That this is definitely not the case, is evidenced by two important aspects of the axiomatic treatment of geometry which will now be briefly considered.

The first of these features is the well-known fact that in the more recent development of mathematics, several systems of geometry have been constructed which are incompatible with euclidean geometry, and in which, for example, the two propositions just mentioned do not necessarily hold. Let us briefly recollect some of the basic facts concerning these *non-euclidean geometries*. The postulates on which euclidean geometry rests include the famous postulate of the parallels, which, in the case of plane geometry, asserts in effect that through every point P not on a given line l there exists exactly one parallel to l, *i.e.*, one straight line which does not meet l. As this postulate is considerably less simple than the others, and as it was also felt to be intuitively less plausible than the latter, many efforts were made in the history of geometry to prove that this proposition need not be accepted as an axiom, but that it can be deduced as a theorem from the remaining body of postulates. All attempts in this direction failed, however; and finally it was conclusively demonstrated that a proof of the parallel principle on the basis of the other postulates of euclidean geometry (even in its modern, completed form) is impossible. This was shown by proving that a perfectly self-consistent geometrical theory is obtained if the postulate of the parallels is replaced by the assumption that through any point P not on a given straight line l there exist at least two parallels to l. This postulate obviously contradicts the euclidean postulate of the parallels, and if the latter were actually a consequence of the other postulates of euclidean geometry, then the new set of postulates would clearly involve a contradiction, which can be shown not to be the case. This first non-euclidean type of geometry, which is called hyperbolic geometry, was discovered in the early 20's of the last century almost simultaneously, but independently by the Russian N. I. Lobatschefskij, and by the Hungarian J. Bolyai. Later, Riemann developed an alternative geometry, known as elliptical geometry, in which the axiom of the parallels is replaced by the postulate that no line has any parallels. (The acceptance of this postulate, however, in contradistinction to that of hyperbolic geometry, requires the modification of some further axioms of euclidean geometry, if a consistent new theory is to result.) As is to be expected, many of the theorems of these non-euclidean geometries are at variance with those of euclidean theory; thus, *e.g.*, in the hyperbolic geometry of two dimensions, there exist, for each straight line l, through any point P not on l, infinitely many straight lines which do not meet l; also, the sum of the angles in any triangle is less than two right angles. In elliptic geometry, this angle sum is always greater than two right angles; no two straight lines are parallel; and while two different points usually determine exactly one straight line connecting them (as they always do in euclidean geometry), there are certain pairs of points which are connected by infinitely many different straight lines. An illustration of this latter type of geometry is provided by the geometrical structure of that curved two-dimensional space which is represented by the surface of a sphere, when the concept of straight line is interpreted by that of great circle on the sphere. In this space, there are no parallel lines since any two great circles intersect; the endpoints of any diameter of the sphere are points connected by infinitely many different "straight lines," and the sum of the angles in a triangle is always in excess of two right angles. Also, in this space, the ratio between the circumference and the diameter of a circle (not necessarily a great circle) is always less than 2π.

Elliptic and hyperbolic geometry are not the only types of non-euclidean geometry; various other types have been developed; we shall later have occasion to refer to a much more general form of non-euclidean geometry which was likewise devised by Riemann.

The fact that these different types of geometry have been developed in modern mathematics shows clearly that mathematics cannot be said to assert the truth of any particular set of geometrical postulates; all that pure mathematics is interested in, and all that it can establish, is the deductive consequences of given sets of postulates and thus the necessary truth of the ensuing theorems relatively to the postulates under consideration.

A second observation which likewise shows that mathematics does not assert the truth of any particular set of postulates refers to *the status of the concepts in geometry*. There exists, in every axiomatized theory, a close parallelism between the treatment of the propositions and that of the concepts of the system. As we have seen, the propositions fall into two classes: the postulates, for which no proof is given, and the theorems, each of which has to be derived from the postulates. Analogously, the concepts fall into two classes: the primitive or basic concepts, for which no definition is given, and the others, each of which has to be precisely defined in terms of the primitives. (The admission of some undefined concepts is clearly necessary if an infinite regress in definition is to be avoided.) The analogy goes farther: Just as there exists an infinity of theoretically suitable axiom systems for one and the same theory—say, euclidean geometry—, so there also exists an infinity of theoretically possible choices for the primitive terms of that theory; very often—but not always—different axiomatizations of the same theory involve not only different postulates, but also different sets of primitives. Hilbert's axiomatization of plane geometry contains six primitives: point, straight line, incidence (of a point on a line), betweenness (as a relation of three points on a straight line), congruence for line segments, and congruence for angles. (Solid geometry, in Hilbert's axiomatization, requires two further primitives, that of plane and that of incidence of a point on a plane.) All other concepts of geometry, such as those of angle, triangle, circle, *etc.*, are defined in terms of these basic concepts.

But if the primitives are not defined within geometrical theory, what meaning are we to assign to them? The answer is that it is entirely unnecessary to connect any particular meaning with them. True, the words "point," "straight line," *etc.*, carry definite connotations with them which relate to the familiar geometrical figures, but the validity of the propositions is completely independent of these connotations. Indeed, suppose that in axiomatized euclidean geometry, we replace the over-suggestive terms "point," "straight line," "incidence," "betweenness," *etc.*, by the neutral terms "object of kind 1," " object of kind 2," "relation No. 1," "relation No. 2," *etc.*, and suppose that we present this modified wording of geometry to a competent mathematician or logician who, however, knows nothing of the customary connotations of the primitive terms. For this logician, all proofs would clearly remain valid, for as we saw before, a rigorous proof in geometry rests on deduction from the axioms alone without any reference to the customary interpretation of the various geometrical concepts used. We see therefore that indeed no specific meaning has to be attached to the primitive terms of an axiomatized theory; and in a precise logical presentation of axiomatized geometry the primitive concepts are accordingly treated as so-called logical variables.

As a consequence, geometry cannot be said to assert the truth of its postulates, since the latter are formulated in terms of concepts without any specific meaning; indeed, for this very reason, the postulates themselves do not make any specific assertion which could possibly be called true or false! In the terminology of modern logic, the postulates are not sentences, but sentential functions with the primitive concepts as variable arguments.—This point also shows that the postulates of geometry cannot be considered as "self-evident truths," because where no assertion is made, no self-evidence can be claimed.

5. Pure and physical geometry. Geometry thus construed is a purely formal discipline; we shall refer to it also as *pure geometry*. A pure geometry, then,—no matter whether it is of the euclidean or of a non-euclidean variety—deals with no specific subject-matter; in particular, it asserts nothing about physical space. All its theorems are analytic and thus true with certainty precisely because they are devoid of factual content. Thus, to characterize the import of pure geometry, we might use the standard form of a movie-disclaimer: No portrayal of the characteristics of geometrical figures or of the spatial properties or relationships of actual physical bodies is intended, and any similarities between the primitive concepts and their customary geometrical connotations are purely coincidental.

But just as in the case of some motion pictures, so in the case at least of euclidean geometry, the disclaimer does not sound quite convincing: Historically speaking, at least, euclidean geometry has its origin in the generalization and systematization of certain empirical discoveries which were made in connection with the measurement of areas and volumes, the practice of surveying, and the development of astronomy. Thus understood, geometry has factual import; it is an empirical science which might be called, in very general terms, the theory of the structure of physical space, or briefly, *physical geometry*. What is the relation between pure and physical geometry?

When the physicist uses the concepts of point, straight line, incidence, *etc.*, in statements about physical objects, he obviously connects with each of them a more or less definite physical meaning. Thus, the term "point" serves to designate physical points, *i.e.*, objects of the kind illustrated by pin-points, cross hairs, *etc.* Similarly, the term "straight line" refers to straight lines in the sense of physics, such as illustrated by taut strings or by the path of light rays in a homogeneous medium. Analogously, each of the other geometrical concepts has a concrete physical meaning in the statements of physical geometry. In view of this situation, we can say that physical geometry is obtained by what is called, in contemporary logic, a semantical interpretation of pure geometry, Generally speaking, a semantical interpretation of a pure mathematical theory, whose primitives are not assigned any specific meaning, consists in giving each primitive (and thus, indirectly, each defined term) a specific meaning or designatum. In the case of physical geometry, this meaning is physical in the sense just illustrated; it is possible, however, to assign a purely arithmetical meaning to each concept of geometry; the possibility of such an arithmetical interpretation of geometry is of great importance in the study of the consistency and other logical characteristics of geometry, but it falls outside the scope of the present discussion.

By virtue of the physical interpretation of the originally uninterpreted primitives of a geometrical theory, physical meaning is indirectly assigned also to every defined concept of the theory; and if every geometrical term is now taken in its physical interpretation, then every postulate and every theorem of the theory under consideration turns into a statement of physics, with respect to which the question as to truth or falsity may meaningfully be raised—a circumstance which clearly contradistinguishes the propositions of physical geometry from those of the corresponding uninterpreted pure theory.—Consider, for example, the following postulate of pure euclidean geometry: For any two objects

x, y of kind 1, there exists exactly one object l of kind 2 such that both x and y stand in relation No. 1 to l. As long as the three primitives occurring in this postulate are uninterpreted, it is obviously meaningless to ask whether the postulate is true. But by virtue of the above physical interpretation, the postulate turns into the following statement: For any two physical points x, y there exists exactly one physical straight line l such that both x and y lie on l. But this is a physical hypothesis, and we may now meaningfully ask whether it is true or false. Similarly, the theorem about the sum of the angles in a triangle turns into the assertion that the sum of the angles (in the physical sense) of a figure bounded by the paths of three light rays equals two right angles.

Thus, the physical interpretation transforms a given pure geometrical theory—euclidean or non-euclidean—into a system of physical hypotheses which, if true, might be said to constitute a theory of the structure of physical space. But the question whether a given geometrical theory in physical interpretation is factually correct represents a problem not of pure mathematics but of empirical science; it has to be settled on the basis of suitable experiments or systematic observations. The only assertion the mathematician can make in this context is this: If all the postulates of a given geometry, in their physical interpretation, are true, then all the theorems of that geometry, in their physical interpretation, are necessarily true, too, since they are logically deducible from the postulates. It might seem, therefore, that in order to decide whether physical space is euclidean or non-euclidean in structure, all that we have to do is to test the respective postulates in their physical interpretation. However, this is not directly feasible; here, as in the case of any other physical theory, the basic hypotheses are largely incapable of a direct experimental test; in geometry, this is particularly obvious for such postulates as the parallel axiom or Cantor's axiom of continuity in Hilbert's system of euclidean geometry, which makes an assertion about certain infinite sets of points on a straight line. Thus, the empirical test of a physical geometry no less than that of any other scientific theory has to proceed indirectly; namely, by deducing from the basic hypotheses of the theory certain consequences, or predictions, which are amenable to an experimental test. If a test bears out a prediction, then it constitutes confirming evidence (though, of course, no conclusive proof) for the theory; otherwise, it disconfirms the theory. If an adequate amount of confirming evidence for a theory has been established, and if no disconfirming evidence has been found, then the theory may be accepted by the scientist "until further notice."

It is in the context of this indirect procedure that pure mathematics and logic acquire their inestimable importance for empirical science: While formal logic and pure mathematics do not in themselves establish any assertions about matters of empirical fact, they provide an efficient and entirely indispensable machinery for deducing, from abstract theoretical assumptions, such as the laws of Newtonian mechanics or the postulates of euclidean geometry in physical interpretation, consequences concrete and specific enough to be accessible to direct experimental test. Thus, *e.g.*, pure euclidean geometry shows that from its postulates there may be deduced the theorem about the sum of the angles in a triangle, and that this deduction is possible no matter how the basic concepts of geometry are interpreted; hence also in the case of the physical interpretation of euclidean geometry. This theorem, in its physical interpretation, is accessible to experimental test; and since the postulates of elliptic and of hyperbolic geometry imply values different from two right angles for the angle sum of a triangle, this particular proposition seems to afford a good opportunity for a crucial experiment. And no less a mathematician than Gauss did indeed perform this test; by means of optical methods—and thus using the interpretation of physical straight lines as paths of light rays—he ascertained the angle sum of a large triangle determined by three mountain tops. Within the limits of experimental error, he found it equal to two right angles.

6. On Poincaré's conventionalism concerning geometry. But suppose that Gauss had found a noticeable deviation from this value; would that have meant a refutation of euclidean geometry in its physical interpretation, or, in other words, of the hypothesis that physical space is euclidean in structure? Not necessarily; for the deviation might have been accounted for by a hypothesis to the effects that the paths of the light rays involved in the sighting process were bent by some disturbing force and thus were not actually straight lines. The same kind of reference to deforming forces could also be used if, say, the euclidean theorems of congruence for plane figures were tested in their physical interpretation by means of experiments involving rigid bodies, and if any violations of the theorems were found. This point is by no means trivial; Henri Poincaré, the great French mathematician and theoretical physicist, based on considerations of this type his famous *conventionalism concerning geometry*. It was his opinion that no empirical test, whatever its outcome, can conclusively invalidate the euclidean conception of physical space; in other words, the validity of euclidean geometry in physical science can always be preserved—if necessary, by suitable changes in the theories of physics, such as the introduction of new hypotheses concerning deforming or deflecting forces. Thus, the question as to whether physical space has a euclidean or a non-euclidean structure would become a matter of convention, and the decision to preserve euclidean geometry at all costs would recommend itself, according to Poincaré, by the greater simplicity of euclidean as compared with non-euclidean geometrical theory.

It appears, however, that Poincaré's account is an oversimplification. It rightly calls attention to the fact that the test of a physical geometry G always presupposes a certain body P of non-geometrical physical hypotheses (including the physical theory of the instruments of measurement and observation used in the test), and that the so-called test of G actually bears on the combined theoretical system $G \cdot P$ rather than on G alone. Now, if predictions derived from $G \cdot P$ are contradicted by experimental findings, then a change in the theoretical structure becomes necessary. In classical physics, G always was euclidean geometry in its physical interpretation, GE; and when experimental evidence required a modification of the theory, it was P rather than GE which was changed. But Poincaré's assertion that this procedure would always be distinguished by its greater simplicity is not entirely correct; for what has to be taken into consideration is the simplicity of the total system $G \cdot P$, and not just that of its geometrical part. And here it is clearly conceivable that a simpler total theory in accordance with all the relevant empirical evidence is obtainable by going over to a non-euclidean form of geometry rather than by preserving the euclidean structure of physical space and making adjustments only in part P.

And indeed, just this situation has arisen in physics in connection with the development of the general theory of relativity: If the primitive terms of geometry are given physical interpretations along the lines indicated before, then certain findings in astronomy represent good evidence in favor of a total physical theory with a non-euclidean geometry as part G. According to this theory, the physical universe at large is a three-dimensional curved space of a very complex geometrical structure; it is finite in volume and yet unbounded in all directions. However, in comparatively small areas, such as those involved in Gauss' experiment, euclidean geometry can serve as a good approximative account of the geometrical structure of space. The kind of structure ascribed to physical space in this theory may be illustrated by an analogue in two dimensions; namely, the surface of a sphere. The geometrical structure of the latter, as was pointed out before, can be described by means of elliptic geometry, if the primitive term "straight line" is interpreted as meaning "great circle," and if the other primitives are given analogous interpretations. In this sense, the surface of a sphere is a two-dimensional curved space of non-euclidean structure, whereas the plane is a two-dimensional space of euclidean structure. While the plane is unbounded in all directions, and infinite in size, the spherical surface is finite in size and yet unbounded in all directions: a two-dimensional physicist, travelling along "straight lines" of that space would never encounter any boundaries of his space; instead, he would finally return to his point of departure, provided that his life span and his technical facilities were sufficient for such a trip in consideration of the size of his "universe." It is interesting to note that the physicists of that world, even if they lacked any intuition of a three-dimensional space, could empirically ascertain the fact that their two-dimensional space was curved. This might be done by means of the method of traveling along straight lines; another, simpler test would consist in determining the angle sum in a triangle; again another in determining, by means of measuring tapes, the ratio of the circumference of a circle (not necessarily a great circle) to its diameter; this ratio would turn out to be less than π.

The geometrical structure which relativity physics ascribes to physical space is a three-dimensional analogue to that of the surface of a sphere, or, to be more exact, to that of the closed and finite surface of a potato, whose curvature varies from point to point. In our physical universe, the curvature of space at a given point is determined by the distribution of masses in its neighborhood; near large masses such as the sun, space is strongly curved, while in regions of low mass-density, the structure of the universe is approximately euclidean. The hypothesis stating the connection between the mass distribution and the curvature of space at a point has been approximately confirmed by astronomical observations concerning the paths of light rays in the gravitational field of the sun.

The geometrical theory which is used to describe the structure of the physical universe is of a type that may be characterized as a generalization of elliptic geometry. It was originally constructed by Riemann as a purely mathematical theory, without any concrete possibility of practical application at hand. When Einstein, in developing his general theory of relativity, looked for an appropriate mathematical theory to deal with the structure of physical space, he found in Riemann's abstract system the conceptual tool he needed. This fact throws an interesting sidelight on the importance for scientific progress of that type of investigation which the "practical-minded" man in the street tends to dismiss as useless, abstract mathematical speculation.

Of course, a geometrical theory in physical interpretation can never be validated with mathematical certainty, no matter how extensive the experimental tests to which it is subjected; like any other theory of empirical science, it can acquire only a more or less high degree of confirmation. Indeed, the considerations presented in this article show that the demand for mathematical certainty in empirical matters is misguided and unreasonable; for, as we saw, mathematical certainty of knowledge can be attained only at the price of analyticity and thus of complete lack of factual content. Let me summarize this insight in Einstein's words:

"As far as the laws of mathematics refer to reality, they are not certain; and as far as they are certain, they do not refer to reality."

A THEOREM IN ELEMENTARY MATHEMATICS*

A. D. WALLACE, Tulane University

The following definition is adopted from a textbook on algebra: An extraneous root of $f(x) = 0$ is a number which is not a root but which satisfies an equation derived from $f(x) = 0$ by permissible operations.

Our purpose is to the prove the

THEOREM: *Any number not a root of $f(x) = 0$ is an extraneous root.*

Proof: The equation $(x-t)f(x) = 0$ is derived from $f(x) = 0$ by a permissible operation.

We have (following the author of the textbook) failed to define "equation" and "permissible operation." Such definitions are beyond the scope of this paper. Their significance will, however, be apparent to any thoughtful reader.

In closing we express the hope that our theorem will, in some measure, indicate the importance to be attached to the concept of extraneous root.

---

* From AMERICAN MATHEMATICAL MONTHLY, vol. 55 (1948), p. 639.

OBJECTIVES IN THE TEACHING OF COLLEGE MATHEMATICS*†

F. S. NOWLAN, University of Illinois

1. Introduction. Any constructive study of objectives in the teaching of mathematics must presuppose an agreement as to what is meant by mathematics. In this paper, we concern ourselves with mathematics in the sense of mathematical thinking and not as an inclusive term applying to the vast body of mathematical truths; also we are not thinking primarily of the mastery of rules and devices for the manipulation of numbers and algebraic expressions. What, then, is mathematics in this sense and what is its value?

It is a truism to state that mathematical thinking is deductive reasoning, but we can profitably examine this statement. Deductive reasoning is based upon assumptions, and these apply to what, for lack of a better word, we will designate elements. The elements may, in part, be accepted without definition, for example, our concept of point in geometry. Any mathematical development begins by assuming that the elements under consideration satisfy certain conditions, that is, are subject to certain rules of behavior. Mathematics consists in the logical deduction of the necessary consequences of the initial assumptions.

It may happen that the assumptions clash, or lead to conclusions which contradict established or accepted theory. One may then modify the assumptions, or start with a new set, or, in the latter contingency, one may be radical and hold to the initial assumptions and ignore the accepted theory. This radical action has occurred frequently in the development of science and has made scientific progress possible.

The procedure that we have outlined is basic in mathematical thinking, and, although not always followed, is useful in every field of human endeavor.

Unfortunately, mathematics, in this sense, will fail to bring about a Utopia. The initial assumptions may include the doctrine that the greatest good results from self-advancement, or that one's chief mission in life is the dissemination of certain creeds or ideologies, or the retention of certain creeds or ideologies, and that the end justifies the means. The individual may then reason logically and honestly, that is, in agreement with his assumptions, but the result may be a Hitler, or worse, whose success or failure, proves or disproves (according to accepted standards of proof) his theories.

All this goes to show that mathematical thinking, in itself, is no cure-all for human ills, but it is essential if these ills are to be alleviated. To repeat, mathematical thinking merely means the study of the necessary consequences of initial assumptions which motivate courses of action. As such it applies even to the social and economic sciences. It is non-moral in character and has no concern with variable standards of rightness or wrongness in behavior.

---

* From AMERICAN MATHEMATICAL MONTHLY, vol. 57 (1950), pp. 73–82.

† Presented to the Illinois Section of the Association at Chicago, Illinois, May 14, 1948. Also presented at the meeting of the National Council, Columbus, Ohio, December 29, 1948.

2. Objectives in teaching mathematics.

By now you are wondering what this has to do with objectives in the teaching of college mathematics. Let me explain, and in doing so you will understand that I am merely presenting my personal views and reactions to conditions which are fairly common in our colleges and universities. Also, I would emphasize at this point that I do not underestimate the value of practical applications in the teaching of mathematics, nor the necessity for the mastery of mechanical skills in numerical and algebraic operations. I assume that these are taken care of, as a matter of course, in our instruction.

There are two main aims in the teaching of college mathematics, whether to liberal arts students or to students of engineering. These, in order of importance, are: (1) Training in precise thinking and a grasp of basic principles. (2) The acquisition of information and a mastery of certain technical skills.

It appears, however, that, in general, a disproportionate stress has been placed upon the second objective to the almost total neglect of the first. The excuse has been that the time at the instructor's disposal is limited and that present day students are poorly prepared and intellectually immature.

In answer to these objections, we point out that mechanical work that is not based upon an understanding of principles has little value, practical or otherwise, and that a student can progress rapidly, and without mental strain, only when he works understandingly.

It is true that in every student body there are those who are incapable of logical thinking but we should hardly permit them to determine our standards. Also, there are those who are so gifted intellectually that they will progress under any system and in spite of handicaps of instruction and textbooks. It happens, however, that the majority of students are in a middle class; they need guidance and can profit from guidance. They are capable of logical thinking but, in many cases, have never undergone that experience. These students constitute the main reason for instructors in our colleges.

One of the first duties of the instructor should be to purge the student's mind of the belief that mathematics is a purely mechanical art and that it consists in the formal application of rules that must be memorized. This concept of mathematics, which is fairly common among both students and instructors, stultifies the intellect and the imagination.

To digress, I recall the case of an ex-marine, in a navy V-12 course, a young man of more than average perseverance and intelligence. This student was asked to show that three points A, B, and C were collinear. He found that the lengths of AB, BC, and AC were $\sqrt{13}$, $\sqrt{13}$, and $\sqrt{52}$, respectively, and remarked, "We must prove $\sqrt{52} = 2\sqrt{13}$." Then follows the deduction:

$$\sqrt{13} = \sqrt{4+9} = 2 + 3 = 5,$$
$$\sqrt{52} = \sqrt{16+36} = 4 + 6 = 10,$$

but

$$10 = 2 \times 5,$$

therefore
$$\sqrt{52} = 2\sqrt{13}.$$

In doing this, the young man showed imagination and some power of analysis and, in the best tradition of his service, he attained his objective, but he had never learned to associate mathematics and thinking.

3. Some suggestions. How then can we remedy matters? Let me presume to suggest a means. Because I believe that it is only by virtue of an intelligent understanding and grasp of principles that a student can obtain lasting profit from his mathematical studies, and also make progress and acquire speed in his later work, and because I believe that algebra is the study that can best give the student the requisite viewpoint and training, I suggest that more time be devoted to that subject. Not only are we obliged to give the student a new point of view but, frequently, we must eradicate the result of earlier training. This takes time and cannot be accomplished overnight; certainly we can not build a lasting structure upon a superficial foundation. I therefore suggest that the instruction in algebra courses, for the first few weeks, be deliberate, and that the stress, in the first few lessons, be upon the nature of deductive reasoning.

In this connection, the instructor might find profit in reading a work such as the recently published book by B. W. Jones, entitled *Elementary Concepts of Mathematics*. Although serving as an introduction to abstract thinking, the discussion and examples which appear in this book are such as to arouse the interest of the student and stimulate his imagination. Unfortunately, on the other hand, we find influential writers on the teaching of mathematics who seem obsessed with the notion that student interest and mental stimulus can be obtained only by a study of problems which deal with dollars and cents or which concern some form of physical activity. A learning process that is so motivated can scarcely broaden the student's intellectual outlook, and would tend to accentuate his adherence to preconceived notions and personal prejudices. It would be of doubtful cultural value.

The instructor might next take up with his class the growth of the number system and indicate the part that postulates play in its development. The treatment would be abstract since it deals with numbers which are abstractions, but it could be presented so as to capture the students' interest. One of the difficulties at this stage of the work is the necessity of convincing the student that rules of operation must be reexamined with every extension of the number system. This can be done, and the student's interest captured, by an illustration somewhat similar to the following: Consider three colored beads, red, white, and blue, upon a wire in the order

$$\underline{\qquad r \qquad\qquad w \qquad\qquad b \qquad}$$

An interchange of the beads to produce a new order, say,

$$\underline{\qquad w \qquad\qquad b \qquad\qquad r \qquad}$$

is called a substitution and may be designated by the symbol

$$s_1 = \begin{pmatrix} r & w & b \\ w & b & r \end{pmatrix},$$

which indicates the replacement of r by w, w by b, and b by r. In case we are dissatisfied with the new color scheme, we may continue with another substitution, say:

$$s_2 = \begin{pmatrix} w & b & r \\ w & r & b \end{pmatrix}$$

to produce the order

w	r	b

We then observe that the single substitution,

$$\begin{pmatrix} r & w & b \\ w & r & b \end{pmatrix},$$

is the resultant, or equivalent, of the two substitutions s_1 and s_2, in the indicated order. For lack of a better name, we call the resultant substitution the product of s_1 and s_2, in that order, and we agree to represent it by the notation s_1s_2. We also call the process of forming the product multiplication. In doing all this we have used the words product and multiplication in a new sense; that is, we have extended the meaning of these words. It is then fairly obvious that rules for multiplication which were established in our earlier work, say for positive numbers, will not carry over automatically to the multiplication of substitutions. In fact, upon trial, we obtain for the substitution s_2 followed by s_1, the product

$$s_2s_1 = \begin{pmatrix} r & w & b \\ r & b & w \end{pmatrix}.$$

Thus it follows for substitutions that

$$s_1s_2 \neq s_2s_1,$$

whereas, in our earlier work with positive numbers,

$$s_1s_2 = s_2s_1.$$

An illustration of this sort should prepare for a discussion of the problems that attend any extension of the number system. It should also suggest that if, by definition, we introduce a new type of number, we should never assume that rules, which were not included in our definitions, apply merely because they were established for more elementary types of numbers. We must inquire into the rules of operation every time we extend the number system.

All this means that we propose beginning the study of college algebra with an attempt to indoctrinate the student in the rudiments of abstract thinking. This has the advantage that the student becomes conscious of the postulational character of mathematics and learns at an early stage to refer back to his assumptions before arriving at conclusions. In other words, he gets training in arriving at conclusions which are not based on prejudice, or preconceived notions, or on the word of some authority. He must think for himself, and work from first principles.

At a later time, the introduction of complex numbers into the number system furnishes another excellent opportunity for training in deductive reasoning.

We have now completed the introductory stage in the study of college algebra. From here on, our remarks apply equally well to any course in college mathematics and merely suggest how the instructor may attain the objective that the student shall understand his mathematics and shall be capable of independent thinking. Of course, as mentioned at the outset, we assume throughout that the student shall receive the usual instruction and drill in manipulative processes, except that this shall not dominate the picture. With all this in mind, it is most important that the student shall be required to express himself precisely, and without ambiguity, in speech, in symbolism, and in writing, especially in the latter since while writing he has more opportunity for deliberation. Information which one cannot impart to another has little intrinsic value. Furthermore, a person's inability to express his ideas clearly is indicative of confusion in thinking, and the converse is equally true.

The student must be aware of the exact meaning of the notation which he employs, and this, of course, involves an awareness of any restriction upon the interpretation of notation. I have in mind, for instance, the symbol $\sqrt{a}$, in which a is positive, which by universal agreement represents the positive square root of a. How many students realize that this agreement will compel him to write

$$\sqrt{x^2} = -x,$$

when x represents the abscissa of a point in the second quadrant? In any case, an example of this sort will prove useful. Also, in this connection, the student should be able to locate the fallacy in the statement

$$1 = \sqrt{1} = \sqrt{i^4} = i^2 = -1.$$

The same necessity for proper interpretation arises in the evaluation of the length of arc in the case of the astroid

$$x^{2/3} + y^{2/3} = a^{2/3}.$$

The usual formula gives the integral

$$\int \sqrt{\frac{a^{2/3}}{x^{2/3}}} \, dx.$$

The student should be able to explain why

$$\int_{-a}^{a} \frac{a^{1/3}}{x^{1/3}}\, dx$$

gives zero and not the length of the portion of the astroid above the x-axis. The difficulty is not in the improper integral but in failing to observe that $-a^{1/3}/x^{1/3}$ is the positive square root for (x, y) a point in the second quadrant.

Since writing the above, I have seen a reference to trigonometric identities in a recent number of a leading mathematical journal. The statement was made: "As a matter of fact, both members of the identity will tolerate any fundamental process, except, and here the caution must be flashed, the extraction of roots. For example,

$$\sin^2(-x) = 1 - \cos^2 x, \qquad \text{for any } x.$$

Extracting roots

$$\sin(-x) = \sqrt{1 - \cos^2 x},$$
$$-\sin x = \sqrt{1 - \cos^2 x}."$$

The author seems to labor under the confusion that we have just pointed out. He should have written

$$\pm \sin x = \sqrt{1 - \cos^2 x},$$

with the proviso that the sign plus or minus must be chosen so as to make the expression on the left side positive. For example, if

$$x = 240°,$$

he would have the true relation

$$\sin(-240°) = -\sin 240° = \sqrt{1 - \cos^2 240°}.$$

Another example in interpretation, one which involves a restriction in a differentiation, is the substitution

$$x = a \sin \theta$$

in the integral

$$\int_{-a}^{a} \sqrt{a^2 - x^2}\, dx.$$

Why must the limits in the resulting integral read from $-\pi/2$ to $\pi/2$, rather than from $3\pi/2$ to $\pi/2$? Attention to such details can mean the difference between the student working mechanically or intelligently.

OBJECTIVES IN THE TEACHING OF COLLEGE MATHEMATICS 51

The word "equals" is one of the important technical terms in elementary mathematics. The student must employ this word in its proper sense and also be aware of the fact that an equality sign is not a punctuation mark. The instructor should not accept the statement: "The equation equals," or, "The formula equals." Quite regardless of the numerical result that he obtains, a student should not receive a passing grade on the solution of a problem that concerns the speeds of two airplanes if he starts out: "Let x equal the first airplane."

I have seen these statements in recent student work, and also the following which illustrate additional misuse of the equality sign:

(1) $$2x - 3y = 7$$
$$= 2x - 3y - 7 = 0.$$

(2) $$x/2 + y/3 = 1$$
$$= 3x + 2y = 6.$$

(3) In finding the seventh term in the expansion of $(3x-2y)^{10}$:

$$(3x - 2y)^{10} = 7\text{th term} = etc.$$

(4) In finding the foci of the ellipse

$$16x^2 + 25y^2 = 400:$$

$$F = ae = \sqrt{a^2 - b^2} = \sqrt{25 - 16} = 3 = F_1(3, 0) \text{ and } F_2(-3, 0).$$

(5) Also,

$$\frac{27.8 \times .0364}{.456} = \frac{\log 27.8 + \log .0364}{\log .456}$$

$$= \frac{1.4440 + 8.5611 - 10}{9.6590 - 10} = \frac{10.0051 - 10}{9.6590 - 10}$$

$$= .3461 = 2.218 \text{ (the correct answer to the problem)}.$$

I suggest that to help the student appreciate the fault in work such as the last, the instructor might assign a simple problem of the type: Determine a number x which is less by 3 than the product of 6 and 7. He could then submit to the class, for their approval or disapproval, the solution.

$$x = 6 + 7 \div 3 = 42 \div 3 = 39.$$

If they disapprove, he could exhibit the logarithmic computation and make comparisons. If this makes no impression, the instructor may conclude that the student is not a prospective mathematical genius.

One final example: The student of college algebra is likely to believe that polynomial equations in x and y can be combined freely by addition, subtraction, and by the substitution of the one into the other and that, provided there have been no numerical errors, the values that are obtained for x and y constitute solutions. The student might, with profit, consider this procedure as applied to the equations

(a) $(x^2 + y^2 + x)^2 = 9(x^2 + y^2)$ and (b) $x^2 + y^2 = 1.$

The substitution of (b) into (a) gives

$$(x + 1)^2 = 9, \quad x + 1 = \pm 3, \quad x = 2, \quad -4.$$

If he substitutes 2 for x in (a), he gets the equation

$$(y^2 + 6)^2 = 9(y^2 + 4),$$

or,

$$y^2(y^2 + 3) = 0,$$

which has the real solution

$$y = 0.$$

The student would likely conclude that (2, 0) is a common solution of the given equations and so a common point of their graphs. It happens, however, that the graphs have no point in common and the equations no common solution.

It is desirable that the student should realize that the propriety of combining the equations as we have done is based on an assumption that there is a common pair of values for x and y which satisfy the two equations and that, as a consequence of such an assumption, even when the numerical work is known to be correct, the values that have been obtained for x and y must be checked in both equations.

The student should learn to employ words in their correct sense and so as to convey his meaning without ambiguity, and he must realize that symbolism constitutes a shorthand notation and must be employed correctly if it is to convey information. He should be expected, in his various written tests, to give precise definitions of terms that are employed in his studies and in doing this and in proving principles he should be encouraged to employ, so far as possible, his own phrasing rather than that of the instructor, or the text, verbatim. It would be interesting, and perhaps surprising, to learn the small percentage of those in classes in college mathematics who have any clear conception of the meaning of the terms which they employ in their daily work, words and phrases such as rational number, polynomial in x, polynomial in x of the nth degree, sum of an infinite series, convergent series, *etc.* Of course, in fairness, we should not place the entire blame for the student's inability to express himself upon his early training in mathematics. His instruction, or lack of instruction, in English must, in part, be held responsible.

We find that in many cases the student is adverse to reading his textbooks and is inclined to depend too much upon the instructor. It would help matters if he were not only required to read the text but were held responsible for interpreting assignments from it to the class.

From all that has been said, one might infer, and correctly so, that the speaker has little sympathy with many of the so-called objective tests and multiple choice questions in examinations. These appear, to him, to defeat the very aim of education, since they require little more of the student than the ability to mark the spot with an x, and surely there is more to a measure of educational attainment than the winning of credits, especially when we simplify the process to this extent.

I realize that these remarks are out of step with some present-day educational theory. However, when, in educational journals, I find the emphasis on mathematical textbooks (and I quote) "being easy," "meaningful to the student," "not a fusion of abstract mathematics," and find references to "hatred towards mathematics" engendered by its difficulty, and this in the very elementary books, then I admit that I am not greatly concerned if I seem out of step with certain theorists. It happens that we are not traveling in the same direction. Also, unlike these people, I am not prepared to grant that American youth is less intelligent than the youth of other countries. It seems desirable to me that a textbook should be sufficiently difficult to require some mental effort on the part of the student. Otherwise, it can scarcely serve as a means of mental stimulus and intellectual development. Furthermore, I feel that we should not be disturbed by adverse criticisms of mathematics which may come from the man on the street, or from non-mathematicians. Their views on mathematics should carry the same weight as their views on relativity or on the technique of atomic research. In general, the so-called popular treatments of mathematics and science are a doubtful blessing and are likely to give rise to misleading concepts.

4. Summary. In summary: We have indicated that mathematics should be taught with the emphasis upon the thinking process. However, we have warned that this statement must not be interpreted as discounting the value of applied mathematics. Applied mathematicians must think. The statement merely implies that there are social values in the study of mathematics, and that these are as important as its applications in industry and in science.

We have shown that proper conclusions are inevitable in a process of reasoning and are a necessary consequence of the initial assumptions, but we have given no recipe for the selection of these assumptions. It is unfortunate, but true, that the initial assumptions may generate intolerance and bigotry, and the danger to our social and economic life is the fact that indoctrination into these premises takes place at an intellectually immature age. In this way they become a part of one's being, and constitute preconceived notions, or prejudices, which limit tolerance. They are accepted as realities and are not looked upon as assumptions.

A social value of the postulational approach to the study of mathematics is that it presupposes an examination and study of the premises. The student comes to realize that in a given study there are alternative sets of premises, all equally possible and proper. Thus, for example, in building an algebra (in the sense of linear algebras) he may start with any one of various sets of postulates, all equally possible and proper, yet one generates an algebra which is rich, and another barren, in its possibilities. The abstract treatment of algebra assures that the student will develop the habit of considering and questioning sets of postulates and this without bias in favor of one set or another. It is then inevitable that he shall acquire habits of reflection and judgment, and that these will affect his reactions to the problems of everyday life. They will make him more tolerant of opinions with which he disagrees.

We are living in a remarkable but dangerous age, one in which new and powerful forces, physical and spiritual, have been let loose. The individual is faced with a greater challenge, and a greater opportunity, than perhaps at any other time in world history. The forces, some good and some evil, must be met with understanding. This calls for independent thinking, sound judgment, and intelligent leadership. It follows that teachers of college mathematics hold a key position. It is their obligation to train our future leaders in habits of sound thinking, in clearness of vision and expression, and in tolerance, which is a by-product of mathematical thinking. However, to do this, teachers must shift the emphasis from the mechanics of mathematics to its understanding. In doing all this, teachers of mathematics can play a part in moulding world destiny.

MATHEMATICS AND INTELLECTUAL HONESTY*†

MOSES RICHARDSON, Brooklyn College

An old legend tells of the explorer who lived for years among the lions and learned to speak their language. When he returned to human society, he reported that all the lions were certain that God is a lion. At one time or another, God has been nominated to honorary membership in the Mathematical Society, the Physical Society, and doubtless many others, as well as to the post of sponsor of many armies. I am aware of this universal tendency to deify the virtues of one's own kind. I am also aware of the claims of other subjects, and of the difficulty of defining Mathematics to the satisfaction of all mathematicians. Despite all this, I propose the following, if not as a definition, then at least as a partial description: *Mathematics is persistent intellectual honesty*. In the remainder of this talk, I wish to amplify this statement and point out some of its implications for teachers.

First, let us examine what Mathematics means to a modern mathematician. The most familiar example of mathematical thinking is doubtless Euclidean

---

* From AMERICAN MATHEMATICAL MONTHLY, vol. 59 (1952), pp. 73–78.

† An address delivered at the first meeting of the Association of Mathematics Teachers of New York State held at Syracuse University on May 12, 1951.

geometry. Let us imagine a conversation between an Unusually Patient Mathematician and an Uncommonly Intelligent Layman. The Layman begins the conversation by asking whether the Pythagorean theorem is true. The Mathematician responds by proving it in a usual way, namely by deducing it from theorems on similar triangles. The Layman perceives that his argument is of the airtight variety of strict deductive logic, but proves merely that if these theorems on similar triangles were true then so would the Pythagorean theorem be true. The Mathematician then deduces these from still simpler propositions, and so on. But since he must avoid both circularity and infinite regress, he must stop ultimately and base the entire discussion on a set of primitive propositions, called postulates, which are left without proof. The Layman then admits that if these postulates were true so would be the Pythagorean theorem, but contends that he is puzzled because he does not understand the meaning of certain terms which were used, such as "triangle." The Mathematician responds by defining "triangle" in the usual way in terms of "point," "line-segment," *etc.* The Layman says he would now understand the meaning of "triangle" if he were not puzzled about the meaning of the terms "line-segment," *etc.* The Mathematician then defines these in terms of still more primitive terms, and so on. But again, since he must avoid both circularity and infinite regress, he must stop ultimately and base all his definitions on a set of primitive terms which are left undefined.

At this point, he has converted Euclidean geometry into an *abstract mathematical science*, that is, a collection of statements of which some (the postulates) are unproved, but all others (the theorems) are logical consequences of these, and in which some terms are undefined but all others are defined in terms of these.

The Layman then asks, "Are there objects in the universe for which these propositions are true?"

The Mathematician then answers, "We have proved that if there were objects such as points, etc., satisfying the postulates, then the theorems would be true of them too. This is all that Pure Mathematics is concerned with. Questions of truth belong to Applied Mathematics. Since the validity of a deductive argument is independent of the meanings of the terms involved, we regard our undefined terms as dummy symbols in Pure Mathematics. Thus we might write 'two x's determine a y' instead of 'two points determine a line' in the abstract mathematical science of Euclidean geometry. *Pure Mathematics* is the totality of abstract mathematical sciences. If concrete meanings are substituted for the undefined terms in an abstract mathematical science, then we have before us a *concrete interpretation* or *application* of it. *Applied Mathematics* is the totality of such concrete interpretations, and *Mathematics* consists of both Pure and Applied Mathematics. If we somehow knew that for a given interpretation the postulates were true, then we would be sure of the truth of the theorems also. But we are seldom certain concerning the truth of our postulates. We try to make them as simple and as plausible as possible, but we cannot guarantee their truth. In Applied Mathematics we use our postulates as good working hypotheses as long as they produce usable theorems, but we are unable to promise that their usefulness will be eternal. If a scientific hypothesis implies a single conclusion that is not in accord with observation, then we know decisively that it is not true as it stands. If all its implications so far drawn do appear to be in accord with observation, we cannot know decisively that it is true, but can only regard it *pro tempore* as a good working hypothesis."

The Layman then asks, "But is not this situation the same in other subjects as well as in Mathematics?"

The Mathematician responds, "I think you are right. If you attempt to organize any body of subject matter logically, you will ultimately cast it in the form of an abstract mathematical science, and if you apply it to reality you will use a concrete interpretation of this abstract mathematical science. It is in this sense that Mathematics is basic to all sciences even more than in the old sense of Mathematics as the science of space and quantity. The mural in the Hall of Science at the Chicago World's Fair, entitled the Tree of Knowledge, exhibits this notion by placing Mathematics at the base of the Tree, the fundamental sciences as the older, better developed branches of the Tree, and the practical arts and applied sciences as the younger, less developed branches which must draw their strength from below. With our definition of Pure and Applied Mathematics, it is true that all deductively organized sciences ultimately become branches of Mathematics. This has been considered unfortunate even by some mathematicians, but attempts to restrict the meaning of the word 'Mathematics' restrict it too much. Rather than throw out the baby with the bath water, we have grudgingly become reconciled to the embracing nature of our subject."

The Layman then says, "You have indeed opened inspiring vistas to me, and I begin to perceive why mathematicians are so enamored of their subject. But you have not yet told me whether the Pythagorean theorem is true."

The Mathematician says, "I have shown you that if there existed objects in reality which had the properties demanded of points, *etc.*, in the postulates of Euclidean geometry, then the Pythagorean theorem would be true of these objects. While in most situations Euclidean geometry seems to work well enough for practical purposes, I cannot assert that there are such objects."

The Layman then asks, "But did not the Greeks say that the axioms of geometry were true because they were self-evident?"

The Mathematician responds, "This was a weakness on their part which modern mathematicians have outgrown. The history of science is a road strewn with the decaying bones of assumptions which were once considered self-evident and later found to be false. In fact, much fundamental progress has been made by questioning assumptions which sometimes were considered so self-evident that they were made tacitly. For the practical purposes of geometry other postulates may be used as well as Euclid's, and these non-Euclidean geometries contradict Euclidean geometry in many places. I am afraid that if you mean the word 'truth' in the somewhat naive sense of absolute truth, then

I cannot assert that Euclidean geometry is true. Pontius Pilate is said to have asked 'What is truth?' long ago, and, so far as I know, no thoroughly satisfactory answer has been given. Fortunately, science progresses without requiring absolute truth. Its successive theories provide better and better approximations to truth in the sense that a new theory must give correct predictions of a more inclusive set of observational facts than the theory it supplants. Every scientist knows how difficult it is to discover truth even in this tentative evolving sense, *a fortiori* in any absolute sense. It might be a better world if others besides scientists were more aware of this difficulty, for it is an excellent antidote for fanatical dogmatism. Throughout history, people who have persuaded themselves that they were already in possession of absolute truth have tended to suppress free inquiry and censor free expression, sometimes on the somewhat surprising grounds that, while the absolute truth was self-evident to them, others would be easily misled by arguments supporting contrary views, and have exhibited at times a distressing willingness to kill those who disagreed with them. A mature individual must learn to live with some unavoidable uncertainty. I recall an old comic strip character who said, 'It ain't ignorance that causes all the trouble—it's the things people know that ain't so'."

Our Layman, who is, you must remember, an Uncommonly Intelligent Layman, then says, "You have shown me that Mathematics is indeed the epitome of intellectual honesty, carried to rather unusual lengths. You do not assert more than you can prove. Even when you deduce your theorems from postulates which you could easily persuade me to believe to be self-evident truths, you warn me against believing this, and point out and investigate the possibility of alternative assumptions. Your insistence on explicit statement of assumptions and undefined terms, on strict logical proof of theorems, and on clarity in definitions is inspiring. Mathematics is the antithesis of dogmatism and fanaticism and authoritarianism, and I suspect it is no coincidence that authoritarians have often fought and even liquidated mathematicians and other scientists. I intend to study mathematics further as soon as I can find the time."

At this happy outcome, let us take leave of our Mathematician and our Layman, because I am afraid of what might happen if we were to follow our Layman's subsequent career in various mathematics classrooms. I fear greatly that he might not find these ideals of intellectual honesty very much in evidence in every single classroom.

Bertrand Russell remarks somewhere that one of the hardest things in the world for a man to do is to stand before an audience and refrain from saying more than he knows. To be sure, we may all sometimes make a hasty statement or an honest mistake. I am not asking for perfection. But I fear that the teaching of mathematics and of natural science does not always live up to the high ideals that our Unusually Patient Mathematician has been expounding. The liberating influence of scientific rationalism seems nowadays to be lost too often, and instruction turns into authoritarian dicta on the part of the teacher, and regurgitated undigested rote responses on the part of the student. A college president once remarked within my hearing that after all, geometry was mostly a matter of memory. Is it possible that he could have been referring to the same study that I have been using as my prototype of mathematics today? I am not at all impugning the honesty of his reaction or his recollection of his own studies. I fear that he may well have been entirely accurate about the way in which he was taught geometry. But I deny that it need be so. I have met teachers of secondary mathematics who apparently taught geometry with due traditional regard for reasoning, but routinized their instruction in algebra and trigonometry until these subjects became nothing more than an elaborate game of tic-tac-toe with symbols. The same thing can be said of some college teachers who turn algebra and trigonometry and even calculus and differential equations into cook book courses. I deny that this need be so.

It is well known that men have a striking ability to compartmentalize their minds. People who are habitually honest critical thinkers in their mathematical work may fall far short of that ideal in other subjects, or in their personal dealings which may be governed by emotional reactions, ingrained prejudices, jealousies, *etc.* Even people who are sound critical thinkers in one branch of mathematics may fall short of that ideal in other branches of mathematics itself. I do not contend that the study of mathematics will automatically produce intellectually honest people. Nevertheless, I do contend that mathematics is the ideal subject in which to point out to the student the virtues of intellectual honesty. In this age of the advertising man and his more vicious cousin, the propaganda technician, is there any more important function for education to perform?

But this can be done only if mathematics is properly taught. Ideals of intellectual honesty can be communicated to the student by instruction and by example only if the teacher is himself saturated with them and will seize the abundant opportunities to stress them in the classroom. However, if mathematics is presented as meaningless rote learning, as mere drill, pure unrationalized technique, routine mechanical manipulation of symbols justified only by the teacher's authority, will these ideals be served? If incorrect arguments are presented as valid proofs whose invalidity may or may not be disclosed in subsequent courses, will these ideals be served? I recall a remark of H. G. Forder, asking if it can be contended that "an unsound proof has an educational value not possessed by a sound one." I do not mean at all that all instruction must be rigorous. I have no objection whatever to omitting too difficult proofs and replacing them by properly labelled informal heuristic discussions. I mean merely that unless a proof is sound, it should not be presented as if it were. Otherwise the teacher is in a somewhat less defensible position than a passer of counterfeit currency whose innocent dupes suffer only in material wealth.

Mathematics should be taught so as to give the student:

1. An appreciation of the natural origin and evolutionary growth of the basic mathematical ideas;

2. A critical logical attitude, and a wholesome respect for correct reasoning, precise definitions, and clear grasp of underlying assumptions;

3. An understanding of the role of mathematics as one of the major branches of human endeavor and its relations with other branches of the accumulated wisdom of the human race;

4. A discussion of some of the important problems of pure mathematics and its applications;

5. An understanding of the nature and practical importance of postulational thinking.

An attempt should be made to emphasize the distinction between familiarity and understanding, between logical proof and routine manipulation, between a critical attitude of mind and habitual unquestioning belief, between scientific knowledge and both encyclopedic collections of facts and mere opinion and conjecture.

These remarks are intended to apply not merely to teachers of mathematics at secondary and collegiate levels, but also to those who are training such teachers. If a teacher is to carry out these purposes, he must have the background for it. If you will pardon me for perverting the modern educational scripture, he must not only know the student, he must also know his subject. By this I do not mean that a secondary teacher needs to know a great deal about differential geometry or topology or the calculus of variations. But he should be well grounded in the fundamental concepts of elementary mathematics. Unfortunately, he is frequently expected to absorb these fundamentals by osmosis. Too often, this process does not take place, and the prospective teacher emerges from his mathematics major with an equipment of heterogeneous techniques, and little or no understanding of the historical development, philosophy, fundamental concepts, nature, spirit, or even *raison d'être* of his subject.

A student in a democracy should learn that the worth of a scientific theory is not to be determined by authoritarian dictates, nor even by popular acclaim, but only by dispassionate and free inquiry into the evidence for and against it and for and against possible alternative theories. This is a difficult task in controversial subjects. What better training can there be for this task than initiation into intellectual honesty through mathematics where controversy is comparatively absent and where standards of intellectual honesty are so high? And although there are certainly other objectives as well involved in the teaching of mathematics, is there a more important aim to be served in this perilous age when freedom of thought and expression is being assailed on all sides?

Bibliography

C. B. Allendoerfer, Mathematics for liberal arts students, this MONTHLY, vol. 54, 1947, pp. 573–578.

E. T. Bell, The Search for Truth, Reynal and Hitchcock, 1934.

C. J. Keyser, The Human Worth of Rigorous Thinking, Columbia University Press, 1925.

E. P. Northrop, Mathematics in a liberal education, this MONTHLY, vol. 52, 1945, pp. 132–135.

F. S. Nowlan, Objectives in the teaching of college mathematics, this MONTHLY, vol. 57, 1950, pp. 73–82.

O. Ore, Mathematics for students of the humanities, this MONTHLY, vol. 51, 1944, pp. 453–458.

M. Richardson, On the teaching of elementary mathematics, this MONTHLY, vol. 49, 1942, pp. 498–505.

M. Richardson, The place of mathematics in a liberal education, Nat. Math. Magazine, vol. 19, 1945, pp. 349–358.

M. Richardson, Fundamentals in the teaching of undergraduate mathematics, this MONTHLY, vol. 58, 1951, pp. 182–186.

B. Russell, Human Knowledge, Its Scope and Limits, Simon and Schuster, 1948.

B. Russell, The Functions of a Teacher, in Unpopular Essays, Simon and Schuster, 1950.

THE PROOF OF THE LAW OF SINES*

ALFRED BRAUER, University of North Carolina

Most mistakes made by mathematicians consist in neglecting to consider a certain possibility. A proof or a theorem can become invalid if two roots of an equation become equal, if a determinant becomes zero, if the factorization in the considered number field is not unique, and so on. Often it is possible to complete the proof for the exceptional case, but sometimes the theorem is wrong. It is therefore of importance that the student learn as early as possible to check if a proof is really complete.

It is surprising that in a large number of textbooks on trigonometry the proof for the law of sines is incomplete.

In general, the proof is given by drawing one of the altitudes. Then two cases are considered, namely the case that each angle is less than 90° and the case that one of the angles is greater than 90°. Two figures illustrate the proof.

The authors of these books seem to believe that it is not necessary to prove the theorem for the case that one of the angles equals 90° since the trigonometric functions can directly be used for right triangles. But this is not correct.

Given two sides a and b, and the angle B opposite to the smaller side b. In this case we are not permitted to use the law of sines if this theorem is not proved for right triangles since it can be that the angle $A = 90°$. But we cannot use the trigonometric functions either, since we do not know that the triangle has a right angle.

In these books the law of sines is used and it is explicitly stated that A can be equal to 90° and that we have one and only one solution in this case. But it is not recognized that here a theorem is used in a case in which it is not proved.

If a, b, and B are given, but not numerically, we cannot determine whether the triangle is a right one or not. If the law of sines is not proved for right triangles, a complete solution must consist of two parts, namely assuming that no angle equals 90°, and assuming that one angle equals 90°. If a, b, and B are given numerically, it can happen that $\sin B = b/a$ if a 4-place table is used, but

---

* From AMERICAN MATHEMATICAL MONTHLY, vol. 59 (1952), p. 319.

$\sin B \neq b/a$ if a 5-place table is used, and the method of solution would depend on the kind of table to be used.

This shows that it is necessary to prove the law of sines for right triangles. Of course, using continuity it is obvious that the theorem is correct for right triangles, too. But also the elementary proof is so simple that there is no reason to omit the proof of this case in any book or class.

ON THE DEFINITION OF FUNCTIONS*†

H. P. THIELMAN, Iowa State College

1. Fundamental concepts. We shall begin with certain undefined concepts such as set and element of a set. We might give some synonyms of these terms, but the synonyms would then have to be left undefined. Thus a set might be described as a collection of definite, distinct objects associated in thought. By the word "definite" we mean here the following. Given a set S and an object, one and only one of the next two statements is true: the object is an element of the given set; the object is not an element of the given set. It is not required that we, nor anyone else, need to know which of these statements is true. Thus the collection of ladies present that are over sixteen years of age at this moment is a set, even though we or anybody may not know whether a given lady belongs to this set or not. The word "distinct" is used here to indicate that no object can be considered as an element of a given set more than once. For example, the collection of letters a, b, a, c, b constitutes the set a, b, c. In contradistinction to the expression "set of elements" (or set of sets) we shall use the terms "collection of elements" (or collection of sets) to indicate that it is not implied that the elements (or the sets) be all distinct.

When a symbol such as a letter, say x, stands for an unspecified element of a set, then this symbol x is said to *vary* over the set, and the symbol x is called a *variable* on the set. If x varies over a set which consists of only one element then x is called a constant.

An *ordered pair of elements* is a collection which consists of two elements one of which has been designated as the first. If an ordered pair consists of the elements a and b of which a has been designated as the first, we indicate this by the symbol (a, b). The word "ordered" refers only to the order in which the elements appear within the parentheses. Two ordered pairs (a, b) and (c, d) are equal if and only if $a = c$, and $b = d$.

A *relation* is a set of ordered pairs. For example, the relation $>$ (greater) is the set of all ordered pairs (x, y) of real numbers such that $x - y$ is a positive

---

* From AMERICAN MATHEMATICAL MONTHLY, vol. 60 (1953), pp. 259–262.

† An excerpt from an address presented under the title *Types of Functions* to the Minnesota Section of the Mathematical Association of America at the invitation of the Executive Committee on May 10, 1952. Another excerpt from this address is published in the March issue of this MONTHLY.

real number. The *domain of definition of a relation* is the set which consists of all the first elements, and the *range of the relation* is the set which consists of all the second elements of the ordered pairs.

2. Function. A *function* is a set of ordered pairs such that no two ordered pairs have the same first element. The set which consists of all the first elements of the ordered pairs of the given function is called the *domain of definition of the function*. The set which consists of all the second elements of the ordered pairs is called the *range of the function*.

Let f be a given function with domain of definition X, and with range Y. If x stands for an unspecified element of X, x is called the *independent variable* of the given function. If y stands for an unspecified element of the range Y, then y is called the *dependent variable* of the given function. For a given ordered pair (x, y) of f, y is called the *image* of x under f, while x is called the *counter image* or *source* of y under f. The image of x under f is also called the *value of the function f at x*, and is denoted by $f(x)$. A function f whose domain of definition is X, and whose range is Y is frequently denoted by $f: X \to Y$, and is referred to as a function *on X onto Y*. If X is a subset of X_1, and Y is a subset of Y_1, then f is said to be a function *from X_1 to Y_1*, or *from X_1 into Y_1*.

3. Inverse function. Let f be a function on X onto Y. Thus f is a set of ordered pairs (x, y) where x is an element of X, and y is an element of Y. If f is such that no two of its ordered pairs have the same second element, then the function obtained from f by interchanging in each ordered pair (x, y) the places of x and y is called the *inverse function* of f. This function is indicated by f^{-1}. Its domain of definition is Y, and its range is X.

The symbols $f(x)$, $f^{-1}(y)$, or $y = f(x)$, $x = f^{-1}(y)$ should be used in referring to a function and its inverse only when it is clear from the context what the domains of definition and the ranges of these functions are.

4. Inverse relations. Let f be a function on X onto Y. If f is such that two or more of its ordered pairs have the same second element, then the set of ordered pairs obtained from f by interchanging in each ordered pair (x, y) the places of y and x does not constitute a function from Y to X, but it does constitute a relation. The symbol $\{f^{-1}(y)\}$ can be used to denote the set which consists of all those elements of X which are counter images of y under f. Thus for every y in the range Y of f, $\{f^{-1}(y)\}$ represents a subset of X. If the inverse function f^{-1} exists, then for each y of Y, $\{f^{-1}(y)\}$ represents the set which consists of the unique element $f^{-1}(y)$.

The definition of function leaves the nature of the sets which constitute the domain of definition and the range of a function unspecified. These sets may be sets of sets, sets of real or complex numbers, sets whose elements are vectors or other arbitrary sets.

If the range of a function is a subset of the set of real numbers, the function is called a *real function*, if the domain of definition of a function is a subset of the real numbers we have a function of a *real variable*. If the domain of definition of a function is a set of sets, we have a *set function*. A set function whose range is a subset of the set of real numbers is called a *real set function*.

5. One-to-one correspondence. If f is a function on X onto Y which has an inverse f^{-1} on Y onto X, then f and f^{-1} are said to constitute a *one-to-one correspondence* between X and Y.

6. Unrestricted functions. Let f be a real function of a real variable. No other restrictions are put on this function. What can be said of such a function? This question has been partially answered in recent years, and the answer is quite different from what had been surmised by some mathematicians. Even E. W. Hobson in his 1921 edition of the Theory of Functions of Real Variables wrote: "No elaborate theory is required for functions which retain their complete generality, . . . since few deductions of importance can be made from the definition which will be valid for all functions." This conjecture was proven to be false by the late Professor Henry Blumberg who has developed an extensive theory which reveals that every unrestricted real function of real variables possesses many properties which are far from obvious and yet have a beauty of simplicity that has attracted the interest of many mathematicians. Since this theory has been presented by its originator in a number of nontechnical addresses before the American Mathematical Society, and since these addresses have been published and are easily accessible we shall not elaborate on this very interesting topic here.*

Some of these results on unrestricted functions of real variables have very recently been extended to relations in general neighborhood spaces.†

It should perhaps be mentioned that the term multiple-valued function which is frequently met in mathematical literature means a relation, that is, it is a name for a set of ordered pairs in which two distinct ordered pairs may have the same first element. If one admits the term multiple-valued function, then the concept *function* as given in section 2 of the present paper would be described as a single-valued function. In recent years the tendency in mathematical literature has been to limit the term function so as to mean a single-valued function. An equation such as $x^2+y^2=1$, where $-1 \leq x \leq 1$, $-1 \leq y \leq 1$, does not define a function. There exist infinitely many functions whose images and counter images satisfy the condition imposed on them by the equation of the preceding sentence. Thus if $y=\sqrt{1-x^2}$ if x is a rational number, and $y=-\sqrt{1-x^2}$ if x is an irrational number, there is defined a function for which $x^2+y^2=1$. By a proper reassignment of the values of y for given values of x,

---

* Blumberg, Henry, Properties of unrestricted real functions, Bull. Amer. Math. Soc., vol. 32, pp. 132–148, 1926; Methods in point sets and the theory of real functions, Bull. Amer. Math. Soc., vol. 36, pp. 809–830, 1930.

† Block, H. D., and Cargal, B., Arbitrary mappings, Proc. Amer. Math. Soc., vol. 3, 1952, pp. 937–941.

arbitrarily many functions can be constructed whose images and counter images satisfy the given equation in x and y.

It is gratifying to see that the unambiguous definition of a function, which is advocated here, has been included in a new elementary textbook.*

A KIND OF PROBLEM THAT EFFECTIVELY TESTS FAMILIARITY WITH FUNCTIONAL RELATIONS†

K. O. MAY, Carleton College

The kind of problem described here probes the student's understanding of the functional concept and notation, tests his familiarity with the functions being studied, and is easy to run off on a duplicator and to correct.

Present the student with the graph of a rather simple function, perhaps piecewise constant or at least piecewise linear. Label this function $f(x)$ and then require the student to graph $-f(x), f(-x), |f(x)|, f(|x|), e^{f(x)}, \sin f(x)$ or whatever functions involve the ideas whose understanding you wish to test. The correct graphs are often decorative and always easily recognized. The student's deviations indicate clearly in many cases the nature of his difficulties. While the resulting graphs are easy to check, it is not hard to construct examples that require considerable thought and challenge the best students. On the other hand, the simpler examples, such as $-f(x)$ or the inverse function of $f(x)$, appeal to the manipulative students. Obvious modifications, such as the use of two functions, will occur to any teacher.

As an example suppose that $y = f(x)$ has the graph:

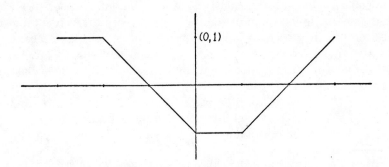

Sketch: $y = |f(x)|$, $y = f(|x|)$, $y = -f(x)$, $y = e^{f(x)}$, $y = \text{Arc sin } f(x)$.

---

* Randolph, John F., Calculus, p. 10, Macmillan Co., 1952.

† From AMERICAN MATHEMATICAL MONTHLY, vol. 60 (1953), p. 624.

"IF THIS BE TREASON..."*

R. P. BOAS, JR., Northwestern University

If I had to name one trait that more than any other is characteristic of professional mathematicians, I should say that it is their willingness, even eagerness, to admit that they are wrong. A sure way to make an impression on the mathematical community is to come forward and declare, "You are doing such-and-such all wrong and you should do it *this* way." Then everybody says, "Yes, how clever you are", and adopts your method. This of course is the way progress is made, but it leads to some curious results. Once upon a time square roots of numbers were found by successive approximations because nobody knew of a better way. Then somebody invented a systematic process and everybody learned it in school. More recently it was realized that very few people ever want to extract square roots of numbers, and besides the traditional process is not really very convenient. So now we are told to teach root extraction, if we teach it at all, by successive approximations. Once upon a time people solved systems of linear equations by elimination. Then somebody invented determinants and Cramer's rule and everybody learned that. Now determinants are regarded as old-fashioned and cumbersome, and it is considered better to solve systems of linear equations by elimination.

We are constantly being told that large parts of the conventional curriculum are both useless and out of date and so might better not be taught. Why teach computation by logarithms when everybody who has to compute uses at least a desk calculator? Why teach the law of tangents when almost nobody ever wants to solve an oblique triangle, and if he does there are more efficient ways? Why teach the conventional theory of equations, and especially why illustrate it with ill-chosen examples that can be handled more efficiently by other methods? As a professional mathematician, I am a sucker for arguments like these. Yet, sometimes I wonder.

There are a few indications that there is a reason for the survival of the traditional curriculum besides the fact that it is traditional. When I was teaching mathematics to future naval officers during the war, I was told that the Navy had found that men who had studied calculus made better line officers than men who had not studied calculus. Nothing is clearer (it was clear even to the Navy) than that a line officer never has the slightest use for calculus. At the most, his duties may require him to look up some numbers in tables and do a little arithmetic with them, or possibly substitute them into formulas. What is the explanation of the paradox?

I think that the answer is supplied by a phenomenon that everybody who teaches mathematics has observed: the students always have to be taught what they should have learned in the preceding course. (We, the teachers, were of course exceptions; it is consequently hard for us to understand the deficiencies

---

* From AMERICAN MATHEMATICAL MONTHLY, vol. 64 (1957), pp. 247–249.

of our students.) The average student does not really learn to add fractions in arithmetic class; but by the time he has survived a course in algebra he can add numerical fractions. He does not learn algebra in the algebra course; he learns it in calculus, when he is forced to use it. He does not learn calculus in the calculus course, either; but if he goes on to differential equations he may have a pretty good grasp of elementary calculus when he gets through. And so on through the hierarchy of courses; the most advanced course, naturally, is learned only by teaching it.

This is not just because each previous teacher did such a rotten job. It is because there is not time for enough practice on each new topic; and even if there were, it would be insufferably dull. Anybody who has really learned to interpolate in trigonometric tables can also interpolate in air navigation tables, or in tables of Bessel functions. He should learn, because interpolation is useful. But one cannot drill students on mere interpolation; not enough, anyway. So the students solve oblique triangles in order (among other things) to practice interpolation. One must not admit this to the students, but one may as well realize the facts.

Consequently, I claim that there is a place, and a use, even for nonsense like the solution of quartics by radicals, or Horner's method, or involutes and evolutes, or whatever your particular candidates for oblivion may be. Here are problems that might conceivably have to be solved; perhaps the methods are not the most practical ones; but that is not the point. The point is that in solving the problems the student gets practice in using the necessary mathematical tools, and gets it by doing something that has more motivation than mere drill. This is not the way to train mathematicians, but it is an excellent way to train mathematical technicians. Now we can understand why calculus improves the line officer. He needs to practice very simple kinds of mathematics; he gets this practice in less distasteful form by studying more advanced mathematics.

It is the fashion to deprecate puzzle problems and artificial story problems. I think that there is a place for them too. Problems about mixing chemicals or sharing work, however unrealistic, give good practice and even have a good deal of popular appeal: witness the frequency with which puzzle problems appear in newspapers, magazines, and the flyers that come with the telephone bill. There was once a story in *The Saturday Evening Post* whose plot turned on the interest aroused by a perfectly preposterous diophantine problem about sailors, coconuts, and a monkey. It is absurd to claim that only "real" applications should be used to illustrate mathematical principles. Most of the real applications are too difficult and/or involve too many side issues. One begins the study of French with simple artificial sentences, not with the philosophical writings of M. Sartre. Similarly one has to begin the study of a branch of mathematics with simple artificial problems.

We may dislike this state of affairs, but as long as it exists we must face it. It would be pleasant to teach only the new and exciting kinds of mathematics;

it would be comforting to teach only the really useful kinds. The traditional topics are some of the topics that once were either new and exciting, or useful. They have persisted partly by mere inertia—and that is bad—but partly because they still serve a real purpose, even if it is not their ostensible purpose. Let us keep this in mind when we are revising the curriculum.

'RYE WHISKEY' IN CONTRAPOSITIVE*

W. P. COOKE, West Texas University
(now at the University of Wyoming)

While discussing some elementary logic in a class in geometry, I found that the students enjoyed those problems which were based on "If-Then" couplets from popular songs. The most fun occurred when we attempted to write them using contraposition. The 'game' was not only to achieve the contrapositive, but also to preserve the rhyme and meter of the song. Following is an only slightly more ambitious undertaking which should illustrate the idea.

In a famous Western 'Classic,' elegantly sung, as I recall, by Tex Ritter, is found (perhaps imperfectly remembered) the verse:

Statements

If the ocean was whiskey and I was a duck,
I'd swim to the bottom and never come up.
But the ocean ain't whiskey and I ain't no duck.
So I'll play Jack-O-Diamonds and trust to my luck.
For it's whiskey, Rye whiskey, Rye whiskey I cry.
If I don't get Rye whiskey I surely will die.

These statements are 'naturals' for contraposition, as follows:

Contrapositives

If I never reach bottom or sometimes come up,
Then the ocean's not whiskey or I'm not a duck.
But my luck can't be trusted or the cards I'll not buck,
So the ocean is whiskey or I am a duck.
For it's whiskey, Rye whiskey, Rye whiskey I cry.
If my death is uncertain then I get whiskey (Rye).

---

* From AMERICAN MATHEMATICAL MONTHLY, vol. 76 (1969), p. 1051.

THE PROFESSOR'S SONG*

Words by Tom Lehrer — Tune: "If You Give Me Your Attention"
from *Princess Ida* (Gilbert and Sullivan)

If you give me your attention, I will tell you what I am.
I'm a brilliant math'matician — also something of a ham.
I have tried for numerous degrees, in fact I've one of each;
Of course that makes me eminently qualified to teach.
I understand the subject matter thoroughly, it's true,
And I can't see why it isn't all as obvious to *you*.
Each lecture is a masterpiece, meticulously planned,
Yet everybody tells me that I'm hard to understand,
   And I can't think why.

My diagrams are models of true art, you must agree,
And my handwriting is famous for its legibility.
Take a word like "minimum" (to choose a random word), †
For anyone to say he cannot read that, is absurd.
The anecdotes I tell get more amusing every year,
Though frankly, what they go to prove is sometimes less than clear,
And all my explanations are quite lucid, I am sure,
Yet everybody tells me that my lectures are obscure,
   And I can't think why.

Consider, for example, just the force of gravity:
It's inversely proportional to something — let me see —
It's r^3 — no, r^2 — no, it's just r, I'll bet —
The sign in front is plus — or is it minus, I forget —
Well, anyway, there *is* a force, of that there is no doubt.
All these formulas are trivial if you only think them out.
Yet students tell me, "I have memorized the whole year through
Ev'rything you've told us, but the problems I can't do."
   And I can't think why!

---

* From AMERICAN MATHEMATICAL MONTHLY, vol. 81 (1974), p. 745.
† This was performed at a blackboard, and the professor wrote: ∧∨∧∨∧∨∧∨∧∨∧

CLUB TOPICS*

H. D. LARSEN (Editor), Albion College

Bibliographies have been published for a considerable number of topics frequently used for club programs. A summary list may be of value to program directors and others.

The bibliographies which follow were published in this MONTHLY, the numbers referring to volume and page.

American mathematicians. 47, 107.
Apportionment of representatives. 47, 484; 49, 115.
Arithmetical prodigies. 25, 91.
The binary scale of notation, etc. 25, 139.
Calculating machines. 43, 99; 46, 233; 47, 106.
The cattle problem of Archimedes. 25, 411.
The Chinese suan p'an. 27, 180.
Codes and ciphers. 26, 409.
Constructions with a double-edged ruler. 25, 358.
Constructions with compasses alone. 47, 107.
La courbe du diable. 33, 273.
Development of present day numerals. 43, 99.
Euler integrals and Euler's spiral, etc. 25, 276.
A Fibonacci series. 25, 235.
Fiedler's cyclography. 32, 517.
Finite geometries. 28, 85.
First printed mathematical books. 47, 108.
Forecasting the population of the U. S. 47, 484.
The four-color problem. 43, 181.
Functional equations. 32, 428.
Geometrography, etc. 25, 37.
Geometry of four dimensions. 25, 316.
Golden section. 25, 232.
Historical items pertaining to the calculus. 46, 233.
History of algebra. 46, 234; 47, 107.
History of American mathematicians. 46, 233; 47, 107.
A home made mathematics exhibit. 40, 555.
The logarithmic spiral. 25, 189.
Magic squares and circles. 47, 106.
Maps and map projection. 46, 650.
Mathematics and art. 47, 108.
Mathematics and defense. 47, 484.
Mathematics and music. 47, 108.
Mathematics in the ancient world. 46, 234.
Mathematics in certain countries. 47, 107.
Nomographs. 47, 106.
Notes on famous mathematicians. 47, 109.
The number π. 26, 209.
Numerals and number systems. 46, 234.
The oldest mathematical work extant. 25, 36.
Origin of various mathematical terms and symbols. 46, 233; 47, 107.

---

* From AMERICAN MATHEMATICAL MONTHLY, vol. 59 (1952), pp. 475–478.

Paper folding. 25, 95.
The pasturage problem of Sir Isaac Newton. 33, 155.
The polar planimeter. 46, 45.
Proportional representation and preferential voting. 47, 484.
Ptolemy's theorem and formulae of trigonometry. 25, 94.
Scholars in other fields—interest in mathematics. 47, 108.
Symbolic logic. 46, 289.
Theorem of Bang. 33, 224.
Women as mathematicians and astronomers. 25, 136.

The following bibliographies appeared in *The Pentagon*.

The abacus. 10, 93.
Amicable numbers, 8, 85.
Apportionment in Congress. 7, 85.
The bee as a mathematician. 6, 19.
Calculating machines. 6, 18.
Calendar problems. 7, 30.
The cattle problem of Archimedes. 5, 67.
Codes and ciphers. 7, 81.
The construction of sundials. 8, 19.
Constructions with limited means. 7, 29.
La Courbe du diable. 10, 94.
Duplication of the cube. 10, 31.
Fermat's last theorem. 9, 33.
Fibonacci series. 8, 20.
The "fifteen" puzzle. 9, 104.
The four-color problem. 8, 86.
The golden section. 11, 103.
History of mathematics in the United States. 10, 33.
Linkages. 7, 84.
The magic number nine. 11, 32.
Magic squares. 10, 28.
Mathematics and music. 7, 31.
Mathematical prodigies. 6, 17.
Nim. 11, 101.
Non-euclidean geometry. 9, 33.
Paper folding. 5, 68.
Perfect numbers. 11, 102.
The planimeter. 6, 69.
Plays. 11, 32.
Proofs of the Pythagorean theorem. 7, 28.
Ptolemy's theorem. 8, 85.
Rational-sided triangles. 9, 102.
Repeating decimal fractions. 10, 95.
Scales of notation. 6, 67.
Solutions of the quadratic equation. 6, 66.
Squaring the circle. 9, 105.
The story of pi. 11, 34.
Trisection of an angle. 8, 21.
Women as mathematicians. 5, 67.

Several comprehensive bibliographies have been prepared by William L. Schaaf and published in *The Mathematics Teacher*. The following selections contain material for clubs.

The correlation of mathematics and science. 44, 340. (1. Curriculum integration in mathematics and science. 2. Biology and mathematics. 3. Chemistry and mathematics. 4. Physics and mathematics. 5. Mathematics and aviation. 6. Mathematics and navigation. 7. Miscellaneous applications of mathematics to science.)

Guidance: the case for mathematics. 44, 130. (1. Why learn mathematics? 2. Guidance in mathematics. 3. Functional competence in mathematics. 4. Mathematics for industry, trade, and business. 5. Mathematics for professional careers. 6. Cultural values of mathematics.)

Laboratory mathematics. 44, 422. (1. Laboratory methods and equipment. 2. Laboratory devices for plane geometry. 3. Laboratory devices for solid geometry. 4. Models for polyhedrons. 5. Laboratory devices for algebra. 6. Laboratory devices for trigonometry. 7. Models for conic sections. 8. Paper folding: dissection of figures. 9. Linkages. 10. Mechanical construction of curves.)

Logarithms and exponentials. 45, 361. (1. History of logarithms. 2. Theory of logarithms. 3. Logarithms and exponential functions. 4. Teaching logarithms.)

Mathematics and art. 43, 423. (1. Art. aesthetics, architecture. 2. Ornament and design; pattern; geometric drawing. 3. Dynamic symmetry. 4. The golden section.)

Mathematics and engineering. 44, 54. (1. Mathematics and engineering education. 2. Engineering and mathematics. 3. Texts and reference works on mathematics for engineers.)

Mathematics and modern science. 43, 294.

Mathematical plays and programs. 44, 526. (1. Plays, skits, and pageants. 2. Programs for assemblies and mathematics clubs.)

Modern calculating machines. 45, 110. (1. General articles. 2. Technical books. 3. Digital machines; electronic calculators. 4. Differential analyzers. 5. Harmonic analyzers and synthesizers. 6. Network analyzers. 7. Algebraic equation solvers.)

The numerical solution of equations. 44, 204. (1. General approximation methods. 2. On the location of roots. 3. Miscellaneous methods of solving cubic and quartic equations.)

The theorem of Pythagoras. 44, 585. (1. Proofs and discussions. 2. Pythagorean numbers: integral and rational right triangles. 3. Special cases of rational right triangles; Heronian triangles. 4. Pythagorean theorem in design. 5. Books and monographs.)

BIBLIOGRAPHIC ENTRIES: PEDAGOGY

The references below are to the AMERICAN MATHEMATICAL MONTHLY.

1. K. L. Nielsen, Industrial experience for mathematics professors, vol. 54, p. 91.

2. C. C. MacDuffee, The scholar in a scientific world, vol. 55, p. 129.

3. R. Schorling, A program for improving the teaching of science and mathematics, vol. 55, p. 221.

4. N. Bourbaki, The architecture of mathematics, vol. 57, p. 221.

An analysis of the procedures of mathematics in order to answer the question of whether we have a mathematic or several mathematics.

5. W. L. Schaaf, Art and mathematics: a brief guide to source materials, vol. 58, p. 167.

Bibliography.

6. H. P. Thielman, On the definition of functions, vol. 60, p. 259.

7. A. Weil, Mathematical teaching in universities, vol. 61, p. 34.

Some important pedagogical considerations for the university mathematics teacher.

8. G. Pólya, On picture-writing, vol. 63, p. 689.

The use of "figurate series", the terms of which are pictures, in the method of generating functions.

9. O. Ore, Pascal and the invention of probability theory, vol. 67, p. 409.

   History of the beginnings of probability theory.

10. H. Levy, Analytic geometry and the calculus, vol. 68, p. 925.

   The importance of analytic geometry as a course separate from calculus.

11. E. T. Parker, A memorable teacher, vol. 72, p. 1127.

12. P. R. Halmos, E. E. Moise and G. Piranian, The problem of learning to teach, vol. 82, p. 466.

2

NUMBERS

(a)

INTEGERS

PASCAL'S TRIANGLE AND NEGATIVE EXPONENTS*

L. V. ROBINSON, University of South Carolina

A most interesting and useful fact apparently overlooked heretofore in the use of Pascal's triangle is that its methods are equally valid for negative, as well as positive, integral exponents. To illustrate this extension in evaluating the binomial coefficients for all integral values of the exponent, it is sufficient to refer to the following scheme:

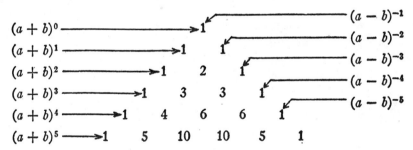

Thus the horizontal numbers are the coefficients for $(a+b)^n$ and the diagonal ones are for $(a-b)^{-n}$. For example, the binomial expansion of $(a-b)^{-2}$ is, accordingly,

$$(a-b)^{-2} = a^{-2} + 2a^{-3}b + 3a^{-4}b^2 + 4a^{-5}b^3 + 5a^{-6}b^4 + \cdots, \qquad b < a.$$

It seems never inappropriate to point out also to students the use of Pascal's triangle in evaluating the probability of a given number of heads, say, from an equal, or greater, number of tosses of a coin.

---

* From AMERICAN MATHEMATICAL MONTHLY, vol. 54 (1947), pp. 540, 541.

ON SOME MATHEMATICAL RECREATIONS*

RICHARD BELLMAN, Rand Corporation, Santa Monica, California

1. Introduction. There are a number of arithmetic puzzles which may be put in the following general form: "Given a sequence of positive integers, $[a_1, a_2, \cdots, a_N]$, and a set of permissible operations upon these integers, such as addition, subtraction, multiplication, division, the taking of a square root, a logarithm, or of a factorial, it is required to obtain representation of all integers in another set, $[b_1, b_2, \cdots, b_M]$, in the form

$$b_i = R_1(a_1 R_2(a_2 \cdots R_N a_N) \cdots),$$

where the R_i belong to the set of admissible operations. A well-known problem of this type is that involving the four 4's, and one which we shall discuss below involves the use of the integers $1, 2, \cdots, 9$ in consecutive order so as to arrive at 100. In this note we wish to show how recurrence techniques, and dynamic programming techniques in particular [1], enable us to obtain the number of representations of different types and minimal representations.

If only addition and multiplication are allowed, a representation due to S. Markowitz, as quoted by Willy Ley [2], is

$$1 + 2 + 3 + 4 + 5 + 6 + 7 + 8 \times 9 = 100.$$

It is natural to ask whether there are other representations, how they might be found, and to look for a minimal representation, say one involving the smallest total number of $+$ signs. We shall consider only this simple problem involving the operations of addition and multiplication. It is easy to see how our methods can be extended to treat more general problems.

2. Number of representations. Let $\{a_i\}$, $i=1, 2, \cdots$, be a given infinite sequence of positive integers and let $f_N(n)$ denote the number of ways in which a positive integer n can be represented in the form

(1) $$n = a_1 * a_2 * \cdots * a_N,$$

where the a_i must be used in the designated order, but where there is a choice of a $+$ or $\times$ for each asterisk. The function $f_N(n)$ is defined for $N=1, 2, \cdots$, $n=1, 2, \cdots$.

Since $6 = 1+2+3 = 1\times 2\times 3$, it is clear that multiple representations exist. Since the last asterisk can be a $+$, or the last r asterisks all $\times$'s, $r=1, 2, \cdots$, $N-2$, we see that

$$f_N(n) = f_{N-1}(n - a_N) + f_{N-2}(n - a_N a_{N-1})$$
$$+ f_{N-3}(n - a_N a_{N-1} a_{N-2}) + \cdots$$
$$+ f_2(n - a_N a_{N-1} \cdots a_4 a_3) + f_1(n - a_N a_{N-1} \cdots a_3 a_2) + g(n),$$

---

* From AMERICAN MATHEMATICAL MONTHLY, vol. 69 (1962), pp. 640–643.

where

$$g(n) = 1 \quad \text{if } n = a_1 a_2 \cdots a_N,$$
$$= 0 \quad \text{if } n \neq a_1 a_2 \cdots a_N.$$

3. Minimal representation. Let us now seek to minimize the number of + signs. Define $F_N(n) =$ minimum total number of + signs in a representation of the form of (1), when at least one representation exists. If the representation is not possible, we define $F_N(n) = \infty$. Arguing as above, we see that we have a choice of using a + sign before a_N, in which case

$$F_N(n) = 1 + F_{N-1}(n - a_N),$$

or we can use $r \times$ signs between $a_N, a_{N-1}, \cdots,$ and a_{N-r}, in which case

$$F_N(n) = F_{N-r-1}(n - a_N a_{N-1} \cdots a_{N-r}).$$

Since we wish to minimize $F_N(n)$, we have

$$F_N(n) = \min\left[1 + F_{N-1}(n - a_N), \min_{1 \leq r \leq N-2} F_{N-r-1}(n - a_N a_{N-1} \cdots a_{N-r})\right].$$

This is a particular application of the principle of optimality [1].

4. An example.* Let us compute $F_9(100)$. We have

(2)
$$F_9(100) = \min[1 + F_8(91), 1 + F_7(28)],$$
$$F_8(91) = \min[1 + F_7(83), 1 + F_6(35)],$$
$$F_7(83) = \min[1 + F_6(76), 1 + F_5(41)],$$
$$F_6(76) = \min[1 + F_5(70), 1 + F_4(46)] = 1 + F_5(70),$$
$$F_5(70) = \min[1 + F_4(65), 1 + F_3(50), 1 + F_2(10)] = \infty,$$
$$F_4(46) = \infty,$$

Hence

$$F_7(83) = 1 + F_5(41) = 1 + \min[1 + F_4(36), 1 + F_3(21)] = \infty.$$

Thus

$$F_8(91) = 1 + F_6(35) = 1 + \min[1 + F_5(29), 1 + F_4(5)]$$
$$= 2 + F_5(29) = 2 + \min[1 + F_4(24), F_3(9)]$$
$$= 3 + F_4(24) = 3,$$

since $24 = 1 \cdot 2 \cdot 3 \cdot 4$.

Consequently, returning to the first equation of (2),

$$F_9(100) = \min[4, 1 + F_7(28)].$$

---

* I wish to thank L. Shapley for correcting my original calculations.

Since
$$F_7(28) = 1 + F_6(21) = 2 + F_5(15) = 3 + F_4(10)$$
$$= 4 + F_3(6) = 4,$$

(since $6 = 1 \cdot 2 \cdot 3$) we see that $F_9(100) = 4$, and, indeed, retracing our steps, we have

$$100 = 1 \cdot 2 \cdot 3 \cdot 4 + 5 + 6 + 7 \cdot 8 + 9.$$

5. Discussion. The method used here can be used to study the design of complex systems, and, particularly, questions of the reliability of multistage systems, (see [3]).

References

1. R. Bellman, Dynamic Programming, Princeton University Press, Princeton, New Jersey, 1957.
2. W. Ley, For your information, Galaxy Magazine, August 1961, p. 141.
3. R. Bellman and S. Dreyfus, Applied Dynamic Programming, Princeton University Press, Princeton, New Jersey, to appear in 1962.

PROBLEMS ABOUT PROBLEMS*

I. D. MACDONALD, University of Newcastle, N. S. W., Australia

The desire to construct elementary numerical problems sometimes gives rise to what may be called second-level problems which have a completely different character. For instance Sholander [1], wishing to find three mutually orthogonal 3-vectors α, β, γ with integral components and the same length D, other than the stock text-book trio

$$\alpha = [1, 2, 2], \quad \beta = [2, 1, -2], \quad \gamma = [2, -2, 1],$$

was led to the Diophantine equations

(1) $$\alpha_3^2 + \beta_3^2 + \gamma_3^2 = D^2 = \gamma_1^2 + \gamma_2^2 + \gamma_3^2.$$

An appeal to Dickson [3] produced the vectors

$$\alpha = [a^2 + b^2 - c^2 - d^2, 2(ac + bd), 2(ad - bc)],$$
$$\beta = [2(ac - bd), b^2 + c^2 - a^2 - d^2, 2(ab + cd)],$$
$$\gamma = [2(ad + bc), 2(cd - ab), b^2 + d^2 - a^2 - c^2],$$

which Reiersöl [2] showed by matrix methods to be essentially unique.

The point in which we are interested here is that this procedure yields a window through which some fragments of advanced mathematics in the theory

---

* From AMERICAN MATHEMATICAL MONTHLY, vol. 72 (1965), pp. 648–651.

of numbers may be seen. It may therefore be worthwhile to indicate a few more such problems with their second-level problems in the hope of strengthening this bond between elementary and advanced topics.

Let a conic in some coordinate system be represented by the equation

(2) $$ax^2 + 2hxy + by^2 = k,$$

which is to be reduced to a standard form by rotating the axes through some angle θ. If

$$x = x_1 \cos \theta - y_1 \sin \theta,$$
$$y = x_1 \sin \theta + y_1 \cos \theta,$$

as usual, substitution gives the equation $Ax_1^2 + 2Hx_1y_1 + By_1^2 = k$, where

$$A = \tfrac{1}{2}(a+b) + \tfrac{1}{2}(a-b) \cos 2\theta + h \sin 2\theta,$$
$$H = h \cos 2\theta - \tfrac{1}{2}(a-b) \sin 2\theta,$$
$$B = \tfrac{1}{2}(a+b) - \tfrac{1}{2}(a-b) \cos 2\theta - h \sin 2\theta.$$

When θ is chosen so that $H=0$ the equation becomes

$$(a+b - \sqrt{(a-b)^2 + (2h)^2})\, x_1^2 + (a+b + \sqrt{(a-b)^2 + (2h)^2})\, y_1^2 = 2k.$$

Now in numerical problems free from troublesome surds we essentially want a, h, b, k and c to be integers, where

$$(a-b)^2 + (2h)^2 = c^2.$$

The integral solutions to this equation are well known (cf. [3], p. 165). If we take

$$a - b = (\alpha^2 - \beta^2)\gamma, \quad 2h = 2\alpha\beta\gamma, \quad c = (\alpha^2 + \beta^2)\gamma,$$

with α, β, γ integers, and put $\delta = a - \alpha^2 \gamma$, (2) becomes

(3) $$\delta(x^2 + y^2) + \gamma(\alpha x + \beta y)^2 = k.$$

The other possibility, namely $a-b = 2\alpha\beta\gamma$, $2h = (\alpha^2 - \beta^2)\gamma$, $c = (\alpha^2 + \beta^2)\gamma$ with $\delta = a - \tfrac{1}{2}(\alpha+\beta)^2\gamma$, produces essentially the same equation as (3):

$$\delta(x^2 + y^2) + \tfrac{1}{2}\gamma((\alpha+\beta)x + (\alpha-\beta)y)^2 = k.$$

Thus if we want numerical examples of (1) to reduce by rotation we will do well to give numerical values to $\alpha, \beta, \gamma, \delta$ in (3) and so avoid problems with awkward surds.

Next, suppose that we are interested in the bisectors of the angles between the straight lines

$$ax + by + p = 0, \quad cx + dy + q = 0.$$

As the text-books prove, the equations of the bisectors are

$$\frac{ax + by + p}{\sqrt{(a^2 + b^2)}} = \pm \frac{cx + dy + q}{\sqrt{(c^2 + d^2)}}.$$

We shall have rational coefficients here provided a, b, p, c, d, q were rational and in addition $\sqrt{(a^2+b^2)}/\sqrt{(c^2+d^2)}$ is rational so that surds cancel out. Thus we are led to the Diophantine equation

(4) $$a^2 + b^2 = c^2 + d^2.$$

Now there is a formula which gives some solutions of this, namely

$$(\alpha\beta + \gamma\delta)^2 + (\alpha\gamma - \beta\delta)^2 = (\alpha\beta - \gamma\delta)^2 + (\alpha\gamma + \beta\delta)^2,$$

and according to Dickson (p. 254) the general solution is similar:

$$2a = \alpha\beta + \gamma\delta, \quad 2b = \alpha\gamma - \beta\delta, \quad 2c = \alpha\beta - \gamma\delta, \quad 2d = \alpha\gamma + \beta\delta,$$

where the integers $\alpha, \beta, \gamma, \delta$ are such that α and δ are even, or β and γ are even, or all four are odd.

The same equation (4) arises from problems in a quite different part of geometry. It may be proved, as it is in all the text-books, that the two tangents to the ellipse

$$\frac{x^2}{a^2} + \frac{y^2}{b^2} = 1$$

from the point (c, d) are perpendicular if and only if (4) holds. On the other hand tangents to the hyperbola

$$\frac{x^2}{a^2} - \frac{y^2}{b^2} = 1$$

give rise to

(5) $$a^2 = b^2 + c^2 + d^2,$$

an equation similar to that treated by Sholander.

The following problem occurs on p. 131 of [4]: "A rectangular sheet of metal has four equal square portions of metal removed at the corners, and the sides are then turned up so as to form an open rectangular box. Show that when the volume contained in the box is a maximum, the depth will be

$$\tfrac{1}{6}\{a + b - \sqrt{(a^2 - ab + b^2)}\},$$

where a, b are the sides of the original rectangle." Thus to make neat numerical examples out of this we want solutions to the Diophantine equation

(6) $$a^2 - ab + b^2 = c^2.$$

There is a geometrical problem which presents a very similar equation. If (2) with $k=0$ represents a pair of lines and if t is the tangent of an angle between them, then it can be proved that

$$t = \frac{2\sqrt{(h^2 - ab)}}{a + b},$$

from which simple algebra gives

(7) $$t^2a^2 + (2t^2 + 4)ab + t^2b^2 = (2h)^2.$$

The most interesting case of (7) has $t=1$. This and (6) and indeed the equation $a^2+b^2=c^2$ are all contained in the general case

(8) $$u^2a^2 + vab + u^2b^2 = c^2,$$

in which u and v are given integers. We show how to solve this by a method of Dickson (p. 406). We have

$$a(u^2a + vb) = (c + ub)(c - ub),$$

so if there is a solution there are coprime integers λ, μ for which

$$\lambda a = \mu(c + ub), \quad \mu(u^2a + vb) = \lambda(c - ub).$$

Elimination of c gives

(9) $$(\lambda^2 - u^2\mu^2)a = (2u\lambda\mu + v\mu^2)b.$$

Now suppose that λ and μ are any coprime integers with $\lambda^2 - u^2\mu^2 \neq 0$ and $2u\lambda\mu + v\mu^2 \neq 0$—we ignore the solutions arising when one of these numbers is 0. Then we can find coprime integers α and β for which

$$(\lambda^2 - u^2\mu^2)\alpha = (2u\lambda\mu + v\mu^2)\beta,$$

and we take $a=\alpha\gamma, b=\beta\gamma$, where γ is arbitrary. Substitution for c in $\lambda a = \mu(c+ub)$ gives

$$c = \frac{\lambda}{\mu} a - ub,$$

and we note that μ divides a. It may be verified that we now have a solution of (8).

Finally we point out a problem stated by Dickson (p. 502) to be unsolved. This is to find a parallelepiped with integral sides, face diagonals and long diagonal; in number-theoretic terms, to find integers a, b, c for which each of the following is a square:

$$b^2 + c^2, \quad c^2 + a^2, \quad a^2 + b^2, \quad a^2 + b^2 + c^2.$$

References

1. Marlow Sholander, Rational Orthogonal Matrices, this MONTHLY, 68 (1961) 350.
2. Olav Reiersöl, Rational Orthogonal Matrices, this MONTHLY, 70 (1963) 63-65.

3. L. E. Dickson, History of the Theory of Numbers, vol. II, Washington, 1919.
4. Sir H. Lamb, An Elementary Course of Infinitesimal Calculus, Cambridge, 1919.
5. A. Sutcliffe, Complete solution of the ladder problem in integers, Mathem. Gazette, 47 (1963) 133–36.
6. Problem E 1633, this MONTHLY, 70 (1963) 1005 and 71 (1964) 795–796.

ANOTHER PROOF OF THE INFINITE PRIMES THEOREM*

M. WUNDERLICH, University of Colorado
(Now at the State University of New York at Buffalo)

Let F_n be the nth Fibonacci number. It is a well-known and easily proved result [1] that

$$(1) \qquad F_{(m,n)} = (F_m, F_n),$$

where (m, n) as usual denotes the greatest common divisor. This property yields another proof of the infinite prime theorem.

THEOREM. *There are infinitely many primes.*

Proof. Suppose $p_1, p_2, \cdots, p_k$ are all the prime numbers. Then consider

$$(2) \qquad F_{p_1}, F_{p_2}, \cdots, F_{p_k}.$$

From (1), the numbers in (2) are pairwise relatively prime, and since there are only k primes, each of the numbers in (2) has only one prime factor. But this contradicts the fact that

$$F_{19} = 4181 = 113 \cdot 37.$$

Reference

1. N. N. Vorob'ev, Fibonacci Numbers, Blaisdell, New York, 1961, p. 30.

MORE ON THE INFINITE PRIMES THEOREM †

R. L. HEMMINGER, Vanderbilt University

Call a sequence $\{F_n\}_{n=1}^{\infty}$ of positive integers acceptable if (1) $F_n \neq F_m$ for $n \neq m$, (2) $(F_n, F_m) = 1$ for $(n, m) = 1$, (3) there exists a prime p for which F_p is not prime, and (4) $F_p \neq 1$ for p a prime.

The purpose of this note is to observe that *any* acceptable sequence can be used to establish the infinitude of the primes in exactly the same manner as the Fibonacci numbers were used in [1].

Let $\{F_n\}_{n=1}^{\infty}$ be an acceptable sequence and suppose that $p_1, p_2, \cdots, p_k$ are all the prime numbers. Then, by (1), (2), and (4), $F_{p_1}, F_{p_2}, \cdots, F_{p_k}$ are pairwise

---

* From AMERICAN MATHEMATICAL MONTHLY, vol. 72 (1965), p. 305.

† From AMERICAN MATHEMATICAL MONTHLY, vol. 73 (1966), pp. 1001–1002.

relatively prime integers greater than 1. Thus, since there are only k primes, each of them has only one prime factor. But this contradicts (3).

One sees immediately that, in addition to the Fibonacci numbers, the Mersenne and Fermat numbers form acceptable sequences since, in each case, the terms of the sequence are relatively prime.

Finally one notes that the essence of the classical proof is recaptured by considering the acceptable sequence $\{F_n\}_{n=1}^{\infty}$, defined recursively by $F_1 = 2$ and $F_{n+1} = F_1 \cdots F_n + 1$.

Reference

1. M. Wunderlich, Another proof of the infinite primes theorem, this MONTHLY, 72 (1965) 305.

ON THE GEOMETRY OF NUMBERS IN ELEMENTARY NUMBER THEORY*

G. E. ANDREWS, Pennsylvania State University

In most elementary texts on number theory, the geometry of numbers is rarely mentioned. This is probably due partly to the feeling that the introduction of convexity, n-dimensional geometry, etc., in an elementary course unnecessarily burdens students; also most of the theorems in the geometry of numbers seem beyond the field of interest of a first course in number theory. The following proof of the theorem on the linear diophantine equation gives the flavor of the geometry of numbers but does not call for any knowledge beyond a little analytic geometry.

We remark that the geometric interpretation of the linear Diophantine equation used here was originally given by H. J. S. Smith [4, p. 147], but Smith's proof is entirely algebraic, depending on the theory of continued fractions. A totally different geometric proof depending on the construction of regular polygons was given by P. Johnson [2].

In the following, $P = (p_1, p_2)$, $Q = (q_1, q_2)$, $\cdots$ will denote points of the plane; $O = (0, 0)$. A lattice point is a point with integral coordinates.

LEMMA. *If there exists no integer $b > 1$ such that $b \mid a_1$ and $b \mid a_2$ (where a_1 and a_2 are integers with $a_1 > 0$, $a_2 > 0$), then all the lattice points on the line $L_0 = \{(x, y); y = a_2 x/a_1\}$ are given by (ta_1, ta_2) with $t = 0, \pm 1, \pm 2, \cdots$.*

Proof. Let R be the lattice point with positive coordinates nearest the origin on L_0. Clearly R exists since there are at most $a_1 a_2$ lattice points (x^*, y^*) with $1 \leq x^* \leq a_1$ and $1 \leq y^* \leq a_2$. Assume that the n lattice points with positive coordinates on L_0 nearest O are $(r_1, r_2), (2r_1, 2r_2), \cdots, (nr_1, nr_2)$. Suppose there is a

---

* From AMERICAN MATHEMATICAL MONTHLY, vol. 74 (1967), pp. 1124–1125.

lattice point P such that $nr_1 < p_1 < (n+1)r_1$ and $nr_2 < p_2 < (n+1)r_2$. Then $0 < p_1 - nr_1 < r_1$ and $0 < p_2 - nr_2 < r_2$; hence, $(p_1 - nr_1, p_2 - nr_2)$ is a lattice point on L_0 nearer to O than R. Hence by mathematical induction (and by the fact that L_0 is symmetric with respect to O), all lattice points on L_0 are given by (tr_1, tr_2) with $t = 0, \pm 1, \pm 2, \cdots$. Thus $(a_1, a_2) = (t_0 r_1, t_0 r_2)$ for some positive integer t_0. Thus by the hypothesis of the lemma, we must have $t_0 = 1$. Hence $A = R$.

For further use note that the distance between (ta_1, ta_2) and $((t+1)a_1, (t+1)a_2)$ is given by $(a_1^2 + a_2^2)^{1/2}$.

THEOREM. *If there exists no integer $b > 1$ such that $b | a_1$ and $b | a_2$ (where a_1 and a_2 are integers with $a_1 > 0$, $a_2 > 0$), then there exist integers k and m such that $ka_1 + ma_2 = 1$.*

Proof. Consider the set Σ of all lines parallel to $L_0 = \{(x, y); y = a_2 x / a_1\}$ and containing lattice points. If $L \in \Sigma$, P is a lattice point on L, and $d(L)$ is the distance between L and L_0, then the area of the triangle $\triangle(OAP)$ is $|\frac{1}{2}(a_1 p_2 - a_2 p_1)| = \frac{1}{2}(a_1^2 + a_2^2)^{1/2} d(L)$. Hence $d(L) = n \cdot (a_1^2 + a_2^2)^{-1/2}$ where n is an integer. It is easy to verify that the adjacent lines of Σ are all the same distance d apart, for translating a lattice point on one line to a lattice point on another line transforms Σ into itself. Consider the square S of side ρ centered on O with two sides parallel to L_0. There are $\rho^2 + O(\rho)$ lattice points inside or on S (a fact easily established by the technique of Gauss which is well described in [1, pp. 29–30]). But there are at most $\rho/d + 1$ lines in S containing lattice points and parallel to L_0, and on each such line there are at most $\rho/(a_1^2 + a_2^2)^{1/2} + 1$ lattice points (by the note following the lemma and by the remarks concerning the translation properties of Σ). Hence

$$\rho^2 + O(\rho) \leq \rho^2 d^{-1} (a_1^2 + a_2^2)^{-1/2} + \rho d^{-1} + \rho \cdot (a_1^2 + a_2^2)^{-1/2} + 1$$

which is true for all positive ρ only if $d \leq (a_1^2 + a_2^2)^{-1/2}$. Since we have already established that $d \geq (a_1^2 + a_2^2)^{-1/2}$, we obtain $d = (a_1^2 + a_2^2)^{-1/2}$. Thus there is a lattice point P on a line a perpendicular distance $(a_1^2 + a_2^2)^{-1/2}$ from L_0. Thus the area of the triangle $\triangle(OAP)$ is

$$\tfrac{1}{2}(a_1^2 + a_2^2)^{1/2}(a_1^2 + a_2^2)^{-1/2} = \tfrac{1}{2} = |\tfrac{1}{2}(a_1 p_2 - a_2 p_1)|.$$

Hence one of the pairs $(k, m) = (p_2, -p_1)$ or $(-p_2, p_1)$ must fulfill the conclusion of the theorem.

Thus a standard technique from the geometry of numbers yields the essential result needed to develop such properties as unique factorization, etc., [3, Chapter 2]. In conclusion, I should like to express my appreciation to a very helpful referee; in particular, he brought references [1] and [4] to my attention.

References

1. D. Hilbert and S. Cohn-Vossen, Anschauliche Geometrie, Dover, New York, 1944.
2. P. B. Johnson, A construction of regular polygons of pq sides leading to a geometrical proof of $rp - sq = 1$, Math. Mag., 38 (1965) 164–165.

3. W. J. LeVeque, Topics in Number Theory, vol. 1, Addison-Wesley, Reading, Mass., 1956.
4. H. J. S. Smith, Collected Works, vol. II, Chelsea, New York, 1965.

MUSIMATICS OR THE NUN'S FIDDLE*†

A. L. LEIGH SILVER, Fellow of the Institute of Musical Instrument Technology, England

1. The divine ratio. "*Abominandum!*" said Cicero as he went a purler over a hidden obstacle—"*quid est quod?*"—and scrabbling in the undergrowth he uncovered an ancient monument. The lettering was illegible but the design—a cylinder circumscribing a sphere—was clearly that which Archimedes, who was killed in the fall of Syracuse 212 B.C., had charged his friends to inscribe on his tombstone. Since Cicero made this discovery about 75 B.C., the tomb has again been lost, probably forever.

Archimedes transformed empirical knowledge into theoretical science and developed the integral calculus which he said would be used by mathematicians "as yet unborn." In keeping with Aristotle's dictum that "it is proper to consider the similar even in things far distant from each other," he considered it highly significant that the cylinder and inscribed sphere, as regards surface area and volume, are in the ratio 3:2 and that the same relationship exists between the frequencies of an important musical interval.

The ear is very sensitive to this interval—the perfect fifth—and it has been used for tuning instruments from the earliest times. A power of 3/2 can never equal a power of 2/1 and superimposed perfect fifths will never arrive at an octave duplication of the fundamental. On a keyboard instrument, however, we find that twelve fifths pass through the twelve semitones of the chromatic scale and finish on the seventh octave of the fundamental note (Figure 1), which means that the fifths are not all perfect and somewhere the difference between 12 perfect fifths and 7 octaves has been lost. This difference $(3/2)^{12} \div 2^7 = 3^{12}/2^{19}$ is called a 'Pythagorean comma.'

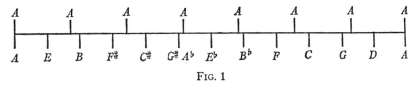

Fig. 1

For tuning purposes the series of fifths is kept within the limits of an octave

---

* From AMERICAN MATHEMATICAL MONTHLY, vol. 78 (1971), pp. 351–357.

† A symbolic title with Chaucerian overtones. This one-stringed instrument, better known as the 'Marine Trumpet', has clarion qualities well suited for trans-Atlantic communication.

A. L. Leigh Silver writes that he is a 3M man: medicine, music, and maths. Son of a professional organist, he is an Oxford and London educated physician and presently is employed by the 7520 U.S.A.F. Hospital. He is a fellow of the British Medical Association, Fellow of the Inst. of Musical Instrument Technology, and Hon. Fellow Mercator Music Foundation. *Editor.*

by descending an octave each time this limit is exceeded (dotted lines Figure 2).

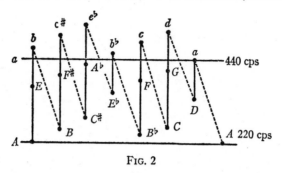

Fig. 2

The ancient Greeks and Chinese calculated the Pythagorean comma which equals about 24 cents (the **cent** being one twelve hundredth part of an octave, or the base two logarithm of the ratio multiplied by 1200).

About 40 B.C., King-Fang, a scholar of the Han dynasty, continued the series of superimposed fifths in order to find a closer approximation to an octave. His first improvement came with the 41st fifth which was less than 24 octaves by about 20 cents. Not content, he carried on until he came to the 53rd fifth, which exceeds 31 octaves by about 3.6 cents. This excellent approximation was later recommended by Mercator and the 53 note octave was incorporated in several instruments including Bosanquet's *Enharmonic Harmonium* which was exhibited in the South Kensington Museum in 1876. In the hope of achieving immortality, I carried on Fang's calculations (without an abacus) and found the next better approximations to be:

$$(3/2)^{306} < 2^{179} \text{ by about 1.8 cents}$$

$$(3/2)^{665} > 2^{389} \text{ by about 0.074 cents.}$$

Leonardo da Vinci *ca.* 1470 observed that "two men shouting together do not seem to produce twice the amount of noise that one man would" [1] and in general we now know that sensations vary as the logarithm of the stimulus (Fechner's Law). We talk and think of two octaves as twice the size of one (as on the piano keyboard), three octaves as three times the size, and so on. Yet the frequencies of these intervals are in the ratio $2:4:8 \cdots$. Base two logarithms are therefore naturally suited for musical purposes and were published in 1670, fifty-six years after Napier's tables [2]. Modern tables are available [3].

2. Lesser divine ratios. Over the centuries musical opinion has been remarkably consistent—

(a) The satisfying intervals are derived from natural harmonics, the frequencies of which are related as the natural number series $1:2:3 \cdots$.

(b) Successive ratios are favoured and are named 'superparticular.' They are an infinite series $2/1, 3/2, 4/3, 5/4, \cdots$.

(c) The lower members are pleasing; the higher members tend to harshness and eventually become unacceptable.

(d) Certain ratios, although within the range of acceptable harshness, are regularly rejected, e.g., 7/6, 8/7, 11/10, 12/11, 13/12, 14/13, $\cdots$.

There is no obvious reason for this last empirical fact. However, an analysis of a large amount of material discloses that *the ear prefers superparticular ratios that are derived from the first three primes*, and when other ratios are omitted we are left with the following *finite* series of well-known intervals:

2/1 octave	9/8 major tone
3/2 perfect fifth	10/9 lesser tone
4/3 perfect fourth	16/15 diatonic semitone
5/4 major third	25/24 chromatic semitone
6/5 minor third	81/80 comma of Didymus

The enthusiast will no doubt relate these intervals (excluding the octave) to the nine Platonic and Kepler-Poinsot regular polyhedra. Since the perfect fifth and the major third contain the first three primes, all other intervals may be compounded from them.

3. Just tunings. Perfection in tuning is an *ignis fatuus* which philosophers and musicians have followed since the beginning of time. They have concentrated on tunings largely composed of primary intervals (the 3/2 fifth and the 5/4 third) and which are loosely termed 'just tunings.' Complexity, vagueness and the absence of a simple method for recording observations have caused confusion and reduplication, but with a simple definition and a geometrical analogy suggested by T. H. O'Beirne [4] I hope to show that there are 118 just tunings and all possess undesirable qualities in varying degree.

A just tuning is one in which every note is related to at least one other note by a primary interval. Such a tuning can be plotted on squared paper. Vertices represent notes, horizontal lines joining them (left-right) perfect fifths, and vertical lines (up-down) major thirds. The problem resolves itself into finding the total number of unbroken patterns that can be formed.

Patterns are easily memorised and each completely defines a tuning. We start with the simplest—Pythagorean tuning—a sequence of eleven perfect fifths which can be plotted on a single horizontal line (Figure 3, i). The twelfth fifth uniting the last note with an octave of the first is left blank, and this indicates that it is imperfect. It is called the **Procrustean fifth** since it is cut to fit, and in this instance it is a perfect fifth less a Pythagorean comma: $2^{18}/3^{11}$. See [5].

Next we list all possible patterns occupying two horizontal lines (Figure 3, ii–xii). The symmetrical pattern (vii) with ten perfect fifths and four major thirds was suggested by Ramis de Pareja in 1482.

There are 43 patterns occupying three horizontal lines, and space will not allow these to be listed. The symmetrical pattern (Figure 4) is of special interest.

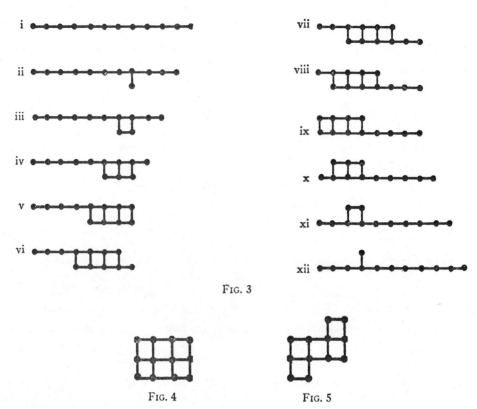

Fig. 3

Fig. 4　　Fig. 5

I regard it as the most perfect of all the just tunings because it contains the maximum number of primary intervals (nine perfect fifths, eight major thirds).

Four horizontal lines give 55 patterns. The following (Figure 5) by Marpurg 1776, is generally spoken of as the 'model form' of just tuning, although it has one less perfect fifth than the symmetrical pattern above.

Five horizontal lines give 8 patterns and this completes the list of 118 just tunings.

Just tunings are pleasing, and each key has a character which can be suited to the mood of the composition (now a forgotten artistic refinement). Their inherent imperfections render them unacceptable for the harmonic and modulatory demands of modern music.

4. Temperaments. Unpleasant intervals cannot be abolished and improvement is only obtained by "tempering" or adjusting, so that the unpleasantness is shared with other intervals. This can be done in an infinite number of ways of which two will be outlined.

EQUAL TEMPERAMENT (ET). The Chinese were concerned with the problem of dividing the octave into twelve equal intervals more than 1000 years B.C. Their music did not require twelve chromatic notes but they realised the need for this number for the purpose of transposition. This meant finding the twelfth root of two which was not an easy problem. The astronomer Ho-Cheng-Tien was accused of "doing violence to figures" when he tried to find a solution *ca.* 420 A.D. Wang-Po, a physician, produced inaccurate results *ca.* 938 A.D., and not until 1598 did Prince Chu-Tsai-Yu "after meditating for days and nights before the light of Truth was revealed" come up with an answer said to be correct to nine places. In Europe the same feat was performed in 1600 by Simon Stevin, an inspector of canals in Holland, author of *La Disme*, and inventor of a sailing barge.

There is no evidence that J. S. Bach (1685–1750) attempted or intended to tune equally. The "48" were written for "*Das wohltemperierte Clavier*"—the *well tempered* clavier, not the equally tempered. It has been pointed out that the frets on ancient instruments appear to be spaced equally (in the logarithmic sense) and the 6-string lute in *The Ambassadors* by Hans Holbein the Younger 1533, has been quoted as an example. This instrument, and a number of curious objects including a German arithmetic book, lie on the lower shelf of the buffet on which the ambassadors are leaning. The finger board is foreshortened by perspective, and all in all the example is not convincing.

Equal temperament was not generally adopted until the beginning of the present century. It is a tedious temperament for the tuner because every interval is "out of tune." Accuracy is seldom achieved and then only by counting twelve different beat rates or by utilising apparatus such as the "Stroboconn."

MEANTONE TEMPERAMENT. In 1523 Aron suggested that fifths should be tempered to produce 5/4 thirds. Four perfect fifths—say C-G-D-A-E—produce a third C-E (plus two octaves) with an unpleasantly large ratio, i.e., $(3/2)^4$ divided by 4 to get rid of the octaves $= 81/64$. In order to give a 5/4 third, the ratio of each fifth must be reduced to $\sqrt[4]{5}$.

Meantone tuning is not just, because the network is broken (Figure 6). The middle note of the major third divides the latter into two equal major seconds, hence the name "meantone."

FIG. 6

This temperament was established by about 1600 and remained popular for a long time. Many organs were still tuned to it at the beginning of the present century. It died a lingering death because musicians were strongly opposed to its replacement by equal temperament.

5. Equal beating scale (EBS). This is evolved in a different manner and is not a tuning or a temperament. It possesses the following advantages:

   i. It can be used for all musical purposes.
   ii. It introduces a soupçon of colour to all keys.

iii. It may represent a close approximation to J. S. Bach's "well-tempered" scale.
iv. It enhances the resonance of stringed keyboard instruments.
v. Above all, ease in tuning is marked and greater accuracy is likely to be achieved.

The principle is simple. All intervals in the tuning series (Figure 2) have the same beating rate. Beats occur between the 3rd partial of the lower note and the 2nd partial of the upper note of an imperfect fifth (Figure 7, a): and between the 3rd partial of the upper note and the 4th partial of the lower note of an imperfect fourth (Figure 7, b).

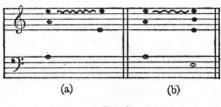

(a)   (b)

Fig. 7

The beat rate is the difference between the frequencies of these partials. In this example, if the frequency of the lower E is half the rate of the upper E (i.e., they are "in tune") then the beat rate is the same in each case.

The common beat rate (β) for the EBS is found by solving the twelve chain equations of the tuning series (Figure 2) in terms of β and a.

Thus

(i) $\beta = 3a - 2e$ or $e = \tfrac{3}{2}a - \tfrac{1}{2}\beta$

(ii) $\beta = 3e - 4b$ $\quad b = \tfrac{3}{4}e - \tfrac{1}{4}\beta$

(iii) $\beta = 3b - 2f\#$ $\quad f\# = \tfrac{3}{2}b - \tfrac{1}{2}\beta$

. . . .

(xii) $\quad\quad\quad\quad 2a = \tfrac{3}{2}d - \tfrac{1}{2}\beta.$

From which $\beta = (7153/1568693)a$. This is the key of all necessary calculations. The ratios of all intervals can be expressed as integral numbers, and with international tuning frequency $A_4 = 440$ cps the beat rate is 1.00317 p/s which, for most practical purposes, may be taken as unity.

6. Apologia. It has been said, with regard to musical problems, that musicians generally give the correct answers supported by illogical argument, but mathematicians arrive at incorrect answers through a process of irrefutable reasoning.

This puts me in a quandary. I would like to be thought of as the operator "little i"—neither one thing nor the other, perhaps imaginary but sometimes useful.

References

1. The Notebooks of Leonardo da Vinci I, London, 1954.
2. Murray Barbour, Musical logarithms, Scripta Mathematica, 3 (1940) 21.
3. Silver and Newby, A Table of Logarithms to Base Two, TN/Math III, Ministry of Aviation, London, W.C.2, 1964.
4. T. H. O'Beirne, Puzzles and paradoxes, New Scientist, 274 (Feb. 15, 1962).
5. A. L. Leigh Silver, Notes on the Duodecimal Division of the Octave, Institute of Musical Instrument Technology, London, (Dec. 1964).
6. ———, Equal beating chromatic scale, J. Acoustical Soc. Amer., 29 (1957) 476–481.
7. ———, Some musico-mathematical curiosities, Math. Gazette, 48 (1964) 1–17.

GALILEO SEQUENCES, A GOOD DANGLING PROBLEM*

KENNETH O. MAY, University of Toronto

1. Galileo's idea. In 1615 Galileo observed that the sequence of odd integers had the property

$$(1) \qquad \frac{1}{3} = \frac{1+3}{5+7} = \frac{1+3+5}{7+9+11} = \cdots.$$

The observation was closely related to his work on freely falling bodies. Indeed, if distance is proportional to time squared and is one in the first time unit, then the total distances at integral times are the perfect squares, and the incremental distances in successive unit time intervals are the odd integers. If we take a new unit of time equal to some multiple of the original, the ratio of the distances travelled in the first two time units should be unchanged. But this is just the significance of (1), since it says that the distance in the first n time units is always one third of the distance in the next n time units.

Galileo observed that the sequence of odd integers is the only arithmetic progression with this property, and he considered this an important argument for this law of free fall. Was Galileo right? What can be said about sequences for which the ratio of the sum of the first n terms to the sum of the next n terms is a constant? [1]

2. Dangling problems. This is a typical dangling problem. It can be presented with little symbolism, is easily understood, has intuitive appeal, and is wide open to student initiative in experimenting, formulating questions, conjecturing, and proving. Dangling such a question before a class may lead to general participation in class discussion, group projects, or individual efforts. At the very least it provides the students with a participatory glimpse of mathematics in the making. At best it may "turn on" a potential mathematician.

The problem of Galileo sequences was dangled before a class of future teachers

---

* From AMERICAN MATHEMATICAL MONTHLY, vol. 79 (1972), pp. 67–69.

at the College of Education at the University of Toronto during 1968–1969. Practically all students participated verbally, and several made significant written contributions [2].

3. Galileo sequences. Let the nth term of a sequence be a_n and the sum of the first n terms S_n. A *Galileo sequence* (*GS*) is a sequence of positive integers satisfying

(2) $$S_{2n} - S_n = pS_n$$

for fixed p ($= 3$ for the odd integers). Equivalent conditions are

(3) $$S_{2n} = qS_n \quad (q = p + 1), \text{ and}$$

(4) $$a_{2n-1} + a_{2n} = qa_n.$$

Experimentation suggests many easily proved results relating to sums, differences, multiples, special hypotheses on a_n, etc. In particular:

(5) *If one sequence of positive integers is a multiple of another, then if either is a GS so is the other and they have the same ratios.*

This suggests defining a *primitive GS* as one that is minimal with respect to multiplication. Then it is easy to prove that the only primitive increasing *GS* in arithmetic progression is the odd integers, but that there are many other primitive increasing *GS*.

An early conjecture might be:

(6) *In a GS the second term must be an integral multiple of the first, i.e., p and q are integers.*

To prove this let $q = h/k$ in lowest terms. Then from (4) every a_n is a multiple of k, and we may form a new *GS* with the same ratio by dividing all terms by k. Repeating the process with the new *GS* and its successors m times, we see that k^m divides a_n for arbitrarily large m, which is the case only if $k = 1$.

The most interesting result of the year was the following:

(7) *A necessary and sufficient condition for the existence of a strictly increasing GS is that $p > 2$.*

The following argument is based on the first proof by D. A. Gautreau.

To prove the impossibility for $p = 2$, we show that for any i there is a $j > i$ such that $d_j < d_i$, where $d_n = a_{n+1} - a_n$. Then it follows that eventually the difference of successive terms will be non-positive. In order to prove the inequality, we use the identity

(8) $$d_{2i+1} + 2d_{2i} + d_{2i-1} = 3d_i,$$

which follows from the definition of d_n and (4) with $q = 3$. Now at least one of the three d's in the left member must be less than d_i, for otherwise the left member would be at least $4d_i$.

Since the sequence of odd numbers has $p=3$, we suppose that p is greater than 3, i.e., $p \geq 4$, $q \geq 5$.

We claim that a strictly increasing GS is given by $a_1 = 1$,

(9) $$a_{2n-1} = \left[\frac{qa_n - 1}{2}\right], \quad a_{2n} = \left[\frac{qa_n}{2}\right] + 1,$$

where the square bracket indicates the greatest integer function. (The choice is suggested by experiments in which one chooses at each stage the nearest pair of numbers that do not violate the requirements.) Since (9) satisfies (4), and a_{2n-1} is obviously less than a_{2n}, it will be sufficient to prove that $a_{2n} < a_{2n+1}$. This can be done recursively by noting that $a_2 < a_3$ and proving that if $a_n < a_{n+1}$, then $a_{2n} < a_{2n+1}$. From (9) and the fact that $q \geq 5$,

(10) $$a_{2n+1} - a_{2n} \geq \frac{qa_{n+1} - 2}{2} - \frac{qa_n + 2}{2}$$

(11) $$\geq \frac{5}{2}(a_{n+1} - a_n) - 2.$$

But if $a_{n+1} > a_n$, their difference is at least 1 and the right member of (11) is greater than 1/2.

Notes

1. The problem was suggested by a conversation with Stillman Drake of the Institute for the History and Philosophy of Science and Technology at the University of Toronto. See his *Galileo Studies* (University of Michigan Press, 1970), pp. 218–219, 228.

2. The most substantial contributors were D. A. Gautreau (an auditor from grade 13 of the University of Toronto Schools), S. K. Pasricha, G. C. Reid, F. Riad, and D. Sale. Paul Erdös, while visiting Toronto, concurred in some conjectures under consideration.

RECURRENT SEQUENCES AND PASCAL'S TRIANGLE*

THOMAS M. GREEN, Contra Costa College

A. Introduction. The Fibonacci sequence can be found by summing the terms on successive "diagonals" of Pascal's Triangle [1]. J. Raab [2] generalized this procedure to show other sets of parallel diagonals generating different recurrent sequences. This generalization is essentially the same as Phase One in what is to follow. The purpose of this paper is to show that there exist infinitely many more recurrent sequences within Pascal's Triangle by summing the terms on diagonals of different slopes. Each sequence shall be of the type such that

---

* From Mathematics Magazine, vol. 41 (1968), pp. 13–21.

each term is the sum of *two* former terms. There is also a unique relationship between just what two terms are involved and the slope of the diagonals being considered.

For this purpose it is convenient to arrange the terms of Pascal's Triangle on the point-lattice determined by the nonnegative integral points of a rectangular coordinate system. (See Figure 1.)

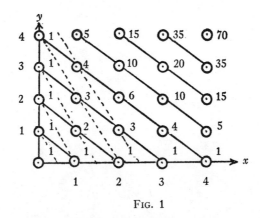

Fig. 1

With this arrangement the coordinates, (x, y), of the lattice point uniquely determine the location and value of a Pascal number $\binom{n}{r}$. The value is seen to be

A(1) $$\binom{n}{r} = \frac{(x+y)!}{x! \cdot y!},$$

since $n = x + y$ and $r = y$ (or x, because of the symmetry involved).

B. Phase one. Consider the linear equation

B(1) $$x + y = n \quad \text{for } n = 0, 1, 2, \cdots.$$

This equation represents the nth row (diagonal) of Pascal's Triangle. If we sum the Pascal numbers on each row determined by B(1) for successive values of n, we obtain the sequence

B(1.1) $$1, 2, 4, 8, \cdots, 2^n, \cdots$$

whose recurrence relation is given by

B(1.2) $$P_n = P_{n-1} + P_{n-1},$$

where $P_0, P_1, \cdots, P_n, \cdots$ denote the terms of the sequence, and the formula for the nth term is given by

B(1.3) $$P_n = 2^n = \sum_{\substack{x=0, y=0 \\ x+y=n}}^{n,n} \frac{(x+y)!}{x! \cdot y!} = \sum_{r=0}^{n} \binom{n}{r}.$$

(*Note*: the nth term is the term formed by summing all of the Pascal numbers on the line $x+y=n$ and, if we were counting the terms, this term would actually be the $(n+1)$th term in the sequence.)

The sum of the first n terms of the sequence is given by

B(1.4) $$\sum_{k=0}^{n-1} P_k = P_n - 1.$$

Now consider the linear equation

B(2) $$2x + y = n \quad \text{for } n = 0, 1, 2, \cdots.$$

This equation represents the nth diagonal referred to above used to obtain the nth Fibonacci number. By summing the Pascal numbers on each diagonal determined by B(2) for successive values of n (see the dotted lines, Figure 1), we obtain the sequence

B(2.1) $$1, 1, 2, 3, 5, \cdots, F_n, \cdots$$

whose recurrence relation is given by

B(2.2) $$F_n = F_{n-2} + F_{n-1}.$$

The formula for the nth term is given by

B(2.3) $$F_n = \sum_{\substack{x=0,y=0 \\ 2x+y=n}}^{[n/2],n} \frac{(x+y)!}{x! \cdot y!} = \sum_{r=0}^{[n/2]} \binom{n-r}{r}$$

where [ ] denotes the greatest integer function and the sum of the first n terms of the sequence is given by

B(2.4) $$\sum_{k=0}^{n-1} F_k = F_{n+1} - 1.$$

Next consider the linear equation

B(3) $$3x + y = n \quad \text{for } n = 0, 1, 2, \cdots.$$

In a way similar to that used above we establish the sequence

B(3.1) $$1, 1, 1, 2, 3, 4, 6, 9, \cdots, G_n, \cdots$$

whose recurrence relation is given by

B(3.2) $$G_n = G_{n-3} + G_{n-1}.$$

The formula for the nth term is given by

B(3.3) $$G_n = \sum_{\substack{x=0,y=0 \\ 3x+y=n}}^{[n/3],n} \frac{(x+y)!}{x! \cdot y!} = \sum_{r=0}^{[n/3]} \binom{n-2r}{r}$$

and the sum of the first n terms of the sequence is given by

B(3.4) $$\sum_{k=0}^{n-1} G_k = G_{n+2} - 1.$$

Now consider the linear equation

B(j) $\quad\quad jx + y = n \quad$ for $n = 0, 1, 2, \cdots$ and $j = 1, 2, 3, \cdots$.

This equation, by the procedure referred to above, establishes a sequence whose recurrence relation is given by

B(j.2) $$T_n = T_{n-j} + T_{n-1}.$$

The formula for the nth term is given by

B(j.3) $$T_n = \sum_{\substack{x=0, y=0 \\ jx+y=n}}^{[n/j], n} \frac{(x+y)!}{x! \cdot y!} = \sum_{r=0}^{[n/j]} \binom{n - (j-1)r}{r}$$

and the sum of the first n terms of the sequence is given by

B(j.4) $$\sum_{k=0}^{n-1} T_k = T_{n+(j-1)} - 1.$$

For a proof of B(j.2) and B(j.4), see E(a.2) and E(a.4).

C. Phase two. In Phase One each linear equation had the coefficient pair $\langle j, 1 \rangle$, giving rise to infinitely many recurrent sequences. We now consider any coefficient pair $\langle j, 2 \rangle$, determining the following equations:

C(1) $\quad\quad x + 2y = n$

C(2) $\quad\quad 2x + 2y = n$

C(3) $\quad\quad 3x + 2y = n \quad$ for $n = 0, 1, 2, \cdots$.

$\quad\quad\quad\quad\quad\quad\quad\vdots$

C(j) $\quad\quad jx + 2y = n$

Several of these are equivalent to cases already discussed, namely,

(i) any coefficient pair $\langle a, b \rangle$ will yield the same sequence and recurrence relation as the pair $\langle b, a \rangle$, because of the symmetry of the Pascal Triangle about the line $y = x$;

(ii) any coefficient pair $\langle a, b \rangle$ presupposes the fact that a and b are relatively prime, since if they are not, then two possibilities occur. Either the equation will be reduced by dividing thru by the greatest common divisor or it will be such that the given value of n will not be divisible by the g.c.d. of a and b. If the equation is reduced, it will have been treated in an earlier phase, and if the equation cannot be reduced, there will be no integral solutions (see [3]), and thus no sequence will be determined;

(iii) even when a and b are relatively prime, there will be cases where $ax+by=n$ will not have nonnegative integral solutions. This means that for those particular values of n, the recurrent sequence derived from $ax+by=n$ will have zero as a value for those nth terms in the sequence, since there will be no Pascal numbers to sum. Thus we establish the following useful

LEMMA. *The equation $ax+by=n$, where a, b, and n are nonnegative integers and $ab \neq 0$ and $(a, b)=1$, will not have nonnegative integral solutions when $n=ab-(ja+kb)$, where $j, k=1, 2, 3, \cdots$.*

Proof. Assume $n=ab-(ja+kb)$ so that $ax+by=ab-ja-kb$ with nonnegative solution (x, y). Thus $a(x+j)+b(y+k)=ab$. Let $X=x+j$ and $Y=y+k$; then $aX+bY=ab$. It is important to note here that both X and Y, as well as both a and b, are greater than or equal to one. We can now transform the above equation to $b=X+(bY)/a$ or $b-X=(bY)/a$. Now $b-X$ is an integer; therefore a divides Y since a and b are relatively prime. Suppose $Y/a=r$ so that $Y=ar$.

Similarly we can show that $a-Y=(aX)/b$ and hence conclude that b divides X. Suppose $X/b=s$ so that $X=bs$. Then, by substitution, we have $abs+abr=ab$ or $ab(r+s)=ab$. Therefore $r+s=1$. But both r and s are greater than or equal to one; therefore we have a contradiction.

Thus, in view of the above discussion, the first new case in Phase Two is C(3):

C(3) $\qquad\qquad 3x + 2y = n \qquad$ for $n = 0, 1, 2, \cdots$.

By summing the Pascal numbers on each diagonal determined by C(3) for successive values of n we find that there is no positive integral solution for $n=1$, since, $1=3 \cdot 2 - (3+2)$ as predicted by the lemma; therefore the sequence is

C(3.1) $\qquad$ 1, 0, 1, 1, 1, 2, 2, 3, 4, 5, 7, 9, 12, $\cdots$, H_n, $\cdots$

whose recurrence relation is given by

C(3.2) $\qquad\qquad H_n = H_{n-3} + H_{n-2}.$

The formula for the nth term is given by

C(3.3) $\qquad\qquad H_n = \sum_{\substack{x=0, y=0 \\ 3x+2y=n}}^{[n/3],[n/2]} \frac{(x+y)!}{x! \cdot y!}.$

A formula for H_n in terms of n and r could also be given; however, it actually requires *two* formulas and, in general, the formula will require b different representations, one for each of the different values in the residue class of $n \pmod{b}$. Phase Three will need three formulas, etc. More will be said about this in the discussion of the general phase.

The formula for the sum of the first n terms of the sequence is given by

C(3.4) $\qquad\qquad \sum_{k=0}^{n-1} H_k = H_{n+2} + H_{n+1} - 1.$

Now by similar considerations of the next new case,

C(5) $\qquad 5x + 2y = n \qquad$ for $n = 0, 1, 2, \cdots,$

we find that there are no solutions for $n=1$ and $n=3$, since $1 = 5 \cdot 2 - (5 + 2 \cdot 2)$ and $3 = 5 \cdot 2 - (5+2)$. Hence, by summing the Pascal numbers on each successive diagonal determined by C(5), we obtain the sequence

C(5.1) $\qquad 1, 0, 1, 0, 1, 1, 1, 2, 1, 3, 2, 4, 4, 5, 7, 7, 11, 11, \cdots, I_n, \cdots$

whose recurrence relation is given by

C(5.2) $\qquad I_n = I_{n-5} + I_{n-2}.$

The formula for the nth term is given by

C(5.3) $\qquad I_n = \sum_{\substack{x=0, y=0 \\ 5x+2y=n}}^{[n/5], [n/2]} \frac{(x+y)!}{x! \cdot y!},$

and the sum of the first n terms is given by

C(5.4) $\qquad \sum_{k=0}^{n-1} I_k = I_{n+4} + I_{n+3} - 1.$

Now consider the general case of Phase Two,

C(j) $\qquad jx + 2y = n \qquad$ for $n = 0, 1, 2, \cdots \quad$ and $j = 1, 2, 3, \cdots.$

This equation establishes a sequence whose recurrence relation is given by

C(j.2) $\qquad T_n = T_{n-j} + T_{n-2}$

The formula for the nth term is given by

C(j.3) $\qquad T_n = \sum_{\substack{x=0, y=0 \\ jx+2y=n}}^{[n/j], [n/2]} \frac{(x+y)!}{x! \cdot y!}$

and the sum of the first n terms of the sequence is given by

C(j.4) $\qquad \sum_{k=0}^{n-1} T_k = T_{n+(j-1)} + T_{n+(j-2)} - 1.$

D. Phase three. Consider the equation

D(j) $\qquad jx + 3y = n \quad$ for $\quad n = 0, 1, 2, \cdots \quad$ and $\quad j = 1, 2, 3, \cdots$

where the pair $\langle j, 3 \rangle$ complies with the remarks made in section C. By summing the Pascal numbers on each diagonal given by D(j) for successive values of n we obtain the sequence whose recurrence relation is given by

D(j.2) $\qquad T_n = T_{n-j} + T_{n-3}.$

The formula for the nth term is given by

D(j.3) $$T_n = \sum_{\substack{x=0, y=0 \\ jx+3y=n}}^{[n/j],[n/3]} \frac{(x+y)!}{x! \cdot y!},$$

and the sum of the first n terms is given by

D(j.4) $$\sum_{k=0}^{n-1} T_k = T_{n+(j-1)} + T_{n+(j-2)} + T_{n+(j-3)} - 1.$$

Proof of the formulas of Phase Two and Phase Three will be covered by the proofs in the general phase that follows.

E. Phase b. In general, the equation

E(a) $\qquad ax + by = n \quad \text{for} \quad n = 0, 1, 2, \cdots \quad \text{and} \quad a, b = 1, 2, 3, \cdots,$

where the pair $\langle a, b \rangle$ complies with the remarks made in section C, will, by summing the Pascal numbers on each diagonal for successive values of n, yield a recurrent sequence whose recurrence relation is given by

E(a.2) $$T_n = T_{n-a} + T_{n-b}.$$

Proof. The first term in the series representing T_n, as defined by E(a.3) below, will be $(x+y)!/(x! \cdot y!)$, where x and y satisfies E(a). In the notation of A(1) this will equal

$$\binom{x+y}{x}.$$

Suppose that this first solution of x and y is the one where x is minimum (and hence y is maximum); then the next solution would be $(x_{\min}+b, y_{\max}-a)$ and the next would be $(x_{\min}+2b, y_{\max}-2a)$, etc., until $y_{\max}-ra$ becomes $y_{\min}$, where r is the greatest integer in the quotient $n/(ab)$. (This is a modified form of a standard result of number theory; see, for example, [3].) Now if we let $k = x_{\min}+y_{\max}$, we can write the first few terms of T_n as follows

(1) $$T_n = \binom{k}{x} + \binom{k+b-a}{x+b} + \binom{k+2b-2a}{x+2b} + \cdots,$$

where the x refers to only $x_{\min}$.

Next we look at T_{n-a}. The first term of this series will be of the form $\binom{k'}{x'}$ where k' and x' are related to k and x above in the following manner. First we note that

(2) $$ax' + by' = n - a, \quad \text{and} \quad x' + y' = k'.$$

Now since a, b and n have all been fixed we find that $x' = x_{\min}-1$ and $y' = y_{\max}$

is a solution, which upon substitution satisfies $ax+by=n$. Furthermore, since y' is the same y_{max} as found in the consideration of T_n, it will also be the maximum y in the consideration of T_{n-a}, since $n-a$ is less than n; hence x' is the corresponding minimum value. This makes $k'=k-1$. Therefore, we can write the first few terms of T_{n-a} as follows:

$$(3) \quad T_{n-a} = \binom{k-1}{x-1} + \binom{k-1+b-a}{x-1+b} + \binom{k-1+2b-2a}{x-1+2b} + \cdots,$$

where, again, x refers to the original x_{min}. If x_{min} is zero to begin with, then for the solution of (2) we choose $x'=x_{min}+b-1$ and $y'=y_{max}-a$ and this choice modifies (3) only to the extent that the first term is omitted in the series for T_{n-a}.

Next we consider T_{n-b}. The first term of this series will be of a form $\binom{k''}{x''}$ where

$$(4) \quad ax'' + by'' = n - b, \text{ and } x'' + y'' = k''.$$

We note here that

$$\binom{k''}{x''} = \binom{k''}{y''}$$

and we find that $x''=x_{min}$ and $y''=y_{max}-1$ is the appropriate solution of (4). Therefore, $k''=k-1$ and we can write the first few terms of T_{n-b} as follows:

$$(5) \quad T_{n-b} = \binom{k-1}{x} + \binom{k-1+b-a}{x+b} + \binom{k-1+2b-2a}{x+2b} + \cdots,$$

where again x refers to the original x_{min}. Now if we add the two series (3) and (5) termwise, we observe the result that we desired, namely, series (1), through the use of Pascal's Rule, which determines the very nature of Pascal's Triangle.

The general nth term of this sequence in terms of Pascal numbers is given by

$$E(a.3) \quad T_n = \sum_{\substack{x=0, y=0 \\ ax+by=n}}^{[n/a] \cdot [n/b]} \frac{(x+y)!}{x! \cdot y!};$$

and the sum of the first n terms of this sequence is given by

$$E(a.4) \quad \sum_{k=0}^{n-1} T_k = T_{n+(a-1)} + T_{n+(a-2)} + \cdots + T_{n+(a-b)} - 1.$$

Proof (by induction). In the development of the phases above a was always greater than or equal to b. The remarks in section C indicate that this is an arbitrary choice because of the symmetry involved; however, one or the other choices must be made, but not both. We will assume here that $a \geq b$. We divide the proof into two parts.

PART 1. We establish the formula for $n=1$. E(a.4) becomes

$$T_0 = T_a + T_{a-1} + T_{a-2} + \cdots + T_{a-b+1} - 1.$$

Now $T_0 = 1$, since $ax+by=0$ has only the single solution $(0, 0)$ and $(0+0)!/(0! \cdot 0!) = 1$. Also we see that $T_a = 1$, since the only solution of $ax+by=a$ is $(1, 0)$ and $(1+0)!/(1! \cdot 0!) = 1$. In considering the other terms, $T_{a-1}, T_{a-2}, \cdots, T_{a-b+1}$, we find that the only solutions to the corresponding equations are, in all cases, $x=0$ and y equal to $(a-1)/b$, $(a-2)/b$, $\cdots$, $(a-b+1)/b$ respectively. Now only one number of this set of values is integral since the set $a, a-1, a-2, \cdots, a-b+1$ forms a residue class modulo b and, since a and b are relatively prime, the value a is omitted from consideration. All other solutions are nonintegral and therefore discarded and T_k for those values equals zero. Let k' be the one value that yields the integral solution and let k'' be that solution. Then

$$T_k = \frac{(0+k'')!}{0! \cdot k''!} = 1$$

Hence $T_0 = T_a + T_{k'} - 1$ or $1 = 1 + 1 - 1$ an identity.

PART 2. We assume the formula is true for n and show that then it is also true for $n+1$. To both sides of E(a.4) add T_n. Thus

$$\sum_{k=0}^{n} T_k = T_{n+(a-1)} + T_{n+(a-2)} + \cdots + T_{n+(a-b)} - 1 + T_n.$$

But from E(a.4) we have

$$\sum_{k=0}^{n} T_k = T_{n+1+(a-1)} + T_{n+1+(a-2)} + \cdots + T_{n+1+(a-b)} - 1.$$

We must show that the right members of the above equations are equal. Upon equating these two members and simplifying we have $T_{n+a-b} + T_n = T_{n+a}$. But we know from E(a.2) that $T_k = T_{k-a} + T_{k-b}$. Thus if $k=n+a$, we have the exact statement above.

References

1. Brother U. Alfred, An Introduction to Fibonacci Discovery, The Fibonacci Association, San Jose State College, 1965, pp. 25-26.
2. J. A. Raab, A generalization of the connection between the Fibonacci sequence and Pascal's Triangle, The Fibonacci Quarterly, vol. 1, 3 (1963) 21-31.
3. T. Nagell, Introduction to Number Theory, Wiley, New York, 1951, p. 29.

PROPERTIES OF A GAME BASED ON EUCLID'S ALGORITHM*

EDWARD L. SPITZNAGEL, JR., Washington University

1. Introduction. In a recent paper [3] Cole and Davie have described a remarkable little game based on the Euclidean algorithm. Appropriately, they have given it the name Euclid. As they point out, Euclid furnishes a simple example of a game with an explicit winning strategy. In [3] the authors were primarily concerned with the exposition of that winning strategy and did not mention any of the other intriguing features of Euclid. In this paper we give an account of some of these other features.

We first give a brief account of the game and winning strategy, thus making this paper self-contained, and then pass on to study the other properties.

2. Rules and strategy. We paraphrase the rules of Euclid from [3]: Let (p, q) be a pair of positive numbers satisfying $p > q$ and let A, B be two players. Each player in turn must move. A move consists of replacing the larger of two numbers given him by any nonnegative number obtained by subtracting a positive multiple of the smaller number from the larger number. The winner of the game is the person who obtains 0 for the (new) smaller number. For example, starting with the pair (51, 30), the successive moves in the game could be:

	(51, 30)	or		(51, 30)	
A:	(30, 21)		A:	(30, 21)	
B:	(21, 9)		B:	(21, 9)	
A:	(9, 3)		A:	(12, 9)	
B:	(3, 0)	B wins	B:	(9, 3)	
			A:	(3, 0)	A wins.

The strategy can be summarized as follows:

DEFINITION. *Let*

$$c = (\sqrt{5} - 1)/2 \doteq .618.$$

(*This is the golden section.*) *Let* (p, q), $p > q$ *be a pair occurring in the game of Euclid. Call* (p, q) *a safe position if* $q/p > c$. *Otherwise call* (p, q) *an unsafe position.*

PROPOSITION 1. *A player moving from an unsafe position is always capable of moving to a safe position.*

---

* From MATHEMATICS MAGAZINE, vol. 46 (1973), pp. 87–92.

PROPOSITION 2. *A player moving from a safe position can make just one move and that move will always be to an unsafe position.*

Combining the above two results, we see that a player A who once is given an unsafe position from which to move can ensure that in all future moves he, A, will always be able to move to safe positions and that the other player, B will always be forced to move to unsafe positions. Since the larger number p strictly decreases as the game progresses and since $p \neq q$ at the start, the game must eventually pass through the unsafe position

(1) $\qquad\qquad (kq, q) \quad k$ an integer > 1.

Therefore the winning strategy is for A always to move to safe positions until B is forced to move to position (1). Then A can produce 0 on his next move and win.

3. Strategy without the golden section. Can a person be taught the winning strategy without his knowing the golden section c? The answer is yes. A close look at the game reveals the following result:

PROPOSITION 3. *Let $p > 2q$ and let p/q be nonintegral. Then the person A about to move can move to at least two positions, but exactly one of all his possible moves will yield a safe position. If p_0 denotes the remainder upon division of p by q, and $p_1 = p_0 + q$, then the safe position is whichever of*

$$(p_1, q) \qquad\qquad (q, p_0)$$

has the larger ratio of second entry to first entry.

Proof. Since $q/p < 1/2 < c$, the position from which A moves is unsafe. By Proposition 1, therefore, A is able to move to a safe position. The various positions to which A can move are (q, p_0), (p_1, q), and all positions of the form $(p_0 + kq, q)$, $k = 2, 3, \cdots$, such that $p_0 + kq < p$. Since for $k \geq 2$, $q/(p_0 + kq) < q/(kq) = 1/k < c$, the only two positions that can possibly be safe are (q, p_0) and (p_1, q). We know at least one must be safe, and we will now show that both cannot be. Suppose (q, p_0) is safe, so that $p_0/q > c$. Then

$$\frac{p_1}{q} = \frac{p_0 + q}{q} = \frac{p_0}{q} + 1 > c + 1 = \frac{1}{c}.$$

Therefore, $q/p_1 < c$, so that (p_1, q) is unsafe. We have now shown that exactly one of (q, p_0), (p_1, q) is safe. That is, exactly one of these pairs will have ratio of second entry to first entry greater than c. Therefore, whichever pair has the larger ratio of second entry to first entry is the unique safe position. This completes the proof.

In Euclid the only two types of position (p, q) at which one has a choice of moves are those in which
  (1) p/q is an integer ≥ 2, or
  (2) $p > 2q$ and p/q is not an integer.
The latter case is the hypothesis of Proposition 3.

Therefore, the strategy can be rephrased as follows, with no mention of the golden ratio c: If there is a choice of moves and p/q is an integer, return 0 and win. If there is a choice of moves and p/q is not an integer, compute

$$p_0 = \text{remainder after } p \text{ is divided by } q$$
$$p_1 = p_0 + q$$

and move to whichever of (p_1, q), (q, p_0) has the larger ratio of second to first entry.

Of course, one need not actually compute the ratios to determine which is the larger. If $p_0 p_1 < q^2$, then q/p_1 is the larger ratio, so (p_1, q) is the safe position; otherwise (q, p_0) is the safe position. For example, when presented with the pair $(70, 11)$, one would compute $p_0 = 4$, $p_1 = 15$. Then since $4 \cdot 15 < 11^2$, he would make the move $(15, 11)$ rather than $(11, 4)$.

4. Relation to Fibonacci series. Once the strategy has been mastered by a person, he is then likely to turn his interest to those parts of the game in which neither player must know the strategy—that is, those parts consisting of a string of moves in which neither player has any choice. In such a string of moves, each player in turn is presented with a pair of numbers $p > q$ such that $p < 2q$ and so can only move to the position $(q, p-q)$.

Of course, eventually such a string must end, with one or the other player being presented with a pair of numbers (p, q) with $p \geq 2q$. In case such a pair arises only once in the game, it occurs right at the end and is of the form (1). In this case, the various numbers in the ordered pairs can be found by working backward. They are:

$$q, kq, (k+1)q, (2k+1)q, (3k+2)q, (5k+3)q, \cdots$$

The above sequence of numbers is q times a Fibonacci sequence. The fact that Fibonacci sequences are closely connected to the game is not surprising, since their relations to the Euclidean algorithm [4] and to the golden section [1, 7] are well known.

Now suppose we have a string of forced moves that ends in a position (p_0, q_0), $p_0 \geq 2q_0$, not necessarily of the form (1). Working backward through the string, we again see a Fibonacci sequence, with the ith move prior to (p_0, q_0) given recursively by

(2) $$(p_i, q_i) = (p_{i-1} + q_{i-1}, p_{i-1}).$$

Then we have the following result, similar to the well-known result [2, 5, 7] on the convergents of a continued fraction.

THEOREM. *Let p_i and q_i be defined recursively by* (2), *starting with the pair* (p_0, q_0). *Then*
  (i) q_{2k}/p_{2k} *is an increasing function of* k, $k = 0, 1, 2, \cdots$
  (ii) q_{2k+1}/p_{2k+1} *is a decreasing function of* k, $k = 0, 1, 2, \cdots$.

Proof. By (2) we have

$$q_{i+1}p_i - p_{i+1}q_i = p_i(p_{i-1} + q_{i-1}) - (p_i + q_i)p_{i-1}$$
$$= -(q_i p_{i-1} - p_i q_{i-1}).$$

Applying this result i times, we obtain

$$(3) \qquad q_{i+1}p_i - p_{i+1}q_i = (-1)^i(q_1 p_0 - p_1 q_0).$$

Let $d = q_1 p_0 - p_1 q_0$. Then from (3) we obtain

$$\frac{q_{i+1}}{p_{i+1}} - \frac{q_i}{p_i} = (-1)^i d \frac{1}{p_{i+1} p_i}$$

$$\frac{q_{i+2}}{p_{i+2}} - \frac{q_{i+1}}{p_{i+1}} = (-1)^{i+1} d \frac{1}{p_{i+2} p_{i+1}}$$

and therefore

$$\frac{q_{i+2}}{p_{i+2}} - \frac{q_i}{p_i} = (-1)^i d \left(\frac{1}{p_{i+1} p_i} - \frac{1}{p_{i+2} p_{i+1}} \right).$$

Since $p_{i+2} > p_i$,

$$\frac{1}{p_{i+1} p_i} - \frac{1}{p_{i+2} p_{i+1}} > 0$$

and since $p_0 \geq 2q_0$,

$$d = q_1 p_0 - p_1 q_0$$
$$= p_0^2 - (p_0 + q_0)q_0$$
$$\geq 2p_0 q_0 - p_0 q_0 - q_0^2 = q_0(p_0 - q_0)$$
$$> 0.$$

Therefore

$$\frac{q_{i+2}}{p_{i+2}} > \frac{q_i}{p_i}$$

for i even, while the reverse inequality is true for i odd. This completes the proof.

Returning to the direction in which the game is played, we can thus say the following about any string of forced moves: The player A who is once presented with two numbers of ratio $q/p < c$ finds thereafter that the ratios of the numbers presented him steadily fall until finally he is given a pair with ratio $\leq 1/2$ and so is enabled to make a decision. If player A always makes his decisions according to the strategy, then player B finds himself completely trapped. During the strings of forced moves B finds the ratios q/p given to him steadily rising to 1 while the ratios he must give to A steadily decrease until one falls below $1/2$. Then A chooses a move that may lower the ratio of the numbers he gives to B, but as soon as the next string of forced moves begins, B finds once again that the ratios q/p given him go the wrong way, increasing toward 1, so that he never has an opportunity to make a decision.

Compared to Nim, then, the workings of the strategy in Euclid lie much closer to the surface. In Nim, the opponent of a player following the strategy usually has more than one move open to him at every stage of the game except the last. He therefore does not get such an obvious warning that he is doomed to lose. In Euclid, on the other hand, the opponent of someone following the strategy is likely to notice his moves are being forced every step of the way, and from this observation it might be possible for him to determine what the strategy must be.

5. Probability of starting with a safe position. Now that the strategy of Euclid has been explicated, there arises the question, What is the probability that an arbitrarily chosen starting position is unsafe?

The following would seem to be a reasonable interpretation of an arbitrarily chosen starting position: Both numbers are chosen from the set S of positive integers less than or equal to some positive integer n. The first number x is chosen with uniform probability from S, and the second is chosen with uniform probability from $S \sim \{x\}$. After the two numbers are chosen, they are labeled p and q so that $p > q$.

Under these assumptions, standard probability arguments, as presented in [6] show that q can be considered as drawn uniformly from the set $\{1, 2, \cdots, p-1\}$. Given p, we have

$$P(q/p < c) = P(q < cp)$$
$$= [cp]/(p-1)$$

where the square brackets denote the integral part. For large p, this fraction approaches c, so for large n, the probability of an arbitrarily chosen starting position being unsafe approaches c. That is, the first to move, player A, is the more likely to have a winning strategy available.

References

1. W. W. R. Ball, Mathematical Recreations and Essays, Chapter 2, Macmillan, New York, 1960.
2. G. Chrystal, Textbook of Algebra, Chapter 32, Dover, New York, 1961.

3. A. J. Cole and A. J. T. Davie, A game based on the Euclidean algorithm and a winning strategy for it, Math. Gaz., 53 (1969) 354–357.

4. J. D. Dixon, A simple estimate for the number of steps in the Euclidean algorithm, Amer. Math. Monthly, 78 (1971) 374–376.

5. A. Ya. Khinchin, Continued Fractions, Chapter 1, University of Chicago Press, Chicago, 1964.

6. E. Parzen, Modern Probability Theory and Its Applications, Wiley, New York, 1960.

7. N. N. Vorob'ev, Fibonacci Numbers, Blaisdell, New York, 1961.

BIBLIOGRAPHIC ENTRIES: INTEGERS

Except for the entries labeled MATHEMATICS MAGAZINE, the references below are to the AMERICAN MATHEMATICAL MONTHLY.

1. C. B. Boyer, An early reference to division by zero, vol. 50, p. 487.

Discussion of early theories on division by zero.

2. E. F. Beckenbach, Interesting integers, vol. 52, p. 211.

Conjecture: There is an interesting fact concerning each of the positive integers.

3. R. Dubisch, Representation of the integers by positive integers, vol. 58, p. 615.

4. P. Franklin, The Euclidean algorithm, vol. 63, p. 663.

Determining an upper bound to the number of divisions required to find the greatest common divisor of two positive integers.

5. M. Sholander, Least common multiples and highest common factors, vol. 68, p. 984.

6. M. O. Le Van, A triangle for partitions, vol. 79, p. 507.

A triangle method, similar to Pascal's triangle, to find the number of ways to partition a positive integer.

7. Z. Usiskin, Perfect square patterns in the Pascal triangle, MATHEMATICS MAGAZINE, vol. 46, p. 203.

Finding patterns of numbers in the Pascal triangle that can be split into subsets with equal products.

8. D. E. Knuth, Representing numbers using only one 4, MATHEMATICS MAGAZINE, vol. 37, p. 308.

9. J. C. Holladay, Some generalizations of Wythoff's game and other related games, MATHEMATICS MAGAZINE, vol. 41, p. 7.

Analysis of some Nim-like games.

(b)

IRRATIONAL NUMBERS

THE LIMIT FOR e*

RICHARD LYON AND MORGAN WARD, California Institute of Technology

The following way of showing that $(1 + 1/n)^n$ tends to e as n tends to infinity seems simpler than the current proofs in elementary texts.

Define e as $\lim_{n\to\infty} \sum_0^n 1/r!$. Then the result is evident from the inequality

$$\sum_0^n \frac{1}{r!} > \left(1 + \frac{1}{n}\right)^n > \sum_0^n \frac{1}{r!} - \frac{3}{2n}, \qquad n \geq 3.$$

This inequality may be proved as follows. Assume $n \geq 3$. Then by the binomial theorem,

$$\left(1 + \frac{1}{n}\right)^n = 2 + \sum_2^n \left(1 - \frac{1}{n}\right)\left(1 - \frac{2}{n}\right) \cdots \left(1 - \frac{r-1}{n}\right)\frac{1}{r!}.$$

Clearly

$$1 > \left(1 - \frac{1}{n}\right)\left(1 - \frac{2}{n}\right) \cdots \left(1 - \frac{r-1}{n}\right).$$

Hence

$$\sum_0^n \frac{1}{r!} = 2 + \sum_2^n \frac{1}{r!} > \left(1 + \frac{1}{n}\right)^n.$$

On the other hand,

$$\left(1 - \frac{1}{n}\right)\left(1 - \frac{2}{n}\right) = 1 - \frac{(1+2)}{n} + \frac{2}{n^2} > 1 - \frac{(1+2)}{n}.$$

Hence

$$\left(1 - \frac{1}{n}\right)\left(1 - \frac{2}{n}\right)\left(1 - \frac{3}{n}\right) > \left(1 - \frac{(1+2)}{n}\right)\left(1 - \frac{3}{n}\right)$$

$$> 1 - \frac{(1+2+3)}{n},$$

---

* From AMERICAN MATHEMATICAL MONTHLY, vol. 59 (1952), pp. 102–103.

and so on. Consequently if $r \geq 2$,

$$\left(1 - \frac{1}{n}\right)\left(1 - \frac{2}{n}\right) \cdots \left(1 - \frac{r-1}{n}\right) \geq 1 - \frac{(1 + 2 + 3 + \cdots + r - 1)}{n}$$

$$= 1 - \frac{r(r-1)}{2n}$$

with equality only when $r=2$. Consequently since $n \geq 3$

$$\left(1 + \frac{1}{n}\right)^n > 2 + \sum_2^n \left(1 - \frac{r(r-1)}{2n}\right)\frac{1}{r!} = \sum_0^n \frac{1}{r!} - \frac{1}{2n}\sum_0^{n-2} \frac{1}{r!}$$

$$> \sum_0^n \frac{1}{r!} - \frac{3}{2n}, \quad \text{since} \quad \sum_0^{n-2} \frac{1}{r!} \text{ is less than 3.}$$

ANOTHER PROOF OF AN ESTIMATE FOR e*

R. F. JOHNSONBAUGH, Chicago State University

At the national meeting of the Mathematical Association of America in January, 1973, Professor Paul Halmos suggested that publication of mathematical folklore would be a worthwhile contribution to mathematical literature. I thus submit an old proof of the monotonicity and boundedness of the sequence $\{(1 + 1/n)^n\}$ which I believe is the simplest that I have ever seen.

The proof uses the simple inequality

$$\frac{b^{n+1} - a^{n+1}}{b - a} < (n + 1)b^n$$

which we rewrite as

$$b^n[(n + 1)a - nb] < a^{n+1}$$

valid for $0 \leq a < b$. Taking $a = 1 + (1/(n + 1))$ and $b = 1 + (1/n)$, the term in brackets reduces to 1 and we have

(*) $$\left(1 + \frac{1}{n}\right)^n < \left(1 + \frac{1}{n + 1}\right)^{n+1}.$$

Taking $a = 1$ and $b = 1 + (1/2n)$, the term in brackets reduces to $\frac{1}{2}$ and we have

$$\left(1 + \frac{1}{2n}\right)^n \frac{1}{2} < 1.$$

---

* From AMERICAN MATHEMATICAL MONTHLY, vol. 81 (1974), pp. 1011–1012.

Multiplying by 2 and squaring we have

(**) $$\left(1 + \frac{1}{2n}\right)^{2n} < 4.$$

Inequalities (*) and (**) imply that the sequence $\{(1 + (1/n))^n\}$ is increasing and bounded above by 4.

In a similar way, one can use the inequality

$$\frac{b^{n+1} - a^{n+1}}{b - a} > (n + 1)a^n$$

with $a = 1 + (1/(n + 1))$ and $b = 1 + (1/n)$ to derive

$$\left(1 + \frac{1}{n}\right)^{n+1} > \left(1 + \frac{1}{n+1}\right)^{n+2} \left[\frac{n^3 + 4n^2 + 4n + 1}{n(n + 2)^2}\right].$$

Since the term in brackets is at least 1, we have shown that $(1 + (1/n))^{n+1}$ is decreasing.

DEPARTMENT OF MATHEMATICS, CHICAGO STATE UNIVERSITY, CHICAGO, IL 60628.

THE IRRATIONALITY OF $\sqrt{2}$ *

EDWIN HALFAR, University of Nebraska

The following proof has some novelty in that it does not use the usual assumption that the fraction n/m which is to represent $\sqrt{2}$ is such that n and m have no common factors.

Suppose $\sqrt{2} = (n/m)$, where n and m are positive integers. Then $n > m$, and there is an integer $p > 0$ such that $n = m + p$, and $2m^2 = m^2 + 2pm + p^2$. This implies $m > p$. Consequently, for some integer $a > 0$, $m = p + a$, $n = 2p + a$ and $2(p+a)^2 = (2p+a)^2$. The last equality implies $a^2 = 2p^2$ so that the entire process may be repeated indefinitely giving $n > m > a > p > \cdots$, but since every non-null set of positive integers has a smallest element, this is a contradiction and $\sqrt{2}$ is irrational.

---

* From AMERICAN MATHEMATICAL MONTHLY, vol. 62 (1955), p. 437.

THE IRRATIONALITY OF $\sqrt{2}$ *

ROBERT GAUNTT, Purdue University, and GUSTAVE RABSON, Antioch College

(The following proof was invented by Robert James Gauntt, in 1952, while he was a freshman at Purdue. I was unable to induce him to write up his proof. G. R.)

$a^2 = 2b^2$ cannot have a non-zero solution in integers because the last non-zero digit of a square, written in the base three, must be 1, whereas the last non-zero digit of twice a square is 2.

ON THE IRRATIONALITY OF CERTAIN TRIGONOMETRIC NUMBERS†

E. A. MAIER, University of Oregon

In this note we present an elementary proof of the following theorem. (See Niven, *Irrational Numbers*, Carus Mathematical Monograph No. 11, MAA, 1956, p. 41, for references to other proofs.)

THEOREM. *If θ is rational in degrees, then* (i) $\cos\theta = 0$, $\pm\frac{1}{2}$, ± 1 *or is irrational*, (ii) $\sin\theta = 0$, $\pm\frac{1}{2}$, ± 1 *or is irrational*, (iii) $\tan\theta = 0$, ± 1, *is undefined or is irrational*.

LEMMA. *If there exists a positive integer n such that $\cos n\theta$ is integral, then $\cos\theta = 0$, $\pm\frac{1}{2}$, ± 1 or is irrational.*

Proof. From the identity $2\cos(n+1)\theta = (2\cos\theta) 2\cos n\theta - 2\cos(n-1)\theta$, it follows by induction that if n is a positive integer then there exists a monic polynomial of degree n with integral coefficients such that $2\cos n\theta = f(2\cos\theta)$. Thus if $\cos n\theta$ is an integer, $2\cos\theta$ is a root of the monic polynomial $f(x) - 2\cos n\theta$. Hence $2\cos\theta$ is an integer or is irrational. Since $|2\cos\theta| \leq 2$, the lemma follows.

Proof of the theorem. If θ is rational in degrees then for appropriate integers a and $b > 0$ we have $\theta = \pi(a/b)$ in radian measure. Then $\cos n\theta$, $\cos n(\pi/2 - \theta)$ and $\cos n(2\theta)$ are integers for $n = b$. Thus (i) follows directly from the lemma and (ii) and (iii) follow from the lemma and the identities $\sin\theta = \cos(\pi/2 - \theta)$, $\tan^2\theta = (1 - \cos 2\theta)/(1 + \cos 2\theta)$. Note that the latter identity implies that if $\tan\theta$ is rational, then $\cos 2\theta$ is rational.

---

* From AMERICAN MATHEMATICAL MONTHLY, vol. 63 (1956), p. 247.

† From AMERICAN MATHEMATICAL MONTHLY, vol. 72 (1965), pp. 1012–1013.

A SIMPLE IRRATIONALITY PROOF FOR QUADRATIC SURDS*

M. V. SUBBARAO, University of Missouri

Let N be a positive integer which is not a square of another integer. If $\sqrt{N}$ is rational, we will obtain contradictions in three ways, thus providing three different proofs for the irrationality of $\sqrt{N}$.

Write $\sqrt{N} = a/b$, where the fraction on the right is chosen so that:

1. The numerator a is the smallest possible positive integer (for the first proof);
2. The denominator b is the smallest possible positive integer (for the second proof);
3. The sum of the numerator and denominator is the smallest possible (for the third proof).

Since $a^2 = Nb^2$ we have $a^2 - Aab = Nb^2 - Aab$, where A is the unique positive integer given by $A < \sqrt{N} < A+1$. Hence, $a(a-Ab) = b(Nb-Aa)$ giving $\sqrt{N} = a/b = (Nb-Aa)/(a-Ab)$. But, in this new expression for $\sqrt{N}$, the numerator $Nb - Aa$ is less than a, the denominator $a - Ab$ is less than b, and the sum of the numerator and denominator is less than $a+b$—three contradictions to complete the three proofs!

Whether this kind of reasoning can be extended to establish the irrationality of the kth root of a non-kth power integer for $k>2$ is an open question; the writer's attempts in this direction ran into difficulties.

IRRATIONAL NUMBERS†

J. P. JONES AND S. TOPOROWSKI, University of Calgary

For the past few years a clever proof has been making the rounds of the various mathematics departments.

THEOREM 1. *An irrational number raised to an irrational power may be rational.*

Proof: Consider the identity

$$[\sqrt{2}^{\sqrt{2}}]^{\sqrt{2}} = 2.$$

If $\sqrt{2}^{\sqrt{2}}$ is rational then we are finished. If not then $\sqrt{2}^{\sqrt{2}}$ is irrational so $(\sqrt{2}^{\sqrt{2}})^{\sqrt{2}}$ is the example.

This proof seems first to have been published by Dov Jarden as a curiosity in [3]. The proof was published again in [2]. Note that while the proof is elementary, it is non-constructive. The non-constructivity enters in the form of the logical

---

* From AMERICAN MATHEMATICAL MONTHLY, vol. 75 (1968), pp. 772–773.

† From AMERICAN MATHEMATICAL MONTHLY, vol. 80 (1973), pp. 423–424.

principle of the excluded middle (*tertium non datur*) which the intuitionists reject.

Actually $\sqrt{2}^{\sqrt{2}}$ is irrational, being the square root of Hilbert's number $2^{\sqrt{2}}$, proved transcendental by Kuzmin [1] in 1930. But this result, which is not elementary, is not used above. Only the irrationality of $\sqrt{2}$ is used.

Consider next the related theorem.

THEOREM 2. *An irrational number raised to an irrational power may be irrational.*

Of course we can use set theoretical principles to prove that a^b is irrational for almost all real numbers b. Or we can use the result of Kuzmin [1] to prove Theorem 2 But does Theorem 2 have an elementary proof?

Proof: Consider the identity $\sqrt{2}^{(\sqrt{2}+1)} = (\sqrt{2}^{\sqrt{2}})\sqrt{2}$.

If $\sqrt{2}^{\sqrt{2}}$ is irrational then we are finished. If not, then $\sqrt{2}^{\sqrt{2}}$ is rational. Hence $(\sqrt{2}^{\sqrt{2}})\sqrt{2}$ is irrational, and $\sqrt{2}^{(\sqrt{2}+1)}$ is the example in this case.

There is also a simple identity by means of which it can be proved that a rational number raised to an irrational power may be irrational. But perhaps the reader would enjoy finding this one himself.

References

1. R. Kuzmin, On a new class of transcendental numbers, Izv. Akad. Nauk SSSR, Ser. Mat., 7 (1930) 585–597.

2. Mathematics Magazine, 39(1966) 111, 134.

3. Scripta Mathematica, 19 (1953) 229.

A METHOD OF ESTABLISHING CERTAIN IRRATIONALITIES*

IVAN NIVEN AND E. A. MAIER, University of Oregon

1. Introduction. H. Steinhaus [1] has given a proof of the irrationality of $\sqrt{2}$ that approaches the question through inequalities rather than divisibility. A variation of this proof is given in section 2; the form given here is more suitable to extension to other questions. Several of these extensions are given in sections 3, 4, and 5. Actually, Theorem 2 of section 5 includes all previous results as special cases. However the chain of results leading to Theorem 2 reveals how we arrived at the proofs. The only background results needed for sections 2 and 3 are a few basic propositions about inequalities; beyond section 3 the division algorithm is also needed.

2. The irrationality of $\sqrt{2}$. Suppose that $\sqrt{2}$ is rational, say

(1) $$\sqrt{2} = a/b,$$

where a and b are positive integers. Suppose that b is the smallest positive

---

* From MATHEMATICS MAGAZINE, vol. 37 (1964), pp. 208–210.

integer for which there is a representation (1). From (1) and the fact that $\sqrt{2}$ lies between 1 and 2 we see that

(2) $\qquad 1 < \dfrac{a}{b} < 2, \qquad b < a < 2b, \qquad 0 < a - b < b.$

Then we note from (1) that

(3) $\quad a^2 = 2b^2, \quad a^2 - ab = 2b^2 - ab, \quad a(a-b) = b(2b-a), \quad \sqrt{2} = \dfrac{a}{b} = \dfrac{2b-a}{a-b}.$

Thus $\sqrt{2}$ has been expressed in another rational form with positive denominator $a-b$, smaller than b, contrary to the minimal property assumed for b. Thus $\sqrt{2}$ is irrational.

3. Extension to $\sqrt{d}$. This proof generalizes at once from $\sqrt{2}$ to $\sqrt{d}$, where d is a positive integer not a perfect square. Thus $\sqrt{d}$ lies between two consecutive integers, say

$$t < \sqrt{d} < t + 1.$$

Then if (1) is replaced by $\sqrt{d} = a/b$, where the positive integer b is minimal, we see that (2) can be replaced by

$$t < \dfrac{a}{b} < t+1, \qquad tb < a < tb + b, \qquad 0 < a - tb < b.$$

Hence the equations leading up to (3) are replaced by

$$a^2 = db^2, \quad a^2 - tab = db^2 - tab, \quad a(a - tb) = b(db - ta), \quad \sqrt{d} = \dfrac{a}{b} = \dfrac{db - ta}{a - tb}$$

and the contradiction is analogous to that in the $\sqrt{2}$ case.

4. The extension to nth roots. The minimum property of the integer b in the foregoing proofs amounts to this: Let α be a rational number, and suppose that $\alpha = a/b$ is the representation of α having minimum positive integer b in the denominator. Then b is seen to be the smallest positive integer such that $b\alpha$ is an integer. In what follows it will be convenient to use this interpretation of b.

To generalize the proof to the case of nth roots, we make use of the division algorithm: Given any integers a and b, with b positive, there exist integers q and r such that $a = bq + r$, $0 \leq r < b$.

LEMMA 1. *Let α be rational, and let b be the least positive integer such that $b\alpha$ is an integer, say $b\alpha = a$. For any nonnegative integer m, if s is the least positive integer such that $sa^m\alpha$ is an integer, then $s = b$.*

Proof. The proof is by induction on m. Note that the result holds for $m = 0$. We assume that b is the least positive integer such that $ba^m\alpha$ is an integer, and that s is the least positive integer such that $sa^{m+1}\alpha$ is an integer, and we prove that $s=b$. Note that $s \leq b$ since $ba^{m+1}\alpha$ is an integer. Apply the division algorithm to sa^{m+1} and b to get integers q and r satisfying

$$sa^{m+1} = bq + r, \qquad 0 \leq r < b.$$

Thus we see that

(4) $$r\alpha = (sa^{m+1} - bq)\alpha = sa^{m+1}\alpha - aq.$$

Thus $r\alpha$ is an integer, and so $r=0$ by the minimal property of b. Hence (4) reduces to

$$0 = sa^{m+1}\alpha - aq, \qquad sa^m\alpha = q.$$

Thus $s \geq b$ by the induction hypothesis, and hence $s=b$.

THEOREM 1. *Let n and d be positive integers greater than 1 such that $\sqrt[n]{d}$ is not an integer. Then $\sqrt[n]{d}$ is irrational.*

Proof. Let $\sqrt[n]{d}$ lie between the consecutive integers t and $t+1$, so that

(5) $$t < \sqrt[n]{d} < t + 1.$$

Assume that $\sqrt[n]{d}$ is rational, and let b be the least positive integer such that $b\sqrt[n]{d}$ is an integer, say $b\sqrt[n]{d} = a$; then (5) implies that

(6) $$t < \frac{a}{b} < t+1, \qquad tb < a < tb + b, \qquad 0 < a - tb < b.$$

Furthermore we observe that (writing α for $\sqrt[n]{d}$)

$$(a - tb)a^{n-2}\alpha = a^{n-1}\alpha - ta^{n-2}b\alpha = b^{n-1}d - ta^{n-1},$$

an integer. But in view of the last part of (6) this contradicts Lemma 1, where m is here $n-2$.

5. Algebraic integers. A number α is said to be an algebraic integer if it satisfies some equation of the type

(7) $$\alpha^n + c_{n-1}\alpha^{n-1} + c_{n-2}\alpha^{n-2} + \cdots + c_1\alpha + c_0 = 0,$$

with integer coefficients. For example the number $\sqrt[n]{d}$ of the preceding section is an algebraic integer. We confine attention to those algebraic numbers that are real numbers.

THEOREM 2. *Let α be a real algebraic integer. If α is not a rational integer (i.e., one that satisfies an equation of type (7) of degree $n=1$), then α is irrational.*

Proof. Suppose that α satisfies (7), and that α is rational. Let b be the least positive integer such that $b\alpha$ is an integer, say $b\alpha = a$. Then, as in the proof of

Theorem 1, there exists an integer t such that

(8) $$0 < a - tb < b.$$

Also by (7) we note that $b^{n-1}\alpha^n$ is an integer:

$$b^{n-1}\alpha^n = b^{n-1}[-c_{n-1}\alpha^{n-1} - c_{n-2}\alpha^{n-2} - \cdots - c_1\alpha - c_0]$$
$$= -c_{n-1}(b\alpha)^{n-1} - c_{n-2}b(b\alpha)^{n-2} - \cdots - c_1 b^{n-2}(b\alpha) - c_0 b^{n-1}.$$

But then

$$(a - tb)a^{n-2}\alpha = a^{n-1}\alpha - ta^{n-2}b\alpha = b^{n-1}\alpha^n - ta^{n-1}$$

is an integer which, in view of (8), contradicts Lemma 1.

Reference

1. H. Steinhaus, Mathematical snapshots, Stechert-Hafner, New York, 1938, p. 132.

BIBLIOGRAPHIC ENTRIES: IRRATIONAL NUMBERS

Except for the entry labeled AMERICAN MATHEMATICAL MONTHLY, the references below are to the MATHEMATICS MAGAZINE.

1. L. L. Pennisi, Elementary proof that e is irrational, AMERICAN MATHEMATICAL MONTHLY, vol. 60, p. 474.

2. W. R. Ransom, Approximations to square roots, vol. 23, p. 54.

3. R. J. Dowling, A radical suggestion, vol. 36, p. 59.

A simple proof of the irrationality of $\sqrt{10}$.

4. R. T. Hood, A base suggestion?, vol. 36, p. 218.

Comments and extension of the Dowling article (MATHEMATICS MAGAZINE, vol. 36, p. 59).

5. D. V. Weyers, A new twist to an old problem, vol. 38, p. 106.

Proof that $\sqrt{2}$ is irrational.

6. L. J. Lange, A simple irrationality proof for nth roots of positive integers, vol. 42, p. 242.

(c)

COMPLEX NUMBERS

A NOTE ON GEOMETRICAL APPLICATIONS OF COMPLEX NUMBERS*

S. A. SCHELKUNOFF, Bell Telephone Laboratories Inc.

The writer was greatly pleased with Professor L. L. Smail's suggestion of geometric problems for illustrating the use of complex numbers.† On several occasions in his teaching (at the State College of Washington) the present writer used similar illustrations, although he favored a somewhat more direct application of the algebra of complex quantities. The purpose of this note is partly to point out the difference in the two methods of approach but mainly to say another word for the cause. To make the comparison of the two methods effective, we shall discuss some of the problems treated by Professor Smail.

It is hoped that in this way we can illustrate the effectiveness of the algebra of complex numbers in certain types of geometric problems. We wish to emphasize the fact that in the following proofs we do not separate the real and imaginary parts of complex numbers and that thereby we obtain simple and elegant demonstrations. It is readily granted that in some problems such separation, evidently preferred by Professor Smail, is advantageous. However, even in treating problems involving distances, it is often convenient to use conjugate complex numbers to avoid the separation into real and imaginary parts.

In this article we shall make use of the following simple propositions:

(a) *If k is real, then kz and z represent parallel vectors whose lengths are in the ratio $k:1$;*

(b) *If z_1 and z_2 are the complex numbers representing respectively points A and B, then $\frac{1}{2}(z_1+z_2)$ represents the mid-point of the line-segment AB.*

We shall now prove a few geometric theorems:

THEOREM 1. *The line-segment joining the mid-points of two sides of a triangle is equal to half the third side and is parallel to it.*

Proof. Let the complex numbers corresponding to the vertices of any triangle in reference to some origin be as shown in Fig. 1. Then, the complex numbers representing the mid-points E and F are immediately found, and we have: $EF = \frac{1}{2}(z_2 - z_1)$. But, we have also: $AB = z_2 - z_1$. Hence, the theorem follows.

---

* From AMERICAN MATHEMATICAL MONTHLY, vol. 37 (1930), pp. 301–303.

† Smail, Lloyd L., *Some geometrical applications of complex numbers*, in this Monthly, vol. 36 (1929), pp. 504–511.

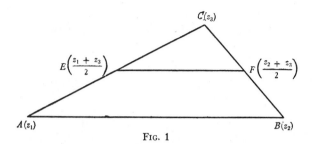

Fig. 1

THEOREM 2. *The line-segments joining the mid-points of opposite sides of any quadrilateral bisect each other.*

Proof. The procedure is similar to the above. Thus, in the notation of Fig. 2, the mid-point of EF is represented by the complex numbers,

$$\tfrac{1}{2}\{\tfrac{1}{2}(z_1+z_4)+\tfrac{1}{2}(z_2+z_3)\}=\tfrac{1}{4}(z_1+z_2+z_3+z_4),$$

and the mid-point of MN is represented by

$$\tfrac{1}{2}\{\tfrac{1}{2}(z_1+z_2)+\tfrac{1}{2}(z_3+z_4)\}=\tfrac{1}{4}(z_1+z_2+z_3+z_4),$$

which is equal to the above. Hence the theorem is proved.

The complex number representing P can be also written as follows:

$$\tfrac{1}{4}(z_1+z_2+z_3+z_4)=\tfrac{1}{2}\{\tfrac{1}{2}(z_1+z_3)+\tfrac{1}{2}(z_2+z_4)\},$$

where $\tfrac{1}{2}(z_1+z_3)$ evidently represents the mid-point K of the diagonal AC and $\tfrac{1}{2}(z_2+z_4)$ the mid-point L of BD; hence P is also the mid-point of KL. Thus, we have the following:

THEOREM 3. *The mid-point of the line-segment joining the mid-points of the diagonals of a quadrilateral coincides with the point of intersection of the line-segments joining the mid-points of opposite sides of the quadrilateral.*

THEOREM 4. *The lines joining mid-points of adjacent sides of a quadrilateral form a parallelogram.*

Again, referring to Fig. 2, we have: $EM=\tfrac{1}{2}(z_3-z_1)$, and $NF=\tfrac{1}{2}(z_3-z_1)$, which proves that EM and NF are parallel and equal in length. Hence, the theorem follows.

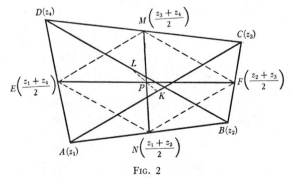

Fig. 2

COMPLEX NUMBERS AND VECTOR ALGEBRA*

D. E. RICHMOND, Williams College

1. Introduction. Vectors in the plane are frequently represented by complex numbers and treated by the corresponding rules. However, in many applications they are treated by vector algebra and multiplied to form vector and scalar products. The student is sometimes puzzled about the relation between these two schemes of representation. The literature appears to contain no explicit discussion of this point. The present paper supplies this lack and also presents complex numbers in a somewhat new light.

Let us imagine that we are faced with the task of inventing an algebra of directed quantities, that is, quantities representable by arrows or vectors in the plane of the paper. By parallel displacement, any such vector may be drawn from the origin of a rectangular coordinate system to some point P in the plane (see Figure 1). It is natural to use letters, say $\alpha, \beta, \gamma, \cdots$, to represent such vectors and to try to give meanings to sums, differences and products like $\alpha+\beta$, $\alpha-\beta$, $\alpha\beta$. Keeping in mind simple physical applications like the composition of forces or velocities, it is easy to arrive at the usual geometric interpretations of the addition and subtraction of two vectors and of the product of a vector by a scalar. The corresponding algebra is also simple.

Let $\mathbf{H}$ denote a unit vector along the horizontal axis and $\mathbf{V}$ a unit vector along the vertical axis (see Figure 1). In terms of vector addition and of multiplication by a scalar, we may represent an arbitrary vector in the form

$$\alpha = a_1 \mathbf{H} + a_2 \mathbf{V}$$

with real numbers a_1 and a_2. We may also define the magnitude of α, written $|\alpha|$, as $\sqrt{a_1^2+a_2^2}$.

The difficulty is to find an appropriate definition for the product of two vectors α and β. It is natural to attempt to define the product in such a way that magnitudes shall multiply, that is, so that $|\alpha\beta| = |\alpha| |\beta|$. This stipulation leads to an essentially unique result if we add the requirement that $\mathbf{H}$ shall act like unity, so that specifically.

$$\mathbf{HV} = \mathbf{VH}(=\mathbf{V}) \quad \text{and} \quad \mathbf{HH} = \mathbf{H} \text{ (i.e., } \mathbf{H}^2 = \mathbf{H}\text{)}.$$

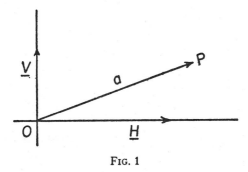

Fig. 1

---

* From AMERICAN MATHEMATICAL MONTHLY, vol. 58 (1951), pp. 622–628.

We proceed to derive this unique result.

To begin with, it is necessary to evaluate $\mathbf{V}^2$. Let
$$\mathbf{V}^2 = a\mathbf{H} + b\mathbf{V}$$
where a and b are to be determined.

Since $|\mathbf{V}| = 1$, we have, equating the squares of the magnitudes,
$$a^2 + b^2 = 1.$$

Now consider
$$(\mathbf{H} + \mathbf{V})(\mathbf{H} - \mathbf{V}) = \mathbf{H}^2 + \mathbf{VH} - \mathbf{HV} - \mathbf{V}^2 = \mathbf{H}^2 - \mathbf{V}^2$$
$$= \mathbf{H} - a\mathbf{H} - b\mathbf{V} = (1-a)\mathbf{H} - b\mathbf{V}.$$

Equating squares of magnitudes,
$$4 = (1-a)^2 + b^2.$$

Combining with
$$1 = a^2 + b^2$$
we have
$$a = -1, \quad b = 0.$$

Hence
$$\mathbf{V}^2 = -\mathbf{H}.$$

If
$$\boldsymbol{\alpha} = a_1\mathbf{H} + a_2\mathbf{V}$$
and
$$\boldsymbol{\beta} = b_1\mathbf{H} + b_2\mathbf{V}$$
are two vectors, their product is easily found to be

(1) $$\boldsymbol{\alpha\beta} = (a_1b_1 + a_2b_2)\mathbf{H} + (a_1b_2 + a_2b_1)\mathbf{V}.$$

Then
$$|\boldsymbol{\alpha\beta}|^2 = (a_1b_1 - a_2b_2)^2 + (a_1b_2 + a_2b_1)^2$$
$$= (a_1^2 + a_2^2)(b_1^2 + b_2^2) = |\boldsymbol{\alpha}|^2|\boldsymbol{\beta}|^2$$

for arbitrary a_1, a_2, b_1, b_2. Hence the stipulation $|\boldsymbol{\alpha\beta}| = |\boldsymbol{\alpha}||\boldsymbol{\beta}|$ can be carried through in general.

The scheme just discussed is, except for notation, that of complex numbers. In fact, if we write $\boldsymbol{\alpha} = a_1 + ib_1$ and $\boldsymbol{\beta} = a_2 + ib_2$ and multiply as usual, replacing i^2 by -1, the result corresponds to (1). The scheme of complex numbers is

COMPLEX NUMBERS AND VECTOR ALGEBRA

therefore essentially the only one for which the magnitude of the product equals the product of the magnitudes.

2. Vector algebra. Vector algebra uses a somewhat different scheme of representation of vectors in the plane. The vector α of section 1 would be written

$$\alpha = a_1 \mathbf{i} + a_2 \mathbf{j}$$

replacing **H** by **i** and **V** by **j**.

So long as one deals only with addition, subtraction and multiplication by scalars, no difference occurs other than this trivial notational one. But as soon as one speaks of multiplying vectors α and β, an essential difference comes in.

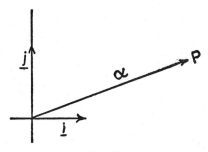

Fig. 2

As is well known, there are two types of product used in vector algebra, the scalar product and the vector product. If

$$\alpha = a_1 \mathbf{i} + a_2 \mathbf{j}$$
$$\beta = b_1 \mathbf{i} + b_2 \mathbf{j}$$

these products are

(2) $\qquad \alpha \cdot \beta = a_1 b_1 + a_2 b_2 \qquad$ (scalar product)

(3) $\qquad \alpha \times \beta = (a_1 b_2 - a_2 b_1) \mathbf{k} \qquad$ (vector product)

where **k** is a unit vector perpendicular to the plane of α and β.

The choice of these expressions to represent products is dictated by physical applications (work done by a force, moment of a force). Here we are concerned only with the formal relations among the definitions (1), (2) and (3).

The relations are the following.

$$\alpha \cdot \beta = \operatorname{Re}(\bar{\alpha}\beta)$$
$$\alpha \times \beta = \operatorname{Im}(\bar{\alpha}\beta)\mathbf{k}$$

where Re stands for "real part of" and Im for "imaginary part of" and where $\bar{\alpha}$

is the conjugate of $\boldsymbol{\alpha}$, that is, $a_1 - a_2 i$. In fact,

$$\bar{\alpha}\beta = (a_1 b_1 + a_2 b_2) + (a_1 b_2 - b_1 a_2)i,$$
$$\text{Re}(\bar{\alpha}\beta) = a_1 b_1 + a_2 b_2 = \boldsymbol{\alpha} \cdot \boldsymbol{\beta},$$
$$\text{Im}(\bar{\alpha}\beta) = a_1 b_2 - b_1 a_2, \text{ the coefficient of } \mathbf{k} \text{ in } \boldsymbol{\alpha} \times \boldsymbol{\beta}.$$

To illustrate this relationship, consider two examples from geometry.

EXAMPLE 1. To prove that an angle inscribed in a semicircle is a right angle.

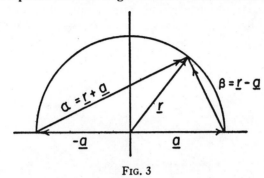

FIG. 3

Vector algebra proof:

$$\boldsymbol{\alpha} \cdot \boldsymbol{\beta} = (\mathbf{r} + \mathbf{a}) \cdot (\mathbf{r} - \mathbf{a})$$
$$= \mathbf{r}^2 - \mathbf{a}^2$$
$$= 0, \quad \text{since} \quad |\mathbf{r}| = |\mathbf{a}|.$$

Hence $\boldsymbol{\alpha} \perp \boldsymbol{\beta}$.

Complex number proof:

$$\bar{\alpha}\beta = [(x + a) - iy][(x - a) + iy]$$
$$= x^2 - a^2 + y^2 + 2aiy$$
$$= 2aiy, \quad \text{since} \quad |x + iy| = a.$$
$$\text{Re}(\bar{\alpha}\beta) = 0.$$

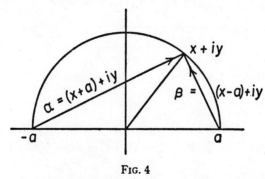

FIG. 4

Hence $\boldsymbol{\alpha} \perp \boldsymbol{\beta}$.

EXAMPLE 2. To derive the difference formulas of trigonometry.
Vector algebra proof:

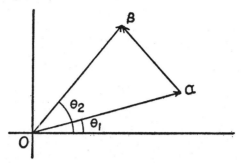

FIG. 5

$$\alpha = \cos\theta_1 \mathbf{i} + \sin\theta_1 \mathbf{j}$$
$$\beta = \cos\theta_2 \mathbf{i} + \sin\theta_2 \mathbf{j}$$
$$\alpha \cdot \beta = \cos\theta_1 \cos\theta_2 + \sin\theta_1 \sin\theta_2.$$

But
$$\alpha \cdot \beta = |\alpha||\beta| \cos(\theta_2 - \theta_1) = \cos(\theta_2 - \theta_1).$$

Hence
$$\cos(\theta_2 - \theta_1) = \cos\theta_1 \cos\theta_2 + \sin\theta_1 \sin\theta_2$$
$$\alpha \times \beta = (\sin\theta_2 \cos\theta_1 - \cos\theta_2 \sin\theta_1)\mathbf{k}$$
$$= |\alpha||\beta| \sin(\theta_2 - \theta_1)\mathbf{k} = \sin(\theta_2 - \theta_1)\mathbf{k}.$$

Hence
$$\sin(\theta_2 - \theta_1) = \sin\theta_2 \cos\theta_1 - \cos\theta_2 \sin\theta_1.$$

Note:

Since the formulas $\alpha \cdot \beta = |\alpha||\beta| \cos(\theta_2 - \theta_1)$ and $\alpha \times \beta = |\alpha||\beta| \sin(\theta_2 - \theta_1)\mathbf{k}$, used in this proof, are often established by way of the difference formulas, the proof may give the appearance of circularity. However, there is no difficulty in avoiding circularity. In fact,

$$(\beta - \alpha)^2 = \beta^2 + \alpha^2 - 2\alpha \cdot \beta.$$

By the law of cosines

$$(\beta - \alpha)^2 = |\beta - \alpha|^2 = \beta^2 + \alpha^2 - 2|\alpha||\beta| \cos(\theta_2 - \theta_1).$$

Equating these two results gives the first formula. The second follows by equating two expressions for the area of the triangle $O\alpha\beta$.

Complex number proof:

$$\alpha = \cos\theta_1 + \sin\theta_1 i$$

$$\beta = \cos\theta_2 + \sin\theta_2 i$$

$$\bar{\alpha}\beta = (\cos\theta_2\cos\theta_1 + \sin\theta_2\sin\theta_1) + (\sin\theta_2\cos\theta_1 - \cos\theta_2\sin\theta_1)i$$

$$= \cos(\theta_2 - \theta_1) + \sin(\theta_2 - \theta_1)i.$$

Conclusions as above.

Note that

$$\alpha \cdot \beta = \operatorname{Re}(\bar{\alpha}\beta), \quad \alpha \times \beta = \operatorname{Im}(\bar{\alpha}\beta)\mathbf{k}.$$

3. Dimensions higher than two. It is of interest to show that there exists no method of multiplication which yields $|\alpha\beta| = |\alpha||\beta|$ for arbitrary vectors α and β in *three* dimensions, if we make the additional assumption that the associative law holds, $(\alpha\beta)\gamma = \alpha(\beta\gamma)$.

Let

$$\alpha = a_1 \mathbf{u}_1 + a_2 \mathbf{u}_2 + a_3 \mathbf{u}_3$$

where $\mathbf{u}_1, \mathbf{u}_2, \mathbf{u}_3$ are unit vectors along the axes of a rectangular coordinate system. Assume that $\mathbf{u}_1$ acts like unity so that

$$\mathbf{u}_1^2 = \mathbf{u}_1, \quad \mathbf{u}_1\mathbf{u}_2 = \mathbf{u}_2\mathbf{u}_1(=\mathbf{u}_2), \quad \mathbf{u}_1\mathbf{u}_3 = \mathbf{u}_3\mathbf{u}_1(=\mathbf{u}_3).$$

If

$$\mathbf{u}_2^2 = a\mathbf{u}_1 + b\mathbf{u}_2 + c\mathbf{u}_3$$

with undetermined a, b and c,

$$(\mathbf{u}_1 - \mathbf{u}_2)(\mathbf{u}_1 + \mathbf{u}_2) = \mathbf{u}_1^2 - \mathbf{u}_2^2 = \mathbf{u}_1 - (a\mathbf{u}_1 + b\mathbf{u}_2 + c\mathbf{u}_3)$$

$$= (1-a)\mathbf{u}_1 - b\mathbf{u}_2 - c\mathbf{u}_3.$$

Equating squares of magnitudes

$$4 = (1-a)^2 + b^2 + c^2.$$

Since $a^2 + b^2 + c^2 = 1$, we have

$$a = -1, \quad b = c = 0.$$

Thus

$$\mathbf{u}_2^2 = -\mathbf{u}_1.$$

Similarly,

$$\mathbf{u}_3^2 = -\mathbf{u}_1.$$

Now let
$$u_2 u_3 = r u_1 + s u_2 + t u_3.$$
Then
$$u_2(u_2 + u_3) = u_2^2 + u_2 u_3$$
$$= (r - 1) u_1 + s u_2 + t u_3.$$
Equating squares of magnitudes
$$2 = (r - 1)^2 + s^2 + t^2.$$
Since $r^2 + s^2 + t^2 = 1$, $r = 0$ and
$$u_2 u_3 = s u_2 + t u_3.$$
Multiply on the left by u_2. Since
$$u_2(u_2 u_3) = (u_2^2) u_3 = -u_1 u_3 = -u_3,$$
$$-u_3 = -s u_1 + t(s u_2 + t u_3).$$

Hence $s = 0$, $t^2 = -1$. This result is inconsistent with the fact that t must be real. It also conflicts with the necessary condition $t^2 + s^2 = 1$.

For arbitrary vectors in 4 dimensions, it is possible to define the product so that magnitudes *do* multiply. In fact the methods of this section lead directly to quaternions which have this property. The same methods show that if multiplication is associative, the condition $|\alpha\beta| = |\alpha| |\beta|$ is impossible to satisfy for arbitrary vectors in a space of more than 4 dimensions. The proofs of both statements are well within the ability of any good student and should prove illuminating.

INTRODUCTION OF COMPLEX NUMBERS AS VECTORS IN THE PLANE*

L. FUCHS, Budapest, and T. SZELE, Debrecen, Hungary

It is a well-known elementary fact that the complex number field is the only possible field extension of degree 2 of the field of the real numbers. This statement may be sharpened to the assertion that *if a (not necessarily commutative) associative algebra of order 2 over the real number field has no divisors of zero, then it is isomorphic to the field of complex numbers*. A vector space of dimension 2 over the real numbers has a simple geometric interpretation as the set of the vectors of the Euclidean plane, this fact being used several times for introducing the complex numbers. Hence it seems to be natural to raise the problem of prov-

---

* From AMERICAN MATHEMATICAL MONTHLY, vol. 59 (1952), pp. 628–631.

ing the cited theorem in a purely elementary geometric way. Such a proof would help beginners to get a deeper insight into the kernel of this fact. Our present purpose is to give such a proof which seems to have some interest in itself and to be new.

In what follows let the vectors of the plane be denoted by bold-face and real numbers by Greek letters. We shall prove the following

THEOREM. *Suppose we have defined addition and multiplication of any two vectors of the plane as well as multiplication of a vector by a scalar (i.e., a real number) with the following properties*:

(1) *addition is vector-addition*; it is therefore commutative, associative, and unique subtraction is possible;

(2) *multiplication by scalars is performed as usual*; therefore we have $\alpha(\mathbf{a}+\mathbf{b}) = \alpha\mathbf{a}+\alpha\mathbf{b}$, $(\alpha+\beta)\mathbf{a} = \alpha\mathbf{a}+\beta\mathbf{a}$, $(\alpha\beta)\mathbf{a} = \alpha(\beta\mathbf{a})$ for arbitrary real numbers α, β and vectors $\mathbf{a}$, $\mathbf{b}$;

(3) *multiplication of vectors is associative*: $\mathbf{a}(\mathbf{bc}) = (\mathbf{ab})\mathbf{c}$;

(4) $(\alpha\mathbf{a})\mathbf{b} = \alpha(\mathbf{ab}) = \mathbf{a}(\alpha\mathbf{b})$;

(5) *divisors of zero do not exist*: $\mathbf{ab} = 0$ implies either $\mathbf{a} = 0$ or $\mathbf{b} = 0$;

(6) *distributivity holds*: $\mathbf{a}(\mathbf{b}+\mathbf{c}) = \mathbf{ab}+\mathbf{ac}$ and $(\mathbf{b}+\mathbf{c})\mathbf{a} = \mathbf{ba}+\mathbf{ca}$.

Then, apart from an affine transformation, our vector system is just the set of all vectors of the complex plane with the well-known operations. In other words, if (1)–(6) hold, then the vectors of the plane form a field isomorphic to the complex number field.

Proof. Clearly, only the statement concerning multiplication needs a verification.

First we show that multiplication of vectors is commutative. Let us consider a vector $\mathbf{x}$ not parallel to its own square $\mathbf{x}^2$. Such an $\mathbf{x}$ certainly exists, for if $\mathbf{y} \neq 0$ is any vector parallel to $\mathbf{y}^2$, *i.e.* $\mathbf{y}^2 = \lambda\mathbf{y}$ (plainly $\lambda \neq 0$), then each vector $\mathbf{x} \neq 0$ not parallel to $\mathbf{y}$ has the required property. In fact, if we had $\mathbf{x}^2 = \mu\mathbf{x}$, then from the equations

$$(\mu\mathbf{y} - \mathbf{yx})\mathbf{x} = 0 \quad \text{and} \quad \mathbf{y}(\lambda\mathbf{x} - \mathbf{yx}) = 0$$

we would get

$$\mu\mathbf{y} = \mathbf{yx} = \lambda\mathbf{x}$$

on account of (5). But this would mean that $\mathbf{x}$ and $\mathbf{y}$ are parallel, contrary to our hypothesis. Choosing a vector $\mathbf{x}$ with the prescribed property, it is evident that the vectors $\mathbf{x}$ and $\mathbf{x}^2$ span the plane. Hence each vector $\mathbf{a}$ may be represented (uniquely) in the form

$$\mathbf{a} = \alpha\mathbf{x} + \beta\mathbf{x}^2$$

with some real numbers α, β. For vectors of this type it is obvious, in view of the above axioms, that they commute. Therefore we obtain

$$\mathbf{ab} = \mathbf{ba}$$

for all vectors $\mathbf{a}$, $\mathbf{b}$.

INTRODUCTION OF COMPLEX NUMBERS AS VECTORS IN THE PLANE   125

For the square of the vector $\mathbf{a} = \alpha \mathbf{x} + \beta \mathbf{x}^2$ we have;

$$\mathbf{a}^2 = \alpha^2 \mathbf{x}^2 + 2\alpha\beta \mathbf{x}^3 + \beta^2 \mathbf{x}^4.$$

Hence it is seen that if we let $\mathbf{a}$ (*i.e.* α and β) vary a little, then $\mathbf{a}^2$ can vary but a little. Thus we get that if $\mathbf{a}$ varies continuously, then so does $\mathbf{a}^2$.

Now let $\mathbf{u}$ be an arbitrary unit vector and turn $\mathbf{u}$ through the angle π until it arrives at the position $-\mathbf{u}$. Meanwhile, according to what has been said above, the square vector has to rotate in one and the same direction, for, on the contrary, there would exist two non-parallel unit vectors $\mathbf{k}$ and $\mathbf{l}$ such that $\mathbf{k}^2 = \lambda \mathbf{l}^2$ with $\lambda > 0$. Hence, by commutativity, we have $(\mathbf{k} + \sqrt{\lambda}\mathbf{l})(\mathbf{k} - \sqrt{\lambda}\mathbf{l}) = 0$ and thus from (5) we are led to either of the equations

$$\mathbf{k} + \sqrt{\lambda}\,\mathbf{l} = 0 \quad \text{and} \quad \mathbf{k} - \sqrt{\lambda}\,\mathbf{l} = 0$$

contradicting the assumption on $\mathbf{k}$ and $\mathbf{l}$. Since $(-\mathbf{u})^2 = \mathbf{u}^2$, we see that the square vector turns through the angle 2π, or -2π.

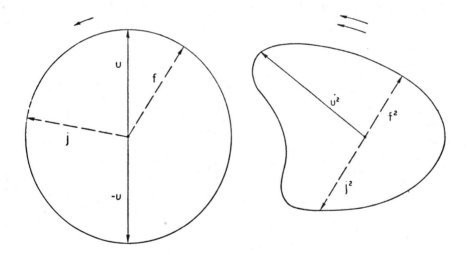

Let us now consider simultaneously the rotating of the unit vector and of its square (Figure). It is evident that during the rotation of the unit vector $\mathbf{u}$ through an angle 2π ($\mathbf{u}^2$ will then turn through an angle $\pm 4\pi$) $\mathbf{u}$ coincides with a vector $\mathbf{f}$ which is parallel to its own square in the same direction: $\mathbf{f}^2 = \rho \mathbf{f}$ with $\rho > 0$, and with a vector $\mathbf{j}$ whose square lies in the same direction as the vector $-\mathbf{f}$, that is, $\mathbf{j}^2 = -\sigma \mathbf{f}$ with $\sigma > 0$. Put $\mathbf{e} = (1/\rho)\mathbf{f}$ and $\mathbf{i} = (1/\sqrt{\rho\sigma})\mathbf{j}$, then we obtain

$$\mathbf{e}^2 = \frac{1}{\rho^2}\mathbf{f}^2 = \frac{1}{\rho^2}\rho\mathbf{f} = \mathbf{e}$$

and
$$\mathbf{i}^2 = \frac{1}{\rho\sigma}\mathbf{j}^2 = -\frac{1}{\rho\sigma}\sigma\mathbf{f} = -\mathbf{e}.$$

The vectors **e** and **i** are surely not parallel, since otherwise their squares would have the same direction. Hence it follows that **e** and **i** span the plane and therefore we may write each vector **a** in the plane in the form

$$\mathbf{a} = \xi\mathbf{e} + \eta\mathbf{i}$$

with suitable real numbers ξ and η. We have still to know which vector of the plane is the vector **ei**.

We proceed to prove that $\mathbf{ei} = \mathbf{i}$, or, more generally, $\mathbf{ea} = \mathbf{a}$ for all vectors **a**. The equation $\mathbf{e}^2 = \mathbf{e}$ implies $\mathbf{e}^2\mathbf{a} = \mathbf{ea}$ or $\mathbf{e}(\mathbf{ea} - \mathbf{a}) = 0$ for all **a**. Since there are no divisors of zero, it follows that $\mathbf{ea} - \mathbf{a} = 0$, *i.e.* $\mathbf{ea} = \mathbf{a}$, as stated.

From the results obtained so far we may conclude that multiplication of two vectors is to be performed as though **e** and **i** were equal to the real number 1 and the imaginary unit $\sqrt{-1}$, respectively, *i.e.* we multiply according to the rules of multiplication of complex numbers.

Now use an affine transformation of the plane which carries the vectors **e** and **i** into two unit vectors perpendicular to each other; such a transformation clearly always exists. Then define the operations in the new vector system as induced by this affine transformation. It is obvious that we arrive just at the vectors of the complex plane with the well-known definitions of operations.

This completes the proof of the theorem.

Remark 1. As the proof above shows, the requirement of associativity can be replaced by commutativity. Also, it is worth while noticing that if we suppose neither associativity nor commutativity, the theorem is no longer valid. It is readily checked that if for two non-parallel vectors **a** and **b** we put

$$\mathbf{a}^2 = \mathbf{b}^2 = \mathbf{a}, \quad \mathbf{ab} = -\mathbf{ba} = \mathbf{b}$$

(associativity fails to hold: $\mathbf{b}^2\mathbf{b} \neq \mathbf{bb}^2$), then we obtain an algebra without divisors of zero, but not isomorphic to the complex number system.

Remark 2. In the above proof we have made use of continuity; this can, however, be avoided, as is shown by the following argument.

For an arbitrary vector $\mathbf{u} \neq 0$ the vectors **au** and **bu** span the plane provided that the same holds for **a** and **b** (zero divisors do not exist!), therefore for any given vector **v** we have

$$\mathbf{v} = \alpha\mathbf{au} + \beta\mathbf{bu} = (\alpha\mathbf{a} + \beta\mathbf{b})\mathbf{u} = \mathbf{xu}$$

with $\mathbf{x} = \alpha\mathbf{a} + \beta\mathbf{b}$ and α, β suitable real numbers. Hence right-division and, similarly, left-division by any non-zero vector exists. We conclude that the vectors not equal to 0 form a group under multiplication with, say, $\mathbf{e}$ as identity. Then each vector has the form $\gamma\mathbf{e} + \delta\mathbf{d}$ with $\mathbf{d}$ an arbitrary vector not parallel to $\mathbf{e}$. Hence the multiplication is commutative.

Now from $\mathbf{d}^2 = \lambda\mathbf{e} + \mu\mathbf{d}$ we have $(\mathbf{d} - \frac{1}{2}\mu\mathbf{e})^2 = (\lambda + \frac{1}{4}\mu^2)\mathbf{e}$ where the real number $\lambda + \frac{1}{4}\mu^2 = \kappa$ must be negative, since $\kappa \geq 0$ would imply that the vectors $\mathbf{d} - \frac{1}{2}\mu\mathbf{e} \pm \sqrt{\kappa}\mathbf{e}$ are divisors of zero. Consequently, for the vector

$$\mathbf{i} = \frac{\mathbf{d} - \frac{1}{2}\mu\mathbf{e}}{\sqrt{-\kappa}}$$

we get $\mathbf{i}^2 = -\mathbf{e}$, q.e.d.

AN ISOMETRY OF THE PLANE*

EDWARD D. GAUGHAN, New Mexico State University

It is well known that any distance-preserving map of the plane into the plane that fixes the origin is either a rotation or a rotation followed by reflection about the x-axis. The purpose of this note is to give a short rigorous proof of this theorem from elementary principles with no recourse to any notions from geometry.

Suppose f is such a map. Since f is distance preserving and $f(0) = 0$, f preserves moduli. We have

$$|f(z) - f(1)|^2 = |f(z)|^2 + |f(1)|^2 - 2 \operatorname{Re} f(z)\overline{f(1)}$$
$$|z - 1|^2 = |z|^2 + |1|^2 - 2 \operatorname{Re} z.$$

Thus we see for any z,

(1) $$\operatorname{Re} f(z)\overline{f(1)} = \operatorname{Re} z.$$

For $z = i$, this gives $\operatorname{Re} f(i)\overline{f(1)} = \operatorname{Re} i = 0$, hence $\operatorname{Im} f(i)\overline{f(1)} = \pm 1$ since $|f(i)| = |f(1)| = 1$. Consequently, $f(i) = \pm if(1)$. Similarly,

$$|f(z) - f(i)|^2 = |f(z)|^2 + |f(i)|^2 - 2 \operatorname{Re} f(z)\overline{f(i)},$$
$$|z - i|^2 = |z|^2 + |i|^2 - 2 \operatorname{Re} z\bar{i}.$$

---

* From AMERICAN MATHEMATICAL MONTHLY, vol. 69 (1962), p. 546.

Now

(2)
$$\pm \operatorname{Im} f(z)\overline{f(1)} = \pm \operatorname{Re} f(z) i \, \overline{f(1)}$$
$$= \operatorname{Re} f(z) \, \overline{f(i)} = \operatorname{Re} z\, \bar{\imath} = \operatorname{Im} z.$$

Combining (1) and (2), we obtain

$$f(z)\overline{f(1)} = z \text{ or } \bar{z},$$

hence

$$f(z) = f(1)z \text{ or } f(1)\bar{z}.$$

THE PROOF OF THE TRIANGLE INEQUALITY*

M. F. SMILEY, University of California, Riverside

The writer has been perhaps overexposed to proofs of the triangle inequality which are based on the discriminant of a quadratic or other lengthy calculations. Some of these exposures were sponsored by the Mathematical Association of America. In addition to a concise proof, a discussion of this inequality should include a clear statement of the case of equality. (Cf. K. Urbanik and F. B. Wright, Absolute-valued algebras, Proc. Amer. Math. Soc., 11 (1960) 861–866; especially their proof of Lemma 3.) The purpose of this note is to provide a geometric proof of the triangle inequality. Our proof leads at once to an equally simple proof of the quadrilateral inequality and its consequent polygonal inequalities. These matters we propose, however, to discuss in a separate publication.

It is a fact that if z and w are complex numbers then

(1) $|z+w| \leq |z| + |w|$ and equality holds iff $|z|w = |w|z$.

The fact (1) is called the triangle inequality.

A logical proof of (1) is not only possible but almost canonical. We have

$$|z+w| \leq |z| + |w| \text{ iff } (z+w)(\bar{z}+\bar{w}) \leq z\bar{z} + 2|z||w| + w\bar{w}$$

$$\text{iff } z\bar{w} + w\bar{z} \leq 2|z||w|$$
$$\text{iff } z\bar{w}|z||w| + w\bar{z}|z||w| \leq 2|z|^2|w|^2$$
$$\text{iff } 0 \leq (|w|z - |z|w)(|w|\bar{z} - |z|\bar{w}).$$

Observing that all of these implications hold if "$\leq$" is everywhere replaced by "$=$", we have proved (1).

---

* From AMERICAN MATHEMATICAL MONTHLY, vol. 70 (1963), pp. 546–547.

The statement (1) remains a fact if we require only that z and w are vectors in a vector space X over the complex field which is equipped with a sesquilinear inner product which is positive definite, i.e., there is a map of $X \times X$ into the complex field denoted by (z, w) such that $(w, z) = \overline{(z, w)}$, $(z_1+z_2, w) = (z_1, w)+(z_2, w)$, $(\alpha z, w) = \alpha(z, w)$ for all z, z_1, z_2, w in X and complex α; and $(z, z) > 0$ unless $z = 0$. In this situation we have $|z| = (z, z)^{1/2}$ and

$$|z+w| \leq |z| + |w|$$

iff $(w, z) + (z, w) \leq 2|z||w|$ (or $\operatorname{Re}(z, w) \leq |z||w|$)
iff $(|z|w, |w|z) + (|w|z, |z|w) \leq 2|z|^2|w|^2$
iff $0 \leq (|w|z - |z|w, |w|z - |z|w)$.

We have not only proved (1), but our first statement yields

(2) $|(z, w)| \leq |z||w|$ *and equality holds iff z, w are linearly dependent.*

We obtain (2) by replacing z by $e^{-i\theta}z$ where $(z, w) = |(z, w)|e^{i\theta}$ (θ real) when we realize that $|w|ze^{-i\theta} = |z|w$ iff z and w are linearly dependent. The fact (2) is often labeled "the inequality of Schwarz."

Moral: What geometry hath joined together, let neither algebra nor analysis put asunder.

BIBLIOGRAPHIC ENTRIES: COMPLEX NUMBERS

The references below are to the AMERICAN MATHEMATICAL MONTHLY.

1. G. A. Bingley, The complex variable in the solution of problems in elementary analytic geometry, vol. 33, p. 418.

2. L. L. Smail, Some geometrical applications of complex numbers, vol. 36, p. 504.

Geometry proofs in the complex plane.

(d)
INFINITY

INFINITE AND IMAGINARY ELEMENTS IN ALGEBRA AND GEOMETRY*

R. M. WINGER, University of Washington

1. Introduction. Imaginary points in analytic geometry. The traditional treatment of imaginary and infinite elements in algebra and geometry has curiously resisted the reform movement in American textbook writing. While imaginary numbers are receiving perhaps more attention in the algebra books, the imaginary point is still an outlaw.** The equation $x^2+y^2=0$ "represents a single point or point circle"† while the equation $x^2+y^2+1=0$ "represents no locus whatever." This is all the more remarkable in view of the historical fact that it was the geometric representation of imaginaries that gave these numbers their algebraic standing. Probably these same writers would not hesitate to speak of a point in space of four dimensions as a geometric image of a tetrad of numbers although such a point is as impossible to plot as one whose Cartesian coördinates in the plane are imaginary. Now that modern pure geometry has found a way to introduce imaginary points independent of algebra, it would seem absurd not to utilize the algebraic approach which is more elementary.

A reciprocal custom prevails with respect to infinity. Modern geometry has found a line at infinity indispensable as a bond of union between projective and metric properties, to render the principle of duality universally valid and to preserve a $(1,1)$ correspondence between a figure and its projection. Geometry has also found the number ∞ serviceable in the analytic theory of collineations, the parametric representation of curves and in studying the behavior of curves at infinity. But algebra steadfastly refuses to avail itself of these conveniences. True in the theory of equations in one variable the conditions for infinite roots are frequently given, but no mention is made of infinite roots of simultaneous equations where the idea is most valuable.

---

* From AMERICAN MATHEMATICAL MONTHLY, vol. 29 (1922), pp. 290–297.

**Exception must be made of the excellent book by Snyder and Sisam, *Analytic Geometry of Space*. Even here after discussing imaginary elements the authors lapse into the old form of statement,—that a tangent plane cuts an ellipsoid in a "point ellipse," *e.g.*, rather than in a pair of imaginary lines. Crawley and Evans, *Analytic Geometry*, is the only freshman text that has come to my notice which recognizes imaginary points.

† This statement always recalls a bizarre work on *Algeometry* in which the author declares that while $x^2+y^2=0$ is the equation of a *round* point no one has been able to write the equation of a *square* point!

Authors of college algebras may hesitate to increase the formidable multiplicity of topics or they may regard the difficulties insuperable to freshmen. But these authors do not scruple to encroach on the fields of analytic geometry and even calculus to an extent that important algebraic material is crowded out altogether * On the other hand, projective geometry has usually been presented synthetically so that any algebraic discussion would appear artificial. Careful writers on both subjects follow the sound practice—under the limitations they have prescribed—of telling the truth and nothing but the truth even if it is impossible to tell the whole truth. An occasional author however fails to tell the strict truth, as for example when he says that a linear and a quadratic equation in two variables *always* have two solutions or that two quadratics always have four. And I am not sure that the writer who says that the locus of the equation $x^2+y^2=0$ is a single point does not convey the impression that he is telling a general truth rather than a truth restricted to the real domain. For he invariably uses "point" and "locus" to signify "real point" and "real locus" without troubling himself to explain his meaning.

Certainly in analytic geometry where the streams of algebra and geometry merge there would seem to be the least excuse for these half-truths. For nowhere is the interplay between algebra and geometry more beautifully exemplified than in the theories of infinite and imaginary elements. The college student should come to the study of analytic geometry with a knowledge of imaginary numbers which, thanks to such books as Wilczynski's,† have been freed of their stigma and made genuinely real in his experience. Imaginary solutions of equations have been accepted on the same terms as real solutions. He has been taught that a quadratic equation in one variable always has two roots, real or imaginary, distinct or equal. Why should he be told that certain equations have no loci and that equations with imaginary solutions represent curves which do not intersect, if indeed they represent curves at all? Why not take advantage of the ground already gained in algebra and say that a line and a circle in the plane always meet in two points, real and distinct, real and coincident or conjugate imaginary? The new statement not only serves every purpose of the old "two points, one point, no point" terminology but is actually more descriptive of the relation of the line and circle. A generator of a circular cone is a true example of a line that meets the circle of the base in but one point while the axis of the cone meets the circle in no point.

Absolute coördinates. It will be instructive to consider briefly a system of metrical coördinates in which the conventional rôles of real and imaginary are varied. We refer to the absolute coördinates used with such signal success by Professor Morley and his students in discussing metrical properties. In this system the bilinear axes are the circular rays through the Cartesian origin. If X and Y are

---

* One recent book of 250 pages devotes nearly a third of its space to calculus, both differential and integral.

† *College Algebra with Applications*, Boston, 1916.

the rectangular coördinates of a point, the absolute coördinates $x, \bar{x}$ are defined to be

(1) $$x = X + iY, \quad \bar{x} = X - iY, \quad i = \sqrt{-1}.$$

The absolute coördinates, being complex, are represented geometrically in the usual manner on the complex plane. But if plotted separately x and $\bar{x}$ would in general determine two points. To avoid ambiguity we plot only the x, although the $\bar{x}$ cannot be ignored. When the Cartesian coördinates are real the absolute coördinates are in general imaginary. The converse is however not generally true. Neither does the reality of the absolute coördinates imply reality of the corresponding point. Thus on the line

(2) $$x + \bar{x} = 2, \quad \text{or} \quad X = 1,$$

those points are real whose x-coördinates are of the form $1 + iY$. The corresponding $\bar{x}$-coördinates are then $1 - iY$. A single real point of the line has real absolute coördinates, viz., the point $(1,1)$. But while the number pairs $(2,0)$, $(3,-1)$, $(1+\sqrt{2}, 1-\sqrt{2})$, $(i, 2-i)$ satisfy the equation of the line, they cannot be represented. For if we plot the x alone the first three points would all lie on the line $x - \bar{x} = 0$ whereas this line cuts (2) only in the point $(1,1)$. If the algebra and geometry are to correspond, we must say that these pairs of numbers are coördinates of *imaginary* points on the line. In this system we have the following criterion:

A point is real or imaginary according as its absolute coördinates are or are not conjugate complex numbers.[*]

Again the parametric representation of curves in absolute coördinates differs from that in Cartesian coördinates. In the latter system real parameters are assigned to real points. In absolute coördinates, on the other hand, if the coefficients in the equations are real, the parameters of real points will be imaginary in general. Instead of thinking of the parameter as running along a line and ranging through all values, we may suppose the parameter to run around the unit circle in the complex plane, *i.e.*, the parameter t is a complex number of absolute value 1. The conjugate of t is then $1/t$. Thus the parametric equations of the unit circle are

(3) $$x = t, \quad \bar{x} = 1/t,$$

for the x-coördinate of a point is the same as the parameter. The only real parameters attached to real points of the circle are ± 1 which are cut out by the line $x - \bar{x} = 0$.

We need go no farther in this development here. Enough has been said perhaps to indicate the danger of dismissing imaginary coördinates as representing no points. The introduction of imaginary points into analytic geometry obviates a

---

[*] Complex refers to either real or imaginary numbers, a real number being self-conjugate.

multitude of exceptions which are distasteful to a geometer who believes that his subject is every whit as good as algebra.*

2. The number infinity in algebra. On the other hand, the algebraist who believes his subject is as good as geometry will be distressed at the failure of his fraternity to appropriate the advantage won by the geometer in the introduction of infinite elements. The geometer has found it convenient to postulate a point at infinity on every line or a line at infinity in every plane.** The most immediate consequence is that parallel lines may then be defined as lines which meet at infinity.† How can we make use of this definition and postulate in algebra? Perhaps the best approach is through homogeneous equations when infinite roots are placed on the same basis as zero roots. But we need not resort to homogeneous equations. We might begin by postulating a number ∞—call it an improper number if you like—which shall be the coördinate, in a Cartesian coördinate system along a line, of the "improper" point at infinity on the line.‡ This number is now clothed with properties to conform to those of its geometric counterpart. Thus the x-intercept of the line

(4) $$y = mx - k$$

is

(5) $$x_0 = k/m.$$

---

* One has but to glance through the recent book of Osgood and Graustein, *Plane and Solid Analytic Geometry*, to see how many such exceptions must be noted by an exacting writer. A conspicuous example is the important theorem (pp. 167–8) relating to a pencil of curves $u + kv = 0$. They divide the theorem into two parts according as $u = 0$ and $v = 0$ do or do not intersect. The first part has an exception when $u + kv = 0$ represents a single point (or a finite number of points). The second part reads: "Let $u = 0$ and $v = 0$ be the equations of two non-intersecting curves. Then the equation $u + kv = 0$ ($k \neq 0$) represents, in general, a curve not meeting either of the two curves." The beautiful positive theorem (part one) is thus replaced by a weak and purely negative theorem devoid of its chief interest. Moreover two exceptions are given (in addition to $k \neq 0$), viz., $u + kv = 0$ may represent (a) a single point, (b) no locus. Even then (b) must stand for such diverse cases as constant $= 0$ and $x^2 + y^2 + 2 = 0$. Finally a special case is given when $u = 0$ and $v = 0$ represent parallel lines. How much more satisfactory, having annexed the infinite and imaginary domain, to be able to say: $u + kv = 0$ (*always*) represents a curve through all the common points of $u = 0$ and $v = 0$, and, if $k \neq 0$ or ∞, through no other point of either curve. The trivial exception here applies only to the parameter whereas the exceptions in the text restrict the base curves themselves. Again one objects to a definition (p. 312) which confines the polar of a point outside a conic to the line segment cut off by the conic—especially when its equation is given, for the equation certainly represents the whole line. These and kindred exceptions due to the refusal of the authors to countenance imaginary and infinite elements and to assign values to indeterminates mar, in my judgement, an otherwise admirable book.

** I wish to insist that this is not projective geometry. Any geometry that isolates a line at infinity and invests it with exceptional properties is not projective. The line at infinity is considered in projective geometry because it serves as a bridge between projective and Euclidean metric properties.

† This definition suffices for either the plane or space since two intersecting lines necessarily lie in the same plane.

‡ This is precisely what is done in Veblen and Young, *Projective Geometry*, vol. 1, chapter 7—EDITOR.

Now if $m=0$, the line is parallel to the x-axis and cuts it therefore at infinity. Accordingly in virtue of (5) we attribute to ∞ the property

(6) $$k/0 = \infty, \qquad k \neq 0,$$

for if $k=0$ at the same time, the line (1) coincides with the x-axis and intersects it at every point.* It is then but a step to the theorem that the equation

(7) $$a_0 x^n + a_1 x^{n-1} + \cdots + a_r x^{n-r} + \cdots + a_{n-1} x + a_n = 0$$

has r infinite roots if $a_0 = a_1 = \cdots = a_{r-1} = 0$, so that an equation of apparent degree r may be regarded as an equation of degree n with $n-r$ roots infinite.

Although some writers explain that repeated and imaginary roots must be counted if an equation of degree n is to have n roots, few authors include infinite roots in the enumeration on the ground that the degree of the equation has been reduced. The recognition of infinite roots is however no less essential. For example, the student has learned to expect two solutions of a linear and a quadratic equation even though the roots be coincident or imaginary. Then he encounters two equations like

(8) $$x^2 - 9y^2 = 7,$$
(9) $$x = 3y + 1,$$

which on combination lead to

(10) $$9y^2 - 9y^2 + 6y = 6,$$

whence $y=1$, $x=4$. Again if he attempts to solve (8) with

(11) $$x = 3y,$$

he will get

(12) $$9y^2 - 9y^2 = 0 = 7,$$

a traditional absurdity. Do (8) and (9) then have a single solution while (8) and (11) have none? This will be the view of his textbook. The orthodox algebraist must say —if he does not avoid the cases in question—that pairs of linear and quadratic equations in two variables and with real coefficients fall into five classes as follows: those with two solutions which are (*a*) real and distinct, (*b*) real and equal, (*c*) conjugate imaginary, (*d*) those with one real solution, and (*e*) those with no solution whatever,† while his hyper-orthodox geometrical brother would say that a line cuts a conic in two points, one point or no point. In a sense the algebraic statement is above criticism—it is largely a question of the point of view. But the geometrical statement is not adequate since it affords no criterion for distinguishing a tangent from a line parallel to an asymptote, nor an asymptote from other lines which "do not cut the conic." Furthermore, it places tangent and asymptote

---

* Other properties may be assigned by considering the intercepts of the parabola $y = ax^2 + bx + c$, but this will suffice for our present purpose.

† If imaginary coefficients are taken into account, he would have even more varieties.

in entirely different categories whereas an asymptote might at least be defined as the limit of a tangent. The heterodox but progressive analytic geometer would say that *a linear equation and a quadratic always have two solutions and that a line always cuts a conic in two points*—real or imaginary, distinct or coincident, finite or infinite—thus placing algebra and geometry on an equal footing. Only by recognizing infinite as well as imaginary roots and both imaginary and infinite intersections can this theorem be generalized to two equations (or curves) of degrees m and n.

But we need not go to simultaneous equations and the geometry of two dimensions nor to the infinite region of the plane to find a sound reason for introducing infinite elements. The recognition of infinite elements in the geometry on a curve, which is essentially a geometry of one dimension, is extremely useful. For example, the equation of a circle with radius 1 and center at the origin can be written parametrically in rectangular coördinates

(13) $$x = \cos\theta = (1-t^2)/(1+t^2),$$
$$y = \sin\theta = 2t/(1+t^2),$$

where $t = \tan(\theta/2)$. To find the intersections with the x-axis, we set $y = 0$ whence $t = 0$, $x = 1$. But obviously the x-axis cuts the circle in a second point $(-1, 0)$. We must say then (*a*) there is one point on the circle with no parameter, (*b*) equations (13) represent a circle exclusive of the point $(-1, 0)$ or (*c*) the x-axis cuts the circle in a second point with parameter $t = \infty$. Is there anyone who will not admit that on considerations of continuity, generality and elegance the last statement is superior? Isn't it more in harmony with the spirit of modern mathematics and is it any less intelligible to the student?

It frequently happens in Cartesian geometry when curves are written parametrically that the parameter ∞ is attached to a point in the finite part of the plane. Another familiar instance is the folium of Descartes,

$$x = \frac{at}{t^3+1}, \quad y = \frac{at^2}{t^3+1},$$

the double point of which is located at the origin but has the parameters 0 and ∞. If θ is the parameter on a curve and x and y are rational functions of $\sin\theta$, $\cos\theta$, $\tan\theta$, then the coördinates of the points on the curve can be expressed as rational functions of $\tan\theta/2$ by means of the relations in (13). There will be one point on the curve with the parameter ∞ and this point is as likely to be in the finite as in the infinite region.

3. Infinite roots and inconsistent equations. The solution of the simultaneous equations:

(14) $$a_1 x + b_1 y + c_1 = 0,$$
$$a_2 x + b_2 y + c_2 = 0,$$

may be written

(15) $$x = \frac{\begin{vmatrix} b_1 & c_1 \\ b_2 & c_2 \end{vmatrix}}{\begin{vmatrix} a_1 & b_1 \\ a_2 & b_2 \end{vmatrix}} \equiv \frac{A}{C}, \qquad y = \frac{\begin{vmatrix} c_1 & a_1 \\ c_2 & a_2 \end{vmatrix}}{\begin{vmatrix} a_1 & b_1 \\ a_2 & b_2 \end{vmatrix}} \equiv \frac{B}{C}.$$

Obviously the solution is unique unless $C=0$. If we examine the possibilities when $C=0$, we recognize two cases: (I) A and B not both $=0$ and (II) $A=B=0$.

Case I. $C=0$, $A \neq 0$. The equations are commonly said to have no solution. There are three characteristic hypotheses depending on the way in which C becomes zero.

(a) $a_1 = b_1 = 0$. Then $A = -b_2 c_1$, $B = c_1 a_2$. The intercepts of the first line are $-c_1/a_1$ and $-c_1/b_1$ which under our hypothesis become infinite and the equation of the first line takes the form

(16) $$0x + 0y + 1 = 0.$$

In other words the line has moved off to infinity and (16) may be regarded as the equation of the line at infinity. Since A is different from zero $b_2, c_1 \neq 0$ and the other line remains finite. The two lines thus meet at infinity,—at the infinitely distant point on the second line.

(b) $a_1 = a_2 = 0$. Then $A = b_1 c_2 - b_2 c_1$, $B = 0$ and the two lines reduce to $b_1 y + c_1 = 0$ and $b_2 y + c_2 = 0$. Since $A \neq 0$ the two lines exist and remain distinct. They are manifestly parallel to the x-axis (or one is parallel to the x-axis and the other coincident with it) and they meet therefore at the infinitely distant point of that axis.

(c) $a_1/a_2 = b_1/b_2 \neq c_1/c_2$. Barring hypotheses (a) and (b) the equations assume the form

$$a_1 x + b_1 y + c_1 = 0, \qquad a_1 x + b_1 y + c_2 = 0,$$

whose solution is $x = \infty$, $y = \infty$. The lines are parallel and meet at infinity.

Under each hypothesis of case I we have found that the two equations may be interpreted as representing distinct lines which meet in a point at infinity. The coördinates of this point will be a solution of the equations and we no longer have pairs of inconsistent linear equations. Or we may retain the term "inconsistent," introducing the definition: The two linear equations (14) are inconsistent when and only when they represent lines which meet at infinity.

Under case II, $A = B = C = 0$ when the lines coincide and the equations have a single infinity of solutions. If however $a_1 = a_2 = b_1 = b_2 = 0$, the lines coincide with the line at infinity and the equations would then be "inconsistent."

Three linear equations. Similarly the solutions of the three equations

(17) $$a_i x + b_i y + c_i z + d_i = 0, \qquad i = 1,2,3,$$

may be written $x = A/D$, $y = B/D$, $z = C/D$. If $D = 0$ but not all the numerators are zero, the values of one, two, or all of the unknowns will be infinite. The equations will then be "inconsistent" as before. Geometrically, the three planes meet in a unique point on the plane at infinity.*

If $A = B = C = D = 0$, the equations are dependent and have ∞^1 or ∞^2 solutions according as the planes have a common line or coincide. Such equations however may have "no solution" according to the orthodox view, as, e.g., the three equations

(18) $$ax + by + cz = d_i, \qquad i = 1,2,3, \qquad d_1 \neq d_2 \neq d_3.$$

These three planes are parallel and hence have their infinitely distant lines in common. We are led to extend our definition of inconsistent equations as follows: Three non-homogeneous linear equations in three variables, whether dependent or independent, are inconsistent when and only when they represent planes which meet only at infinity.

A general definition of inconsistent equations. These results may be generalized for a set of n linear equations in n variables.† Indeed the definition is valid when there are fewer equations than variables. Thus the equations of two parallel planes would be inconsistent. Furthermore any set of equations in linear form will be inconsistent under similar conditions. For example, the equations

(19) $$\begin{aligned} ax^2 + by^2 + c_1 &= 0, \\ ax^2 + by^2 + c_2 &= 0, \end{aligned} \qquad c_1 \neq c_2,$$

are linear in x^2 and y^2. Subtracting we should get constant $= 0$ so that the equations are inconsistent in the usual sense. But using formula (15) we find $x^2 = \infty$, $y^2 = \infty$ and we may say that the equations have two pairs of equal roots, i.e., the curves have double contact at infinity.

Two equations need not be of the same degree to be inconsistent. An instance of this already noticed is the equation of a hyperbola and that of its asymptote. Another example is furnished by the equations

(20) $$\begin{aligned} y^3 - xy^2 - x^2 + 2xy &= 0, \\ y^2 - y + x + 1 &= 0, \end{aligned}$$

which, on elimination of x, yield

(21) $$0y^4 + 0y^3 + 0y^2 + 0y - 1 = 0.$$

The equations which should have six solutions thus have no finite solutions and the corresponding curves meet wholly at infinity. We may now formulate a general criterion for inconsistent equations:

---

* There are three possibilities: (1) one plane cuts the other two and is parallel to their line of intersection, (2) two of the planes are parallel and the third a finite plane, (3) two planes meet in a finite line while the third is the plane at infinity.

† This demands of course the assumption that the locus of points at infinity in a linear space S_n of n dimensions is an S_{n-1}.

r non-homogeneous (*dependent or independent*) *equations in n variables*, $r \leq n$, *are inconsistent when and only when the loci of the equations intersect, whether in real or imaginary points, wholly at infinity*.

The purpose of this discussion has been not merely to criticize the shortcomings of our current elementary textbooks in respect to imaginary and infinite elements but to illustrate how these fruitful ideas can be employed to enrich both algebra and geometry. The suggestions here embodied are consonant with the spirit that pervades modern mathematics and with sound European tradition. It is my conviction that they could be incorporated into our elementary texts without sacrificing either clearness or rigor and at no greater cost of space than is required to detail the exceptions which they eliminate.

INFINITE AND IMAGINARY ELEMENTS IN ALGEBRA AND GEOMETRY*

TOMLINSON FORT, University of Alabama

Since the things advocated by Professor R. M. Winger in his article bearing the above title and appearing in a recent number of the MONTHLY (*1922*, 290–296) are at variance with the general practice followed by me in my teaching of freshman and sophomore classes, I am prompted to comment on what he says.

In my opinion one of the duties of the teacher of elementary mathematics is to attempt to rid the subject of mysticism and to make it appear at all times in conformity with the student's experience. Like most teachers I welcome elementary uses for imaginary numbers but can not believe the first course in analytic geometry the place for them. To the freshman, points and lines are more or less the pictures that he sees and are not ideals. His analytic geometry is truly a geometry of reals and for the instructor to confine his teaching to the "real domain" seems to me sound mathematically and to rid the subject of that element of mystery which I do not believe it possible to eliminate when trying to teach geometry with imaginary elements to such immature students. I never feel that I am telling the freshman a half truth when I tell him that $(0,0)$ is the only point satisfying $x^2+y^2=0$ or that $x^2+y^2+1=0$ has no locus. I think that imaginary elements should be introduced at a later time when the whole subject can be subjected to a more critical examination. The same is true of infinite regions of the plane. To bring a class of freshmen to understand and remember their ideal nature seems to me out of the question. I simply never mention the subject: Parallel lines do not meet. The parabola is an open curve and division by zero is never permitted.

It is with a kind of horror that I read where the author advocates the postulation of "*The Number Infinity*" defined by $k/0=\infty$, $k \neq 0$. How I fight division by zero in my classes! How many times have I told my students that 90° has no tangent and have forbidden this same infinity! To describe the behavior of

---

* From AMERICAN MATHEMATICAL MONTHLY, vol. 30 (1923), pp. 255–256.

the roots of an equation,

$$a_0 x^n + \cdots + a_n = 0; \qquad a_0 \neq 0,$$

when $a_0, \ldots, a_{r-1}$ approach zero seems to me sound teaching, but to speak of the equation as having r infinite roots when $a_0 = \cdots = a_{r-1} = 0$ and the equation is only of the $(n-r)$th degree, as the author advocates, is to me very bad. It is just that from which I thought American mathematics was growing. The author states that what he advocates is in accordance with "*sound* European tradition." If I understand him correctly, I am sure in particular that his infinity was not in use in the teaching at Göttingen in 1912 or in Paris in 1913 when I was a student at those universities. Also in my opinion he will find that it has been eliminated from the better teaching in England. With reference to infinity, I think the best policy in elementary teaching is to explain carefully the various unending processes that arise and in particular to make it clear that there is no largest real number. In more advanced courses where infinite elements are introduced care should be taken to explain their ideal character and to show just why it is wise to postulate them in exactly the form that it is done.

THE INFINITE AND IMAGINARY IN ALGEBRA AND GEOMETRY: A REPLY[*][†]

W. L. G. WILLIAMS, Cornell University

"The traditional treatment of imaginary and infinite elements in algebra and geometry," says Professor R. M. Winger in a recent number of this MONTHLY, "has curiously resisted the reform movement in American text-book writing." What this reform movement is, where it exists, and what text-books its adherents have written, we do not know and until we can examine its products we should hesitate to call it a reform movement.

However, the two questions that Professor Winger raises, the question of the introduction into our elementary college courses in algebra of a number ∞ and the question of considering, in our elementary courses in analytic geometry, imaginary as well as real points, are so different that they require separate examination.

1. Professor Winger points out the service of a line at infinity in rendering the principle of duality universally valid and in preserving a one-to-one correspondence between a figure and its projection. This idea of a line at infinity, which Poncelet introduced into geometry just a century ago, has nevertheless given rise to many misconceptions and errors. One form of the equation of the line at infinity is the one which appears in the article under discussion, viz.,

$$0x + 0y + 1 = 0,$$

---

[*] From AMERICAN MATHEMATICAL MONTHLY, vol. 30 (1923) pp. 384–391.
[†] V. Infinite and Imaginary Elements in Algebra and Geometry, this MONTHLY (*1922*, 290).

and another is
$$1 = 0,$$
equations satisfied by no real or imaginary points. For mathematics as it now exists these two equations are identical and equally meaningless as equations of lines or of any other loci. The second of these equations is evidently quite independent of x and y, but it has been proposed by Professor Winger and others to introduce into algebra one or more infinite numbers with the property that their products by zero are different from zero in order to give a meaning to the first of these equations.

Professor Winger is right in believing that if infinite numbers can render a service in analytic geometry they can also do so in algebra, since analytic geometry is an application of algebra. For their use in either we should require a logically constructed theory of the system of numbers obtained by adjoining to the real system or the complex system one or more infinite numbers.

The most important properties of the real and complex number systems are, perhaps, that addition, subtraction, and multiplication in them are always possible and always unique, that the same is true of division when the divisor is different from zero, and that the result is always a number of the system. These properties are not inherent in the nature of a number system, but the systems which do not have them seem so bizarre to all except the professional mathematician and are so unimportant relatively that we find it hard to believe that such systems should have any part in the curriculum of our colleges.

Professor Winger proposes to create a number system in which there shall exist a single infinite number ∞ defined by the relation

$$\frac{k}{0} = \infty, \qquad (k \neq 0).$$

What meaning is to be attached to the words, "k divided by zero"? By the words, "six divided by two," we mean the number x such that $2x = 6$, so that division is the operation inverse to multiplication. In every number system familiar to the present writer in which division has a meaning it is the inverse of multiplication, which is an operation with a definite meaning. If Professor Winger intends to follow the analogy he will say

$$k = 0 \cdot \infty = \infty \cdot 0, \qquad (k \neq 0),$$

but we find it difficult to imagine what this means. It certainly does not mean

$$k = 0 + 0 + 0 + \cdots,$$

for $k \neq 0$. But whatever it may mean we have, for example (operating formally), $0 \cdot \infty = 1, 0 \cdot \infty = 2, \ldots$. In fact, $0 \cdot \infty = $ any number of the system except 0 and possibly ∞.

Such a system of numbers, if it has any uses, is so much more difficult to deal with than quaternions or even linear non-associative algebras that we can hardly consider its study fitting for elementary students.

After laying down the above definition, but without enunciating any other laws of operation, Professor Winger states: "It is then but a step to the theorem that the

THE INFINITE AND IMAGINARY IN ALGEBRA AND GEOMETRY: A REPLY

equation
$$a_0 x^n + \cdots + a_n = 0$$
has r infinite roots if
$$a_0 = a_1 = \cdots = a_{r-1} = 0$$
so that an equation of apparent degree $(n-r)$ may be regarded as an equation of degree n."

Granting that the phrase "apparent degree $(n-r)$" has a meaning, I do not know what it means to say that this equation has r infinite roots, but I am unable to prove that it has one infinite root. In fact there is one simple case in which it has none.

In making this proof I assume that a root of an equation is a number which satisfies it. Consider the equation
$$0 \cdot x + 0 = 0$$
which satisfies all of Professor Winger's conditions. Substituting ∞ for x we have $k + 0 \neq 0$ since by the definition $k \neq 0$. If any coefficient is different from zero the theorem cannot be verified, e.g., consider the equation
$$0 \cdot x^2 + x = 0.$$
This equation is satisfied if and only if $0 \cdot \infty^2 + \infty = 0$, i.e., if and only if $k \cdot 0 + \infty = 0$, i.e., if and only if $(k+1)\infty = 0$.* We do not know the value of k except that it is not 0; if $k = -1$ this necessary and sufficient condition reduces to $0 \cdot \infty = 0$, which is untrue since $0 \cdot \infty = k \neq 0$ by hypothesis. If $k \neq -1$ or 0 we have no means of knowing whether or not the equation is satisfied. Thus we see that it is impossible to verify Professor Winger's theorem even in an extremely simple case.

A useful conception in respect to points at infinity is that there is one and only one on every straight line, and that all straight lines parallel to a given straight line pass through the same point at infinity.† Let us see how this fits in with Professor Winger's number system. In his system there exist the following points at infinity: $(c, \infty)(\infty, c)$ and (∞, ∞) where $c \neq \infty$. A point (c, ∞) lies on the line $x + y = 0$ if and only if $c + \infty = 0$ and a point (∞, c) lies on this line if and only if $\infty + c = 0$. Since $c \neq \infty$, no points with the coördinates (∞, c) or (c, ∞) lie on $x + y = 0$. The point (∞, ∞) lies on it if and only if $\infty + \infty = 0$.‡ Whether or not this is true Professor Winger's definition of ∞ does not enable us to say. If, however, $\infty + \infty = 0$, and if $\infty^2 = \infty$ (since formally $\infty^2 = (1/0)^2 = (1/0) = \infty$), then ∞ is a root of the equation $x^2 + x = 0$, and, consequently, this equation has three roots, 0, 1, and ∞. Accordingly, either we have a line (and indeed an infinite number of lines) which contains no point at infinity or we have an equation of the second degree with three roots, so that Professor Winger's system leads us even in the simplest kinds of cases to

---

* Formally, $(k+1)\infty = (k+1)\dfrac{1}{0} = \dfrac{k+1}{0} = \infty$, when $k \neq -1$.

† See Winger, loc. cit., page 296.

‡ Probably $\infty + \infty = \infty$, for $\infty + \infty = \dfrac{1}{0} + \dfrac{1}{0} = \dfrac{2}{0} = \infty$; possibly $\infty + \infty = \dfrac{1}{0} + \dfrac{-1}{0} = \dfrac{0}{0}$.

results inconsistent with the general theorems and postulates whose universality he seeks to maintain.

The author in the incompleteness and vagueness of his statements about his number system is but typical of those who talk about a number ∞. Turn to a work on the theory of functions of a real variable like that of Pierpont or de la Vallée Poussin and you find that the greatest care is taken to lay the foundations of the real number system; turn to a work on the complex variable like that of Burkhardt or Pringsheim and you find no vagueness as to this system; quaternions and other unusual systems receive an exact treatment in Dickson's Cambridge Tract; but those who talk of the number ∞ act as if all that was needed was to define it in a formal way.

Since, for the purposes of algebra and analytic geometry there is no logically constructed system in which ordinary real and infinite numbers exist side by side, we must reject not only from our teaching but from all our mathematical work any such "system" until and unless such a system shall be created, and thus avoid the paradoxes which arise in the works of those who attempt to make use of infinite numbers.*

A great number of text-books contain such misstatements and misconceptions regarding the so-called infinite that the honest student is bewildered and discouraged by them. These errors and deceptive statements arise from the use of infinite numbers, from imperfect realization of the fact that the infinite regions of different geometries† are different, and from a desire to claim more for mathematics than is justified by the facts.

Listen, for example, to Salmon, that most inspiring of romanticists: "Let A and B be both $=0$, then the intercepts become infinite and the line is altogether situated at an infinite distance from the origin. Now it was proved (Art. 63) that the equation under consideration is equivalent to $0x + 0y + C = 0$, and though it cannot be satisfied by any finite values of the coördinates, it may be by infinite values since the product of nothing by infinity may be finite. ******; and that the equation of an infinitely distant right line, in Cartesian coördinates, is $0x + 0y + C = 0$. We shall, for shortness, commonly cite the latter equation in the less accurate form $C = 0$."

Another author, whose book is in many respects interesting and excellent, solves the equations

$$y = \pm \left(\frac{b}{a} - \epsilon\right)x$$

---

* Bôcher: *Bulletin of the American Mathematical Society*, vol. 11 (1904–05), page 134. "Finally there is what may perhaps be called the method of optimism which leads us either wilfully or instinctively to shut our eyes to the possibility of evil. Thus the optimist who treats a problem in algebra or analytic geometry will say, if he stops to reflect on what he is doing: 'I know that I have no right to divide by zero; but there are so many other values which the expression by which I am dividing might have that I will assume that the Evil One has not thrown a zero in my denominator this time.' This method, if a proceeding often unconscious can be called a method, has been of great service in the rapid development of many branches of mathematics, though it may well be doubted whether in a subject as highly developed as is ordinary algebra it has not now survived its usefulness."

† Bôcher: Infinite Regions of Various Geometries, *Bulletin of the American Mathematical Society*, vol. 20 (1913–14), page 185.

THE INFINITE AND IMAGINARY IN ALGEBRA AND GEOMETRY: A REPLY

with
$$\frac{x^2}{a^2} - \frac{y^2}{b^2} = 1$$
and finds that each of these lines cuts the hyperbola in points whose x coördinates satisfy the relation
$$x^2 = \frac{1}{\frac{2\epsilon}{ab} - \frac{\epsilon^2}{b^2}}.$$

He then adds: "These two lines then tend as ϵ becomes smaller and smaller to become tangents to the hyperbola, for the points in which either of them meets the hyperbola are at a very great distance and they are on opposite sides of the centre. And it is one of the paradoxes of geometry with which the student is probably acquainted that the points on a line in two opposite directions and at a very great distance tend to become the same point."

This treats geometry as an experimental science in which the physical phenomenon of points moving in opposite directions and yet eventually almost colliding has already startled the young student.

Another author writes: "Thus of the coördinates x, y, z, certainly two and possibly all three are infinite and the point x, y, z is at infinity. We see then that two lines may be so situated that instead of a finite intersection they have their intersection at infinity."

There is not a hint of axiom, definition, postulate; in books of this character infinite points, a line at infinity, circular points at infinity, are as real, as much a part of the physical universe, as thrilling, as Robinson Crusoe to a boy of ten. Salmon would no more have given away the game than would Defoe have printed on the titlepage of his immortal work: "Small boys are hereby warned that no such person as Robinson Crusoe ever existed. Those who think he did, do so at their own risk."

The algebraist who reads these works, the last of which was printed in the present century by one of the world's greatest presses, will hardly cry out in envy of the geometer: "My subject is every whit as good as his; why can't I take over a little of this wonderful thing called infinity and make my subject fascinating too?"

At the top of page 294 Professor Winger says: "For example, the student has learned to expect two solutions of a linear and a quadratic equation even though the roots be coincident or imaginary." If the student has learned to expect any such thing the business of his teacher is to show him that he has been misled, not to try to invent some system of numbers which he will not understand in order apparently to justify a totally false expectation. Neither in the domain of real numbers nor in the domain of complex numbers nor in the system obtained by Professor Winger do we know that
$$x^2 - 9y^2 = 7,$$
$$x = 3y + 1$$
have two solutions. In the real domain they have but one, in the complex domain they have one and only one, and we have no means of knowing whether (∞, ∞) is

another solution in Professor Winger's domain, but if so it is of very little interest for algebra in the ordinary sense or for analytic geometry.

Professor Winger seems to imply that without taking his point of view it is impossible to tell which of

$$x = 3y + 1,$$
$$x = 3y$$

is an asymptote to

$$x^2 - 9y^2 = 7.$$

It is true as Professor Winger states (page 294) that an asymptote may be defined as the limit of a tangent (if it exist), but the method of treatment of asymptotes which he suggests has nothing to do with limits. It is in various ways easy to lay down a rule of thumb method for finding asymptotes, but modern mathematics has nothing to do with rule of thumb methods. The principles of finding asymptotes, whether we define one as Professor Winger suggests, or as a line whose distance from a point on the curve$\to 0$ as one or both of its coördinates$\to\infty$, depend upon a knowledge of limits and of infinity in the sense of all analysts since Cauchy.

With the word "infinity" are bound up some of the most important ideas in mathematics. There are points at infinity in analytic geometry, but their coördinates are not infinite numbers; there is a line at infinity, but its equation is not $0x + 0y + 1 = 0$; there are infinite collections of objects, but the sum of an infinite number of zeros is not a number different from zero. The experience of many generations of mathematicians has shown that the ideas of limit and infinity present important psychological difficulties. A Bernoulli could say: "If there are an infinite number of terms (in an infinite sequence) there must be an infinitieth, for the same reason that if there are ten there must be a tenth." An Euler could say that

$$\tfrac{1}{2} = 1 - 1 + 1 - 1 + 1 - 1 + \cdots$$

and the haziness of his notions is not excused by modern developments in divergent series. Though Abel at twenty called divergent series an invention of the devil even his great genius could not bring order out of the chaos, a chaos which was not dispelled until Cauchy and Weierstrass had devoted long lives to reconstructing the foundations of mathematics.

With regard to the teaching of these difficult ideas to students G. H. Hardy in the preface to that excellent book, "A Course of Pure Mathematics," says: "It has been my good fortune during the last eight or nine years to have a share in the teaching of a good many of the ablest candidates for the Mathematical Tripos; and it is very rarely indeed that I have encountered a pupil who could face the simplest problem involving the ideas of infinity, limit or continuity with a vestige of the confidence with which he would deal with questions of a different character and of far greater intrinsic difficulty. I have indeed in an examination asked a dozen candidates, including several future Senior Wranglers, to sum the series $1 + x +$

$x^2\cdots$ and not received a single answer that was not practically worthless—and this from men quite capable of solving difficult problems connected with the curvature and torsion of twisted curves.

"I cannot believe that this is due solely to the nature of the subject. There are difficulties in these ideas, no doubt; but they are not so great as many other difficulties inherent in mathematics that every young mathematician completely overcomes. The fault is not that of the subject or of the student but of the textbook and the teacher. It is not enough for the latter, if he wishes to drive sound ideas on these points well into the mind of his pupils, to be careful and exact himself. He must be prepared not merely to tell the truth but to tell it elaborately and ostentatiously. He must drill his pupils in 'infinity' and 'continuity' with an abundance of written exercises and examples as he drills them at present in poles and polars or symmetric functions or the consequences of De Moivre's theorem. Then and only then he may hope that accurate thought in connection with these matters will become an integral part of their ordinary mathematical habit of mind."

Mathematics is a division of knowledge which is worthy of the best efforts of the ablest student. It embraces a body of facts which are interesting and definite, and it does not need either to claim what it cannot do in order to make itself popular or in elementary teaching to bring into question statements as to whose validity doubt exists. Mathematics, as Whitehead says, has nothing to do with mysteries except to explain them. The duty of the teacher is not to introduce mysteries, but to make plain the great simple facts of mathematics. Instead of an introduction into our teaching of the number infinity our great task is to get rid of it. Half of our freshmen have already learned in high school that $(1/0) = \infty$, that an equation $ax^2 + bx + c = 0$ in which $a = 0$ has the root infinity, and many other things about "infinity" that are either meaningless or false, but which tickle their fancy, excite their imagination, and make it almost impossible for them to learn mathematics. No *deus ex machina* in the form of infinite numbers can replace the hard honest work which is necessary for teacher and student in order that the latter may acquire clear and logical ideas with respect to these fundamental concepts of mathematics.

2. In dealing with imaginary numbers we are on entirely different ground from that traversed in the previous section. There is a perfectly logical method of introducing ordinary complex numbers into algebra and geometry, just as there is a logical method of introducing quaternions or numbers of any linear algebra. The methods which Professor Morley has used are particularly interesting, but the most interesting thing about them is that they give through the use of complex numbers facts about figures in a real geometry. Now that is exactly the opposite of the state of affairs that confronts us in the use of imaginary numbers in the ordinary Cartesian system.

The only questions that arise are those of interest and point of view. Why, for example, should Professor Winger criticize those who call

$$x^2 + y^2 = 0$$

a "point-circle" or single point rather than those who say that

$$x^2+y^2+1=0$$

is an imaginary circle, when they might be more original and entertaining and say that it represents the two quaternion straight lines

$$x+yi+j=0,$$
$$x-yi-j=0?$$

To most people, of whom the writer is one, as long as we lack a space of four dimensions to enable us to plot all the points, real and imaginary, which satisfy an equation in two variables, it will be more interesting to confine themselves to points in the real plane, and they will not consider that they are telling only part of the truth when they say that

$$x^2+y^2=0$$

represents only a single point.

INFINITE AND IMAGINARY ELEMENTS IN ALGEBRA AND GEOMETRY: REPLY TO CRITICISMS*

R. M. WINGER, University of Washington

Two criticisms of my article** have been printed in the MONTHLY, one by Professor Tomlinson Fort,[†] the other by Professor W. L. G. Williams.[‡] Both these gentlemen express emphatic disagreement with my suggestions, partly on the grounds of pedagogy, but chiefly as it seems to me because they regard half of the program as mathematically unsound. Professor Williams indeed by a detailed analysis of several hypothetical cases attempts to prove the mathematics faulty. But his proofs, as I shall point out, are based on a misinterpretation of my paper. For he quotes me as *defining* a number infinity by the relation $k/0=\infty$, where k cannot be zero.

To answer his argument, I must review a part of my original statement. I first recall briefly the service of infinite elements in modern geometry and, assuming the geometry as known, suggest that the advantages be carried over into algebra. I say that perhaps the best approach is through homogeneous equations but that we might begin by postulating an "improper number" ∞ which should play in algebra a rôle analogous to that played in geometry by the "improper point" at infinity on the line. The next few lines involve the heart of the matter at issue: "This number is now clothed with properties to conform to those of its geometric counterpart. Thus

---

* From AMERICAN MATHEMATICAL MONTHLY, vol. 31 (1924), pp. 383–387.
**This MONTHLY, September, 1922, p. 290.
† *Ibid.*, July–August, 1923, p. 255.
‡ *Ibid.*, November, 1923, p. 384.

the x-intercept of the line

(4) $$y = mx - k$$

is

(5) $$x_0 = k/m.$$

Now if $m = 0$, the line is parallel to the x-axis and cuts it therefore at infinity. Accordingly, in virtue of (5) we attribute to ∞ the property

(6) $$k/0 = \infty, \quad k \neq 0,$$

for if $k = 0$ at the same time, the line (4) coincides with the x-axis and intersects it at every point."

I cannot see how Professors Fort and Williams find in this a proposal to define the symbol ∞ by the single relation (6).* For, attributing a property to a symbol is quite a different thing from defining it. I gave in fact no formal definition at all. I only postulated the existence of the symbol ∞, attributed to it one property and stated a theorem about infinite roots of equations—for this was all I had occasion to use in the illustrations considered in my paper. Other properties than (6) are however implied by the context, expecially by the phrase "*clothed with properties to conform to those of its geometric counterpart.*" Further I remark in a footnote that other properties may be assigned by considering the intercepts of the parabola $y = ax^2 + bx + c$.

Indeed relation (6) does not even say that k cannot be zero, a restriction that Professor Williams uses at least five times.† What it does say is that $k/0 = \infty$, if $k \neq 0$. If $k = 0$ the expression is indeterminate and might have any value including ∞, since the line (4) then coincides with the x-axis and meets it at each of its points, including the point at infinity. In other words $0 \cdot \infty$ is also indeterminate and might have the value 0, contrary to Professor Williams' assertion. This disposes of the first "contradiction" that arises when he seeks to show that $0 \cdot x + 0 = 0$ cannot have an infinite root.

Again when he finds by formal algebraic manipulation that there is doubt whether a certain line contains a point at infinity, a result in conflict with the geometry, what has he learned? Merely that the alleged "definition" (6) is inadequate. But why does Professor Williams ignore the qualification in my paper that the algebra *must conform to the geometry?* He might have found, I should think, quite as much mental recreation in ascertaining what additional property must be ascribed to the symbol ∞ in this case to make it fit the geometry as in speculating on what certain expressions mean in order to catch me in a logical inconsistency.

Since Professor Williams objects that my little sketch falls short of the treatment of number systems in books on the theory of functions, it will be instructive

---

* Professor Williams says that I propose "to create a number system in which there shall exist a single infinite number ∞ defined by the relation (6)" while Professor Fort states: "It is with a kind of horror that I read where the author advocates the postulation of 'The Number Infinity' defined by (6)."

† In such statements as : "for $k \neq 0$, since by definition $k \neq 0$, we do not know the value of k except that it is not 0, which is untrue since $k \neq 0$ by hypothesis."

to see how an authority of his own choosing handles the question under discussion. Turning to Burkhardt,* §12, I find this statement: "*In addition to the complex numbers and their symbols already introduced we introduce now a new one, 'infinity,' with the symbol ∞, which is to be regarded as the result of the division $1/0$.*" This he parallels with a geometrical definition of a point at infinity.** He then goes on to qualify the symbol by several other conventions and points out that $\infty \pm \infty$, $0 \cdot \infty$, and ∞/∞ as well as $0/0$ are indeterminate forms. A few lines farther on he says: "According to conventions of this kind, certain words and symbols previously defined are assigned a wider meaning. That this procedure is permissible we have repeatedly stated in the first chapter; that it is useful is justified by results." Still later he reconciles this view with the meaning of the symbol as used in the theory of limits and proves that: "*These two views of the symbol ∞ as applied to rational functions are not contradictory; and every such equation $f(z_0) = \infty, f(\infty) = w_0, f(\infty) = \infty$ which is true from the one point of view is also true from the other.*"

Now let Professor Williams turn to the strictly logical development of projective geometry by Veblen and Young† and he will find no vagueness about the symbol ∞ as used in my article. When establishing the non-homogeneous coordinate system in one dimension they say: "For the exceptional point P_∞, let us introduce a special symbol ∞ with exceptional properties, *which will be assigned to it as the need arises*. Some five or six properties, which are essentially the same as those of Burkhardt, are then assigned from the theory of collineations."

Professor Fort challenges the statement that my suggestions are in harmony with sound European tradition.‡ I trust that I have in a measure divested "my infinity" of its "horror" for Professor Fort and I must remind him that my remark *applied to the imaginary as well as the infinite*. Let me adduce briefly the evidence.

First I find this statement emanating from Göttingen which all but convicts me of what Sylvester once called unconscious plagiarism:§ "These two equations [a linear and a quadratic homogeneous in three variables] have then precisely two common solutions [*Lösungstripel $x:y:z$*], *i.e., a curve of the first and one of the second order always intersect in exactly two points, which indeed may be real or complex, finite or infinite, distinct or coincident."* "

Again in a textbook originating in Paris†† and intended for pupils from fourteen to seventeen years old, after a very careful discussion of the equation $ax = b$ when $a = 0$, $b \neq 0$,—a discussion to which the most meticulous cannot take ex-

---

* *Theory of Functions of a Complex Variable*, Rasor Translation.

** He uses the convention of a point instead of a line at infinity in the plane, but the *principle* of introducing corresponding concepts into analysis and geometry is the same as I advocated.

† *Projective Geometry*, vol. I, p. 54 ff. They note however that the symbol ∞ does not represent a number of a *field* as defined earlier. The Editor in a footnote to my paper called attention to this treatment but Professor Williams omits to mention it.

‡ He says: "I am sure in particular that his infinity was not in use in the teaching at Göttingen in 1912 or in Paris in 1913 when I was a student at those universities."

§ See my paper, this MONTHLY (*1922*, 294).

" Klein, *Elementarmathematik vom höheren Standpunkte aus*, Teil II, p. 244.

†† E. Borel-P. Stäckel, *Die Elemente der Mathematik*, I: *Arithmetik und Algebra*.

ception,—they summarize (p. 214) as follows: "*We say then that for $a=0$, x is infinitely great and indicate this solution by the symbol ∞.*" The authors do not consider imaginary points but neither do they use imaginary numbers in the solution of quadratics. They do however introduce the point at infinity on a line (p. 177) and assign to it a coördinate and they denote one point on the line by the symbol ∞—and this in a book on arithmetic and algebra.

O. Staude (*Analytische Geometrie*) introduces the point at infinity on a line as well as imaginary points, discusses infinite roots of a quadratic equation and assigns the label ∞ to one point of the line.

Chrystal* devotes ten pages to the interpretation of infinite solutions of equations. For the equation $ax+b=0$, when $a=0$, $b\neq 0$ he obtains an infinite root much as we assigned the property (6) to the symbol ∞. Again in treating the two linear equations

$$ax+by+c=0, \qquad a'x+b'y+c'=0$$

he obtains as the condition for parallelism of the lines $ab'-a'b=0$ and proceeds: "We have thus fallen on the excepted case of §§ 4 and 5. If we assume that the results of the general formulae obtained for the case $ab'-a'b\neq 0 \cdots$ hold also when $ab'-a'b=0$, we see that in the present case neither of the numerators can vanish. $\cdots$ It follows then that

$$x=\frac{bc'-b'c}{0}=\infty, \qquad y=\frac{ca'-c'a}{0}=\infty;$$

and the analytical result agrees with the graphical."

These quotations are to be read like all other mathematical statements,—not excluding those occurring in discussion columns,—in the light of the context and applied as Chrystal warns "with a proper regard to accompanying circumstances."

But half of my proposal,—and not the lesser in importance,—concerned the imaginary in geometry. Instead of calling $x^2+y^2+1=0$ an imaginary circle as I advocate, Professor Williams thinks it would be "more original and entertaining" to call it a pair of quaternion lines. Perhaps he is right—I had not considered the entertaining possibilities. Again he tells us that "to most people, of whom the writer is one,... it will be more interesting to confine themselves to points in the real plane, and they will not consider that they are telling only part of the truth when they say that $x^2+y^2=0$ represents only a single point."

To many geometers the question of which elements of a curve or configuration are real and which imaginary is a matter of comparative indifference—the whole figure is their concern. Certain it is that no distinction is made in such fundamental formulas as Plücker's equations relating to the singularities of curves. Some of the most beautiful configurations in geometry, as for example that associated with the flexes of the plane cubic, are largely imaginary. And I regard it as a kind of hyper-esthetic sense that enables the mathematician to contemplate the intricate symmetry and beauty of a geometric figure which transcend the ordinary senses. I

---

* *Algebra*, vol. I, Second Edition, pp. 385–395.

cannot profess to speak for most geometers, to say nothing of most mathematicians or most persons, but such geometers as Poncelet, Chasles, Darboux, Plücker, von Staudt, Lie and Study have concerned themselves with the imaginary in geometry so that one must conclude that they did not find it wholly devoid of interest. In particular, Salmon, Cayley, Clebsch, Klein, Pascal's *Repertorium* and the mathematical encyclopaedias are on record as referring to a conic of the elliptic type and vanishing discriminant as a pair of imaginary lines. I am content to stand in such geometric company as this.

Finally, not to lose sight of the purpose of my original paper, let me recall the opening lines of Klein's *Higher Geometry*:* "We distinguish in general two kinds of geometry, the synthetic geometry, which considers the figures themselves, and the analytic geometry which builds up its system essentially with the aid of analysis. Besides these two kinds of geometry, we may construct still a third kind which in a sense is the inverse of the two and which will form the subject of the present course of lectures. Thus whereas we ordinarily apply analysis to geometry, it shall be the purpose of this exposition inversely to apply geometry to analysis, to get acquainted with analytic relations in a geometric manner, or somewhat more precisely expressed, *with the aid of geometry to gain an insight into the theory of functions of several variables*." If the dualism here indicated between algebra and geometry is to be without exception, we shall require on the one hand imaginary points in geometry to correspond to imaginary solutions of equations while on the other we shall need a symbol ∞ to correspond in the non-homogeneous coördinate system to the point at infinity on every line. In accordance with a cardinal principle of mathematical development, I advocated both steps to the end that algebra and geometry appear as merely different aspects of the same abstract truth. This concept of algebra and geometry as dualistic or isomorphic in my view leads to a more adequate picture of analytic geometry and is at the same time a distinct advantage in teaching.

---

* *Einleitung in die höhere Geometrie*, vol. I, p. 1.

3

INDUCTION, IDENTITIES, AND INEQUALITIES

(a)

INDUCTION

ON MATHEMATICAL INDUCTION*

J. W. A. YOUNG, University of Chicago

1. Its function and place. In the secondary school, the pupil in mathematics is becoming familiar with this fundamental type of thought by working in it and applying it. He thinks mathematics, but not *about* mathematics. In college, when the student begins to philosophize, he may, in addition to working in and with mathematics, also begin to think about it, to analyze and classify its processes of thought, to seek its essential characteristics and the lines of demarcation between it and other subjects.

He will learn that many mathematicians see the distinctive marks of their science, not in its subject matter, not in numbers, points, lines and symbols, but in the mode of thinking which is used.** He will find the definition, "Mathematics is the science which draws necessary conclusions,"† a good expression of this conception of the subject. It identifies mathematics with deductive reasoning, and accounts for the peculiar certainty and accuracy which, in his experience, has been its distinctive characteristic.

But when he thinks of the way in which he works out mathematics, solves problems, does "original" work of any description, he sees that induction plays an important part in this aspect of mathematics.

Further, as his acquaintance with the subject matter of mathematics widens, he will find numerous instances in which concepts are extended and theorems generalized. He will frequently find that the course of development is from the particular

---

* From AMERICAN MATHEMATICAL MONTHLY, vol. 15 (1908), pp. 145–153.

**For an instructive discussion of various definitions of mathematics, see Bocher, *Bulletin of American Mathematical Society*, 1904, p. 115.

† Peirce, *American Journal of Mathematics*, Vol. IV.

to the general, and in such cases he will often find employed a method of reasoning called *mathematical induction*,* which shares with non-mathematical induction the peculiarity of generalizing from particular instances, but which nevertheless, like other mathematics, produces that unhesitating confidence in the absolute accuracy of the result which is not felt as to the results of non-mathematical reasoning. This two-fold property leads one of the most acute thinkers of the day† to see in mathematical induction the sole instrument whereby the mathematician enlarges the sum-total of mathematical knowledge (at least in pure mathematics or arithmetic, as distinguished from geometry and infinitesimal analysis). Whether or not we concur‡ without reserve in Poincaré's thesis that "we can advance only by means of mathematical induction, which alone can teach us anything new," all will agree that explicitly it produces a large body of the mathematician's most valued results, and that implicitly it lurks unsuspected at the bottom of the reasoning by which a far greater body of mathematical truth is established.

The process of mathematical induction is exceptionally well fitted to introduce the beginner to the philosophic study of mathematical thinking, since it deals neither with the delicate and abstract questions relative to the foundations of mathematics, whose successful treatment requires extensive experience in mathematical reasoning, nor with such concepts from the borderland of mathematics and philosophy as have in the past proved themselves vague and elusive (for example, "continuity" or "infinity"). While a careful discussion of the logical nature and function of mathematical induction should perhaps be deferred to a period later than the early collegiate years, practical acquaintance with the process itself and a certain amount of readiness in its use may well be acquired at that time. To this end, a considerable body of material is requisite, both to illustrate the range and fertility of the method and also to supply a fund of exercises sufficiently ample to meet the student's need of considerable practice, and to permit variety in the same class and in successive classes.

2. Its character. I give some of the exercises which I have collected for such use from many sources, preceded by a discussion of the method itself based on a particular example.

By trial:

$$1 = 1,$$
$$1 + 2 = 3,$$
$$1 + 2 + 3 = 6,$$
$$1 + 2 + 3 + 4 = 10,$$
$$1 + 2 + 3 + 4 + 5 = 15.$$

---

* Also, *Complete Induction*, and, in German, *der Kaestnerische Schluss*, although published by Pascal in 1662 (Cantor, *Gesch. d. Math.*, II. p. 684) long before the time of Kaestner (1719–1800).

† Poincaré, *Sur la nature du raisonnement mathématique, Revue de métaphysique et morale*, 1894, p. 371.

‡ I have discussed elsewhere (*The Teaching of Mathematics*, p. 25, note) the manner in which deductive reasoning enlarges the sum-total of mathematical knowledge.

Letting S_n denote the sum of the first n positive integers, the equations can be written:

$$S_1 = \frac{1.2^*}{2},\ S_2 = \frac{2.3}{2},\ S_3 = \frac{3.4}{2},\ S_4 = \frac{4.5}{2},\ S_5 = \frac{5.6}{2}.$$

Examining these equations, we see that they are all of the type:

$$S_n = \frac{n(n+1)}{2}.$$

By taking a few additional ones, we find that the formula continues to hold. Thus

$$S_6 = 1 + 2 + 3 + 4 + 5 + 6 = 21 = \frac{6.7}{2}.$$

Trial of additional instances increases the "moral" certainty that the formula is true for every positive integer n, but no matter how many instances we may have the patience to try, mathematical certainty is not achieved thereby. This is attained by means of the mind's power of operating with mathematical certainty upon unspecified numbers.

The invariable method of mathematical induction is to prove that whenever the formula in question holds in any particular instance, it also holds in the next following instance.†

Recurring to our example, we must show that if the formula $S_n = \dfrac{n(n+1)}{2}$ holds for any particular value of n, say $n = k$, it also holds for the next value of n or $k + 1$.

(A) That is, we must show that, whenever $S_k = \dfrac{k(k+1)}{2}$ holds, then, $S_{k+1} = \dfrac{(k+1)(k+2)}{2}$ holds also.

Proof. By definition, $S_{k+1} = S_k + k + 1$.
Substituting the assumed value of S_k,

$$S_{k+1} = \frac{k(k+1)}{2} + k + 1 = \frac{(k+1)(k+2)}{2}.$$

The proof required above is thus made, but the result is purely hypothetical. We pass to actually valid results as follows.

(B) Trial has already shown that the formula is true for S_1 to S_6. Since the formula is true for S_6, we know, without trial but solely by the proof at (A), that the formula holds for S_7. Since it is true for S_7, we know similarly without trial, that it is true for S_8. Similarly we know that the formula is true for S_9, for S_{10}, and so on to any S_n whatever.

---

* 1.2 means 1×2.

† The application of the method presupposes that the cases to be considered have in some way been arranged in a definite order and numbered consecutively, and that after each case follows another.

The success of the proof hinges upon three things:

1. The ability to express any instance in terms of the next preceding, even when the latter is not specified.

2. The ability to make the proof (A) without specifying the particular instance that is being considered.

3. The power of the mind to see with certainty that repetition of the steps at (B) would lead to any given n, without actually going through all these steps.

These three things are but various phases of the mind's power of operating with certainty upon unspecified numbers.

In any proof by mathematical induction, the following two parts must unfailingly be present:

1). The proof that *if* the statement holds in any particular instance, it also holds in the next, and

2). The proof that the statement actually holds in the first instance (that is, in *some* particular case).

These two proofs are quite independent and may therefore be made in either order, but the relation to true induction is most apparent when the formula to be proved is actually discovered by true induction (as in the example above), thus amply making the second proof at the outset.

Beginners are prone to regard one or the other of these proofs as sufficient in itself. The need of both may be made clear by non-mathematical illustrations and by mathematical examples.

As a non-mathematical illustration we may consider a row of bricks so arranged that whenever any brick is knocked over, it will in its fall knock over its neighbor on the right. But this is only potential. In order actually to knock over the whole row, it is necessary and sufficient actually to knock over the first brick. If this cannot be done the whole row cannot be knocked over.

A second illustration is that of a ladder.* "We must have a ladder by which to climb from any round (the kth) to the next round (the $k+1$st); but the ladder must rest on a solid basis so that we can get on to the ladder (the $k=1$ or $k=2$ rounds)."

Mathematical examples can also be given in which one of the parts can be proved but not the other, and hence the statement in question is not proved.

1. Consider the series,

$$1 \cdot 1!, 2 \cdot 2!, 3 \cdot 3!, \ldots .$$

It is readily proved that if the sum of the first k terms is $(k+1)!$, then the sum of the first $k+1$ terms is $(k+1+1)!$ Or, denoting the sum of the first n terms of any series by S_n, if, for this series, the formula $S_n = (n+1)!$ holds for any particular value of n, it also holds for the next following value of n. But there is no value of n for which it can be proved to hold, and therefore the formula is not proved. (If it could be proved for any particular value of n after the first, the formula would of course be proved from that value of n on.)

---

* Dickson, *College Algebra*, p. 100.

2. Considering the series,

$$1, \frac{3}{2}, \frac{5}{4}, \frac{7}{8}, \frac{9}{16}, \ldots,$$

it can be proved that if the formula

$$S_n = 1 - \frac{2n+3}{2^{n-1}}$$

holds for any particular value of k, it also holds for $k+1$. But no value k can be found for which the formula holds and hence it is not proved.

On the other hand it may be possible to verify a certain statement in many consecutive instances, without leading to the conclusion that it is true in every instance.

1. The expression, $1 - 2^{n+2} + 2 \cdot 3^{n+1} - 4^{n+1} + 5^n$ is zero for $n = 0, 1, 2, 3$, but not zero for $n = 4$.
2. $(4^n - 6 \cdot 3^n + 14 \cdot 2^n - 14)[1 - (-1)^n]$ is zero for $n = 0, 1, \ldots, 6$, but not zero for $n = 7$.
3. $n^2 + n + 17$ is prime for $n = 0, 1, 2, \ldots, 16$, but not for $n - 17$.
4. $2n^2 + 29$ is prime for $n = 0, 1, 2, \ldots, 28$, but not for $n = 29$.
5. $n^2 - n + 41$ is prime for $n = 0, 1, 2, \ldots, 40$, but not for $n = 41$.
6. The theorem that if an odd prime be increased by 3, the result is the product of an odd prime and a power of two, holds for the odd primes to 43, but does not hold for 47.

3. Exercises. In the exercises that follow, a formula for S_n is to be proved, unless otherwise specified. The finding of the formula by true induction is one of the most fascinating activities, and offers, in proportion to the difficulty of the task, more or less opportunity for the application of mathematical ability and the enjoyment of mathematical inspiration. Not to rob the reader of the possibility of the pleasures, the answers are not as a rule given with the exercises, but are collected at the end.

The exercises are not arranged in order of difficulty. This will naturally vary with different minds. Of those whose answers are reserved, the following are among the easiest, arranged somewhat in order of supposed difficulty: 21, 10, 13, 1, 32, 23, 12, 11, 4, 9. The degree of difficulty of the exercises may be diminished by giving a part of the result. Thus, No. 3 becomes very easy if it is given that $4n^2 + 6n - 1$ is a factor of the result, and No. 7 becomes moderately easy if it is given that $2n+1$ is a factor of the result.

In each of the following the sum of the series is to be found by mathematical induction.

(1) 1.2, 2.3, 3.4,...
(2) 1.2, 3.4, 5.6,...
(3) 1.3, 3.5, 5.7,...
(4) 2.3, 4.6, 6.8,...
(5) $1^2, 2^2, 3^2, \ldots$
(6) $1^2, 3^2, 5^2, \ldots$
(7) $2^2, 4^2, 6^2, \ldots$
(8) $4^2, 7^2, 10^2, 13^2, \ldots$
(9) 0, 2.3, 3.8, 4.15,..., $k(k^2 - 1), \ldots$

(10) $1^3, 2^3, 3^3, \ldots$
(11) $1^3, 3^3, 5^3, \ldots$
(12) $2^3, 4^3, 6^3, \ldots$
(13) $2.3.4, 3.4.5, 4.5.6, \ldots$
(14) $1.3.5, 3.5.7, 5.7.9, \ldots$
(15) $1.1^2, 2.3^2, 3.5^2, 4.7^2, \ldots$
(16) $1.2^2, 2.3^2, 3.4^2, 4.5^2, \ldots$
(17) $1, 4, 10, 20, \ldots, \dfrac{k(k+1)(k+2)}{3!}, \ldots$
(18) $1^4, 2^4, 3^4, 4^4, 5^4, \ldots$
(19) $1.2.3.4, 2.3.4.5, 3.4.5.6, \ldots$
(20) $1, 5, 15, 35, \ldots, \dfrac{k(k+1)(k+2)(k+3)}{4!}, \ldots$
(21) $\dfrac{1}{1.2}, \dfrac{1}{2.3}, \dfrac{1}{3.4}, \ldots$
(22) $\dfrac{1}{1.3}, \dfrac{1}{2.4}, \dfrac{1}{3.5}, \ldots$
(23) $\dfrac{1}{3.5}, \dfrac{1}{5.7}, \dfrac{1}{7.9}, \ldots$
(24) $\dfrac{1}{1.2.3}, \dfrac{1}{2.3.4}, \dfrac{1}{3.4.5}, \ldots$
(25) $\dfrac{1}{1.2.3}, \dfrac{3}{2.3.4}, \dfrac{5}{3.4.5}, \ldots$
(26) $\dfrac{1}{2.3.4}, \dfrac{2}{3.4.5}, \dfrac{3}{4.5.6}, \ldots$
(27) $\dfrac{1}{1.3.5}, \dfrac{1}{3.5.7}, \dfrac{1}{5.7.9}, \ldots$
(28) $\dfrac{1}{2.5.8}, \dfrac{2}{5.8.11}, \dfrac{3}{8.11.14}, \ldots$
(29) $\dfrac{1}{1.2.3.4}, \dfrac{1}{2.3.4.5}, \dfrac{1}{3.4.5.6}, \ldots$
(30) $1^5, 2^5, 3^5, 4^5, \ldots$
(31) $1^6, 2^6, 3^6, 4^6, \ldots$*
(32) $1, 2.2, 3.2^2, 4.2^3, \ldots$
(33) $1, 2.3, 3.3^3, 4.3^3, \ldots$
(34) $1, 2.5, 3.5^2, 4.5^3, \ldots$
(35) $1, 3.2, 5.2^2, 7.2^2, \ldots$
(36) $1, 4.3, 7.3^2, 10.3^3, \ldots$
(37) $1, 5.4, 9.4^2, 13.4^3, \ldots$
(38) $1, 4.2, 9.2^2, 16.2^3, \ldots$
(39) $1, 4.3, 9.3^2, 16.3^3, \ldots$
(40) $1, 8.2, 27.2^2, 64.2^3, \ldots$

---

* On the series $1^n, 2^n, 3^n, \ldots$ see Chrystal's *Alpha*, Vol. I, p. 486.

(41) $1, 3.2, 6.2^2, 10.2^3, \ldots, \dfrac{k(k+1)}{2} 2^{k-1}, \ldots$

(42) $1, \dfrac{4}{3}, \dfrac{9}{3^2}, \dfrac{16}{3^3}, \ldots$

(43) $a, 2(a+1), 3(a+2), 4(a+3), \ldots$

Show that:

(44) $2.6.10.14, \ldots, (4n-6)(4n-2) = (n+1)(n-2), \ldots, (2n-1)2n$.

(45) $n(n+1)(n+2), \ldots, (2n-2) = 1.3.5.7, \ldots, (2n-3).2^{n-1}$.

(46) $(1+x)(1+x^2)(1+x^4)(1+x^8)\cdots(1+x^{2n}) = 1 + x + x^2 + \cdots + x^{2^{n+1}-1}$

(47) Show that $x^{2n} - y^{2n}$ is divisible by $x + y$.

(48) Show that $n^{13} - n$ is divisible by 13, for every positive integer n.

(49) If p is a prime number, show that $n^p - n$ is divisible by p for every positive integer n.

(50) Show that $2.7^n + 3.5^n - 5$ is divisible by 24 for every positive integer n.

Suggestion: Assume, 1) $2.7^k + 3.5^k - 5 =$ multiple of 24.
  To show, $2.7^{k+1} + 3.5^{k+1} =$ multiple of 24.
  Multiple 1) by 7, $2.7^{k+1} + 3.5^k.7 - 5.7 =$ multiple of 24.
  Hence, $2.7^{k+1} + 3.5^{k+1} - 5 - 30 + 3.5.2^k =$ multiple of 24.

To prove the last assertion, we must show that $6.5^k - 30 =$ multiple of 24, which is easily done.

(51) Show that $3.5^{2n+1} + 2^{3n+1}$ is divisible by 17 for every positive integer n.

(52) If n is a positive integer, show that $(a+b)^n = a^n + na^{n-1}b + \dfrac{n(n-1)}{2!}a^{n-2}b^2 + \cdots + \dfrac{n(n-1)(n-2)\cdots(n-k+1)}{h!}a^{n-k}b^k + \cdots$.

(53) If all the positive integers of n and fewer digits be written, show that the numbers of times any digit (other than zero) occurs is $n.10^{n-1}$.

(54) If the positive integers are grouped as follows: $[(1,2),(3)]; [(4,5,6),(7,8)]; [(9,10,11,12),(13,14,15)]; \ldots$ and these groups taken in pairs as indicated, prove that the sum of the numbers in the two groups of any pair is the same.

(55) $\sum_{k=1}^{n}(a+kb)^2 = n[a^2 + ab(n+1) + \dfrac{b^2}{6}(2n^2+3n+1)]$.

(56) $\sum_{k=1}^{n} \dfrac{(2k)!}{k!2^k} = 1.3.5\ldots(2n-1)$.

(57) Letting t_k denote the kth term of a series, show that, if $t_k = k(k+1)\cdots(k+q-2), S_n = \dfrac{n(n+1)\cdots(n+q-1)}{q}$.

(58) If $t_k = \dfrac{1}{k(k+1)\cdots(k+p)}$, $S_n = \dfrac{1}{p}\left[\dfrac{1}{p!} - \dfrac{1}{(n+1)\cdots(n+p)}\right]$.

(59) Assuming that the formulas,

$$\sin(x \pm y) = \sin x \cos y \pm \cos x \sin y,$$
$$\cos(x \pm y) = \cos x \cos y \mp \sin x \sin y,$$

have been proved for all acute angles x, y, prove that they hold for all posiangles x, y. (In this proof, the quadrants are to be regarded as numbered consecutively, 5, 6, 7, 8, etc., as the angle increases beyond 360°, and the formulas of the type $\sin(90 + x) = \cos x$, are to be accepted as proved for every angle x.)

(60) $(\cos a + i \sin a)^n = \cos na + i \sin na$. DeMoivre's formula.

Sum the following:

(61) $\sin a, \sin 2a, \sin 3a, \ldots$
(62) $\sin a, \sin 3a, \sin 5a, \ldots$
(63) $\cos a, \cos 2a, \cos 3a, \ldots$
(64) $\cos a, \cos 3a, \cos 5a, \ldots$

4. Answers.

(1) $\dfrac{n(n+1)(n+2)}{3}$.

(2) $\dfrac{n(n+1)(4n+1)}{3}$.

(3) $\dfrac{n(4n^2+6n-1)}{3}$.

(4) $\dfrac{4n(n+1)(n+2)}{3}$.

(5) $\dfrac{n(n+1)(2n+1)}{6}$.

(6) $\dfrac{n(2n-1)(2n+1)}{6}$.

(7) $\dfrac{2n(n+1)(2n+1)}{3}$.

(8) $\dfrac{n(6n^2+15n+11)}{2}$.

(9) $\dfrac{(n-1)n(n+1)(n+2)}{4}$.

(10) $\left[\dfrac{n(n+1)}{2}\right]^2$.

(11) $n^2(2n^2-1)$.

(12) $2n^2(n+1)^2$.

(13) $\dfrac{(n+1)(n+2)(n+3)}{3}$.

(14) $n(2n^3+8n^2+7n-2)$.

(15) $\dfrac{n(n+1)(6n^2-2n-1)}{6}$.

(16) $\dfrac{n(n+1)(n+2)(3n+5)}{12}$.

(17) $\dfrac{(n+3)!}{4!(n-1)!}$.

(18) $\dfrac{n(n+1)(6n^3+9n^2+n-1)}{30}$.

(19) $\dfrac{n(n+1)(n+2)(n+3)(n+4)}{5}$.

(20) $\dfrac{(n+4)!}{5!(n-1)!}$.

(21) $\dfrac{n}{n+1}$.

(22) $\dfrac{n(3n+5)}{4(n+1)(n+2)}$.

(23) $\dfrac{n}{3(2n+3)}$.

(24) $\dfrac{1}{2}\left[\dfrac{1}{2} - \dfrac{1}{(n+1)(n+2)}\right]$.

(25) $\dfrac{n(3n+1)}{4(n+1)(n+2)}$.

(26) $\dfrac{n(n+1)}{4(n+2)(n+3)}$.

(27) $\dfrac{n(n+2)}{3(2n+1)(2n+3)}$.

(28) $\dfrac{n(n+1)}{4(3n+2)(3n+5)}$.

(29) $\dfrac{1}{3}\left[\dfrac{1}{3!} - \dfrac{1}{(n+1)(n+2)(n+3)}\right]$.

(30) $\dfrac{n^2(n+1)^2(2n^2+2n-1)}{3.4}$.

(31) $\dfrac{n(n+1)(2n+1)(3n^4+6n^3-3n+1)}{6.7}$.

(32) $(n-1)2^n + 1$.

(33) $\dfrac{(2n-1)3^n + 1}{4}$.

(34) $\dfrac{(4n-1)5^n + 1}{16}$.

(35) $(2n-3)2^n + 3$.

(36) $\dfrac{(6n-7)3^n + 7}{4}$.

(37) $\dfrac{(12n-13)4^n + 13}{9}$.

(38) $(n^2 - 2n + 3)2^n - 3$.

(39) $\dfrac{(n^2 - n + 1)3^n - 1}{2}$.

(40) $[(n-1)^3 - 6(n-2)]2^n + 13$.

(41) $2^{n-1}(n^2 - n + 2) - 1$.

(42) $\dfrac{3^{n+1} - (n^2 + 3n + 3)}{2 \cdot 3^{n-1}}$.

(43) $\dfrac{n(n+1)(3a + 2n - 2)}{6}$.

(61) $\dfrac{\sin\dfrac{na}{2}\sin\dfrac{(n+1)a}{2}}{\sin\dfrac{a}{2}}$.

(62) $\dfrac{\sin^2 na}{\sin a}$.

(63) $\dfrac{\sin\dfrac{na}{2}\cos\dfrac{(n+1)a}{2}}{\sin\dfrac{a}{2}}$.

(64) $\dfrac{\sin 2na}{2\sin a}$.

ON PROOFS BY MATHEMATICAL INDUCTION*

E. T. BELL, University of Washington

1. The best way to cure oneself of a crotchet is to confide it to some sympathetic listener. The crotchet in this note is one which has worried me since school days when I was induced to repeat the proof of the binomial theorem by mathematical induction. The same crotchet seems to trouble successive generations of freshmen, for occasionally one has obstinacy enough to balk at the magic formula "*and therefore the theorem is always true*," with which many authors conclude their proofs by recurrence. I hold that mathematical induction has no place in elementary teaching, particularly when such teaching strains at mathematical gnats, as in pseudo-rigorous presentations of the elementary theory of limits, the better to swallow logical camels such as some proofs of the binomial theorem or their equivalent quoted presently from Poincaré. This is the crotchet. In short, elementary teaching would be more convincing if it left rigor to that logistics which was Poincaré's bête noir.

---

* From AMERICAN MATHEMATICAL MONTHLY, vol. 27 (1920), pp. 413–415.

2. It would be difficult to find a balder statement of the logical vice which characterizes many proofs by mathematical induction in the current text books, than the following extract from Poincaré's essay *On the Nature of Mathematical Reasoning*, in *Science and Hypothesis* (Halsted's translation, page 36, section IV). Having considered three examples of "proof by recurrence," and having drawn a false conclusion in each of them, Poincaré says:

"Here I stop this monotonous series of reasonings. But this very monotony has the better brought out the procedure which is uniform and is met again at each step.

"This procedure is the demonstration by recurrence. We first establish a theorem for $n=1$; then we show that if it is true of $n-1$, it is true of n, and thence conclude that it is true for all whole numbers."

It is only fair to state that in the next section (V) of his essay, Poincaré gives an unobjectionable form of "this procedure," which he calls "mathematical reasoning *par excellence*." This, however, may have been an inadvertence, as the closing paragraph of the essay again exhibits the circularity of the reasoning in all its viciousness:

"Observe finally that this induction is possible only if the same operation can be repeated indefinitely."

From this we should expect a rich crop of subtle fallacies when "this induction" is applied to prove that a certain assemblage contains an infinity of members, or when it is used to demonstrate the universal truth of a proposition.

3. To exhibit the logical defect in this "mathematical reasoning par excellence," let us separate Poincaré's summary into its three constituents:

(1) "We first establish a theorem for $n=1$;

(2) "Then we show that if it is true of $n-1$, it is true of n;

(3) "And thence (we) conclude that it is true for all whole numbers."

In (3) "thence," if it means anything definite, must refer to (2) and (1). That is, the conclusion that the theorem is true for *all* whole numbers is to follow from (1) and (2) *only*. It does not so follow. In order to draw the conclusion (3), we need, either as a postulate, or, if it can be proved from simpler assumptions, the proposition:

(4) If a theorem is true for $n=1$, and if its truth for $n-1$ implies that it is true for n, then the theorem is true for all whole numbers.

Without (4), all that (1) and (2) give is the means for *step by step* assertions that the theorem is true in successive cases. Thus, supposing (1) and (2) established, if it be required to see whether the theorem is true for $n=5$, we must, for all that (1) and (2) prove, take the steps 1 to 2, 2 to 3, 3 to 4, 4 to 5, omitting none. Or, again, if (1) and (2) are established, and we wish to know whether the theorem is true for $n=9^{9^9}$, say, we must take $9^{9^9}-1$ steps in order to find out, for neither (1) nor (2) permits us to take more than one step at a time. In this case we might never gratify our curiosity, and the truth or falsity of the theorem for $n=9^{9^9}$ would remain as inaccessible to our knowledge as is the other side of the moon. Nor could we assert that the theorem *most probably* is true for this value of n, for we cannot predicate probabilities in the absence of data. Our belief that the theorem is true for this value of *n* might be strong; but belief belongs to the realm of emotional experiences, and is seldom conspicuous for the reasonableness of its tenets. Until (4) or its equivalent is proved or admitted as a postulate, it would seem to be advisable to banish such phrases as "all whole numbers," and "the same operation can be repeated indefinitely," from elementary texts which make pretensions to rigor.

4. If reasoning by recurrence is, as Poincaré claims, mathematical reasoning *par excellence*, and if the objections put forth in § 3 are not groundless, it would seem to follow that mathematics is like any other science in that the conclusions which it legitimately draws are no more "general" or "universal" than those of other sciences. This contradicts what seems to be a current valuation of mathematical truth in the minds of laymen and some others who hold that mathematics has a timeless, eternal aspect, independent of all the empiricism which characterizes the conclusions of physical sciences.

There is one way of escape which is so obvious that it need only be pointed out. We can beat the mathematical devil round the logical bush by saying that (4) of § 3 is the rule, or law, of inference. But it would be a wise logician indeed who recognized (4) as one of his legitimate children. For where is either a proof of it or its explicit statement as a postulate of logic to be found?

DISCUSSIONS*

W. A. HURWITZ, Cornell University

As the first discussion this month we present a paper which was read at the last annual meeting of the Mathematical Association of America, as part of a program devoted to the consideration of the sort of training in mathematics most useful for students specializing in fields in which mathematics finds frequent application. Professor Reed represents the point of view of the biometrist. His recommendations, briefly summarized, are: the usual courses in algebra, trigonometry, and analytic geometry; a short course in the calculus, emphasizing principles rather than technique; and a course in probability, with stress on statistical theory and the adjustment of curves to given data. It would seem, therefore, that the student of biometry will generally find ready at hand, in most of our colleges, courses agreeing reasonably well with the plan outlined by Professor Reed. Probably, too, the textbooks in elemenatry subjects are no longer so restricted in their treatment as Professor Reed implies. Few, if any, trigonometries of recent date, for example, fail to give applications to mechanics.

The brevity of the course in calculus in Professor Reed's scheme should be considered rather carefully. Is a student of statistics to accept Stirling's formula and the value of the probability integral on faith, or is he to receive demonstrations? The proofs, if given with logical precision, require more thorough treatment of the

---

* From AMERICAN MATHEMATICAL MONTHLY, vol. 27 (1920), pp. 407–409.

behavior of series and improper integrals than is usually found in even rather extensive first courses in calculus.

In the second discussion, Mr. Webb outlines a treatment of complex numbers, including the definition of exponential and trigonometric functions and their relationship. His scheme is similar to that found in some textbooks in trigonometry; seldom, if ever, is so full a discussion included in the ordinary course in algebra. Two items in his outline call for special comment. No. 6 prescribes "the customary development of e and of e^x by the binomial theorem as the limits of $(1+1/n)^n$ and $(1+1/n)^{nx}$ as $n \doteq \infty$." This customary scheme involves either a scandalously inaccurate treatment of the double limit process (as in most textbooks on calculus) or a degree of logical precision scarcely within the reach even of the average college graduate. The question involved is much more delicate than that of mere convergence. Either the properties of uniform convergence, or else some equivalent special process to avoid this concept must be used. It would be desirable, if possible, to find some less difficult approach to the exponential function. No. 8 implies that there must exist some k such that $\cos 1 + i \sin 1 = e^k$. Since the series for $\cos \theta$ and $\sin \theta$ are determined on the hypothesis that such a k exists, and then by use of the series the value $k = i$ is obtained, it is not easy to see how to avoid the hypothesis.

Professor Bell points out that the method of proof known as mathematical induction is valid only by virtue of a distinct assumption, which he formulates very clearly thus: *If a theorem is true for $n=1$, and if its truth for $n-1$ implies that it is true for n, then the theorem is true for all whole numbers.* This discussion should be helpful to our readers, inasmuch as the necessity of this explicit assumption is not always clearly recognized by teachers. However, the author's implication that the rôle of such an assumption is overlooked by specialists in the logic of mathematics seems to be unwarranted. Writers on the foundations of mathematics generally recognize that the validity of mathematical induction is the result of a pertinent hypothesis. A common alternative to Professor Bell's form of statement is the following: *In any set of positive integers there is a least integer.* From this hypothesis, the other form of statement can be proved by considering the set of all numbers n (if any) for which the theorem under consideration is false.

Poincaré himself, in the sections of *Science and Hypothesis* criticized by Professor Bell, states that the principle of mathematical induction can not be obtained either from syllogisms based on the other fundamental axioms or from experience. For example*: "We may readily pass from one enunciation to another, and thus give ourselves the illusion of having proved that reasoning by recurrence is legitimate. But we shall always be brought to a full stop—we shall always come to an indemonstrable axiom, which will at bottom be but the proposition we had to prove translated into another language." Russell has long regarded mathematical induction as a *definition* of the *natural numbers*.† "We *define* the natural numbers as

---

* *Science and Hypothesis*, Part I, Chapter VI.
† *Introduction to Mathematical Philosophy*, Chapter III.

those to which proofs by mathematical induction can be applied." Since the number system of elementary arithmetic and algebra contains no other *integers* except these "natural numbers," it is in this special case only a matter of terminology to say which of the fundamental hypotheses shall be called *definitions* rather than merely *axioms*. Many logicians would prefer to regard the whole set of fundamental hypotheses underlying any mathematical system as collectively *defining* the system.

Professor DeCou gives an instance of a problem in maxima and minima arising in the printing office. The solution is simple; and in fact it may be shown by a slight alteration in notation that the problem belongs to the familiar type in which a sum of two variables is to be minimized while the product remains constant. It is published here on account of the value for the class room of problems arising from practical sources. How many similar instances may occur in trade, industry, and every day life, in which a little knowledge of mathematics would save, as here, "an added cost of $100 to $200," or an equivalent in energy or convenience?

One detail of the problem suggests a further interesting question. As the calculus refuses to discriminate between integers, fractions, and irrationals, the formal solution of the problem generally gives an irrational result for the number of electrotypes needed. Professor DeCou says, "Of course only the nearest integral value of x is used." It is fairly clear, since the relationship of C and x is represented graphically by a hyperbola, that the minimum C for integral x must be given by one of the two integers enclosing the value of x for which the actual minimum occurs. But it is not evident that of these two, the *nearest* will give the value desired. In fact, this turns out not to be precisely correct, and our readers may be interested in verifying the true result. If $\sqrt{PR/ES} = \alpha$, then α should be replaced by the next smaller integer n or the next greater integer $n+1$, not according as α is less or greater than the arithmetic mean $n+\frac{1}{2}$, but according as α is less or greater than the geometric mean $\sqrt{n(n+1)}$. For large values of n, the distinction between the two criteria is slight. Of course in any actual case it would be simple enough to substitute each of the two values in the expression for C in order to determine which was the better.

A PARADOX RELATING TO MATHEMATICAL INDUCTION*

R. G. ALBERT, Providence, R. I.

The axiom of mathematical induction as included in the set of axioms of Peano which characterize the system of natural numbers asserts:

If S be a set of natural numbers containing 1 and containing the successor x' of every natural number x which S contains, then S is the set of *all* natural numbers.

The following proof by mathematical induction (by Landau), that every natural number except 1 has an immediate predecessor, suggested an interesting paradox which may aid in gaining an insight into this important method of proof:

Let x' denote the successor of x (i.e., $x+1$).

Let S be the set of natural numbers including 1 and including all n.n.'s u such that there exists a n.n. x with $x'=u$. Then, obviously, S contains 1. If S contains $u=k$, then S contains $u=k'$, since $x'=k'$ has the evident solution $x=k$. Hence S is the set of all n.n.'s. Since S was defined to be the set comprised of 1 and of all n.n.'s u which have predecessors, it follows that every n.n. except 1 has an immediate predecessor.

Now, examine the following proof which seems to trace the same lines:

We shall try to show that every n.n. except 1 equals its own successor.

Let S be the set of n.n.'s containing 1 and containing all n.n.'s u such that $u=u'$ (u' denotes $u+1$, as before). S contains 1. If S contains $u=k$, then $k=k'$. Since every n.n. has a unique successor (Axiom 2, Peano), then $k'=(k')'$. The latter verifies $u=u'$ for $u=k'$. Thence, S contains $u=k'$. Thus, S is the set of all n.n.'s. Since S was defined to contain 1 and all n.n.'s equal to their own successors, it follows that every n.n. with the possible exception of 1 is its own successor.

This procedure is of course generalizable and seems to suggest that many theorems or formulas which are known to be false can be "proved" for all n.n.'s except 1, by this device of consigning 1 to S a priori. It appears as if tossing 1 in bodily from the start is a kind of circumvention of the intention in the induction hypothesis, raising doubts regarding the validity of arbitrary inclusion of 1.

It has been of interest to me to note how many of my students have failed to "see through" the error in the last proof and why the first proof is still valid.

Of course, the error in the second demonstration lies in the step:

If S contains $u=k$, then $k=k'$.

From S containing $u=k$, follows: *Either* $k=1$ *or* $k=k'$. And in the first of these alternative cases, we can no longer assert that S contains k', since neither qualification for admission to S is satisfied by k', i.e., $1'\neq 1$ and $1'\neq (1')'$.

In the first proof, S containing k *did* imply *invariably* that S contained k', for $x'=k'$ does in all cases have a solution, $x=k$.

---

* From AMERICAN MATHEMATICAL MONTHLY, vol. 57 (1950), pp. 31–32.

MATHEMATICAL INDUCTION AND RECURSIVE DEFINITIONS*

R. C. BUCK, University of Wisconsin and Institute of Defense Analyses

Many students first encounter mathematical induction during a beginning course in algebra, either in secondary school or in college. For some of these students, this can become their first introduction to mathematical ideas, turning their attention away from computational exercises to notions of structure and proof. Their initial distrust is gradually replaced by an appreciation for its power; with reference to dominoes or inductive sets, they will usually become convinced of its reasonableness. Some students will retain a cautious attitude toward certain types of applications of induction; in this paper, I wish to increase the number of these students by discussing some of these problems.

Suppose that we have agreed upon a workable definition of the notion of function. We will deal only with the set $I = \{0, 1, 2, \cdots\}$ of nonnegative integers, or with the set $I \times I$ of pairs of integers, or more generally, with the set $I^k = I \times I \times \cdots \times I$ for some specific k. A function will always be defined on a subset of such a set, and will take its values in I. If we identify a function f with its graph, then a function f on I to I will become a specific set of pairs $\langle n, f(n) \rangle$ for all $n \in I$. More generally, we would say that any nonempty subset S of $I \times I$ that is univalent is a function; its domain will be the projection of S into the first coordinate space.

We now present a student with the following definition of a function f on I to I:

(1)
$$f(0) = 1$$
$$f(1) = 2$$
$$f(n + 1) = f(n) + f(n - 1) \quad \text{all } n = 1, 2, 3, \cdots.$$

I think that it will be quite convincing to the student that there is such a function f, and that it is uniquely defined by formula (1). He has no difficulty seeing that $f(2) = 3$, $f(3) = 5$, $f(4) = 8$, and so on. In short, he believes in the existence of f. However, there often remains a certain unhappiness in the mind of the student; he may say that he wants a formula for f. If he is pressed to explain, it will be found that he feels that functions ought to be described by means of certain allowed operations such as addition and composition, and ought to be built up from simpler functions. Students are constructivists at heart.

In the present example, of course, a formula can be given and should be given. He will express astonishment that so much complexity is needed for such a simple appearing function.

(2)
$$f(n) = \frac{5 + 3\sqrt{5}}{10} \left(\frac{1 + \sqrt{5}}{2}\right)^n + \frac{5 - 3\sqrt{5}}{10} \left(\frac{1 - \sqrt{5}}{2}\right)^n.$$

---

* From AMERICAN MATHEMATICAL MONTHLY, vol. 70 (1963), pp. 128–135.

Of course, you can convince him that this is correct, first by computing several values to check it, and then by using mathematical induction. We observe that

$$f(0) = \frac{5 + 3\sqrt{5}}{10} + \frac{5 - 3\sqrt{5}}{10} = \frac{10}{10} = 1,$$

and that

$$f(1) = \frac{5 + 3\sqrt{5}}{10} \frac{1 + \sqrt{5}}{2} + \frac{5 - 3\sqrt{5}}{10} \frac{1 - \sqrt{5}}{2}$$

$$= \frac{5 + 8\sqrt{5} + 15}{20} + \frac{5 - 8\sqrt{5} + 15}{20}$$

$$= 2.$$

We have verified the correctness of (2) for $n=0$ and $n=1$. Suppose that we have verified it for $n=0, 1, 2, \cdots, k$, can we be sure that it holds for $n=k+1$? For simplicity, set $A = (5 + 3\sqrt{5})/10$, $B = (5 - 3\sqrt{5})/10$, $\alpha = (1 + \sqrt{5})/2$ and $\beta = (1 - \sqrt{5})/2$. Then, formula (2) can be written as

(3) $$f(n) = A\alpha^n + B\beta^n.$$

By (1), we have the right to express $f(k+1)$ as $f(k) + f(k-1)$, so that the inductive hypothesis gives us

$$f(k + 1) = [A\alpha^k + B\beta^k] + [A\alpha^{k-1} + B\beta^{k-1}]$$
$$= A\alpha^{k-1}[1 + \alpha] + B\beta^{k-1}[1 + \beta].$$

Observing that $\alpha^2 = 1 + \alpha$, and $\beta^2 = 1 + \beta$, we obtain

$$f(k + 1) = A\alpha^{k+1} + B\beta^{k+1},$$

verifying (3) and thus (2). Or, if the instructor has been more formal in his presentation, he can point out that this argument has shown that the set $E \subset I$ of those integers n for which (2) holds is an inductive set; since it contains 0, it must, perhaps by an axiom rather than a theorem, be the whole set I.

At this point, the line of development is sure to be interrupted by a clamor to know "where formula (2) came from." This is handled either by suppressive measures, or by embarking upon a brief discussion of difference equations with constant coefficients.

But all of this has, to a certain extent, detoured the real and valid question which was in the student's minds. Is it in fact legitimate to *define* a function by a dodge like that of formula (1)? This certainly does not describe f either as a mapping or as a class of pairs. In this particular example, we were lucky enough to have found a formula. Is this always the case? Moreover, an alert and cautious student may also raise the general problem of how one can tell whether a relation similar to (1) admits a solution. For example, is there a function f which satisfies this relation?

(4)
$$f(1) = 1$$
$$f(2n + 1) = n^2 - n + 1$$
$$f(3n + 1) = 2n + f(2n + 1)$$
$$n \geq 1.$$

[Ans. No; try computing $f(13)$; then, replace n^2-n+1 by $4n+1$ and see what happens.]

Perhaps the following example, which deserves to be better known, will bring the matter more clearly to a head. Suppose we wish to define a function F on the set $I \times I$, which we can for convenience picture as the first quadrant. Suppose we write down the relation:

(5) $$F(m + 1, n + 1) = F(F(m, n + 1), n) \qquad m, n = 0, 1, 2, \cdots.$$

Suppose that we assign the values of F on the edges of the quadrant. Then, a little experimentation leads us to believe that there is such a function, that it is uniquely determined, and that we can calculate any desired value of F.

First, by assumption, we have available a complete knowledge of $F(0, n)$ for each n, and of $F(m, 0)$ for each m; we can calculate their values for any specific choice of m or n. Set $n=0$ in (5), obtaining a simple recursion akin to formula (1):

(6) $$F(m + 1, 1) = F(F(m, 1), 0) \qquad m = 0, 1, 2, \cdots.$$

From this, and the knowledge of the boundary values of the supposed F, we can generate the value of $F(x, 1)$ for any desired x; for example, $F(1, 1) = F(F(0, 1), 0)$ which is computable since we know the number $F(0, 1)$, and can compute the specific value of $F(x, 0)$ which results from setting $x = F(0, 1)$. Proceeding, we next have

$$F(2, 1) = F(F(1, 1), 0)$$

which is now computable since we know $F(1, 1)$, and so on.

We can thus regard $F(x, 1)$ as known for any specific integer x. Now, put $n=1$ in (4), obtaining

(7) $$F(m + 1, 2) = F(F(m, 2), 1) \qquad m = 0, 1, 2, \cdots.$$

Since we know the initial value $F(0, 2)$, we could proceed in the same fashion to compute $F(1, 2)$, $F(2, 2)$, and any later value of $F(x, 2)$. Notice in particular that at any stage, we have needed to know only a finite number of the values of F at *earlier* points; we did not need to know *all* the values of $F(x, 1)$ in order to compute $F(2, 2)$. In Figure 1, we have attempted to make this point clear by shading the region that might be needed in order to compute $F(4, 3)$.

We have thus reached the same spot with this example that we encountered with equations (1). Apparently, we can compute any desired value of F; intuitively, we are therefore convinced that relation (4) admits a solution which is a function defined on the entire set $I \times I$. However, the behavior illustrated by formula (4) may lead a student to seek assurance that the process outlined above leads to a consistent answer, that the value ascribed to $F(4, 3)$ does not depend upon his mode of procedure; again, he would be much happier if we were to exhibit F as an explicit class of ordered pairs, constructed from the relation (5) by standard set operations, for this would demonstrate existence in a much more satisfying way. We shall in fact do this at the end of this paper, and at the same time show that (5) cannot have two different solutions with the same assigned boundary values; the situation is analogous to the study of the Dirichlet problem in partial differential equations.

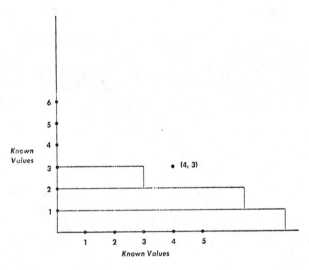

Fig. 1

Before doing this, however, we can gain some appreciation for the latent strength of the scheme (5) by examining the results of a specific choice of boundary values:

(8)
$$F(m, 0) = m + 1 \qquad m = 0, 1, \cdots.$$
$$F(0, 1) = 2$$
$$F(0, 2) = 0$$
$$F(0, n) = 1 \qquad n = 3, 4, \cdots.$$

Formula (6) becomes
$$F(m + 1, 1) = F(F(m, 1), 0)$$
$$m = 0, 1, 2, \cdots.$$
$$= F(m, 1) + 1.$$

With the initial value $F(0, 1) = 2$ from (8), this is easily seen to have the solution
(9) $$F(m, 1) = 2 + m \qquad m = 0, 1, \cdots.$$

In the same manner, (7) becomes
$$F(m + 1, 2) = F(F(m, 2), 1)$$
$$= 2 + F(m, 2) \qquad m = 0, 1, 2, \cdots$$
with the initial condition $F(0, 2) = 0$. From this, we obtain
$$F(1, 2) = 2 + 0 = 2$$
$$F(2, 2) = 2 + 2 = 2(2)$$
$$F(3, 2) = 2 + (2)(2) = 2(3)$$
and in general,
(10) $$F(m, 2) = 2m.$$

Continuing in the same way, we have
$$F(0, 3) = 1$$
$$F(m + 1, 3) = 2F(m, 3)$$
from which we deduce that $F(1, 3) = 2$, $F(2, 3) = 2^2$, $F(3, 3) = 2^3$, and in general
$$F(m, 3) = 2^m.$$

What happens when we go to the next stage? Our simple recursion becomes
$$F(m + 1, 4) = 2^{F(m, 4)}$$
with $F(0, 4) = 1$. We are able to compute the values of F as before:
$$F(1, 4) = 2$$
$$F(2, 4) = 2^2$$
$$F(3, 4) = 2^{2^2} = 16$$
$$F(4, 4) = 2^{2^{2^2}} = 2^{16} = 65536$$
$$F(5, 4) = 2^{2^{2^{2^2}}} = 2^{65536}$$
and by using dots, we can fake a general formula
$$F(m, 4) = 2^{2^{\cdot^{\cdot^{\cdot^2}}}} \qquad \text{[with } m \text{ two's]}.$$

Let us try the next case; we have the simple recursion
$$F(0, 5) = 1$$
$$F(m + 1, 5) = F(F(m, 5), 4) \qquad m = 0, 1, 2, \cdots$$

and using our value for $F(x, 4)$, we write this as

$$F(m + 1, 5) = 2^{2^{2^{\cdot^{\cdot^{\cdot^{2}}}}}} \quad [\text{with } F(m, 5) \text{ two's}].$$

This will suffice to compute some of the early values of F, so that we have for example

$$F(1, 5) = 2$$
$$F(2, 5) = 2^2 = 4$$
$$F(3, 5) = 2^{2^{2^2}} = 65536$$
$$F(4, 5) = 2^{2^{\cdot^{\cdot^{\cdot^{2}}}}} \quad \text{with } 65536 \text{ two's}$$
$$F(5, 5) = 2^{2^{\cdot^{\cdot^{\cdot^{2}}}}} \quad [\text{with } 2^{2^{\cdot^{\cdot^{\cdot^{2}}}}} \quad [\text{with } 65536 \text{ two's}] \text{ two's}].$$

However, I think that it is quite clear that we do not have a suitable way to write down any nonbogus general formula for $F(m, 5)$ within the notational schemes of the standard terminology.

Still less, then, will this be true for $F(m, 6)$, and manifestly more so for the function ψ of one variable which is now definable by the equation

$$\psi(x) = F(x, x) \quad \text{for } x = 0, 1, 2, \cdots.$$

However, it is also clear that $\psi(x)$ can be computed for any specific value of x, granting the necessary time and paper—which undoubtedly exceeds both the estimated size of the universe and its duration. Indeed, $\psi(0) = 1$, $\psi(1) = 3$, $\psi(2) = 4$, $\psi(3) = 8$, $\psi(4) = 65536$, and $\psi(5) = F(5, 5)$, which we have written down just above.

The existence of functions such as ψ yields an unexpected dividend. The following personal illustration may be amusing. I have found that most beginning analysis students seem to accept as plausible the conjecture that, given any increasing sequence of integers $\{c_n\}$, one could find an entire function f such that $f(n) > c_n$ for $n = 1, 2, \cdots$. If you suggest $c_n = 2^n$, they counter with $f(z) = \exp(z)$. If you suggest $c_n = n!$, they suggest $f(z) = \exp(\exp(z))$. However, once they have been shown the construction of the special function ψ, and have come to appreciate its stupendous rate of growth, and the obvious possibility of creating functions which grow even more rapidly, their confidence in the conjecture seems to fade; analyticity is too delicate a phenomenon to match such catastrophic growth. Indeed, in one instance, the only student in the class who was able to overcome this feeling and find the simple general proof was one who had been absent the day before, and did not know about the function ψ. (Proof: Put $f(z) = \sum c_n (z/n)^{c_n}$, convergent for all z.)

The power of recursive definitions is now plain to the student; he will not find it hard to modify (5) for a function of three variables, generalizing Peano's recursion, so that:

$$F(x, y, 1) = x + y$$
$$F(x, y, 2) = xy$$
$$F(x, y, 3) = x^y$$

thus obtaining all the usual arithmetic operations at once. (This and the preceding example are slightly modified versions of examples given originally by Ackermann; see [3] or [4].) At this point, the student is also prepared to see the point of general theorems which deal with the more subtle aspects of existence, definability, and computability of functions.

As an illustration, let us re-examine the recursion schemes we have used, and prove that the solution of (5) is unique and can be exhibited as a set of ordered pairs. Let us start from the simple recursion relation:

(11)
$$f(0) = a$$
$$f(m + 1) = g(f(m)) \qquad m = 0, 1, 2, \cdots,$$

where a is an integer, and g is a previously defined function on I to I. Introduce a special mapping S of $I \times I$ into itself defined by:

$$\text{if } p = \langle u, v \rangle, \quad \text{then} \quad S(p) = \langle u + 1, g(v) \rangle.$$

Let us say that a subset $A \subset I \times I$ is *admissible* if it obeys the pair of conditions

(12)
$$\langle 0, a \rangle \in A$$
$$\text{if } p \in A, \quad \text{then} \quad S(p) \in A.$$

There are admissible sets, for example $I \times I$. More to the point, if there is a function f that obeys (11), then its graph is an admissible set.

Let A_0 be the *smallest* admissible set, e.g. the intersection of all the admissible sets. We show that A_0 is the graph of a function. Observe first that if A is any admissible set, and q is any point in A other than $\langle 0, a \rangle$, and if q is not of the form $S(p)$ for any $p \in A$, then we can remove q from A and still have an admissible set, since neither condition of (12) will be violated. Consequently, since A_0 is minimal, every point in it, except $\langle 0, a \rangle$, is of the form $S(p)$ for some $p \in A_0$. Let π be the projection of $I \times I$ onto the first factor, sending $\langle u, v \rangle$ into u; it is then immediate that π maps A_0 onto I one-to-one, so that A_0 is a function with domain I, and the desired solution of the recursion (11).

We have therefore produced a new function Φ of two variables, one an integer and the other a function, whose values are functions, and which is described by saying that $\Phi(a, g) = f$, where f is the (unique) solution of (11). If we let $\mathfrak{F}$ denote the class of all functions on I to I, then Φ is a mapping on $I \times \mathfrak{F}$ to $\mathfrak{F}$. Suppose now that α and β are in $\mathfrak{F}$, and let us attempt to define a sequence of functions $F_n \in \mathfrak{F}$ by the format

(13)
$$F_0 = \alpha$$
$$F_{n+1} = \Phi(\beta(n), F_n) \qquad n = 0, 1, 2, \cdots.$$

The first line means that $F_0(m) = \alpha(m)$ for $m = 0, 1, \cdots$. The second line is harder to interpret; if we set $f = \Phi(\beta(n), F_n)$ then, by (11) which describes Φ,

$$f(0) = \beta(n)$$
$$f(m+1) = F_n(f(m)) \qquad m = 0, 1, 2, \cdots.$$

Since (13) identifies f with F_{n+1}, these conditions amount to asking that

$$F_{n+1}(0) = \beta(n)$$
$$F_{n+1}(m+1) = F_n(F_{n+1}(m)) \qquad m = 0, 1, 2, \cdots.$$

If we now write $F(x, y)$ for $F_y(x)$, we see that all together, we have recaptured the form of the double recursion (5) exactly:

$$F(m, 0) = \alpha(m) \qquad m = 0, 1, 2, \cdots$$
$$F(0, n) = \beta(n-1) \qquad n = 1, 2, \cdots$$
$$F(m+1, n+1) = F(F(m, n+1), n) \qquad m, n = 0, 1, 2, \cdots.$$

Thus, we have shown that a multiple recursion of the complicated type which we used to create the function F, and then ψ, can in fact be reduced to a primitive recursion format, provided we allow function valued functions. Does (13) have a solution? If so, it will be a sequence of functions F_n, that is, a function F on I to $\mathfrak{F}$; can we show its existence by exhibiting it as a subset of $I \times \mathfrak{F}$? The pattern used earlier can be repeated exactly. Introduce a special mapping S of $I \times \mathfrak{F}$ into itself by:

$$\text{if } p = \langle u, \gamma \rangle, \text{ then } S(p) = \langle u+1, \Phi(\beta(u), \gamma) \rangle.$$

Again, say that $A \subset I \times \mathfrak{F}$ is admissible if A contains the point $\langle 0, \alpha \rangle$ and is mapped into itself by S. Then, in exactly the same fashion, the unique minimal admissible set turns out to be the graph of the sought-for function F.

It is clear how this can be continued, basing the study of multiple recursions on that of primitive recursions with more elaborate function valued functions. At this point, we are within range of the concept of a general recursive function and the related notions of computability and constructability. The reader is herewith referred to the bibliography.

References

1. Martin Davis, Computability and Unsolvability, McGraw-Hill, 1958.
2. Leon Henkin, On mathematical induction, this MONTHLY 67 (1960), 323–337.
3. S. C. Kleene, Introduction to Metamathematics, Van Nostrand, 1952.
4. Rózsa Péter, Rekursive Funktionen, Akad. Kiado, Budapest, 1951.
5. J. B. Rosser, Logic for Mathematicians, McGraw-Hill, 1953.

A REMARK ON MATHEMATICAL INDUCTION*

V. L. KLEE, JR., University of Virginia

Suppose we wish to prove a certain theorem, T. It may happen that the simplest way of proving T is to establish a stronger theorem, T^*, of which T is a simple corollary. That is, it may be easier to prove T^* than it is to prove T without using T^*. This fact is particularly useful in framing a proof which employs mathematical induction, and since, moreover, it is an important notion in many mathematical proofs, it deserves early mention in the classroom treatment of mathematical induction.

Interesting in this connection are the remarks of Felix Bernstein [Bull. Amer. Math. Soc., 52 (1946) Abstract 259, p. 622] who suggests that "the four-color theorem may be a simple consequence of a more inclusive theorem which can be proved by complete induction."

A simple example, suitable for elementary instruction, is the following:

Let $S(n) = 1^2 + 2^2 + \cdots + n^2$ for each positive integer n, and let T be the statement that $n+1$ divides $6S(n)$. A simple, direct proof of T is perhaps not immediately apparent. Moreover, the fact that $k+1$ divides $6S(k)$ is not in itself enough to imply that $(k+1)+1$ divides $6S(k+1)$. Hence a proof by induction is evidently not feasible. But now let T^* be the stronger statement that $6S(n) = n(n+1)(2n+1)$. Then T^* can be easily proved by induction and T is obtained as a corollary.

ON THE DANGER OF INDUCTION†

LEO MOSER, University of North Carolina

In pointing out the danger of jumping to general conclusions from a few particular results the Euler polynomial $f(x) = x^2 + x + 41$ which is prime for $x = 1, 2, \ldots, 39$ (but composite for $x = 40$) is often cited. We will give here some other examples of false theorems which are true for the first few cases, and which seem somewhat less artificial than Euler's result.

Let $f(n)$ be the number of regions determined in the inside of a circle by joining n points on the circumference in all possible ways by straight lines, no three of which are concurrent inside the circle. It is easy to check that $f(n) = 1, 2, 4, 8, 16$, for $n = 1, 2, 3, 4, 5$; and it seems natural to suppose that $f(n) = 2^{n-1}$. However if we go to $n = 6$ we find that $f(6) = 31$. We leave it to the reader to show that the correct result for $n > 1$ is actually $f(n) = \sum_{j=0}^{n} \binom{n-1}{j}$.

Our next example deals with prime representing functions. Consider

---

* From MATHEMATICS MAGAZINE, vol. 22 (1948), p. 52.
† From MATHEMATICS MAGAZINE, vol. 23 (1949), p. 109.

$g(n) = 1 + \sum_{j=1}^{n} \phi(j)$, where $\phi(j)$ is Euler's totient function. For $n=1,2,3,4,5,6$, this yields $2,3,5,7,11,13$, which are the first 6 primes. This together with the fact that $\phi(n)$ is an arithmetic function connected with the theory of prime numbers and the fact that both $g(n)$ and p_n have only a single even element, makes the conjecture $g(n)=p_n$ plausible. However $g(7)=19$ shatters the illusion. $g(8)=23$ and $g(9)=19$, so we might still suspect that $g(n)$ is always prime, but this too is false since $g(10)=33$.

Finally there is the somewhat humorous experimental 'proof' that "Every odd number is a prime!" A prime is a number divisible only by 1 and itself. Certainly this is true for $n=1$. We proceed step by step and in each case prove primality by dividing the odd number by all numbers less than it. In this way we find that 3, 5, and 7, are primes. Apparently $9 = 3 \times 3$ but this would spoil the theory which has covered the facts perfectly so far, so we ascribe the discrepancy to experimental error and continue undaunted. We find the theory to hold for 11 and 13 and then we test a few odd numbers chosen at random like $23, 37, 41, \ldots$. Thus the theorem has been proved.

ON THE NUMBER OF SUBSETS OF A FINITE SET*

DAVID S. GREENSTEIN, Northeastern Illinois State College

In Freshman courses in "Modern Mathematics", it is easy to find students who have difficulty grasping the combinatorial reasoning usually employed to show that an n-element set has 2^n subsets. I have found the following way of showing why adding a new element to a set doubles the number of subsets. This approach plus the obvious fact that the empty set has exactly one subset makes it easy to find the number of subsets of an n-element set in n easy steps.

Let A be a finite set and let $b \notin A$. Then each subset S of A "gives birth" to the two subsets S and $S \cup \{b\}$ of $A \cup \{b\}$. Furthermore, each subset of $A \cup \{b\}$ has a unique "birth" from some subset of A. Thus adding b has doubled the number of subsets.

BIBLIOGRAPHIC ENTRIES: INDUCTION

Except for the entry labeled MATHEMATICS MAGAZINE, the references below are to the AMERICAN MATHEMATICAL MONTHLY.

1. W. H. Bussey, The origin of mathematical induction, vol. 24, p. 199.
2. M. Ward, An interesting theorem, vol. 52, p. 540.

Theorem: All real numbers are uninteresting.

3. G. Pólya, On the harmonic mean of two numbers, vol. 57, p. 26.
4. A. Wilansky, An induction fallacy, MATHEMATICS MAGAZINE, vol. 39, p. 305.

An almost flawless induction proof that all nonnegative integers are odd.

---

* From MATHEMATICS MAGAZINE, vol. 43 (1970), p. 36.

(b)

IDENTITIES

SOME TRIGONOMETRIC PRODUCTS*

R. C. MULLIN, University of Waterloo

Products of the form $\prod_{k=1}^{n-1} f(k\pi/n)$, where f is any one of the six basic trigonometric functions, may easily be formed once $\prod_{k=1}^{n-1} \sin(k\pi/n)$ and $\prod_{k=1}^{n-1} \cos(k\pi/n)$ are known. These two latter products are readily found by elementary means.

We begin by noting that $e^{i\theta}$ and $e^{-i\theta}$ are the zeros of $z^2 - 2z\cos\theta + 1$. Also, the polynomial $\sum_{k=0}^{n-1} z^{2k} = (z^{2n}-1)/(z^2-1)$ has $2n-2$ zeros, and these are obviously $e^{ik\pi/n}$ [$k = \pm 1, \pm 2, \cdots, \pm(n-1)$]. Hence we obtain the identity

$$(*) \qquad \prod_{k=1}^{n-1} [z^2 - 2z\cos(k\pi/n) + 1] = \sum_{k=0}^{n-1} z^{2k}.$$

By placing $z = \mp 1$, we at once obtain

$$\prod_{k=1}^{n-1} [1 \pm \cos(k\pi/n)] = n/2^{n-1};$$

multiplying, we find

$$\prod_{k=1}^{n-1} [1 - \cos^2(k\pi/n)] = (n/2^{n-1})^2.$$

Hence

$$\prod_{k=1}^{n-1} \sin(k\pi/n) = n/2^{n-1},$$

since all factors $\sin(k\pi/n)$ are positive.

Now put $z = -i$ in identity (*) to give

$$\prod_{k=1}^{n-1} 2i \cos(k\pi/n) = \tfrac{1}{2}[1 - (-1)^n].$$

If n is even, $\prod_{k=1}^{n-1} \cos(k\pi/n) = 0$, a fact otherwise obvious, since $\cos\tfrac{1}{2}\pi = 0$. If

---

* From AMERICAN MATHEMATICAL MONTHLY, vol. 69 (1962), pp. 217–218.

n is odd, it immediately follows that

$$\prod_{k=1}^{n-1} \cos (k\pi/n) = (-1)^{\frac{1}{2}(n-1)}/2^{n-1},$$

and we can at once deduce that, for n odd,

$$\prod_{k=1}^{n-1} \tan (k\pi/n) = (-1)^{\frac{1}{2}(n-1)}n.$$

A NOTE ON SUMS OF POWERS OF INTEGERS*

DAVID ALLISON, University of Cape Town, South Africa

The identity

$$\sum_{1}^{n} r^3 = \left(\sum_{1}^{n} r \right)^2 \tag{1}$$

is well known, and the question naturally arises, for what positive integers k, m, p, q does the identity

$$\left(\sum_{1}^{n} r^k \right)^p = \left(\sum_{1}^{n} r^m \right)^q \tag{2}$$

hold? Clearly no cases of interest will be lost by supposing that $p > q$ and that p and q are coprime; and with these assumptions it is easy to show that (1) is the only solution of (2).

The proof depends on the well-known facts that $\sum_{r=1}^{n} r^m$ is a polynomial in n, of degree $m+1$, and that the coefficient of n^{m+1} is $1/(m+1)$. Suppose then that (2) holds, i.e., that

$$\left(\frac{n^{k+1}}{k+1} + \cdots \right)^p = \left(\frac{n^{m+1}}{m+1} + \cdots \right)^q,$$

with p and q as described above. Then $p(k+1) = q(m+1)$, $(k+1)^p = (m+1)^q$. Thus $(m+1)^q = (p/q)^q(k+1)^q$, whence $(k+1)^{p-q} = (p/q)^q$, and the R.H.S. of this is an integer because the L.H.S. is. Thus $q=1$, and $(k+1)^{p-1} = p$. But $k+1 \geq 2$; therefore $2^{p-1} \leq p$, so that $p < 3$. Since $p > q$, we have $p = 2$, $k = 1$, $m = 3$.

Reference

1. Sheila M. Edmonds, Math. Gaz., vol. 41, 1957, pp. 187–188.

---

* From AMERICAN MATHEMATICAL MONTHLY, vol. 68 (1961), p. 272.

A COMBINATORIAL PROOF THAT $\sum k^3 = \left(\sum k\right)^2$*

ROBERT G. STEIN, California State College, San Bernardino

How many rectangles can you find in this picture?

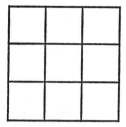

A little counting will show that there are just thirty-six rectangles here. In general, for any positive integer n, a picture of an $n \times n$ square ruled into unit squares will have $(1+2+3+\cdots+n)^2 = [n(n+1)/2]^2$ rectangles in it. Giving two separate counting arguments for this yields a combinatorial proof of the identity

(1) $\qquad 1^3 + 2^3 + 3^3 + \cdots + n^3 = (1 + 2 + 3 + \cdots + n)^2,$

which relates "squares," "triangle numbers," and "cubes" in a most interesting way. (The usual proof of this, by mathematical induction, is quite unenlightening. Toeplitz, in his wonderful *Calculus, A Genetic Approach*, gives an interesting old Arabic proof, so the proof given here is the third known to this writer.)

For the first counting method, notice that the rectangles in question have length $n-a$ and width $n-b$, where a and b are integers between 0 and n. How many such $n-a$ by $n-b$ rectangles are there in our $n \times n$ square? Each such rectangle may be translated $a+1$ units in one direction and $b+1$ in the perpendicular direction, to $(a+1)(b+1)$ positions in all. If $a \neq b$, a 90° rotation of one of our $n-a$ by $n-b$ rectangles produces another such rectangle which is not a translate of the original, and this in turn has $(a+1)(b+1)$ translates. Thus if $a \neq b$, there are just $2(a+1)(b+1)$ rectangles $n-a$ by $n-b$ in the $n \times n$ square. If $a = b$, there are $(a+1)^2$ such rectangles, since rotation by 90°, in the case of a square, does not yield a new rectangle. Then the number of rectangles whose shortest side is $n-a$ is

$$(a+1)^2 + 2(a+1)a + 2(a+1)(a-1) + \cdots + 2(a+1)(1)$$
$$= (a+1)^2 + 2(a+1)[a + (a-1) + (a-2) + \cdots + 1]$$
$$= (a+1)^2 + 2(a+1)\left[\frac{a(a+1)}{2}\right] = (a+1)^3.$$

---

* From MATHEMATICS MAGAZINE, vol. 44 (1971), pp. 161-162.

Since each rectangle has as its shortest side just one of the numbers 1, 2, 3, $\cdots$, n, we may count the rectangles in the $n \times n$ square as

$$\sum_{a=0}^{n-1} (a+1)^3.$$

We now count the rectangles a second way. There are just n^2 points which could serve as the lower left vertex of a rectangle, namely all the points in the picture of the ruled $n \times n$ square where lines meet, except those on the upper or righthand boundaries. If we consider a point c units to the left of the upper right corner of the big square and d units below it, we see this is the lower left corner of just cd of the rectangles we are trying to count. Thus in all there are $\sum_{c=1}^{n} \sum_{d=1}^{n} cd$ rectangles, or $(1+2+3\cdots+n)^2$, which proves (1).

The second counting method generalizes easily to count the number of rectangles in a ruled $n \times m$ rectangle as

$$(1+2+3+\cdots+n)(1+2+3\cdots+m) = \frac{n(n+1)}{2} \frac{m(m+1)}{2}$$

but because the symmetry of n and m is lost, this is no longer an elegant sum of cubes.

BIBLIOGRAPHIC ENTRY: IDENTITIES

1. H. Flanders, A democratic proof of a combinatorial identity, MATHEMATICS MAGAZINE, vol. 44, p. 11.

Proof that $\sum_{k=j}^{i} \binom{k}{j}\binom{i}{k}(-1)^{j+k} = 0$.

(c)

INEQUALITIES

ARITHMETIC, GEOMETRIC INEQUALITY*

D. J. NEWMAN, Brown University

The fact that, for $x_1, \cdots, x_n$ positive,

(1) $$\sqrt[n]{(x_1 \cdots x_n)} \leq \frac{x_1 + \cdots + x_n}{n},$$

---

* From AMERICAN MATHEMATICAL MONTHLY, vol. 67 (1960), p. 886.

has many proofs abounding in the literature. We give a new variant which seems much more direct and which is just the thing for a class equipped with mathematical induction and little else.

(1) is trivial for $n=1$, now suppose it to hold for $n=k$ and assume $x_1 x_2 \cdots x_{k+1} = 1$, $x_i > 0$. It is our burden to prove $x_1 + \cdots + x_{k+1} \geq k+1$.

Since $x_1 x_2 \cdots x_{k+1} = 1$ it follows that one of the x_i's is ≥ 1. We may take this to be x_{k+1} and obtain $x_1 x_2 \cdots x_k = 1/x_{k+1}$. By (1), which is assumed to hold for $n=k$, we obtain $x_1 + x_2 + \cdots + x_k \geq k/\sqrt[k]{x_{k+1}}$ and so $x_1 + x_2 + \cdots + x_{k+1} \geq k/\sqrt[k]{x_{k+1}} + x_{k+1}$. The proof will be complete, then, once we establish

LEMMA. *If $x \geq 1$ then $k/\sqrt[k]{x} + x \geq k+1$.*

(The fact is that, for $x \geq 1$, $k/\sqrt[k]{x} + x$ is an increasing function, as is easily shown by differentiation, but if differentiation is outlawed then the following proof may be used.)

Proof. First recall that, for $a \geq 0$, $(1+a)^{k+1} \geq 1 + (k+1)a$. This is easily established by induction or, for that matter, by the binomial theorem. Hence, with $a = \sqrt[k]{x} - 1$, we obtain $x \cdot \sqrt[k]{x} \geq (k+1)\sqrt[k]{x} - k$, or, transposing and dividing, $k/\sqrt[k]{x} + x \geq k+1$.

That the equality holds in (1) only when $x_1 = x_2 = \cdots = x_n$, can also be obtained by a trivial modification of the above proof.

A PROOF OF THE ARITHMETIC-GEOMETRIC MEAN INEQUALITY*

BENGT ÅKERBERG, Stockholm, Sweden

Suppose that α is a positive number and n a positive integer greater than 1. It is easy to verify that

$$(\alpha - 1)[n - (1 + \alpha + \alpha^2 + \cdots + \alpha^{n-1})] \leq 0$$

or

(1) $$\alpha(n - \alpha^{n-1}) \leq n - 1,$$

and the inequality is strict unless $\alpha = 1$.

If $a_1, a_2, \cdots, a_n$ denote positive numbers and if

$$\alpha = \left(\frac{a_1}{\frac{a_1 + a_2 + \cdots + a_n}{n}} \right)^{1/(n-1)}$$

then it follows from (1) that

$$\left(\frac{a_1 + a_2 + \cdots + a_n}{n} \right)^n \geq a_1 \left(\frac{a_2 + \cdots + a_n}{n - 1} \right)^{n-1}.$$

---

* From AMERICAN MATHEMATICAL MONTHLY, vol. 70 (1963), pp. 997–998.

By repeated applications of this formula we obtain

$$\left(\frac{a_1 + \cdots + a_n}{n}\right)^n \geq a_1\left(\frac{a_2 + \cdots + a_n}{n-1}\right)^{n-1}$$
$$\geq a_1 a_2 \left(\frac{a_3 + \cdots + a_n}{n-2}\right)^{n-2}$$
$$\cdots \cdots \cdots \cdots$$
$$\geq a_1 a_2 \cdots a_n,$$

an inequality which asserts that the arithmetic mean of $a_1, a_2, \cdots, a_n$ is not less than their geometric mean.

The inequality in the first line is strict unless $\alpha = 1$ or $a_1 = (a_1 + \cdots + a_n)/n$, and in the second line unless this holds and also $a_2 = (a_2 + \cdots + a_n)/(n-1)$, or $a_1 = a_2$.

It is clear that we may repeat this argument until we find that the inequality in the final step is strict unless $a_1 = a_2 = \cdots = a_n = (a_1 + \cdots + a_n)/n$.

ON TWO FAMOUS INEQUALITIES*

L. H. LANGE, University of Notre Dame

In December of 1958 the Commission on Mathematics of the College Entrance Examination Board, established in 1955, published its report. The report contains a nine-point program designed to meet the needs of contemporary college-capable students of mathematics. One of these points concerns "treatment of inequalities along with equations" and it is a point well taken.† In the present note we take this point to heart. We discuss briefly several elementary but worthwhile inequality problems and, secondly, we call attention to a rewarding source of material on inequality problems and their solution.

We begin with the simplest special case of the most famous theorem belonging to the subject of inequalities, the *theorem of the arithmetic and geometric means*.‡ If a and b are (positive) real numbers, then $0 \leq (a-b)^2$, with equality taking place if and only if $a = b$. Equivalently, then $2ab \leq a^2 + b^2$. Letting $x = a^2, y = b^2$, we see immediately that what we are really dealing with here is the following relation between the geometric and arithmetic means of the positive numbers x and y:

$\sqrt{xy} \leq \frac{x+y}{2}$, with equality occuring if and only if $x = y$.

---

* From MATHEMATICS MAGAZINE, vol. 32 (1959), pp. 157–160.

† See point number 5 of the list given on pages 773–774 of the *American Mathematical Monthly* for December, 1958.

‡ For several proofs of the *general* statement of this theorem see G. H. Hardy, J. E. Littlewood, and G. Pólya, *Inequalities*, Cambridge, 1952, pp. 16–21.

In its various forms this inequality appears in many contexts, of course. Here is an application which apparently does not occur in the texts. We consider the problem of determining the minimum initial speed, v_0, required to drive a golf ball 300 yards—where we assume the fairway to be level, the atmosphere non-meddling, and a terminal roll of 25 yards. Letting the initial velocity vector of the golf ball have a horizontal component of magnitude $a>0$ and a vertical component of magnitude $b>0$, we have $v_0^2 = a^2 + b^2$. In all cases involving this type of projectile flight, the *range* of the projectile is given by $\frac{2ab}{g}$, where g is the gravitational constant. (This follows from the fact that the parametric equations for the projectile path are $x = at$, $y = -\frac{gt^2}{2} + bt$, and the equation $y=0$ has the positive root $\frac{2b}{g}$. Here x, y, and t are to be assigned their standard meanings.) In our problem the range is a given number, 275 yards. Hence, in particular, the product $2ab$ is a fixed number. Then, since $2ab \leqq a^2 + b^2$, the theorem of the means tells us that the minimum of $a^2 + b^2$ occurs if and only if $a = b$. It follows that we may very easily now compute the minimum, v_0, since we may therefore write

$$275 \text{ yards} = \frac{2ab}{g} = \frac{a^2 + b^2}{g} = \frac{v_0^2}{g}.$$

We may give the approximate answer $v_0 = 160$ feet per second.

We now concern ourselves briefly with another famous inequality, the *Cauchy inequality*. Here too we consider only a special case, though the proof in this case is easily modified to dispose of the general case as well.*

Consider the expression

$$T = (a_1 - tb_1)^2 + (a_2 - tb_2)^2 + (a_3 - tb_3)^2,$$

where t is a real number and $a_i > 0, b_i > 0$, $i = 1, 2, 3$. Unless there exists a constant k such that $a_i = kb_i$ for $i = 1, 2, 3$, we have $T > 0$ for all t. Hence, if the a's and b's are *not* proportional, the following quadratic equation in t,

$$(t^2) \sum b_i^2 + (t)\left(-2 \sum a_i b_i\right) + \sum a_i^2 = 0,$$

must have imaginary roots. Since this implies that the discriminant is negative, we have

$$\left(-2 \sum a_i b_i\right)^2 - 4\left(\sum b_i^2\right)\left(\sum a_i^2\right) < 0.$$

Thus we can arrive at Cauchy's inequality:

$$\left(\sum a_i b_i\right)^2 \leqq \sum a_i^2 \sum b_i^2,$$

with equality occurring if and only if the a's and b's are proportional.

---

* See *Inequalities*, p. 16.

We now take a problem from a book which is rich in inequality problems. It is G. Pólya's *Induction and Analogy in Mathematics*,* and we discuss part (2) of problem 61, page 141, "Given E, the sum of the lengths of the twelve edges of a box, find the maximum (1) of its volume V, (2) of its surface S." (Pólya has earlier replaced the lengthy "rectangular parallelepiped" by "box".)†

Letting x, y, z be the lengths of the three edges drawn from the same vertex of the box, we have $E = 4(x+y+z)$ and we can state our problem this way: *Maximize* the expression $S = 2(xy+yz+zx)$ under the side condition $x+y+z = \dfrac{E}{4} =$ a positive constant. Now, since $(\dfrac{E}{4})^2 = x^2+y^2+z^2+S$, our problem is to *minimize* $S' = x^2+y^2+z^2$ under the same side condition. If we dwell on the fact that Cauchy's inequality contains such a sum of squares, we can make the pleasant discovery that it solves our problem if we set $a_1 = x$, $a_2 = y$, $a_3 = z$; $b_1 = b_2 = b_3 = 1$. Then

$$(x \cdot 1 + y \cdot 1 + z \cdot 1)^2 \leq (x^2+y^2+z^2)(1^2+1^2+1^2);$$

i.e.,

(I) $$(x+y+z)^2 \leq 3(x^2+y^2+z^2).$$

Thus

$$\left(\frac{E}{4}\right)^2 \leq 3S',$$

where S' is a minimum if and only if $x = y = z = \dfrac{E}{12}$. Our box with minimum surface area is a cube.

To be sure, the problem can be solved without an appeal to Cauchy's inequality, either as Pólya does it,‡ or by simply establishing our inequality (I) directly by adding the four inequalities:

$$2xy \leq x^2+y^2$$
$$2yz \leq y^2+z^2$$
$$2zx \leq z^2+x^2$$
$$x^2+y^2+z^2 \leq x^2+y^2+z^2.$$

---

* This is volume I of his two volume set called *Mathematics and Plausible Reasoning* and published by Princeton in 1954. After each chapter in these volumes, Professor Pólya places a sequence of problems. In the spirit of Pólya and Szegö, *Aufgaben und Lehrsätze aus der Analysis* I, II (Berlin, 1925), he later gives solutions to these problems.

† Part (1), which is perhaps easier than part (2), we leave for the reader. Pólya's solution takes but one line on page 257 of that volume.

‡ See pages 257–258 of *Induction and Analogy in Mathematics*.

BIBLIOGRAPHIC ENTRIES: INEQUALITIES

The references below are to the MATHEMATICS MAGAZINE.

1. R. K. Coburn, An analytical method for solving basic inequalities, vol. 34, p. 345.

2. W. P. Cooke, Two-dimensional graphical solution of higher-dimensional linear programming problems, vol. 46, p. 70.

4

TRIGONOMETRY AND TRIGONOMETRIC FUNCTIONS

ON THE REPRESENTATION OF THE TRIGONOMETRIC FUNCTIONS BY LINES*

R. D. CARMICHAEL, Alabama Presbyterian College

The representation of the trigonometric functions dealt with in this note is in some respects different from that usually employed; it has certain advantages which appear to make it superior to other methods of effecting this representation.

Let the circle whose center is O (see figure) have the radius unity; and let O be the origin of the rectangular axes XX' and YY'. Let the angle $MOT = x$ be formed by the radius revolving counter-clockwise from OX to any position OT. Draw TP and TQ perpendicular, respectively, to OX and OY. At T draw a tangent to the circle cutting the X-axis in M and the Y-axis in N. Then in whatever quadrant OT may lie we have

$$\sin x = OQ, \quad \sec x = OM, \quad \tan x = TM,$$
$$\cos x = OP, \quad \csc x = ON, \quad \cot x = TN.$$

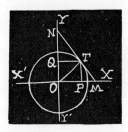

The tangent and the cotangent are measured from the point of tangency; the other functions from the center of the circle. Thus, all functions are measured from an extremity of the revolving radius.

It should be noticed that in any quadrant the tangent is that portion of the tangent line intercepted between the point of tangency and the X-axis, while the cotangent is intercepted between the point of tangency and the Y-axis. *This is the conception of Analytics.* The secant and the cosine are measured along the X-axis,

---

* From AMERICAN MATHEMATICAL MONTHLY, vol. 15 (1908), pp. 199–200.

while the sine and the cosecant are measured along the Y-axis. The following facts are evident from an inspection of the figure for an angle in each quadrant:

The algebraic signs of the sine, cosine, secant, and cosecant are determined by the direction in which each is measured from O in accordance with the usual convention of Analytics. The algebraic sign of the tangent is plus when the tangent is measured to the right of OT (as one looks from O); minus, when measured to the left. The tangent and cotangent have always the same sign.

Any two functions of an angle measured along the same line have unity for their product.

It seems to me that this representation of the functions will make it easier for the student to fix the algebraic sign of any function in any quadrant, and also to remember the group of products each equal to unity.

The method also lends itself very readily to approximate measurements of the functions for rough work. For this purpose the pupil will require a circular protractor and a "square" graduated to tenths of the unit on the inner edges of the angle. This square should have for unit the radius of the protractor. Tangents and cotangents may be read (accurately to tenths, estimated to hundredths) by laying one inner edge of the square along the radius OT which cuts off on the protractor the required angle, and then reading TM or TN according as tangent or cotangent is desired. The sine and the cosine may be read by putting the vertex of the square as at P and reading PO and PT. (It is in measuring tangents and cotangents, of course, that this method has the advantage over the ordinary methods.)

A CURIOUS CASE OF THE USE OF MATHEMATICAL INDUCTION IN GEOMETRY*

J. V. USPENSKY, Carleton College

It is known that the ratio $\sin x/x$ continually decreases when x increases from 0 to π. Ordinarily the proof is given by means of trigonometry or calculus. The purpose of this note is to show how an equivalent statement can be established by purely geometric means using the process of reasoning by induction, which certainly is a rare case in geometry. Following is the fundamental theorem:

THEOREM I. *If α and β denote the angles at the base AB of a triangle ABC (Fig. 1) opposite to the sides CB and CA, then the following inequalities hold*

$$\frac{CA}{BA} > \frac{\beta}{\alpha+\beta} \quad \text{and} \quad \frac{CB}{AB} > \frac{\alpha}{\alpha+\beta}.$$

Proof. (a) First let us consider triangles with commensurable angles at the base, so that we can set

$$\alpha = p\delta \quad \text{and} \quad \beta = q\delta$$

---

* From AMERICAN MATHEMATICAL MONTHLY, vol. 34 (1927), pp. 247–250.

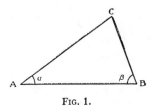

Fig. 1.

where p and q are relatively prime integers and δ is a certain angle. Supposing that the above inequalities have already been proved in every case when the corresponding sum $p+q$ is *less* than a given integer $N>2$, we shall show that they continue to hold when $p+q=N$. Let ABC be a given triangle (Fig. 2) with the angles at the base $p\delta$ and $q\delta$, where p and q are relatively prime integers and $p+q=N>2$. The sides AC and CB cannot be equal, and we can suppose $AC>BC$. Describe from C as center and with CA as radius the circumference which passes through A and cuts the prolonged base AB at the point D. In the triangle ACD,

(1) $\qquad AC=CD \quad \text{and} \quad 2AC>AD=AB+BD.$

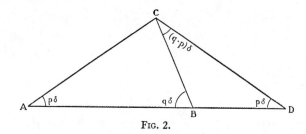

Fig. 2.

Considering CD as the base of the triangle CBD we find that the adjoining angles are $p'\delta$, $q'\delta$, where $p'=p$ and $q'=q-p$, so that

$$p'+q'=q<N.$$

As the sum $p'+q'$ is *less* than N we have by supposition

(2) $\qquad BD > \dfrac{q-p}{q} CD \quad \text{or} \quad BD > \dfrac{q-p}{q} AC.$

Adding (1) and (2) we get

$$2AC+BD > \dfrac{q-p}{q} AC + AB + BD,$$

whence

$$\dfrac{q+p}{q} AC > AB \quad \text{or} \quad \dfrac{AC}{AB} > \dfrac{q}{p+q} = \dfrac{\beta}{\alpha+\beta}.$$

From the same triangle CBD we get

$$CB > \dfrac{p}{q} CA \quad \text{and} \quad CB > \dfrac{p}{p+q} AB \quad \text{or} \quad \dfrac{CB}{AB} > \dfrac{\alpha}{\alpha+\beta}.$$

Now if the sum $p+q$ has the least possible value, namely 2, it must be that $p+q=1$; that is the triangle is isosceles, and for such a triangle it is obvious that

$$\frac{AC}{AB} > \frac{1}{2} \quad \text{and} \quad \frac{BC}{AB} > \frac{1}{2},$$

so that our inequalities hold true when $N=2$. Then they will continue to hold for $N=3,4,5,\ldots$, that is in every possible case, provided the angles α and β are commensurable.

(b) Next suppose α and β incommensurable. Divide α into p equal parts, so that $\alpha = p\delta$ and determine an integer q by the condition $q\delta < \beta < (q+1)\delta$. Then construct the triangle ABC' with the angles $p\delta$ and $q\delta$ at its base AB. As these angles are commensurable we have

$$\frac{AC'}{AB} > \frac{q}{p+q} \quad \text{and} \quad \frac{BC'}{AB} > \frac{p}{p+q}.$$

Now if p increases indefinitely, AC' and BC' approach respectively to AC and BC, and at the same time the ratios $q/(p+q)$ and $p/(p+q)$ converge to the limits $\beta/(\alpha+\beta)$ and $\alpha/(\alpha+\beta)$. Performing the passage to the limit we conclude from the preceding inequalities that

$$\frac{AC}{AB} \geq \frac{\beta}{\alpha+\beta} \quad \text{and} \quad \frac{BC}{AB} \geq \frac{\alpha}{\alpha+\beta}.$$

It is easy to show that the sign $=$ is excluded. Repeat the same construction as in Fig. 1. From the triangle BCD we get

$$BD \geq \frac{\beta-\alpha}{\beta} AC, \quad \text{and furthermore} \quad 2AC > AB + BD,$$

whence

$$\left(2 - \frac{\beta-\alpha}{\beta}\right) AC > AB, \quad \text{or} \quad \frac{AC}{AB} > \frac{\beta}{\alpha+\beta}.$$

Combining this inequality with $\dfrac{CB}{AC} \geq \dfrac{\alpha}{\beta}$, which follows from the same triangle BCD, we get finally

$$\frac{CB}{AB} > \frac{\alpha}{\alpha+\beta}. \qquad\qquad Q.E.D.$$

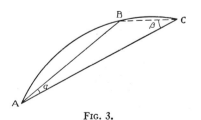

Fig. 3.

THEOREM II. *Two arcs AB and AC, neither exceeding a semicircumference, being taken on the same circle and the latter being the greater of the two, the following inequality holds*:

$$\frac{AB}{AC} > \frac{\operatorname{arc} AB}{\operatorname{arc} AC}.$$

Proof. Applying the preceding theorem to the triangle ABC we have

$$\frac{AB}{AC} > \frac{\beta}{\alpha + \beta}.$$

On the other hand

$$\frac{\operatorname{arc} AB}{\operatorname{arc} BC} = \frac{\beta}{\alpha},$$

whence

$$\frac{\operatorname{arc} AB}{\operatorname{arc} AC} = \frac{\beta}{\alpha + \beta} \quad \text{and} \quad \frac{AB}{AC} > \frac{\operatorname{arc} AB}{\operatorname{arc} AC}. \qquad Q.E.D.$$

COROLLARY I. Taking a certain unit of length to measure distances we can express the lengths of the chords AB and AC by the *numbers* s and s'. Taking a certain arc of our circle, e.g. the whole circumference, as a unit we can express the measures of the arcs AB and AC by two *numbers* σ and σ'. We have then the following relation:

$$\frac{s}{s'} > \frac{\sigma}{\sigma'} \quad \text{or} \quad \frac{s}{\sigma} > \frac{s'}{\sigma'}.$$

if $\sigma' > \sigma$, and this implies that $\sin x / x$ diminishes when x increases from 0 to $\pi/2$.

THEOREM III. *Denote by P and P' the perimeters of two polygons inscribed in the same circle. If the greatest side of the second is less than the smallest side of the first, then*

$$P' > P.$$

Proof. Denote by $s_1, s_2, s_3, \ldots, s_n$ the measures of the sides of the first polygon in the *increasing* order of magnitude, and by $\sigma_1, \sigma_2, \sigma_3, \ldots, \sigma_n$ the measures of the subtended arcs, the whole circumference being taken as a unit. Denote by $s'_1, s'_2, s'_3, \ldots, s'_m$ the measures of the sides of the second polygon in the *decreasing* order of magnitude, $\sigma'_1, \sigma'_2, \sigma'_3, \ldots, \sigma'_m$ being the measures of the subtended arcs. As

$$\sigma_1 < \sigma_2 < \sigma_3 < \cdots < \sigma_n \quad \text{and} \quad \sigma'_1 > \sigma'_2 > \sigma'_3 > \cdots > \sigma'_m,$$

we have by the preceding corollary

$$\frac{s_1}{\sigma_1} > \frac{s_2}{\sigma_2} > \frac{s_3}{\sigma_3} > \cdots > \frac{s_n}{\sigma_n} \quad \text{and} \quad \frac{s'_1}{\sigma'_1} < \frac{s'_2}{\sigma'_2} < \frac{s'_3}{\sigma'_3} < \cdots < \frac{s'_m}{\sigma'_m}$$

whence it follows that

$$\frac{s'_1+s'_2+\cdots+s'_m}{\sigma'_1+\sigma'_2+\cdots+\sigma'_m} > \frac{s'_1}{\sigma'_1} \quad \text{and} \quad \frac{s_1+s_2+\cdots+s_n}{\sigma_1+\sigma_2+\cdots+\sigma_n} < \frac{s_1}{\sigma_1}.$$

By supposition $\sigma'_1 < \sigma_1$, whence by the same corollary $\frac{s_1}{\sigma_1} < \frac{s'_1}{\sigma'_1}$. That is

$$\frac{s_1+s_2+\cdots+s_n}{\sigma_1+\sigma_2+\cdots+\sigma_n} < \frac{s'_1+s'_2+\cdots+s'_m}{\sigma'_1+\sigma'_2+\cdots+\sigma'_m}.$$

Now

$$\sigma_1+\sigma_2+\cdots+\sigma_n = 1 \text{ and } \sigma'_1+\sigma'_2+\cdots+\sigma'_m = 1$$

and consequently

$$s_1+s_2+\cdots+s_n < s'_1+s'_2+\cdots+s'_m.$$

That is
$$P < P'. \qquad Q.E.D.$$

COROLLARY. Perimeters of the regular polygons inscribed in the same circle increase with the increasing number of sides.

THE TRIGONOMETRIC FUNCTIONS OF HALF OR DOUBLE AN ANGLE*

ROSCOE WOODS, State University of Iowa

The following simple method for deriving the formulas connecting the trigonometric functions of double or half an angle is not featured in texts on trigonometry.

For convenience we consider acute angles only. The figure needs no explanation.

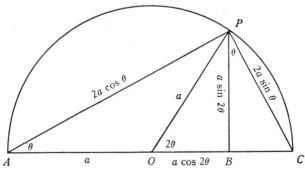

$AB = a(1 + \cos 2\theta)$, and $BC = a(1 - \cos 2\theta)$.

---

* From AMERICAN MATHEMATICAL MONTHLY, vol. 43 (1936), pp. 174–175.

From triangle ABP, we have by definition,

$\sin \theta = a \sin 2\theta / 2a \cos \theta,$ or, $2 \sin \theta \cos \theta = \sin 2\theta,$

$\cos \theta = a(1 + \cos 2\theta)/2a \cos \theta,$ or, $2 \cos^2 \theta = 1 + \cos 2\theta,$

$\tan \theta = a \sin 2\theta / a(1 + \cos 2\theta),$ or, $\tan \theta = \sin 2\theta / (1 + \cos 2\theta).$

From triangle PBC, we have by definition,

$\sin \theta = a(1 - \cos 2\theta)/2a \sin \theta,$ or, $2 \sin^2 \theta = 1 - \cos 2\theta,$

$\cos \theta = a \sin 2\theta / 2a \sin \theta,$ or, $2 \sin \theta \cos \theta = \sin 2\theta,$

$\tan \theta = a(1 - \cos 2\theta)/a \sin 2\theta,$ or, $\tan \theta = (1 - \cos 2\theta)/\sin 2\theta.$

By substitution the relation

$$AP^2 - AB^2 = BP^2 = PC^2 - BC^2$$

becomes

$$\cos^2 \theta - \sin^2 \theta = \cos 2\theta.$$

Editorial Note. The author writes: "I have looked through about fifty texts on trigonometry and have seen the figure in only four, viz., Crockett, p. 73; Smail, p. 105; Granville-Smith-Mikesh, p. 97; Hobson (2nd Ed.), p. 53. The authors use the figure to prove some of the formulas for the half and double angles, but do it in such a way that it appears very hard."

GEOMETRIC PROOFS OF MULTIPLE ANGLE FORMULAS*

WAYNE DANCER, University of Toledo

The following construction for the trisection of an angle, accomplished by means of a marked straight edge, has been well known since the time of Archimedes.* The purpose of this note is to show that this same figure may be utilized to establish and illustrate several common trigonometric identities.

Let the angle GCH, of magnitude $3x$, be placed at the center of a circle of radius unity. Mark on a straight edge a segment AE of unit length. Lay the straight edge on the figure to determine a line through G such that one end of the marked segment, E, will fall on the circumference, and the other end, A, will fall on the diameter extended. Draw the radius CE. Then $\angle EAC = \angle ACE = x$, and $\angle GEC = \angle CGE = 2x$. From points E and G drop perpendiculars EB and GD respectively to the horizontal diameter, and from the center C drop a perpendicular CF to the line EG. It is observed that all segments in the figure

---

* From AMERICAN MATHEMATICAL MONTHLY, vol. 44 (1937), pp. 366–367.

are either of unit length, or equal in length to the sine or cosine of one of the angles x, $2x$, or $3x$, as shown. Application of the Theorem of Pythagoras to proper triangles in the figure illustrates the identity $\sin^2 \alpha + \cos^2 \alpha = 1$ for the three cases where $\alpha = x$, $2x$, $3x$.

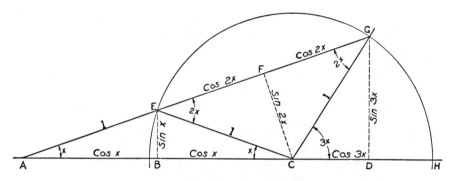

Fig. 1

In the triangle ACF, we have $AF = AC \cos x$, whence $1 + \cos 2x = 2 \cos^2 x$, or, transposing, $\cos 2x = 2 \cos^2 x - 1$; and $CF = AC \sin x$, or $\sin 2x = 2 \sin x \cos x$. Similarly, in the triangle ADG, we have $AD = AG \cos x$, that is,

$$2 \cos x + \cos 3x = (1 + 2 \cos 2x) \cos x,$$

or

$$\cos 3x = 2 \cos x \cos 2x - \cos x$$
$$= 2 \cos x (2 \cos^2 x - 1) - \cos x$$
$$= 4 \cos^3 x - 3 \cos x.$$

In this same triangle, $DG = AG \sin x$, or,

$$\sin 3x = (1 + 2 \cos 2x) \sin x$$
$$= \sin x + 2(1 - 2 \sin^2 x) \sin x$$
$$= 3 \sin x - 4 \sin^3 x.$$

The figure may be extended to include higher multiples of x, but the formulas developed are the fundamental ones.

THE ADDITION FORMULAS FOR THE SINE AND COSINE*

E. J. McSHANE, University of Virginia

Off and on, for some years, I have tried to find a proof of the addition formulas for the sine and cosine which would have the following three properties. (1) It should be valid for all angles, and not involve any discussion of the quadrants in which the angles lie. (2) It should not require previous knowledge of the formulas for the functions of $n \cdot 90° \pm A$ in terms of functions of A. (3) It should not be too difficult for first-year students to follow. The proof below, which I have not seen published, satisfies (1) and (2), and perhaps comes as close to satisfying (3) as any other. It requires a knowledge of the distance formula, of the general definitions of the trigonometric functions, and of the equations

$$\cos^2 \theta + \sin^2 \theta = 1,$$
$$\cos 0° = \sin 90° = 1, \quad \cos 90° = \sin 0° = 0.$$

From the definitions of the sine and cosine we deduce at once that if P is a point whose distance from the origin is r and for which the angle of OP with the positive x-axis is θ, the coördinates of P are $(r \cos \theta, r \sin \theta)$.

Let A and B be any two angles. With a vertex O and a half-line OW as a beginning we construct angles A and B, and on their terminal half-lines we choose points P and Q respectively, each at distance 1 from O. Let d denote the distance from P to Q. We shall now make two computations for d^2, using first OW and then OQ as x-axis.

First using OW as x-axis, we find that the coördinates of P and Q are $(\cos A, \sin A)$ and $(\cos B, \sin B)$ respectively, since they each have distance 1 from the origin O and the half-lines OP, OQ make the respective angles A, B with the positive x-axis. Hence

$$d^2 = (\cos A - \cos B)^2 + (\sin A - \sin B)^2$$
$$= 2 - 2[\cos A \cos B + \sin A \sin B].$$

Next we use OQ as positive x-axis. The half-lines OP, OQ now make the respective angles $A-B$, 0 with the positive x-axis, so P and Q have coördinates $(\cos (A-B), \sin (A-B))$ and $(1, 0)$ respectively. Hence

$$d^2 = (\cos (A - B) - 1)^2 + \sin^2 (A - B)$$
$$= 2 - 2 \cos (A - B).$$

Equating the two expressions for d^2 yields

(1) $\qquad \cos (A - B) = \cos A \cos B + \sin A \sin B.$

If we wish to use the formulas for the functions of $n \cdot 90° \pm A$ in terms of functions of A, the formulas for $\cos (A+B)$ and $\sin (A \pm B)$ can quickly be deduced from (1). However, we do not need to use the formulas for the functions of

---

* From AMERICAN MATHEMATICAL MONTHLY, vol. 48 (1941), pp. 688–689.

$n \cdot 90° \pm A$; the equations mentioned in the introduction, together with (1), are enough.

In (1) we set $A = 0$, obtaining

(2) $$\cos(-B) = \cos B.$$

Again, by setting $A = 90°$ in (1) we find

(3) $$\cos(90° - B) = \sin B.$$

In (3) we set $B = 90° - C$; this yields

(4) $$\sin(90° - C) = \cos C.$$

By (3), and (4) and (1), we have

(5)
$$\begin{aligned}\sin(A + B) &= \cos[90° - (A + B)] \\ &= \cos[(90° - A) - B] \\ &= \cos(90° - A)\cos B + \sin(90° - A)\sin B \\ &= \sin A \cos B + \cos A \sin B.\end{aligned}$$

In (3) we set $B = -A$, obtaining

(6) $$\sin(-A) = \cos(90° + A).$$

Now from (6) and (1), we deduce

(7)
$$\begin{aligned}\sin(-A) &= \cos[A - (-90°)] \\ &= -\sin A.\end{aligned}$$

By (1), (2), and (7) we have

(8)
$$\begin{aligned}\cos(A + B) &= \cos[A - (-B)] \\ &= \cos A \cos(-B) + \sin A \sin(-B) \\ &= \cos A \cos B - \sin A \sin B;\end{aligned}$$

and likewise by (5), (2), and (7) we find

(9) $$\sin(A - B) = \sin A \cos B - \cos A \sin B.$$

Since we have not used the formulas for the functions of $n \cdot 90° \pm A$ in deducing (1), (5), (8), and (9) we can use the latter, with the known values of $\sin(n \cdot 90°)$ and $\cos(n \cdot 90°)$, to deduce the former.

THE ADDITION FORMULAS IN TRIGONOMETRY*

A. S. HOUSEHOLDER, University of Chicago

Cauchy's derivation of the addition formulas in trigonometry, given by Hobson in his *Treatise* and presented recently by McShane, is general and shorter than the standard proof given by most texts. The proof suggested by the accompanying figure (the circle has unit diameter but the figure is otherwise self-explanatory) is admittedly not general and moreover requires a preliminary lemma to the effect that the length of a chord in a circle of unit diameter is equal to the sine of the subtended inscribed angle. Nevertheless, this lemma follows almost immediately from standard high school geometry theorems relating to inscribed angles, and is otherwise useful in providing a simple proof for the law of sines. Once the lemma is established, the addition formula for the sine is immediately evident from this figure, for the case α, β, and $\alpha+\beta$ numerically less than 180°, and no algebraic manipulation is required. The same figure is easily adapted to the cosine formula.

This proof can hardly be new. Its relation to Ptolemy's theorem is too close, and that these formulas are special cases of Ptolemy's theorem is well known. But it does not seem to have found its way into the standard texts, though even if a single general proof is to be demanded it should be pedagogically a worthy adjunct to this.

Note by the Editor. Professor E. F. Beckenbach also has pointed out that McShane's method of deriving the addition formulas is essentially due to

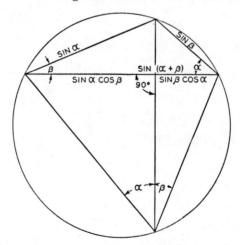

Cauchy, and that this derivation, along with a history of these formulas, appears in Enciclopedia delle Matematiche Elementari, Milano, 1937, vol. 2, part 1, pp. 551–552.
                                                  R. J. W.

---

* From AMERICAN MATHEMATICAL MONTHLY, vol. 49 (1942), pp. 326–327.

VALUES OF THE TRIGONOMETRIC RATIOS OF $\pi/8$ AND $\pi/12$*

H. L. DORWART, Washington and Jefferson College

Although every textbook in trigonometry begins with the determination of the exact values of the trigonometric ratios of $\pi/3$, $\pi/4$ and $\pi/6$ from appropriate triangles, all texts that the writer has seen reserve the exact values of $\pi/8$ and $\pi/12$ until after the functions of sum and difference and half angles have been derived. Since the exact values of the ratios of these latter angles can be

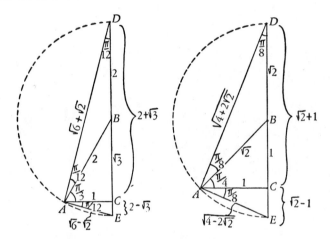

just as useful at the beginning of a trigonometry course as those of the former, the following simple construction* may be of some interest.

In each figure, starting with the basic triangle ABC, a semicircle of radius AB is described on BC (extended) with B as a center. ABD is thus an isosceles triangle and angle DAE is a right angle.

From the right triangle DAE, we have

$$\sin \frac{\pi}{12} = \frac{\sqrt{6} - \sqrt{2}}{4}, \qquad \sin \frac{\pi}{8} = \frac{\sqrt{2 - \sqrt{2}}}{2},$$

$$\cos \frac{\pi}{12} = \frac{\sqrt{6} + \sqrt{2}}{4}, \qquad \cos \frac{\pi}{8} = \frac{\sqrt{2 + \sqrt{2}}}{2},$$

from the right triangle ACE

$$\tan \frac{\pi}{12} = 2 - \sqrt{3}, \qquad \tan \frac{\pi}{8} = \sqrt{2} - 1,$$

$$\sec \frac{\pi}{12} = \sqrt{6} - \sqrt{2}, \qquad \sec \frac{\pi}{8} = \sqrt{4 - 2\sqrt{2}},$$

---

* From AMERICAN MATHEMATICAL MONTHLY, vol. 49 (1942), pp. 324–325.

and from the right triangle ACD

$$\cot\frac{\pi}{12} = 2 + \sqrt{3}, \qquad \cot\frac{\pi}{8} = \sqrt{2} + 1,$$

$$\operatorname{cosec}\frac{\pi}{12} = \sqrt{6} + \sqrt{2}, \qquad \operatorname{cosec}\frac{\pi}{8} = \sqrt{4 + 2\sqrt{2}}.$$

Note by the Editor. Professor Dorwart's construction suggests a simple derivation of the half-angle formulas for angles less than 180°. It is used thus occa-

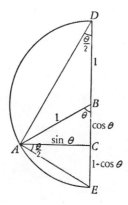

sionally in texts in trigonometry. The accompanying figure is self-explanatory, and we see that

$$\tan \theta/2 = AC/CD = \sin \theta/(1 + \cos \theta)$$
$$= EC/AC = (1 - \cos \theta)/\sin \theta.$$

The other half-angle formulas are easily obtained from these. R. J. W.

A SUBSTITUTION FOR SOLVING TRIGONOMETRIC EQUATIONS*

R. W. WAGNER, Oberlin College

The substitution $t = \tan(\theta/2)$ which is used in the calculus to integrate some expressions containing both $\sin \theta$ and $\cos \theta$ can also be used to advantage in the solution of equations in which the unknown appears only as the argument of one or more trigonometric functions. The usefulness of this substitution arises from the fact that the trigonometric functions of θ are simple rational functions of t. From

$$\tan(\theta/2) = t,$$

---

* From AMERICAN MATHEMATICAL MONTHLY, vol. 54 (1947), pp. 220–221.

one gets, by using the double angle formula,

$$\tan \theta = \frac{2t}{(1-t^2)},$$

and hence that

$$\sin \theta = \frac{2t}{\sqrt{(2t)^2 + (1-t^2)^2}} = \frac{2t}{1+t^2}.$$

There is no ambiguous sign in this last equation. For sin θ and t have the same algebraic sign. Also, one gets

$$\cos \theta = \frac{1-t^2}{1+t^2}.$$

The neatness of this method of solution will be shown by applying it to an example:

EXAMPLE. *Find the angles between* $-\pi$ *and* π *radians such that*

$$5 \sin \theta - 21 \cos \theta = 10.$$

On making the above substitutions one gets

$$\frac{10t}{1+t^2} - 21 \frac{1-t^2}{1+t^2} = 10.$$

When this equation is cleared of fractions and terms collected the result is

$$11t^2 + 10t - 31 = 0.$$

Hence,

$$t = \frac{-10 \pm \sqrt{100 + 1364}}{22}$$

$$= 1.285 \quad \text{or} \quad -2.194.$$

From a table of trigonometric functions in radian measure one finds that

$$\theta/2 = 0.910 \quad \text{or} \quad -1.143,$$

and that

$$\theta = 1.820 \quad \text{or} \quad -2.286 \text{ radians}.$$

The main advantage of this method of solving the given equation lies in the fact that there is no difficulty in determining the quadrant in which an angle may lie. It is easy to bear in mind that negative values of t correspond to negative angles and positive values of t correspond to positive angles.

DERIVATION OF THE TANGENT HALF-ANGLE FORMULA*

F. E. WOOD, University of Oregon

The following derivation of the formula for tan $\theta/2$ appears to be an improvement over standard derivations, for it gives the result directly without a complicated discussion of the appropriate algebraic sign.

From the equation

$$\sin\frac{\theta}{2} = \sin\left(\theta - \frac{\theta}{2}\right) = \sin\theta\cos\frac{\theta}{2} - \cos\theta\sin\frac{\theta}{2}$$

one obtains

$$(1 + \cos\theta)\sin\frac{\theta}{2} = \sin\theta\cos\frac{\theta}{2}.$$

Consequently:

$$\tan\frac{\theta}{2} = \frac{\sin\theta/2}{\cos\theta/2} = \frac{\sin\theta}{1+\cos\theta}.$$

THE LAWS OF SINES AND COSINES†

L. J. BURTON, Bryn Mawr College

Using the distance formula of analytic geometry we may derive the law of cosines in the following simple fashion. For any triangle ABC, choose the coordinate system so that A is at the origin and B is on the positive x-axis. Then B is $(c, 0)$ and C is $(b\cos A, b\sin A)$. From the distance formula:

$$a^2 = (b\cos A - c)^2 + (b\sin A)^2 = b^2 + c^2 - 2bc\cos A.$$

Most textbooks do not point out any connection between the law of cosines and the law of sines. Assuming the former we can prove the latter algebraically as follows:

$$\cos A = \frac{b^2+c^2-a^2}{2bc} \quad \text{and} \quad \sin^2 A = 1 - \cos^2 A = \frac{4b^2c^2 - (b^2+c^2-a^2)^2}{4b^2c^2}$$

or

$$\sin^2 A = \frac{(a+b-c)(a-b+c)(a+b+c)(-a+b+c)}{4b^2c^2}.$$

By symmetry:

$$\sin^2 B = \frac{(a+b-c)(-a+b+c)(a+b+c)(a-b+c)}{4a^2c^2}.$$

---

* From AMERICAN MATHEMATICAL MONTHLY, vol. 56 (1949), p. 103.
† From AMERICAN MATHEMATICAL MONTHLY, vol. 56 (1949), pp. 550–551.

Then $b^2 \sin^2 A = a^2 \sin^2 B$; and since $\sin A$, $\sin B \geq 0$, we have

$$b \sin A = a \sin B.$$

Also, it follows that if $s = \tfrac{1}{2}(a+b+c)$, then

$$\sin A = \frac{2\sqrt{s(s-a)(s-b)(s-c)}}{bc}$$

and since the area of the triangle ABC is clearly $\tfrac{1}{2} bc \sin A$, we have a simple proof that the area is $\sqrt{s(s-a)(s-b)(s-c)}$.

The law of cosines can be proved algebraically from the law of sines as follows. Assuming that

$$\frac{\sin A}{a} = \frac{\sin B}{b} = \frac{\sin C}{c} = k$$

we wish to prove that $a^2 = b^2 + c^2 - 2bc \cos A$, or simply the trigonometric identity

$$\sin^2 A = \sin^2 B + \sin^2 C - 2 \sin B \sin C \cos A$$

where $A = \pi - (B+C)$. Putting $\sin A = \sin B \cos C + \cos B \sin C$ and $\cos A = \sin B \sin C - \cos B \cos C$, the identity is easily verified. In spherical trigonometry, the law of sines follows from the law of cosines by a proof similar to the above, but the law of cosines does not follow from the law of sines.

ANGLES WITH RATIONAL TANGENTS*†

T. S. CHU, National Kunming Teachers College

1. Introduction. The purpose of this note is to show that the class of angles having rational tangents, and the class of angles which are rational multiples of π, intersect only in the obvious cases.

2. Theorem. We shall establish the following result.

THEOREM. *If x is a rational multiple of π, and $\tan x$ is rational, then x is an integral multiple of $\pi/4$.*

3. Proof. Let $\tan x = q/p$. The theorem is trivially satisfied if $q = 0$ or if $|p| = |q|$. Further, without loss of generality, x may be restricted to the first quadrant, so that p and q may be assumed positive, integral, unequal, and coprime.

If $x = m\pi/n$, then $e^{inx} = e^{-inx} = \pm 1$, or

$$(\cos x + i \sin x)^n = (\cos x - i \sin x)^n,$$

---

* From AMERICAN MATHEMATICAL MONTHLY, vol. 57 (1950), pp. 407–408.
† Revised by J. D. Swift.

and
$$(p+iq)^n/(p^2+q^2)^{n/2} = (p-iq)^n/(p^2+q^2)^{n/2}.$$

Thus
$$\begin{aligned}(p-iq)^n &= (p+iq)^n = (p-iq+2iq)^n \\ &= (p-iq)^n + \binom{n}{1}(p-iq)^{n-1}2iq + \cdots \\ &\quad + \binom{n}{n-1}(p-iq)(2iq)^{n-1} + (2iq)^n.\end{aligned}$$

Therefore $(p-iq)$ divides $(2qi)^n$, and p^2+q^2 divides $(2q)^{2n}$. Similarly, p^2+q^2 divides $(2p)^{2n}$ and therefore divides $(2^{2n}p^{2n}, 2^{2n}q^{2n}) = 2^{2n}$. Then $p^2+q^2 = 2^k$; but this is possible in positive coprime integers only when $p=q=k=1$, a contradiction.

4. Corollary. By writing $\tan nx$ as a rational function in terms of $\tan x$, the reader may verify the following corollary.

COROLLARY: *The equations*

$$\sum_{k=0}^{a}(-1)^k\binom{n}{2k+1}x^{2k} = 0, \qquad a = \left[\frac{n-1}{2}\right], \quad n>2,\ n\neq 4,$$

and

$$\sum_{k=0}^{b}(-1)^k\binom{n}{2k}x^{2k} = 0, \qquad b = \left[\frac{n}{2}\right], \qquad\qquad n>2,$$

have no rational roots.

THE GENERAL SINE AND COSINE CURVES*

E. L. EAGLE, University of Tennessee

In many textbooks on analytic geometry, and trigonometry, the equations of the general sine and cosine curves are given expression, canonically, in the form:
$$y = k \sin(bx+c)$$
$$y = k \cos(bx+c).$$

However, there is an advantage to writing such equations in the following form:
$$y = a \sin n(x-\alpha)$$
$$y = a \cos n(x-\alpha).$$

---

* From AMERICAN MATHEMATICAL MONTHLY, vol. 57 (1950), pp. 685–686.

Then, a will be the amplitude and $2\pi/n$ the period (as usual), but the lag or lead will be indicated, definitely, by the point $(\alpha, 0)$. For example: Given $y = 2/3 \sin(3x + \pi/3)$, we convert to canonical form:

$$y = \frac{2}{3} \sin 3\left(x - \left[-\frac{\pi}{9}\right]\right).$$

Therefore, the amplitude is 2/3 and the period is $2\pi/3$; and, to sketch the curve, we may conveniently start a cycle at $(-\pi/9, 0)$.

This approach removes any uncertainty, when the curve is out of phase, as to whether we have a lag or a lead.

It is gratifying to note that the taking of α from behind a minus sign is a procedure familiar to the student for he has already done the same thing in dealing with the circle, parabola, ellipse, and hyperbola.

A DERIVATION OF THE FORMULAS FOR SIN $(\alpha + \beta)$ AND COS $(\alpha + \beta)$*

A. K. BETTINGER, The Creighton University

The following procedure for deriving the addition formulas of plane trigonometry has been used for some time by the author. To the best of our knowledge it has not previously appeared in print.

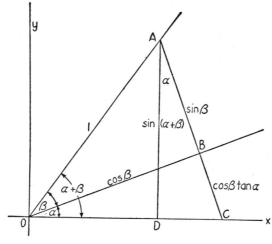

Fig. 1

We shall derive the formulas for the case when both α and β are acute angles and $\alpha + \beta < 90°$.

Figure 1 will be used to prove the formula for $\sin(\alpha + \beta)$.

---

* From AMERICAN MATHEMATICAL MONTHLY, vol. 60 (1953), pp. 108–110.

In Figure 1, let $OA = 1$, draw AC perpendicular to OB, and AD perpendicular to OX. Write the lengths of the sides on the figure as shown.

Now by definition the cosine of angle DAC equals DA/AC, that is,

$$\cos \alpha = \frac{\sin (\alpha + \beta)}{\cos \beta \tan \alpha + \sin \beta}.$$

Solving for $\sin (\alpha+\beta)$, we have

$$\sin (\alpha + \beta) = \cos \alpha (\cos \beta \tan \alpha + \sin \beta)$$
$$= \sin \alpha \cos \beta + \cos \alpha \sin \beta.$$

Figure 2 is found more convenient for deriving the formula for $\cos (\alpha+\beta)$ than the preceding one. Hence we let $OA = 1$, draw CA perpendicular to OB, and AD perpendicular to OY.

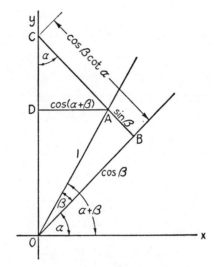

Fig. 2

By definition the cosine of angle DAC equals $\sin \alpha = DA/AC$, hence

$$\sin \alpha = \frac{\cos (\alpha + \beta)}{\cos \beta \cot \alpha - \sin \beta}.$$

Solving for $\cos (\alpha+\beta)$, we have

$$\cos (\alpha + \beta) = \sin \alpha (\cos \beta \cot \alpha - \sin \beta)$$
$$= \cos \alpha \cos \beta - \sin \alpha \sin \beta.$$

The construction is readily extended to other quadrants for demonstrations in different cases. There is no need, however, for showing the construction in the numerous cases that may occur since the generality of the formulas does not depend on the particular construction used to derive the formulas.

The validity of the formulas in general may be shown by a method similar to that in Todhunter's *Plane Trigonometry*, page 53.

TERSE TRIGONOMETRY*

C. M. FULTON, University of California, Davis

This is an attempt at formulating a new approach to trigonometry. Its conciseness and rigor should make it suitable for good students. The basic idea is a truly general definition of the trigonometric functions.

Let $P(x, y)$, $x^2+y^2=1$, be on the terminal side of an angle θ and $P'(x', y')$, $x'^2+y'^2=1$, on its initial side. The angle with its vertex at the origin O is considered positive or negative according as the rotation involved is counterclockwise or clockwise. We define

(1) $$\cos \theta = xx' + yy', \quad \sin \theta = yx' - xy'.$$

Interchanging the sides of the angle it is seen that

(2) $$\cos(-\theta) = \cos \theta, \quad \sin(-\theta) = -\sin \theta.$$

The following identities are easily established:

(3) $$(xx' + yy')^2 + (yx' - xy')^2 = 1,$$

(4) $$(x - x')^2 + (y - y')^2 = 2 - 2(xx' + yy').$$

The meaning of (3) is obvious. Identity (4), on the other hand, is crucial in that it shows the invariance of the definition of cosine: if the angle is rotated, the distance PP' remains unchanged and so does the cosine. Hence, because of (3) the same is true for the square of the sine. Moreover, a continuity argument will take care of the sine itself. It should be apparent to the reader that, in order to save space, we are using technical rather than pedagogical language in this discussion.

We now use the invariance demonstrated above and place the angle in standard position letting $x'=1$, $y'=0$. The definitions (1) reduce to

(5) $$\cos \theta = x, \quad \sin \theta = y.$$

Let us remind ourselves at this stage that the much-stressed trigonometric identities involving one angle only could all be written as algebraic identities by using (5). Returning to definitions (1) we denote by α and α' the angles in standard position whose terminal sides are OP and OP', respectively. Then θ and $\alpha-\alpha'$ differ by some integral multiple of 360° and with the aid of (5) we

---

* From AMERICAN MATHEMATICAL MONTHLY, vol. 65 (1958), pp. 522–523.

obtain the formulas for $\cos(\alpha-\alpha')$ and $\sin(\alpha-\alpha')$. On account of (2) the corresponding formulas for $\alpha+\alpha'$ are easily derived. Eventually, the so-called reduction formulas follow as special cases. All these derivations are, of course, completely general.

Clearly, the definitions (5) can be changed to the usual form $\cos\theta = x/r$, $\sin\theta = y/r$ by taking a point P at a distance r from the origin. This leads directly to the solution of right triangles. Furthermore, identity (4) is generalized to

$$(x - x')^2 + (y - y')^2 = r^2 + r'^2 - 2rr' \cos\theta,$$

which is precisely the law of cosines. May we be allowed to suggest at this point that for practical purposes the latter be rewritten as a haversine formula, namely

$$c^2 = (a - b)^2 + 4ab \text{ hav } \gamma.$$

Finally, to find a trigonometric formula for the area of triangle $OP'P$, let $x' = r'$, $y' = 0$, $y > 0$. Then the area $K = \frac{1}{2}r'y = \frac{1}{2}r'r \sin\theta$. We can now prove the law of sines without difficulty, using the formula for K.

ELEMENTARY PROOFS FOR THE EQUIVALENCE OF FERMAT'S PRINCIPLE AND SNELL'S LAW*

MICHAEL GOLOMB, Purdue University

In a recent SMSG monograph by Beckenbach and Bellman [1] Fermat's minimum time principle for the path of a reflected ray of light is verified by means of an elementary inequality, and also by the well-known geometric construction ascribed to Heron. For the path of a refracted ray the authors state that verification of the minimum time principle does not seem to be obtainable in a similar fashion and that recourse must be made to the differential calculus. Below we give an arithmetic and a geometric proof for the law of refraction which are of a similar character to those given in [1] for the law of reflection.

Suppose that a plane F divides two homogeneous media M_1 and M_2 of a different optical density so that in M_1 light rays travel with velocity v_1 and in M_2 they travel with velocity v_2. Fermat's principle says that the path of a ray from a point P_1 in M_1 to a point P_2 in M_2 is such that it consumes less time than any other path connecting P_1 and P_2. It is clear that only those paths need to be considered which lie in a plane G perpendicular to F and which consist of two line segments P_1Q, QP_2 with $Q \in F \cap G$. Make $F \cap G$ the x axis of a Cartesian coordinate system and let (x_1, y_1), (x_2, y_2), $(x, 0)$ be the coordinates of P_1, P_2, Q, respectively, with $x_1 < x_2$, $y_1 > 0$, $y_2 < 0$. The problem is to minimize the function

$$t(x) = \frac{1}{v_1}\sqrt{\{(x-x_1)^2 + y_1^2\}} + \frac{1}{v_2}\sqrt{\{(x_2-x)^2 + y_2^2\}}.$$

Let θ_1, θ_2 be any real numbers. Then, by Cauchy's inequality,

$$\pm(x_i - x)\sin\theta_i \pm y_i \cos\theta_i \leq \sqrt{\{(x_i-x)^2 + y_i^2\}} \qquad (i = 1, 2).$$

---

* From AMERICAN MATHEMATICAL MONTHLY, vol. 71 (1964), pp. 541–543.

THE EQUIVALENCE OF FERMAT'S PRINCIPLE AND SNELL'S LAW

Therefore

(1) $$t(x) \geqq \frac{(x - x_1) \sin \theta_1 + y_1 \cos \theta_1}{v_1} + \frac{(x_2 - x) \sin \theta_2 - y_2 \cos \theta_2}{v_2}.$$

If we now require that

(2) $$\frac{\sin \theta_1}{\sin \theta_2} = \frac{v_1}{v_2}$$

then the lower bound in (1) is independent of x, and it is attained for $x = x^*$, i.e. $t(x^*)$ is the minimum time for any path, if equality holds in (1) for $x = x^*$. This is the case if and only if

(3) $$\sin \theta_1 = \frac{x - x_1}{\sqrt{\{(x - x_1)^2 + y_1^2\}}}, \quad \cos \theta_1 = \frac{y_1}{\sqrt{\{(x - x_1)^2 + y_1^2\}}}$$

$$\sin \theta_2 = \frac{x_2 - x}{\sqrt{\{(x_2 - x)^2 + y_2^2\}}}, \quad \cos \theta_2 = \frac{-y_2}{\sqrt{\{(x_2 - x)^2 + y_2^2\}}}.$$

It is seen that as x varies in (3) from x_1 to x_2, $\sin \theta_1$ increases monotonically from 0 to $u_1 = (x_2 - x_1)/\sqrt{\{(x_2 - x_1)^2 + y_1^2\}}$ and $\sin \theta_2$ decreases monotonically from $u_2 = (x_2 - x_1)/\sqrt{\{(x_2 - x_1)^2 + y_2^2\}}$ to 0. Thus there is exactly one set of values $x = x^*$, $\theta_1 = \theta_1^*$, $\theta_2 = \theta_2^*$, with $x_1 < x^* < x_2$, $0 < \theta_1^* < \arcsin u_1$, and $0 < \theta_2^* < \arcsin u_2$, for which (2) and (3) are satisfied. If Q_* denotes the point $(x^*, 0)$ then by (3) θ_i^* is the angle between $P_i Q_*$ and the normal to F ($i = 1, 2$), and (2) expresses Snell's law of refraction.

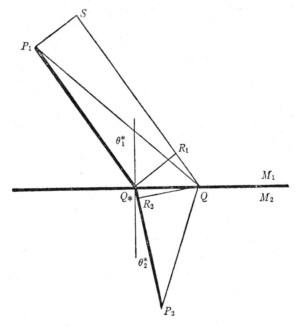

Fig. 1

The following geometric proof goes back to Huygens [2]. Referring to the Figure (with $v_1 > v_2$ and $x > x^*$), let SQ be parallel to P_1Q_*, SP_1 and R_1Q_* perpendicular to P_1Q_*, and R_2Q perpendicular to P_2Q_*. Then

$$\frac{R_1Q}{R_2Q_*} = \frac{\sin \theta_1^*}{\sin \theta_2^*} = \frac{v_1}{v_2}.$$

Hence, if $|AB|$ denotes the time to traverse the segment AB,

$$|R_1Q| = |R_2Q_*|$$

and

$$\begin{aligned}|P_1Q_*| + |Q_*P_2| &= |SR_1| + |R_1Q| - |R_2Q_*| + |Q_*P_2| \\ &= |SQ| + |R_2P_2| \\ &\leq |P_1Q| + |QP_2|.\end{aligned}$$

Equality holds here if and only if $Q = Q_*$. In a similar way one obtains this inequality for $x < x^*$. Thus the minimum time property of the path $P_1Q_* \cup Q_*P_2$ is proved.

References

1. E. Beckenbach and R. Bellman, An Introduction to Inequalities, New Mathematical Library, Random House, New York, 1961.

2. C. Huygens, Treatise on Light, translated by S. P. Thompson, Dover, New York, 1962.

A GEOMETRIC PROOF OF THE EQUIVALENCE OF FERMAT'S PRINCIPLE AND SNELL'S LAW*

DANIEL PEDOE, Purdue University

Let l be the line separating the media, and let $P_1Q^*P_2$ be the actual path of the ray according to Snell's Law, where $\sin \theta_1 / \sin \theta_2 = v_1/v_2$. We wish to show that for any other point Q on the line

$$P_1Q/v_1 + P_2Q/v_2 > P_1Q^*/v_1 + P_2Q^*/v_2,$$

so that the time taken for traversing the actual path is a minimum. Draw the circle through the points P_1, Q^* and P_2, and let the perpendicular to l through Q^* intersect this circle again at the point A. Then we note that $AP_1 = 2R \sin \theta_1$, and $AP_2 = 2R \sin \theta_2$, where R is the radius of the circle $P_1Q^*P_2$. Therefore $AP_1/AP_2 = v_1/v_2$; that is, $AP_1 = k/v_2$ and $AP_2 = k/v_1$, where k is a constant.

Applying the theorem of Ptolemy to the four concyclic points P_1, Q^*, P_2 and A, we have the equality

$$P_1P_2 \cdot AQ^* = P_1Q^* \cdot AP_2 + P_2Q^* \cdot AP_1,$$

---

* From AMERICAN MATHEMATICAL MONTHLY, vol. 71 (1964), pp. 543–544.

whereas if Q is any point $\neq Q^*$ on the line l, the extension of the Ptolemy theorem which arises naturally by inverting the triangle inequality [1] gives the inequality

$$P_1P_2 \cdot AQ < P_1Q \cdot AP_2 + P_2Q \cdot AP_1.$$

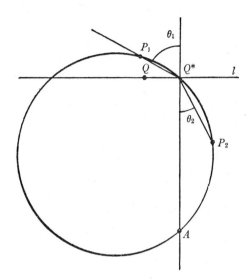

If we substitute for AP_1 and AP_2, we obtain the equality

$$k(P_1Q^*/v_1 + P_2Q^*/v_2) = P_1P_2 \cdot AQ^*,$$

and the inequality

$$k(P_1Q/v_1 + P_2Q/v_2) > P_1P_2 \cdot AQ.$$

Hence the Fermat principle of minimum time is established for a ray which satisfies Snell's Law, since $AQ > AQ^*$.

Reference

1. D. Pedoe, Circles, Pergamon, London, 1957.

SIN $(A+B)$*

F. H. YOUNG, Montana State College

In most trigonometry texts the derivation of $\sin(a+b)$, etc., is either lacking in generality or lacking in simplicity. The following derivation is suggested as a means of avoiding both Scylla and Charybdis.

First, let us suppose that the sine and cosine functions have been defined in terms of ordinate and abscissa in the unit circle. Then, let us assume that the normal form of the straight line has been developed as an application of these functions. We are now prepared to expand $\sin(a+b)$. As in the figure, let the angle b be drawn positively and a negatively in the unit circle. Point P, the terminus of angle b, then, has coordinates $(\cos b, \sin b)$. Let line (1) be the line coinciding with

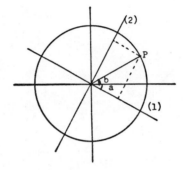

the terminal side of angle a. The equation of (1), in normal form, is

$$x \sin a + y \cos a = 0.$$

The distance from P to (1) is

$$d = \sin(a+b) = \pm(\sin a \cos b + \cos a \sin b).$$

Since this is to be an identity for all a and b, a consideration of the result when $a = 0°$ shows that the $+$ sign should be used.

Similarly, the equation of line (2), normal to (1), is

$$x \cos a - y \sin a = 0.$$

The distance from P to (2) is then

$$d = \cos(a+b) = \cos a \cos b - \sin a \sin b,$$

with the sign determined as before.

If angle a were drawn in a positive direction, the same technique would yield the expansions of $\sin(b-a)$ and $\cos(b-a)$.

Clearly, no restriction is placed on either the magnitude or the sign of a or b.

---

* From MATHEMATICS MAGAZINE, vol. 27 (1954), pp. 208–209.

CONCERNING SIN $(A + B)$*
(F. H. YOUNG-Vol. 27, No. 4)

H. V. CRAIG, University of Texas, Austin
(Editor of *Current Papers and Books*)

Dear Mr. James:

I was very interested in F. H. Young's method for deriving the formulae for sin $(A + B)$ etc., in the March-April, 1954 Mathematics Magazine. There is one snag which I should very much like to have cleared up:

$$\text{"}\sin(A+B) = \pm(\sin A \cos B + \cos A \sin B).$$

Since this is to be an identity for all a and b...."

Assuming the order of treatment suggested, I can't see why this should be an identity — might it not be $+$ sometimes and $-$ others? (cf. $\sin A/2 = \pm\sqrt{(1-\cos A)/2}$.)

I expect the author has a simple answer to this, but if it were a question of dragging in sign-conventions, his method would be no quicker than the traditional one.

I very much hope there is a short answer to this, as his method is so good.

Yours truly,

G. Matthews
St. Dunstan's College
Catford, England

Dear Mr. Matthews:

Suppose that both signs were needed. Then for some A and B, $\sin(A+B) = -(\sin A \cos B + \cos A \sin B)$ with $A + B \not\equiv 0 \pmod{\pi}$. From the continuity of $\sin x$, either A or B, say A, can be changed to $\equiv 0 \pmod{\pi}$ without changing the sign of $\sin(A+B)$. Since the $+$ sign is obviously required when $A \equiv 0 \pmod{\pi}$ for all B, a contradiction is reached. Recall that the sine and cosine functions are assumed defined as coordinates on the unit circle. This continuity argument was implied but perhaps too briefly or too enigmatically stated in the paper as published.

Sincerely yours,

F. H. Young
Montana State College

---

* From MATHEMATICS MAGAZINE, vol. 28 (1955), pp. 51-52.

A SIMPLE PROOF OF THE FORMULA FOR SIN $(A+B)$*

NORMAN SCHAUMBERGER, Bronx Community College of the City University of New York

The following proof was communicated by Professor Jesse Douglas of the City College of the City University of New York, who has been using it for many years in his classes. It is composed of the following four ingredients, each a standard theorem of trigonometry:

(1) The projection law for triangles (each side equals the algebraic sum of the projections upon it of the other two): $c = a \cos B + b \cos A$;

(2) The sine law:
$$\frac{\sin C}{c} = \frac{\sin A}{a} = \frac{\sin B}{b};$$

(3) $A+B+C=\pi$;
(4) $\sin(\pi-\theta) = \sin\theta$.

Multiplying the terms in (1) by the corresponding ones in (2), we get: $\sin C = \sin A \cos B + \cos A \sin B$. By (3) and (4), $\sin C = \sin(A+B)$; hence

(*) $\qquad\qquad \sin(A+B) = \sin A \cos B + \cos A \sin B.$

The addition theorem is thus established for all A, B that may occur as two angles of a triangle, i.e., $A+B<\pi, A>0, B>0$. The case A or $B=0$ and the case $A+B=\pi$ are immediate by the formulas $\sin 0 = 0$, $\cos 0 = 1$, $\cos(\pi-\theta) = -\cos\theta$, and (4). The extension to the case $0<A<\pi$, $0<B<\pi$, $A+B>\pi$ is established by applying (*) to $\pi-A, \pi-B$, since $(\pi-A)+(\pi-B) = 2\pi-(A+B)<\pi$. (*) is now proved for $0 \leqq A <\pi$, $0 \leqq B <\pi$.

We may then extend (*) to all A, B such that $0 \leqq A < 2\pi$, $0 \leqq B < 2\pi$, by applying (*), as so far proved, to $A-\pi, B$; $A, B-\pi$; $A-\pi, B-\pi$; according as A, or B, or both, are $\geqq \pi$. One uses $\sin(\theta-\pi) = -\sin\theta$, $\cos(\theta-\pi) = -\cos\theta$, $\sin(\theta-2\pi) = \sin\theta$.

(*) now follows universally by periodicity, since for arbitrary A, B and appropriate integers m, n, the angles $A+2m\pi, B+2n\pi$ belong to the interval $0 \leqq \theta < 2\pi$.

---

* From MATHEMATICS MAGAZINE, vol. 35 (1962), p. 229.

MATRICES IN TEACHING TRIGONOMETRY*

A. R. AMIR-MOÉZ, University of Florida

In the September-October 1958 issue of this MAGAZINE in a brief article we described a use of vectors in teaching trigonometry [1]. In this note we would like to add to that a few ideas concerning the use of matrices in obtaining trigonometric functions of $t+s$ in terms of the ones of t and s. In [1] a formula for $\cos(t-s)$ was obtained using the inner product of vectors. Here we would like to obtain formulas for $\sin(t+s)$ and $\cos(t+s)$. The advantage of the use of matrices is that we can obtain the formulas in general without referring to complicated diagrams.

1. Linear transformations. A function f on the set of vectors of the plane into the set of vectors of the plane is called a linear transformation if:

I. $f(\mathbf{A} + \mathbf{B}) = f(\mathbf{A}) + f(\mathbf{B})$,

II. $f(a\mathbf{A}) = af(\mathbf{A})$,

where a is a real number and $\mathbf{A}$ and $\mathbf{B}$ are any two vectors. Indeed this idea may be studied in any book in linear algebra, for example [2].

2. The matrix of a linear transformation. Consider the unit vectors $\mathbf{U}$ on the x-axis and $\mathbf{V}$ on the y-axis (Fig. 1). Suppose $f(\mathbf{U}) = \mathbf{A}$, $f(\mathbf{V}) = \mathbf{B}$. If (a_{11}, a_{12}) is the set of coordinates of the point A and (a_{21}, a_{22}) is the set of coordinates of the point B, then the matrix of f is defined to be the array of numbers

$$\begin{pmatrix} a_{11} & a_{12} \\ a_{21} & a_{22} \end{pmatrix}.$$

3. Matrix of a rotation. For any vector $\mathbf{A}$ we define $\mathbf{B} = f(\mathbf{A})$ in such a way that $|\mathbf{A}| = |\mathbf{B}|$ and directed angle from $\mathbf{A}$ to $\mathbf{B}$ is α. Then we call f the rotation of the plane through an angle α.

To obtain the matrix of this rotation we again consider the unit vectors $\mathbf{U}$ and $\mathbf{V}$ respectively on the x and y axes (Fig. 2). Let $f(\mathbf{U}) = \mathbf{U}_1$ and $f(\mathbf{V}) = \mathbf{V}_1$. Then we observe that the set of coordinates of $\mathbf{U}_1$ is $(\cos \alpha, \sin \alpha)$ and the set of coordinates of V_1 is $(-\sin \alpha, \cos \alpha)$. Thus the matrix of the rotation through an angle α is

$$\begin{pmatrix} \cos \alpha & \sin \alpha \\ -\sin \alpha & \cos \alpha \end{pmatrix}.$$

4. Product of linear transformations. For two linear transformations f and g we define $f \cdot g$ as follows. Let $f(\mathbf{A}) = \mathbf{B}$ and $g(\mathbf{B}) = \mathbf{C}$. Then $(f \cdot g)(\mathbf{A}) = g[f(\mathbf{A})] = \mathbf{C}$ (note the order).

---

* From MATHEMATICS MAGAZINE, vol. 37 (1964), pp. 78–81.

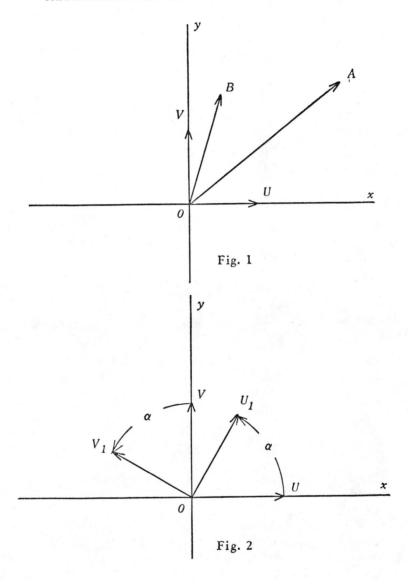

Fig. 1

Fig. 2

Now for the matrix of $f \cdot g$ we suppose that the matrices of f and g are respectively

$$\begin{pmatrix} a_{11} & a_{12} \\ a_{21} & a_{22} \end{pmatrix} \text{ and } \begin{pmatrix} b_{11} & b_{12} \\ b_{21} & b_{22} \end{pmatrix}.$$

This means that

$$f(\mathbf{U}) = a_{11}\mathbf{U} + a_{12}\mathbf{V}, \quad f(\mathbf{V}) = a_{21}\mathbf{U} + a_{22}\mathbf{V}, \quad g(\mathbf{U}) = b_{11}\mathbf{U} + b_{12}\mathbf{V}, \quad g(\mathbf{V}) = b_{21}\mathbf{U} + b_{22}\mathbf{V}.$$

Since the transformations f and g are linear we can write

$$(f \cdot g)(\mathbf{U}) = g[f(\mathbf{U})] = a_{11}g(\mathbf{U}) + a_{12}g(\mathbf{V})$$
$$= a_{11}(b_{11}\mathbf{U} + b_{12}\mathbf{V}) + a_{12}(b_{21}\mathbf{U} + b_{22}\mathbf{V})$$
$$= (a_{11}b_{11} + a_{12}b_{21})\mathbf{U} + (a_{11}b_{12} + a_{12}b_{22})\mathbf{V}.$$

That is, the set of coordinates of $(f \cdot g)(\mathbf{U})$ is

$$(a_{11}b_{11} + a_{12}b_{21}, \; a_{11}b_{12} + a_{12}b_{22}).$$

Similarly the set of coordinates of $(f \cdot g)(\mathbf{V})$ can be obtained as

$$(a_{21}b_{11} + a_{22}b_{21}, \; a_{21}b_{12} + a_{22}b_{22}).$$

Thus the matrix of $f \cdot g$ is

$$\begin{pmatrix} a_{11}b_{11} + a_{12}b_{21} & a_{11}b_{12} + a_{12}b_{22} \\ a_{21}b_{11} + a_{22}b_{21} & a_{21}b_{12} + a_{22}b_{22} \end{pmatrix}.$$

Thus we define this matrix to be the product of

$$\begin{pmatrix} a_{11} & a_{12} \\ a_{21} & a_{22} \end{pmatrix} \quad \text{and} \quad \begin{pmatrix} b_{11} & b_{12} \\ b_{21} & b_{22} \end{pmatrix}.$$

5. Trigonometric functions of $t+s$. Consider the rotation f of the plane through an angle t. The matrix of this rotation is

$$\begin{pmatrix} \cos t & \sin t \\ -\sin t & \cos t \end{pmatrix}.$$

The matrix of the rotation g of the plane through an angle s will be

$$\begin{pmatrix} \cos s & \sin s \\ -\sin s & \cos s \end{pmatrix}.$$

The transformation $f \cdot g$ means the rotation of the plane through an angle t first and then the rotation of the plane through an angle s, i.e., the rotation of the plane through an angle $t+s$ whose matrix is

$$\begin{pmatrix} \cos(t+s) & \sin(t+s) \\ -\sin(t+s) & \cos(t+s) \end{pmatrix}.$$

On the other hand the matrix of $f \cdot g$ is

$$\begin{pmatrix} \cos t \cos s - \sin t \sin s & \cos t \sin s + \sin t \cos s \\ -\sin t \cos s - \cos t \sin s & -\sin t \sin s + \cos t \cos s \end{pmatrix}.$$

This implies that

$$\cos(t+s) = \cos t \cos s - \sin t \sin s$$

and
$$\sin(t+s) = \cos t \sin s + \sin t \cos s.$$
Other formulas can be obtained from these.

References

1. Ali R. Amir-Moéz, Teaching trigonometry through vectors, this MAGAZINE, 31 (1958) 19–23.
2. A. R. Amir-Moéz, and A. L. Fass, Elements of linear spaces, Pergamon Press, 1962, pp. 16, 67, 123.

THE LAW OF SINES AND LAW OF COSINES FOR POLYGONS*

R. B. KERSHNER, The Johns Hopkins University

In connection with my investigations of polygons [1] I had need of a convenient analytic formulation of the restrictions imposed on a set of numbers by the fact that they were the angles and sides of a convex polygon. I found appropriate expressions in the form of straightforward generalizations of the law of sines and law of cosines for triangles. These laws are so elegant, useful, and elementary that they can hardly be new but a modest search has failed to discover them in the literature. Accordingly I am here making them available. I hope that others may find them as useful as I have.

The laws will be stated and proved here for pentagons and hexagons, thus illustrating the slight difference between the even and odd cases; the extension to any number of sides is completely obvious. Let the angles of a pentagon be A, B, C, D, E successively and the sides a, b, c, d, e in such a way that a and b are the sides of angle A. Place this pentagon on the Cartesian plane as in Figure 1, so that side a occupies the interval from $(0, 0)$ to $(a, 0)$. Now the inclination of side b is clearly the exterior angle $\pi - A$. The direction of side c is obtained from that of side b by a further rotation through $\pi - B$, hence side c has inclination $2\pi - (A+B)$; similarly the inclination of side d is $3\pi - (A+B+C)$, the inclination of side e is $4\pi - (A+B+C+D)$, and the inclination of side a is $5\pi - (A+B+C+D+E) = 2\pi$. The last equality expresses the very well known fact that the sum of the interior angles of a pentagon is 3π, i.e.,

(1) $$A + B + C + D + E = 3\pi.$$

The fact that the pentagon (Figure 1) is a closed figure implies that the sum of the horizontal projections of the five sides is zero, i.e.,

(2) $$\begin{aligned}&a + b\cos(\pi - A) + c\cos(2\pi - (A+B))\\&\quad + d\cos(3\pi - (A+B+C)) + e\cos(4\pi - (A+B+C+D)) = 0.\end{aligned}$$

---

* From MATHEMATICS MAGAZINE, vol. 44 (1971), pp. 150–153.

Using (1) and elementary trigonometry, (2) yields

(3) $\qquad a - b \cos A + c \cos(A + B) = e \cos E - d \cos(D + E).$

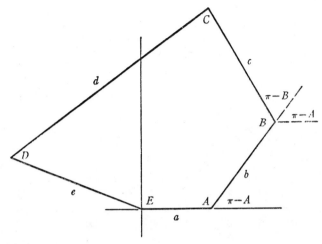

FIG. 1.

In the same way it is seen that the sum of the vertical projections of the five sides in Figure 1 is also zero, i.e.,

(4) $\quad b \sin(\pi - A) + c \sin(2\pi - (A + B)) + d \sin(3\pi - (A + B + C))$
$\qquad\qquad + e \sin(4\pi - (A + B + C + D)) = 0,$

or, from (1),

(5) $\qquad b \sin A - c \sin(A + B) = e \sin E - d \sin(D + E).$

Squaring each of the equations (3) and (5) and adding yields, immediately,

(6) $\quad a^2 + b^2 + c^2 - 2ab \cos A - 2bc \cos B + 2ac \cos(A + B)$
$\qquad\qquad = d^2 + e^2 - 2de \cos D.$

The equation (5) is easily seen to be a direct generalization of the ordinary law of sines for triangles (allowing for the difference between the nomenclature scheme of Figure 1 and the usual nomenclature for a triangle). It clearly can be written in five different ways by relabelling the sides, namely,

Law of sines for pentagons.

(a) $b \sin A - c \sin(A + B) = e \sin E - d \sin(D + E)$
(b) $c \sin B - d \sin(B + C) = a \sin A - e \sin(E + A)$
(c) $d \sin C - e \sin(C + D) = b \sin B - a \sin(A + B)$
(d) $e \sin D - a \sin(D + E) = c \sin C - b \sin(B + C)$
(e) $a \sin E - b \sin(E + A) = d \sin D - c \sin(C + D)$

216 TRIGONOMETRY AND TRIGONOMETRIC FUNCTIONS

Like the law of sines for triangles, each of these equations involves all but one of the sides and all but one of the angles of the polygon.

In the same way equation (6) is a direct generalization of the law of cosines for triangles. It also can be written in five different ways, namely,

Law of cosines for pentagons.

(a) $a^2 + b^2 + c^2 - 2ab \cos A - 2bc \cos B + 2ac \cos(A + B)$
$$= d^2 + e^2 - 2de \cos D$$

(b) $b^2 + c^2 + d^2 - 2bc \cos B - 2cd \cos C + 2bd \cos(B + C)$
$$= e^2 + a^2 - 2ea \cos E$$

(c) $c^2 + d^2 + e^2 - 2cd \cos C - 2de \cos D + 2ce \cos(C + D)$
$$= a^2 + b^2 - 2ab \cos A$$

(d) $d^2 + e^2 + a^2 - 2de \cos D - 2ea \cos E + 2da \cos(D + E)$
$$= b^2 + c^2 - 2bc \cos B$$

(e) $e^2 + a^2 + b^2 - 2ea \cos E - 2ab \cos A + 2eb \cos(E + A)$
$$= c^2 + d^2 - 2cd \cos C$$

Like the law of cosines for triangles each of these equations involves all of the sides and all but two of the angles of the polygon.

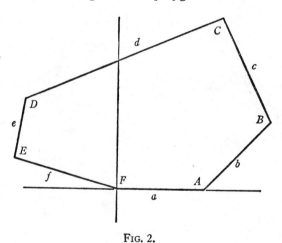

FIG. 2.

The derivation of the equivalent laws for a hexagon is now quite obvious. Let the hexagon be represented by Figure 2. Then the fact that the horizontal and vertical projections add to zero can be written as

(7) $a - b \cos A + c \cos(A + B)$
$$= f \cos F - e \cos(E + F) + d \cos(D + E + F)$$

and

(8) $\quad b \sin A - c \sin(A + B) = f \sin F - e \sin(E + F) + d \sin(D + E + F).$

Squaring (7) and (8) and adding gives

(9) $\quad a^2 + b^2 + c^2 - 2ab \cos A - 2bc \cos B + 2ac \cos(A + B)$
$\quad\quad = d^2 + e^2 + f^2 - 2de \cos D - 2ef \cos E + 2df \cos(D + E).$

Equation (8) will be called the law of sines for hexagons and may be written in six different ways, namely,

Law of sines for hexagons.

(a) $\quad b \sin A - c \sin(A + B) = f \sin F - e \sin(E + F) + d \sin(D + E + F)$
(b) $\quad c \sin B - d \sin(B + C) = a \sin A - f \sin(F + A) + e \sin(E + F + A)$
(c) $\quad d \sin C - e \sin(C + D) = b \sin B - a \sin(A + B) + f \sin(F + A + B)$
(d) $\quad e \sin D - f \sin(D + E) = c \sin C - b \sin(B + C) + a \sin(A + B + C)$
(e) $\quad f \sin E - a \sin(E + F) = d \sin D - c \sin(C + D) + b \sin(B + C + D)$
(f) $\quad a \sin F - b \sin(F + A) = e \sin E - d \sin(D + E) + c \sin(C + D + E)$

Again, equation (9) will be called the law of cosines for hexagons. In this case, because of symmetry the six different formulations consist of three identical pairs so that there are only three forms of the law of cosines for hexagons, namely,

Law of cosines for hexagons.

(a) $\quad a^2 + b^2 + c^2 - 2ab \cos A - 2bc \cos B + 2ac \cos(A + B)$
$\quad\quad = d^2 + e^2 + f^2 - 2de \cos D - 2ef \cos E + 2df \cos(D + E)$
(b) $\quad b^2 + c^2 + d^2 - 2bc \cos B - 2cd \cos C + 2bd \cos(B + C)$
$\quad\quad = e^2 + f^2 + a^2 - 2ef \cos E - 2fa \cos F + 2ea \cos(E + F)$
(c) $\quad c^2 + d^2 + e^2 - 2cd \cos C - 2de \cos D + 2ce \cos(C + D)$
$\quad\quad = f^2 + a^2 + b^2 - 2fa \cos F - 2ab \cos A + 2fb \cos(F + A)$

Again, the equalities in the law of sines involve all but one side and all but one angle, those in the law of cosines involve all sides and all but two angles.

As stated earlier, the generalization of these laws to those for polygons of any number of sides is straightforward.

Reference

1. R. B. Kershner, On paving the plane, Amer. Math. Monthly, 75 (1968) 839–844.

BIBLIOGRAPHIC ENTRIES: TRIGONOMETRY AND TRIGONOMETRIC FUNCTIONS

The references below are to the AMERICAN MATHEMATICAL MONTHLY.

1. L. E. Dickson, Graphical methods in trigonometry, vol. 12, p. 129.

Suggests graphical methods to aid in deriving and understanding trigonometric identities.

2. L. J. Paradiso, A check formula for the first case of oblique triangles, vol. 34, p. 318.

3. H. A. Bender, Line values of powers of trigonometric functions and their use in constructing curves, vol. 34, p. 481.

4. J. W. Wrench, Jr., On the derivation of arctangent equalities, vol. 45, p. 108.

A collection of identities involving π.

5. D. H. Lehmer, On arccotangent relations for π, vol. 45, p. 657.

List of relations with discussion of relative merits for computational purposes.

6. R. E. Greenwood, Jr., On the discovery of certain trigonometric identities, vol. 47, p. 99.

7. J. S. Frame, A trigonometric approximation, vol. 53, p. 454.

An improvement on J. C. Ray's approximation of the trigonometric ratio for fractions of a right angle.

8. W. R. Ransom, Solutions of a trigonometric equation, vol. 56, p. 402.

Introduction of an auxiliary angle to avoid extraneous roots in the solution of $a\sin x + b\cos x = c$ in general form.

9. L. J. Burton and E. A. Hedberg, Proofs of the addition formulae for sines and cosines, vol. 56, p. 471.

10. A. D. Bradley, Trigonometry of right spherical triangles and the gnomonic projection, vol. 58, p. 34.

11. J. M. Thomas, Geometrical solution of spherical triangles, vol. 58, p. 151.

Employs coordinates on the sphere rather than spherical triangles; replaces three circles used by Dutton by single circle; and gives a construction for finding the latitude and longitude of a point from which the altitudes of two stars have been simultaneously observed.

12. S. L. Thompson, Note on the law of cosines, vol. 58, p. 698.

5

ELEMENTARY ALGEBRA

(a)

POLYNOMIALS

ON THE IRREDUCIBILITY OF CERTAIN POLYNOMIALS*

JACOB WESTLUND, Purdue University

The object of the following note is to determine whether the two polynomials

$$f_1(x) = (x - a_1)(x - a_2) \cdots (x - a_n) - 1 \text{ and}$$
$$f_2(x) = (x - a_1)(x - a_2) \cdots (x - a_n) + 1,$$

where $a_1, a_2, \ldots, a_n$ are distinct integers, are reducible or irreducible.

Let us first consider $f_1(x)$. If $f_1(x)$ were reducible we would have

$$f_1(x) = \phi(x)\psi(x),$$

where $\phi(x)$ is irreducible and of a lower degree than n. Then since

$$\phi(a_i)\psi(a_i) = -1, \; i = 1, 2, \ldots, n,$$

we must have

$$\phi(a_i) = \pm 1 \quad \text{and} \quad \psi(a_i) = \mp 1.$$

Hence,

$$\phi(a_i) + \psi(a_i) = 0, \; i = 1, 2, \ldots, n;$$

and hence the equation

$$\phi(x) + \psi(x) = 0,$$

whose degree is less than n, has n distinct roots, which is impossible. Hence, $f_1(x)$ is always irreducible.

Let us next consider $f_2(x)$. If $f_2(x)$ were reducible, we would have

$$f_2(x) = \phi(x)\psi(x),$$

---

*From AMERICAN MATHEMATICAL MONTHLY, vol. 16 (1909), pp. 66–67.

where $\phi(x)$ is irreducible and of a lower degree than n. Then reasoning in the same way as in the first case we find that the equation

$$\phi(x) - \psi(x) = 0,$$

which is of a lower degree than n, has n distinct roots. But this is impossible, unless $\phi(x)$ and $\psi(x)$ are identically equal. Hence the only case when $f_2(x)$ is reducible is when it is a perfect square, in which case n of course must be even.

CONSTRUCTION OF AN ALGEBRAIC EQUATION WITH AN IRRATIONAL ROOT APPROXIMATELY EQUAL TO A GIVEN VALUE*†

L. S. DEDERICK, Princeton University

In teaching Horner's method it is often desirable to have an equation which shall exhibit a given peculiarity in the sequence of digits in the computed root. Thus it may be desired to illustrate the treatment of a zero, or the unreliability of the trial divisor for a large digit occurring early in the computation. The following method furnishes a means of finding an equation of any given degree above the first, with relatively small integral coefficients, in fact a large number of such equations, which shall have a root beginning with any given sequence of digits. No very definite meaning can be attached to the phrase "relatively small." The size of the coefficients will depend not only upon the number of digits given and the degree of the required equation, but also upon the closeness with which the given root happens to approximate to some simple algebraic irrationality. For example, if it were required to find an algebraic equation with a root beginning with 1.2679, it would not be especially obvious that this number is approximately equal to $3 - \sqrt{3}$, and hence satisfies the equation $x^2 - 6x + 6 = 0$.

The method can best be explained in connection with an example. As a certain interest attaches to an algebraic equation with a root approximately equal to π, let it be required to find an equation of the fourth degree having a root beginning with 3.14159. We have then the following values for the powers of this root:

$$x^4 = 97.40909, x^3 = 31.00628, x^2 = 9.86960, x = 3.14159, x^0 = 1.$$

We may now diminish the largest of these five numbers by subtracting from it a suitable multiple of one of the others, then diminish the largest remaining one in the same way, and so continue. The process is similar to Euclid's algorithm for the

---

* From AMERICAN MATHEMATICAL MONTHLY, vol. 23 (1916), pp. 69–71.

† This note may be taken as an answer to algebra problem number 447, proposed in the December issue, though it was presented simultaneously with the problem.—EDITOR.

CONSTRUCTION OF AN ALGEBRAIC EQUATION

greatest common divisor, except that at each stage there is a choice of numbers whose multiples may be used. Thus

$$
\begin{aligned}
A &= x^4 - 3x^3 & &= 4.39025\\
B &= x^3 - 3x^2 & &= 1.39747\\
C &= x^2 - 3x & &= .44482\\
A - 3B = D &= x^4 - 6x^3 + 9x^2 & &= .19784\\
E &= x - 3 &=& .14159\\
B - 3C = F &= x^3 - 6x^2 + 9x & &= .06301\\
1 - 2C = G &= -2x^2 + 6x + 1 &=& .11036\\
C - 3E = H &= x^2 - 6x + 9 &=& .02005\\
D - 3F = I &= x^4 - 9x^3 + 27x^2 - 27x & &= .00881\\
E - G = J &= 2x^2 - 5x - 4 &=& .03123\\
2F - G = K &= 2x^3 - 10x^2 + 12x - 1 &=& .01566\\
F - 2J = L &= x^3 - 10x^2 + 19x + 8 &=& .00055\\
2K - J = M &= 4x^3 - 22x^2 + 29x + 2 &=& .00009\\
H - K = N &= -2x^3 + 11x^2 - 18x + 10 &=& .00439\\
2I - K = P &= 2x^4 - 20x^3 + 64x^2 - 66x + 1 &=& .00196\\
I - 2N = Q &= x^4 - 5x^3 + 5x^2 + 9x - 20 &=& .00003\\
N - 2P = R &= -4x^4 + 38x^3 - 117x^2 + 114x + 8 &=& .00047\\
4L - P = S &= -2x^4 + 24x^3 - 104x^2 + 142x + 31 &=& .00024\\
L - R = T &= 4x^4 - 37x^3 + 107x^2 - 95x & &= .00008\\
2S - R = U &= 10x^3 - 91x^2 + 170x + 54 &=& .00001\\
3T - S = V &= 14x^4 - 135x^3 + 425x^2 - 427x - 31 &=& 0
\end{aligned}
$$

Here then is an equation with a root beginning 3.14159. If we desire only one such equation we need go no further. In fact we might have stopped after finding Q, and have written at once $3Q - M = 0$. We may, however, go on to get other equations. Thus

$$
\begin{aligned}
3Q - M = W &= 3x^4 - 19x^3 + 37x^2 - 2x - 62 = 0\\
3Q - T = X &= -x^4 + 22x^3 - 92x^2 + 122x - 60 = .00001\\
Q - 3X = Y &= 4x^4 - 71x^3 + 281x^2 - 357x + 160 = 0\\
U - X = Z &= x^4 - 12x^3 + x^2 + 48x + 114 = 0
\end{aligned}
$$

V, W, Y, and Z, are four linearly independent polynomials, each having one root approximately equal to 3.14159. Therefore, any linear combination of these will also have such a root. We may thus get a large number of equations satisfying the conditions of the problem. Of these there may be some having coefficients smaller than those first found. To obtain these we may try to diminish the coefficients in the same way that we diminished the values of the polynomials, that is by getting rid of the largest, and then the next largest, and so on. Thus

$$V - Y = A_1 = 10x^4 - 64x^3 + 144x^2 - 70x - 191 = 0$$
$$A_1 - Y = B_1 = 6x^4 + 7x^3 - 137x^2 + 287x - 351 = 0$$

These are improvements upon V and Y respectively. But it soon becomes difficult to diminish one coefficient without increasing another. The four following independent equations can be obtained after a little manipulation:

$$-64W + 6Z + 19A_1 + 3B_1 = C_1 = 22x^4 - 51x^3 - 37x^2 - 53x - 30 = 0$$
$$W + Z = D_1 = 4x^4 - 31x^3 + 38x^2 + 46x + 52 = 0$$
$$-21W + 2Z + 6A_1 + B_1 = E_1 = 5x^4 - 2x^3 - 48x^2 + 5x + 33 = 0$$
$$-20W + 2Z + 6A_1 + B_1 = F_1 = 8x^4 - 21x^3 - 11x^2 + 3x - 29 = 0$$

The last of these not only has remarkably small coefficients but also has a root more nearly equal to π than we had any right to expect, namely $x = 3.1415925$.

The last four equations may be combined to satisfy various further conditions. For example, $F_1 - 2D_1$ gives us the third degree equation

$$41x^3 - 87x^2 - 89x - 133 = 0.$$

Or we may obtain an equation making a still closer approximation. Thus, $3D_1 - F_1$ gives the equation

$$4x^4 - 72x^3 + 125x^2 + 135x + 185 = 0,$$

which has a root equal to 3.1415926557, the value of π being 3.1415926536. Or various other conditions might be imposed.

For the degree of accuracy used here the computation is rather laborious. If, however, only a single third degree equation is desired and not more than four or five significant figures in the root, the work is not long, especially if the powers of x at the beginning are found by logarithms.

ON NONNEGATIVE POLYNOMIALS*

LOUIS BRICKMAN,[†] Yale University, and LEON STEINBERG, Remington Rand Univac

1. Introduction. It is rather well known and easy to prove that a polynomial which is nonnegative on the whole real line can be written as the sum of the squares of two real polynomials. Here we discuss the analogous question for polynomials nonnegative in an interval, finite or half infinite. All polynomials mentioned are understood to have real coefficients.

Beginning with the interval $[0, \infty)$, we quote a result appearing in [3] and [5; 5].

---

* From AMERICAN MATHEMATICAL MONTHLY, vol. 69 (1962), pp. 218–221.

[†] The first author was supported by the United States Air Force through the Office of Scientific Research of the Air Research and Development Command under Contract No. AF49(638)224.

THEOREM 1. *Let the polynomial $f(x)$ be nonnegative for nonnegative x. Then there exist polynomials $p(x)$, $q(x)$, $r(x)$, $s(x)$ such that*

(1) $$f(x) = p^2(x) + q^2(x) + [r^2(x) + s^2(x)]x.$$

In this note we improve Theorem 1 and obtain

THEOREM 2. *Let the polynomial $f(x)$ be nonnegative for nonnegative x. Then there exist polynomials $p(x)$ and $q(x)$ such that*

(2) $$f(x) = p^2(x) + q^2(x)x.$$

The method of proof is quite elementary. Moreover the technique can be applied to obtain a simple proof of the following somewhat deeper result of F. Lukács [5; 4] and [2; 35].

THEOREM 3. *Let the polynomial $f(x)$ be nonnegative in the finite interval $[a, b]$. Then $f(x)$ can be written*

(3) $$f(x) = p^2(x)(x - a) + q^2(x)(b - x)$$

if the degree of $f(x)$ is odd, and

(4) $$f(x) = p^2(x) + q^2(x)(x - a)(b - x)$$

if the degree of $f(x)$ is even. $p(x)$ and $q(x)$ can be chosen so that the degree of each term of (3) and (4) does not exceed the degree of $f(x)$.

These theorems find current application in [1] and [4]. Their analogs for entire and meromorphic functions will appear in a future paper of the first named author.

2. Proof of Theorem 2. The conclusion is obviously correct for polynomials of degree 0 and 1. Suppose $f(x)$ is of degree 2. Then

$$f(x) = a(x^2 + bx + c), \quad a > 0, \quad c \geq 0,$$

or

$$f(x) = a[(x - \sqrt{c})^2 + (b + 2\sqrt{c})x].$$

The conclusion will follow for quadratic polynomials if $b + 2\sqrt{c} \geq 0$. If $b < 0$, the only nontrivial possibility, we have $0 \leq f(-\frac{1}{2}b) = \frac{1}{4}a(4c - b^2)$. Therefore $|b| \leq 2\sqrt{c}$, and the theorem is again verified. In order to complete the proof by mathematical induction, we observe that the class of polynomials (2) is closed under multiplication. This fact is displayed by the identity

(5) $$(p^2 + q^2x)(r^2 + s^2x) = (pr - qsx)^2 + (ps + qr)^2x.$$

Consequently it is sufficient to show that every polynomial $f(x)$ of degree greater than 2 which satisfies the hypothesis can be split into two nonconstant factors which also satisfy the hypothesis. In fact it is enough to prove the existence of one such factor, for the complementary factor then necessarily has the same property. If $f(x)$ has a simple real root r, then r cannot be positive, and so $(x-r)$ is as desired. If r is not simple, we can choose $(x-r)^2$. On the other hand, if $f(x)$ has a pair of complex conjugate roots, r and $\bar{r}$, then $(x-r)(x-\bar{r})$ is actually positive for all real x, and the theorem is proved.

3. Proof of Theorem 3. The theorem is obvious for constant polynomials. For $f(x)$ of degree 1 we have

$$f(x) = \frac{f(b)}{b-a}(x-a) + \frac{f(a)}{b-a}(b-x),$$

evidently of the form (3). Next let $f(x)$ be of degree 2. To verify (4) we define $p(x)$ by

$$p(x) = \frac{1}{b-a}[\sqrt{f(a)} \cdot (x-b) + \sqrt{f(b)} \cdot (x-a)].$$

Then $p^2(a) = f(a)$, $p^2(b) = f(b)$. Hence

$$f(x) - p^2(x) = c(x-a)(b-x)$$

for some real number c, and

$$-p^2(x) \leqq c(x-a)(b-x), \qquad\qquad a \leqq x \leqq b.$$

Since $p(a) \leqq 0$, $p(b) \geqq 0$, we can write $p(x) = d(x-r)$, $a \leqq r \leqq b$. Therefore

$$-d^2(x-r)^2 \leqq c(x-a)(b-x), \qquad\qquad a \leqq x \leqq b.$$

Choosing $x = r$ (or near r if $r = a$ or b) we deduce that $c \geqq 0$ as required.

To conclude the proof we observe some rather interesting multiplication properties of the classes (3) and (4). Let us write

(6) $\qquad\qquad (3)(3) \subset (4), \qquad (3)(4) \subset (3), \qquad (4)(4) \subset (4)$

to indicate that the product of two polynomials of the class (3) belongs to the class (4), etc. The last of these relations follows immediately from (5) with x replaced by $(x-a)(b-x)$. The first and second are consequences of the identities

(7) $\qquad\qquad (p^2y + q^2z)(r^2y + s^2z) = (pry - qsz)^2 + (ps + qr)^2 yz$

and

(8) $\qquad\qquad (p^2y + q^2z)(r^2 + s^2yz) = (pr - qsz)^2 y + (psy + qr)^2 z$

respectively. By means of the relations (6), the induction step can be handled by an argument almost identical with that used in the proof of Theorem 2. As for the assertion concerning degree, we observe in connection with (7) that if degree $p^2y \leqq m$, degree $q^2z \leqq m$, degree $r^2y \leqq n$, degree $s^2z \leqq n$, then degree $(pry - qsz)^2 \leqq m+n$, degree $(ps+qr)^2yz \leqq m+n$, and a similar computation applies to (8). But we have seen that the degree property is valid for linear and quadratic polynomials. Hence this property is established by the induction, and the proof is complete.

We note in closing that representations (2), (3), and (4) need not be unique. For example, we have

$$(2x - 1)^2 + (x - 2)^2 x = 1 + x^2 x,$$
$$(5x + 3)^2(x - 2) + (x + 4)^2(3 - x) = (7x - 3)^2(x - 2) + (5x - 4)^2(3 - x),$$
$$(3x^2 - 11x + 4)^2 + (6x - 1)^2(x - 2)(3 - x)$$
$$= (3x^2 - 7x - 4)^2 - (6x + 1)^2(x - 2)(3 - x).$$

However, Professor S. Karlin has communicated to us the following interesting fact: If $f(x)$ is positive on any of the intervals considered in this paper, then there is one and only one representation of the appropriate type (2), (3), or (4) in which the zeros of $p(x)$ and $q(x)$ are all in the interior of the given interval and strictly interlace.

The referee has kindly informed us that the methods and results of this article have appeared in a Russian book by Achiezer, published in Moscow, 1961, and entitled *Classical Moment Problems*. Nevertheless, an account in English, particularly of Theorem 2, seems worthwhile.

References

1. L. Brickman, A convex class of polynomials and an associated locus problem, Jour. of Math. Analysis and Applications, (to appear).
2. S. Karlin and L. S. Shapley, Geometry of moment spaces, Memoirs of the Amer. Math. Soc., no. 12.
3. G. Polya and G. Szegö, Aufgaben und Lehrsätze aus der Analysis, vol. 2, p. 82, problem 45.
4. I. J. Schoenberg and G. Szegö, An extremum problem for polynomials, Compositio Math., vol. 14, fasc. 3, pp. 260–268.
5. G. Szegö, Orthogonal Polynomials, Amer. Math. Soc. Colloquium Publ., vol. 23.

BIBLIOGRAPHIC ENTRIES: POLYNOMIALS

Except for the entry labeled MATHEMATICS MAGAZINE, the references below are to the AMERICAN MATHEMATICAL MONTHLY.

1. H. L. Dorwart, Irreducibility of polynomials, vol. 42, p. 369.
2. B. A. Hausmann, A new simplification of Kronecker's method of factorization of polynomials, vol. 44, p. 574.

   An old factorization method which might be of interest in modern minicalculator days.

3. F. E. Hohn, The number of terms in a polynomial, vol. 48, p. 686.

   Proof that a complete, homogeneous polynomial of degree n in $r + 1$ variables has $_{n+r}C_r$ distinct terms.

4. C. S. Ogilvy, The binomial coefficients, vol. 57, p. 551.
5. C. M. Fulton, A simple proof of the binomial theorem, vol. 59, p. 243.
6. L. O. Kattsoff, Polynomials and functions, MATHEMATICS MAGAZINE, vol. 33, p. 157.

(b)

VECTOR ALGEBRA

FUNDAMENTAL IDENTITY OF VECTOR ALGEBRA*

C. J. COE and G. Y. RAINICH, University of Michigan

One of the most valuable features of vector analysis as applied to geometry and physics is the fact that the discussion and results can be free from any reference to a particular coordinate system. This makes it desirable to define the vector operations in geometric form and only develop the corresponding coordinate formulas later as aids in the application. In this way the question of the invariance of these quantities under transformation of coordinates does not arise.

However, simple geometric proofs of the fundamental identity,

(1) $$\mathbf{a} \times (\mathbf{b} \times \mathbf{c}) = (\mathbf{c}\cdot\mathbf{a})\mathbf{b} - (\mathbf{a}\cdot\mathbf{b})\mathbf{c}$$

would seem to be lacking, since the proofs offered in the texts are extremely various and usually unsatisfactory from the above point of view. The following geometric proof is suggested, as resting directly on the geometric definitions of the operations involved and as clearly showing the geometric significance of the various terms.

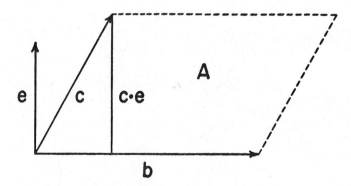

Consider first the case in which **a** is a unit vector **e**, coplanar with **b** and **c** and perpendicular to **b** on the same side as **c**. Inspection of the figure, in which the plane of the paper is that of **b**, **c**, **e** shows that $\mathbf{e} \times (\mathbf{b} \times \mathbf{c})$ has the direction

---

* From AMERICAN MATHEMATICAL MONTHLY, vol. 56 (1949), pp. 175–176.

and sense of **b** and thus that the equality,

$$(2) \qquad \mathbf{e} \times (\mathbf{b} \times \mathbf{c}) = (\mathbf{c} \cdot \mathbf{e})\mathbf{b}$$

is true in direction and sense. Furthermore, the length of the left member is the area A of the parallelogram formed on **b** and **c**, while the length of the right member is the product of the altitude and base of this parallelogram. Thus equation (1) holds for $\mathbf{a} = \mathbf{e}$, since the second term of its second member vanishes in this case.

Likewise, in the case where **a** is a unit vector **f**, coplanar with **b** and **c**, but perpendicular to **c** on the same side as **b**, we have merely to replace **e** in equation (2) by **f** and interchange **b** and **c** to find,

$$(3) \qquad \mathbf{f} \times (\mathbf{b} \times \mathbf{c}) = -(\mathbf{f} \cdot \mathbf{b})\mathbf{c},$$

and thus equation (1) is satisfied for $\mathbf{a} = \mathbf{f}$, since the first term of its second member vanishes in this case. And finally, if **a** is a unit vector **g**, perpendicular to both **b** and **c**, equation (1) holds since every term is zero.

Except in the trivial case in which **b** and **c** are parallel, any vector **a** is of the form $\mathbf{a} = k\mathbf{e} + l\mathbf{f} + m\mathbf{g}$, where k, l, m are scalars, and it follows that equation (1) holds in every case, since the equation involves **a** linearly and has been seen to hold for $\mathbf{a} = \mathbf{e}, \mathbf{f}$ and $\mathbf{g}$.

COORDINATE GEOMETRY FROM THE VECTOR POINT OF VIEW*

SAMUEL BOURNE, University of Connecticut

Introduction. From a mathematical point of view, the vector approach to coordinate geometry is superior to the conventional slope approach, in that concepts do not depend on the dimensionality of the space, and proofs are simplified. From a practical point of view, its superiority rests in the early introduction of vectors into the student's mathematical education, thereby enabling him to utilize the algebra of vectors in his study of the physical sciences. The above opinions are supported by years of teaching this subject at the Johns Hopkins University, under Professor F. D. Murnaghan. We give a few examples to illustrate them.

Vectors. In defining a vector, we are cognizant of the mathematical immaturity of the student and we constantly appeal to his geometric experience. If $P_1 = (x_1, y_1)$ and $P_2 = (x_2, y_2)$, we define the direction numbers of $\overrightarrow{P_1P_2}$ to be $\Delta x = x_2 - x_1$ and $\Delta y = y_2 - y_1$, and we write $\overrightarrow{P_1P_2} = (\Delta x, \Delta y)$. Conversely, each or-

---

* From AMERICAN MATHEMATICAL MONTHLY, vol. 59 (1952), pp. 245–248.

dered pair of numbers determines directed line segments having these numbers as direction numbers. Hence, we may define a *vector as the collection of directed line segments having equal direction numbers*. This definition is adequate at this early date. Each directed line segment is called a representative line segment of the vector and the length of such a line segment is defined to be the magnitude of the vector. We denote by $\vec{v}(P_1 \to P_2)$ a vector, a representative segment of which is $\overrightarrow{P_1P_2}$. If $\overrightarrow{P_1P_2} = (\Delta x, \Delta y)$ and $r = +\sqrt{(\Delta x)^2 + (\Delta y)^2} \neq 0$, then we define the direction cosines (l, m) of P_1P_2 to be $l = \Delta x/r = \cos\theta$ and $m = \Delta y/r = \sin\theta$, where θ is an angle from the positive direction of the x-axis to the direction $\overrightarrow{P_1P_2}$. Obviously, $l^2 + m^2 = \cos^2\theta + \sin^2\theta = 1$.

Angle. We let θ be an angle from the vector $\vec{v_1}$ to the vector $\vec{v_2}$; then we have

THEOREM 1. *If the direction cosines of $\vec{v_1}$ are (l_1, m_1) and if the direction cosines of $\vec{v_2}$ are (l_2, m_2), then $\cos\theta = l_1l_2 + m_1m_2$.*

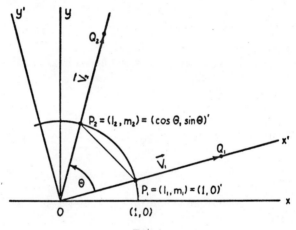

FIG. 1

We let $\overrightarrow{OQ_1}$ and $\overrightarrow{OQ_2}$ be representative segments of the vectors $\vec{v_1}$ and $\vec{v_2}$, respectively, and P_1 and P_2 the points in which the unit circle, at the origin, intersects these segments. Then, $P_1 = (l_1, m_1)$ and $P_2 = (l_2, m_2)$ relative to xy-axes, and $P_1 = (1, 0)$ and $P_2 = (\cos\theta, \sin\theta)$, relative to $x'y'$-axes. We have

$$|\overrightarrow{P_1P_2}|^2 = (l_2 - l_1)^2 + (m_2 - m_1)^2 = 2 - 2(l_1l_2 + m_1m_2),$$

$$|\overrightarrow{P_1P_2}|^2 = (\cos\theta - 1)^2 + \sin^2\theta = 2 - 2\cos\theta;$$

therefore

$$\cos\theta = l_1l_2 + m_1m_2.$$

Since
$$l_1 = \frac{\Delta_1 x}{r_1}, \quad l_2 = \frac{\Delta_2 x}{r_2}, \quad m_1 = \frac{\Delta_1 y}{r_1} \quad \text{and} \quad m_2 = \frac{\Delta_2 y}{r_2}, \quad r_1 r_2 \neq 0,$$
we have that
$$\cos \theta = \frac{(\Delta_1 x)(\Delta_2 x) + (\Delta_1 y)(\Delta_2 y)}{r_1 r_2}.$$

This equality motivates us to define *the symmetric product* $\vec{v_1} \cdot \vec{v_2}$ *of the vector* $\vec{v_1} = (x_1, y_1)$ *by the vector* $\vec{v_2} = (x_2, y_2)$ *to be* $x_1 x_2 + y_1 y_2$. We prove, with ease, the well-known properties of this product with their corresponding geometric interpretations.

Straight line. We use this symmetric product to prove

THEOREM 2. *An equation of a straight line is an equation of the form* $ax + by + c = 0$. *Conversely, the graph of the equation* $ax + by + c = 0$ *is a straight line.*

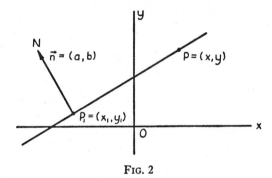

FIG. 2

We let $P_1 = (x_1, y_1)$ be a fixed point on the given line, $\vec{n} = (a, b)$ a vector normal to this line, and $P = (x, y)$ any point on this line. Then
$$\vec{v}(P_1 \to N) \cdot \vec{v}(P_1 \to P) = 0, \quad P \neq P_1;$$
therefore
$$a(x - x_1) + b(y - y_1) = 0, \quad \text{for all} \quad P.$$
Therefore
$$ax + by + c = 0, \quad \text{where} \quad c = -ax_1 - by_1.$$
The proof of the converse follows by a retracing of steps.

Distance from a line to a point. Again, by a judicious employment of the symmetric product, we may prove

THEOREM 3. *The distance d from the line $ax+by+c=0$ to the point $P_1 = (x_1, y_1)$ is given by*
$$\frac{|ax_1 + by_1 + c|}{\sqrt{a^2 + b^2}}.$$

Proof:

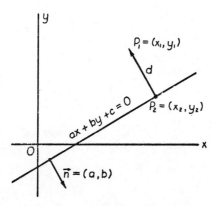

FIG. 3

We let $\vec{n} = (a, b)$ be a normal to the line $ax+by+c=0$ and $P_2 = (x_2, y_2)$ the foot of the perpendicular from $P_1 = (x_1, y_1)$ to this line. Then $\vec{n}$ and $\overrightarrow{P_1P_2}$ are parallel, and if θ is an angle from $\vec{n}$ to $\overrightarrow{P_1P_2}$ we have:

$$(a, b) \cdot (x_1 - x_2, y_1 - y_2) = \sqrt{a^2 + b^2}\, d \cos \theta;$$

therefore

$$a(x_1 - x_2) + b(y_1 - y_2) = \pm d\sqrt{a^2 + b^2},$$

and

$$d = \frac{|ax_1 + by_1 + c|}{\sqrt{a^2 + b^2}}.$$

Conclusion. The above few examples, in the plane, illustrate adequately the effectiveness of this approach. We may point out that no trigonometry need be presupposed. The definition of the vector product of two space vectors is motivated by the necessity to find direction numbers of the line of intersection of two planes. The laws of trigonometry are merely a by-product of the relationships of the algebra of plane vectors, while the laws of spherical trigonometry spring from the corresponding relationships of the algebra of space vectors.

A VECTOR PROOF OF EULER'S THEOREM ON ROTATIONS OF E^3*

M. K. FORT, JR., University of Georgia

A well known theorem of Euler states that every orientation preserving isometry T of Euclidean 3-space that has a fixed point O is a rotation about a line through O. The most difficult part of the proof of Euler's theorem is in showing that there exists a nonzero vector $\mathbf{P}$ such that $T(\mathbf{P}) = \mathbf{P}$. We give a proof of this last statement which does not use the theory of matrices and determinants, and which may be presented to a class that is studying elementary vector algebra.

Let O be the origin, and let $\mathbf{i}, \mathbf{j}, \mathbf{k}$ be a right hand orthogonal system of unit vectors. We define $\mathbf{i}' = T(\mathbf{i})$, $\mathbf{j}' = T(\mathbf{j})$, $\mathbf{k}' = T(\mathbf{k})$. Since T is an isometry, $\mathbf{i}', \mathbf{j}', \mathbf{k}'$ is an orthogonal system of unit vectors. Since T is orientation preserving, $\mathbf{i}', \mathbf{j}', \mathbf{k}'$ is also a right hand system.

The following computation shows that the triple scalar product $[\mathbf{i}-\mathbf{i}', \mathbf{j}-\mathbf{j}', \mathbf{k}-\mathbf{k}']$ vanishes.

$$[\mathbf{i} - \mathbf{i}', \mathbf{j} - \mathbf{j}', \mathbf{k} - \mathbf{k}']$$
$$= [\mathbf{i},\mathbf{j},\mathbf{k}] - [\mathbf{i},\mathbf{j},\mathbf{k}'] - [\mathbf{i},\mathbf{j}',\mathbf{k}]$$
$$+ [\mathbf{i},\mathbf{j}',\mathbf{k}'] - [\mathbf{i}',\mathbf{j},\mathbf{k}] + [\mathbf{i}',\mathbf{j},\mathbf{k}'] + [\mathbf{i}',\mathbf{j}',\mathbf{k}] - [\mathbf{i}',\mathbf{j}',\mathbf{k}']$$
$$= 1 - \mathbf{i} \times \mathbf{j} \cdot \mathbf{k}' - \mathbf{j}' \cdot \mathbf{k} \times \mathbf{i} + \mathbf{i} \cdot \mathbf{j}' \times \mathbf{k}' - \mathbf{i}' \cdot \mathbf{j} \times \mathbf{k} + \mathbf{j} \cdot \mathbf{k}' \times \mathbf{i}' + \mathbf{i}' \times \mathbf{j}' \cdot \mathbf{k} - 1$$
$$= 1 - \mathbf{k} \cdot \mathbf{k}' - \mathbf{j}' \cdot \mathbf{j} + \mathbf{i} \cdot \mathbf{i}' - \mathbf{i}' \cdot \mathbf{i} + \mathbf{j} \cdot \mathbf{j}' + \mathbf{k}' \cdot \mathbf{k} - 1 = 0.$$

It follows that the vectors $\mathbf{i}-\mathbf{i}'$, $\mathbf{j}-\mathbf{j}'$, $\mathbf{k}-\mathbf{k}'$ are coplanar, and that there exists a non-zero vector $\mathbf{P}$ which is orthogonal to each of them. Thus, $\mathbf{P} \cdot \mathbf{i} = \mathbf{P} \cdot \mathbf{i}'$, $\mathbf{P} \cdot \mathbf{j} = \mathbf{P} \cdot \mathbf{j}'$, $\mathbf{P} \cdot \mathbf{k} = \mathbf{P} \cdot \mathbf{k}'$.

Since T is an isometry and $T(O) = O$, T is clearly linear. That is, if x, y and z are real numbers, then $T(x\mathbf{i} + y\mathbf{j} + z\mathbf{k}) = x\mathbf{i}' + y\mathbf{j}' + z\mathbf{k}'$. Thus,

$$T(\mathbf{P}) = T((\mathbf{P} \cdot \mathbf{i})\mathbf{i} + (\mathbf{P} \cdot \mathbf{j})\mathbf{j} + (\mathbf{P} \cdot \mathbf{k})\mathbf{k}) = (\mathbf{P} \cdot \mathbf{i})\mathbf{i}' + (\mathbf{P} \cdot \mathbf{j})\mathbf{j}' + (\mathbf{P} \cdot \mathbf{k})\mathbf{k}'$$
$$= (\mathbf{P} \cdot \mathbf{i}')\mathbf{i}' + (\mathbf{P} \cdot \mathbf{j}')\mathbf{j}' + (\mathbf{P} \cdot \mathbf{k}')\mathbf{k}' = \mathbf{P}.$$

---

* From AMERICAN MATHEMATICAL MONTHLY, vol. 64 (1957), p. 428.

ON THE VECTOR TRIPLE CROSS PRODUCT IDENTITY*

DANIEL T. DWYER, Indiana Technical College

In the Classroom Note, *A proof of the vector triple cross product identity*, (this MONTHLY, vol. 67, 1960, pp. 574–578), Robert C. Wrede presented a proof of the relation

(1) $$A \times (B \times C) = B(A \cdot C) - C(A \cdot B),$$

which avoided the expansion of the two members of equation (1) but which utilized tensor notation.

Another approach is to prove the "Lagrange Identity"

(2) $$(A \times B) \cdot (C \times D) = (A \cdot C)(B \cdot D) - (A \cdot D)(B \cdot C),$$

from which the vector triple product relation follows directly.[†]

Proof of relation (2).

$$(A \times B) \cdot (C \times D) = (AB \overline{C \times D}) = (\overline{A \times B} CD).$$

This follows from the definition of the triple scalar product;

$$(XYZ) = \begin{vmatrix} X_1 & X_2 & X_3 \\ Y_1 & Y_2 & Y_3 \\ Z_1 & Z_2 & Z_3 \end{vmatrix} = \begin{vmatrix} X_1 & Y_1 & Z_1 \\ X_2 & Y_2 & Z_2 \\ X_3 & Y_3 & Z_3 \end{vmatrix}.$$

Then,

(3) $$[(A \times B) \cdot (C \times D)]^2 = \begin{vmatrix} A_1 & A_2 & A_3 \\ B_1 & B_2 & B_3 \\ \overline{C \times D_1} & \overline{C \times D_2} & \overline{C \times D_3} \end{vmatrix} \cdot \begin{vmatrix} \overline{A \times B_1} & C_1 & D_1 \\ \overline{A \times B_2} & C_2 & D_2 \\ \overline{A \times B_3} & C_3 & D_3 \end{vmatrix}.$$

From the rule for multiplying determinants, the right member of (3) equals

$$\begin{vmatrix} (AAB) & A \cdot C & A \cdot D \\ (BAB) & B \cdot C & B \cdot D \\ \overline{C \times D} \cdot \overline{A \times B} & (CDC) & (CDD) \end{vmatrix}.$$

Since the scalar triple products in the above determinant are equal to zero from the definition of (XYZ),

$$[(A \times B) \cdot (C \times D)]^2 = [(\overline{C \times D}) \cdot (\overline{A \times B})][(A \cdot C)(B \cdot D) - (A \cdot D)(B \cdot C)]$$

or $(A \times B) \cdot (C \times D) = (A \cdot C)(B \cdot D) - (A \cdot D)(B \cdot C)$, and Lagrange's identity is established.

---

* From AMERICAN MATHEMATICAL MONTHLY, vol. 68 (1961), p. 910.
† David V. Widder, Advanced Calculus, New York, 1947, pp. 51–52.

VECTOR PROOFS IN SOLID GEOMETRY

M. S. KLAMKIN, Ford Motor Company

1. Introduction. In this paper, we give vector proofs of a number of theorems in solid geometry. Although, as to be expected, one can usually give synthetic proofs, one should always be prepared to use any alternate representations, especially if they are simpler. In many cases, it will turn out that not only is the vector approach simpler, but it is also more direct. Admittedly, however, in a great many other cases, the vector approach will not be quite as simple, but it still may be more direct. This does not mean that the author eschews the use of synthetic geometry, in fact the reverse is true. However, the main difficulty with elegant geometric solutions is that frequently there is no general method to indicate the first few proper steps or constructions which should be made. But once the first few proper steps have been made, the rest of the solution is usually very easy and one sees what is behind the problem. This is probably one of the reasons why analysts and algebraists tend to shun geometry; they have not given themselves enough practice to acquire the necessary geometric intuition to make the first few proper steps. Contrast this with analytic geometry. Here the first few steps are usually routine. You coordinatize everything in sight and then write down the appropriate equations. However, the subsequent equation solving can often be tedious, extremely difficult or unmanageable. On the other hand, vector proofs frequently appear to "lie between" synthetic and analytic proofs in regards to both simplicity and directness. The point to observe here is that one should be ready to use *any* representation, be it synthetic geometry, analytic geometry, analysis, vectors, algebra, etc., which lead to results. An exception to this, of course, is when one is learning some particular representation. This point is explored much further with many illustrative examples in [1] and [2].

Most of the theorems given here are known results. However, some of the proofs are believed to be new. Although some of them are not simpler than their synthetic counterparts, they still can be used to provide nontrivial and, we hope, interesting exercises to be used in classes on vector analysis. A number of these results are inequalities and, as to be expected, their proofs will involve either the triangle or Cauchy's inequality.

2. Angles in tetrahedra.

THEOREM 1: *The sum of the measures of any two face angles of a trihedral angle is greater than the measure of the third face angle* [3, p. 65].

Proof: Since $\cos x$ is decreasing in $[0, \pi]$, we equivalently have to show that

$$\cos AOC > \cos AOB \cos BOC - \sin AOB \sin BOC \text{ (see Fig. 1)}.$$

---

*From AMERICAN MATHEMATICAL MONTHLY, vol. 77 (1970), pp. 1051–1065.

Vectorially, this is equivalent to establishing the inequality
$$(B \cdot B)(A \cdot C) > (B \cdot C)(B \cdot A) - |B \times C| |B \times A|.$$
Since
$$(B \cdot C)(B \cdot A) - (B \cdot B)(A \cdot C) = B \cdot \{C(B \cdot A) - B(A \cdot C)\}$$
$$= B \cdot [A \times (C \times B)] = (B \times A) \cdot (C \times B),$$
the inequality immediately follows. There is equality only if O is in the plane of A, B, C.

THEOREM 2. *The sum of the areas of any three faces of a tetrahedron is greater than the area of the fourth face* [4, p. 106].

Proof: Referring to Fig. 1, we have to show vectorially that
$$|A \times B| + |B \times C| + |C \times A| > |(A - B) \times (C - B)|.$$
Since $|(A-B) \times (C-B)| = |A \times B + B \times C + C \times A|$, the desired result follows from the triangle inequality. Again there is equality only if O is in the plane of A, B, C.

THEOREM 3. *In any trihedral angle, the sines of the dihedral angles are proportional to the sines of the opposite face angles* [3, p. 67].

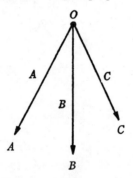

FIG. 1

Proof: If we let A, B, C, denote the dihedral angles containing the edges OA, OB, OC (which can now all be assumed to be of unit length), respectively, in Fig. 1, we have to show that
$$\frac{\sin A}{\sin BOC} = \frac{\sin B}{\sin COA} = \frac{\sin C}{\sin AOB}.$$
Since
$$\frac{\sin B}{\sin COA} = \frac{|(A \times B) \times (B \times C)|}{|A \times B| \cdot |B \times C| \cdot |C \times A|} = \frac{[A \cdot B \times C]}{|A \times B| \cdot |B \times C| \cdot |C \times A|}$$
is symmetric in A, B, C, the result is established.

The synthetic proof given in [3] does not include the proportionality constant which enables one to compute the dihedral angles from the face angles. To accomplish this, we need an expression for $[A \cdot B \times C]$ (a neater way but not quite so direct is given in [5, p. 35]) which gives the volume of a parallelepiped having A, B, C, as three coterminal edges in terms of the face angles. To this end, we let

$$A = i,$$
$$B = i \cos AOB + j \sin AOB,$$
$$C = pi + qj + rk, \quad \text{where} \quad p^2 + q^2 + r^2 = 1.$$

Since

$$A \cdot C = \cos AOC = p,$$
$$B \cdot C = \cos BOC = p \cos AOB + q \sin AOB,$$

(1) $\quad V = [A \cdot B \times C] = r \sin AOB = (1 - p^2 - q^2)^{1/2} \sin AOB$

$$= \{(1 - \cos^2 AOC) \sin^2 AOB - (\cos BOC - \cos AOC \cos AOB)^2\}^{1/2}$$
$$= \{1 - \cos^2 AOB - \cos^2 BOC - \cos^2 COA$$
$$+ 2 \cos AOB \cos BOC \cos COA\}^{1/2}.$$

Thus $\sin A = V/(\sin AOB \sin COA)$, etc.

The symmetrical function $V/2$ of the face angles of the given trihedral angle is called the norm of the sides of the spherical triangle ABC and is denoted by n in spherical trigonometry [6, p. 498]. Also, V or equivalently $2n$ is sometimes called the sine of the solid angle subtended by the trihedral angle.

Reciprocally, the face angles as functions of the dihedral angles are obtained by letting

$$\sin BOC = \lambda \sin A, \text{ etc.}$$

and solving for λ as a function of A, B, C. Whence,

$$\frac{\sin AOB}{\sin A} = \frac{\sin BOC}{\sin B} = \frac{\sin COA}{\sin C} = \frac{2N}{\sin A \sin B \sin C},$$

where $2N = \{1 - \cos^2 A - \cos^2 B - \cos^2 C - 2 \cos A \cos B \cos C\}^{1/2}$. N is called the norm of the angles of the spherical triangle ABC. Alternate derivations by means of spherical trigonometry for the latter formulae are given in [6, pp. 497, 503].

Since V ranges between 0 and 1, we also have the following inequality:

$$3 \geq \sin^2 \theta_1 + \sin^2 \theta_2 + \sin^2 \theta_3 + 2 \cos \theta_1 \cos \theta_2 \cos \theta_3 \geq 2,$$

where

$$\pi \geq \theta_i \geq 0, \quad \theta_1 + \theta_2 + \theta_3 \leq 2\pi,$$
$$\theta_1 + \theta_2 \geq \theta_3, \quad \theta_2 + \theta_3 \geq \theta_1, \quad \theta_3 + \theta_1 \geq \theta_2.$$

The upper bound is achieved for $\theta_i = \pi/2$ and the lower bound when the sum of two angles equals the third angle.

A determinantal representation for V^2 is

$$V^2 = \begin{vmatrix} 1 & \cos AOB & \cos COA \\ \cos AOB & 1 & \cos BOC \\ \cos COA & \cos BOC & 1 \end{vmatrix}.$$

The latter form is equivalent to the following result given in [7, p. 48]:

THEOREM 4. *If α, β, γ are the angles between each pair of the triad of directions OA, OB, OC, then*

$$\begin{vmatrix} 1 & \cos \gamma & \cos \beta \\ \cos \gamma & 1 & \cos \alpha \\ \cos \beta & \cos \alpha & 1 \end{vmatrix} = \sin^2 \gamma \sin^2 \phi,$$

where ϕ is the angle OC makes with the plane OAB.

Proof: Referring to the previous figure again, $\sin \phi = (A \times B \cdot C)/|A \times B|$ or $V^2 = \sin^2 \phi \sin^2 AOB$.

THEOREM 5. *If the edges of the base of a tetrahedron are a, b, c, and each of the lateral edges is equal to d, then the volume v of the tetrahedron is given by* [4, p. 108]

$$v = \{16p(p-a)(p-b)(p-c)d^2 - a^2b^2c^2\}^{1/2},$$

where $2p = a+b+c$.

Proof: It follows from (1), that $v = Vd^3/6$. By substituting $\cos BOC = 1 - a^2/2d^2$, etc., in V in (1) and simplifying we get the desired result.

THEOREM 6: *If a line makes congruent angles with each of three lines in a plane, the line is perpendicular to the plane* [3, p. 45].

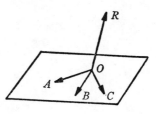

FIG. 2

Proof: If A, B, C denote unit vectors along the three lines in the plane and R is a vector along the other line, we have to show equivalently that (see Fig. 2)

$$R \cdot A = R \cdot B = R \cdot C \Rightarrow R \cdot A = R \cdot B = 0.$$

Let $R \cdot A = \lambda$. Then since A, B, C are coplanar, $C = aA + bB$ and

$$R \cdot C = (a+b)\lambda = \lambda \text{ or } (a+b-1)\lambda = 0.$$

But $a+b=1$ is impossible, since this implies that A, B, C are collinear. Thus $\lambda = 0$. (A simple geometric proof follows by considering the perpendicular bisecting planes of AB and BC. Their intersection which is normal to the plane contains both O and R.)

3. Altitudes. The next set of theorems (7–12) relate to the altitudes of a tetrahedron. While the proofs of all of them are not as simple as the corresponding synthetic ones, they may be more appealing (especially to non-geometers) by virtue of their directness. All of them except Theorem 12 are given in [4, pp. 61–65].

All the proofs will refer to Fig. 3. Here **A**, **B**, **C**, **D**, etc., will denote vectors from a common origin to the points A, B, C, D, etc.

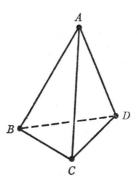

FIG. 3

THEOREM 7. *If a pair of opposite edges of a tetrahedron are rectangular (normal to each other), the two altitudes of the tetrahedron issued from the end of each of these two edges intersect.*

Proof: Equivalently, we have to show that given $(A-B)\cdot(C-D) = 0$, there exists a vector H_1 such that

(2) $\qquad (H_1 - A)\cdot(B - C) = (H_1 - A)\cdot(C - D) = 0,$

(3) $\qquad (H_1 - B)\cdot(C - D) = (H_1 - B)\cdot(D - A) = 0.$

If H is any point on h_a, then (2) is satisfied by $H_1 = A + \lambda(H - A)$ for any λ. Since

$$(A - B)\cdot(C - D) = 0 \Rightarrow (H_1 - B)\cdot(C - D) - (H_1 - A)\cdot(C - D) = 0,$$

the first part of (3) is satisfied. To satisfy the second part of (3), we must have

$$(A - B)\cdot(D - A) + \lambda(H - A)\cdot(D - A) = 0.$$

Since $(H-A)\cdot(D-A) \neq 0$ a unique λ exists.

THEOREM 8 (converse of Th. 7). *If two altitudes of a tetrahedron intersect, the edge joining the two vertices from which these altitudes issue is perpendicular to the opposite edge of the tetrahedron.*

Proof: We wish to show equivalently that

(4) $(H - A) \cdot (C - D) = (H - A) \cdot (B - C) = 0$
(5) $(H - B) \cdot (C - D) = (H - B) \cdot (A - C) = 0$ $\Rightarrow (A - B) \cdot (C - D) = 0.$

On subtracting the first part of (4) from the first part of (5), we get the desired result. Since we did not use the second parts of (4) and (5), we can make the stronger statement:

THEOREM 8′. *If two lines from A and B are normal to CD and intersect, then $AB \perp CD$.*

COROLLARY. *If two altitudes of a tetrahedron intersect, the remaining two altitudes intersect.*

THEOREM 9. *If two pairs of opposite edges of a tetrahedron are rectangular, then the remaining pair of opposite edges is also rectangular.*

Proof: We wish to show equivalently that

(6) $(A - B) \cdot (C - D) = 0$
(7) $(B - C) \cdot (D - A) = 0$ $\Rightarrow (A - B) \cdot (B - D) = 0.$

The result immediately follows by subtracting (6) from (7).

THEOREM 10. *If three altitudes of a tetrahedron are concurrent, then the four altitudes are concurrent.*

Proof: Our first proof is similar to the vector proof for the concurrency of three altitudes of a triangle given in [8, p. 32]. A normal to the plane of A, B, C, is given by

$$[A \times B + B \times C + C \times A].$$

Also,

$$A \times (B \times C) + B \times (C \times A) + C \times (A \times B) = 0.$$

Using the latter identity, it follows immediately that

(8) $\begin{aligned}&(H - A) \times (B \times C + C \times D + D \times B) \\ &+ (H - B) \times (C \times A + A \times D + D \times C) \\ &+ (H - C) \times (D \times A + A \times B + B \times D) \\ &+ (H - D) \times (A \times C + C \times B + B \times A) = 0\end{aligned}$

is also an identity. By choosing H to be the point of concurrency of three of the altitudes, our result follows.

Although this proof is rather neat, it admittedly is not very direct. Consequently, we now give a more direct one.

We wish to show equivalently that

$$(H - A) \cdot (B - C) = (H - A) \cdot (C - D) = 0$$
(9) $\quad (H - B) \cdot (C - D) = (H - B) \cdot (D - A) = 0$
$$(H - C) \cdot (D - A) = (H - C) \cdot (A - B) = 0$$
$$\Rightarrow (H - D) \cdot (A - B) = (H - D) \cdot (B - C) = 0.$$

As in Theorem 8, we then have $(A-B) \cdot (C-D) = (B-C) \cdot (D-A) = 0$. Since
$$C - D = (H - D) - (H - C),$$
$$(H - D) \cdot (A - B) = (H - C) \cdot (A - B) = 0.$$

Similarly, $(H-D) \cdot (B-C) = 0$.

THEOREM 11. *If ABCD is a tetrahedron whose altitudes are concurrent in the point H (orthocenter), then each of the five points is the orthocenter of the tetrahedron formed by the other four points.*

Proof: Follows immediately from (9).

THEOREM 12: *If one altitude of a tetrahedron intersects two other altitudes, then the four altitudes are concurrent.*

Proof: It follows from Theorems 8 and 9 that
$$(B - C) \cdot (A - D) = (A - C) \cdot (B - D) = (A - B) \cdot (C - D) = 0.$$
If H denotes the point of intersection of h_a and h_b, then
$$(H - A) \cdot (B - C) = (H - A) \cdot (B - D) = 0,$$
$$(H - B) \cdot (A - C) = (H - B) \cdot (A - D) = 0.$$
$$(B - D) \cdot (A - C) = (B - D) \cdot (H - C) - (B - D) \cdot (H - A) = 0,$$
$$(A - D) \cdot (B - C) = (A - D) \cdot (H - C) - (A - D) \cdot (H - B) = 0.$$

Whence, $(H-C) \cdot (A-D) = (H-C) \cdot (B-D) = 0$, implying that h_a, h_b and h_c are concurrent. Then by Theorem 10, all the altitudes are concurrent.

4. Tangencies. The next two theorems, which on first glance may seem unrelated, will be shown to be essentially equivalent.

THEOREM 13. *If an ellipsoid is tangent to the four edges of a skew quadrilateral, then the points of tangency are coplanar* [5, p. 180].

THEOREM 14. *If four spheres touch in succession, each one touching two others (the number of external contacts being even), then the four points of tangency lie on a circle* [5, p. 44].

We first establish Carnot's theorem and its converse [4, p. 111].

THEOREM 15. *If P, Q, R, S are four coplanar points on the four sides, respectively, of a skew quadrilateral ABCD, then*

(10) $$\frac{AP}{PB}\cdot\frac{BQ}{QC}\cdot\frac{CR}{RD}\cdot\frac{DS}{SA} = 1$$

and conversely (here AP is taken to be positive if P lies on the ray AB, otherwise it is negative). (See Fig. 4.)

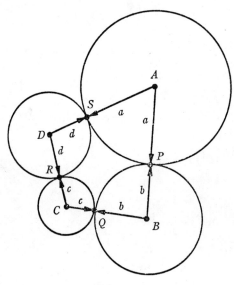

Fig. 4

Proof: Since P is on line AB etc., we have
$$P = aA + (1-a)B, \qquad Q = bB + (1-b)C,$$
$$R = cC + (1-c)D, \qquad S = dD + (1-d)A.$$

Since P, Q, R, S are coplanar, there must exist four constants p, q, r, s, where $p+q+r+s=0$, $pqrs \neq 0$, such that
$$pP + qQ + rR + sS = 0.$$

Since the origin of the vectors is taken outside the 3-space of the quadrilateral, A, B, C, D are linearly independent. Thus the coefficients of A, B, C, D in the latter equation are all zero, i.e.,
$$p(1-d) + qa = 0, \qquad q(1-a) + rb = 0,$$
$$r(1-b) + sc = 0, \qquad s(1-c) + pd = 0.$$

In order that this latter set of equations be consistent, it is necessary that
$$(1-a)(1-b)(1-c)(1-d) = abcd.$$

On solving for p, q, r in terms of s and substituting back in $p+q+r+s$, it follows

after some simplification that the condition is also sufficient. Since
$$AP = (1-a)|\mathbf{A} - \mathbf{B}|, \quad PB = a|\mathbf{A} - \mathbf{B}|,$$
etc., we get the desired result.

For the converse theorem, we are given that (10) holds. Let the plane determined by P, Q, R, intersect AD in S'. Then by the previous theorem,
$$\frac{AP}{PB} \cdot \frac{BQ}{QC} \cdot \frac{CR}{RD} \cdot \frac{DS'}{S'A} = 1.$$

Whence S' coincides with S, and P, Q, R, S are coplanar.

We digress to sketch out an analytic geometry proof and a proof using centroids. Then we shall compare the proofs, including the synthetic one (using Menelaus' theorem) which was referred to in [4], to illustrate the differences in simplicity and directness as mentioned in the introduction.

For the analytic proof, we start out by coordinatizing the points A, B, C, D, i.e., let them be (x_i, y_i, z_i), where $i = 1, 2, 3, 4$. Points P, Q, R are then chosen arbitrarily on the lines AB, BC, and CD, respectively. S is then determined as the intersection of the plane through P, Q, R with the line DA. Finally, the desired relation is verified.

For the centroid proof, assume that masses
$$M_a = \frac{k}{AP}, \quad M_b = \frac{k}{BP}, \quad M_c = \frac{l}{CR}, \quad M_d = \frac{l}{DR}$$
are placed at the points A, B, C, D, respectively, such that k, l satisfy
$$k\frac{AS}{AP} = l\frac{DS}{DR}.$$

These masses have been so chosen that the centroid of M_a and M_b is at P, the centroid of M_c and M_d is at R, and the centroid of M_a and M_d is at S. It now follows immediately that the centroid G of the four masses is on the segment $\overline{PR}$. If the centroid of M_b and M_c is at Q' on $\overline{BC}$, then G is also on SQ'. Thus, $\overline{BC}$ must intersect SQ' at G. Unless Q' and Q coincide A, B, C, D will be coplanar, contradicting the hypothesis. Then,
$$k\frac{BQ}{BP} = l\frac{CQ}{CR}$$
which together with the previous condition in k, l gives the desired result.

On comparing the different proofs it is seen that the analytic geometry one is very direct but involves a lot of arithmetic. The vector proof is somewhat less direct but the arithmetic is much less onerous. The computations in the centroid proof are few and easy but the proof requires the very indirect step of introducing appropriate masses and using centroids. The synthetic proof (see reference)

242 ELEMENTARY ALGEBRA

is also very easy provided that you invoke Menelaus' theorem. To any decent geometer, this would be a fairly direct step. However, to many others, it will not be.

We now return to the proofs of Theorems 13 and 14. Since tangency and planarity are preserved under affine transformations it suffices in Theorem 13 to prove the result for a sphere. Also, since tangents from a point to a sphere are congruent, the configuration we obtain is the same as in the first figure for the four possible configurations, Figs. 4–7 of Theorem 14.

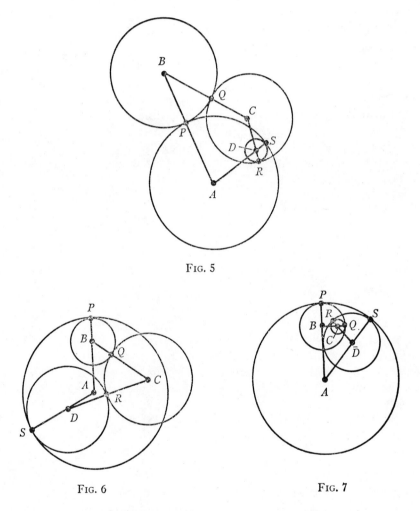

Fig. 5

Fig. 6   Fig. 7

In three of the cases (Figs. 5, 6, 7), some of the points of tangency P, Q, R, S do not lie in the interior of their corresponding segments AB, BC, CD, DA. This occurs when a pair of corresponding spheres are tangent internally, e.g., P in

Figs. 6, 7. However, in all the cases,

$$\frac{AP}{PB} = \pm \frac{a}{b}, \quad \frac{BQ}{QC} = \pm \frac{b}{c}, \quad \frac{CR}{RD} = \pm \frac{c}{d}, \quad \frac{DS}{SA} = \pm \frac{d}{a},$$

where the number of minus signs is even. Then by the converse of Carnot's theorem, P, Q, R, S are coplanar. Since in Theorem 13, P, Q, R, S are on a sphere, they also must lie on a circle. To prove the circular part for Theorem 14 requires an extension of the proof that a sufficient condition (which is also necessary) for a plane quadrilateral $ABCD$ to have an inscribed circle is that $AB+CD=BC+DA$ [9, p. 28]. Our proof is geometric as we do not have a simple enough vectorial one. It will be in two parts, one for Figs. 4, 5 and the other for Figs. 6, 7.

Since P, Q, R, S have been shown to be coplanar, it suffices to show also that they are cospherical.

In Figs. 4, 5, $AB+CD=BC+DA$. Assuming without loss of generality that $AB \geq DA$, then $BC \geq CD$. Points X, Y are chosen on AB and BC, respectively, such that $AX=DA$, $YC=CD$ and then $XB=BY$. It is to be noted that although the configuration in Fig. 8 is taken from Fig. 4, it also applies to Fig. 5. The only change is that R and S are then exterior points of their corresponding segments. The perpendicular bisecting planes of DX, XY, YD will intersect in a line through the circumcenter of $\triangle DXY$ and perpendicular to its plane. By the symmetry of the configuration, each point on this perpendicular line will be equidistant from the four sides of $ABCD$ and also from P, Q, R, S. Thus, P, Q, R, S are cospherical and, additionally, the skew quadrilateral has a family of "inspheres."

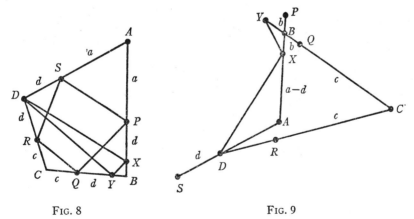

Fig. 8   Fig. 9

In Figs. 6, 7, $AB+BC=CD+DA$. In the case of a plane quadrilateral, this will turn out to be the necessary and sufficient condition for the existence of a circle tangent to the four sides (2 internal and 2 external points of contact). Since the previous argument now applies also for Fig. 9, our proof is completed.

5. Higher dimensions. In the next four theorems, we consider n-dimensional figures. For problems in higher dimensions, the vector approach is particularly useful and often a necessity.

THEOREM 16. *Corresponding to any n-dimensional simplex, there exists a skew $(n+1)$-gon whose sides are congruent and parallel to the $n+1$ medians of the simplex. Furthermore, the volume of the simplex spanned by the $n+1$ vertices of the skew-gon is $(1+1/n)^{n-1}/n$ times the volume of the initial simplex. This generalizes the known results for a triangle and tetrahedron* [4, p. 54].

Proof: Let the $n+1$ vertices of the given simplex be O, A_1, A_2, $\cdots$, A_n and let A_i denote the set of n linearly independent vectors from O to A_i. The volume of the given simplex is then

$$V = [A_1, A_2, \cdots, A_n]/n! = \frac{1}{n!} \begin{vmatrix} a_{11} & a_{12} & \cdots & a_{1n} \\ a_{21} & a_{22} & \cdots & a_{2n} \\ \cdots & \cdots & \cdots & \cdots \\ a_{n1} & a_{n2} & \cdots & a_{nn} \end{vmatrix}.$$

Here $(a_{i1}, a_{i2}, \cdots, a_{in})$ denotes the rectangular components of A_i. The $n+1$ medians of the simplex are given by

$$nM = A_1 + A_2 + \cdots + A_n, \qquad M - \frac{n+1}{n} A_i \; (i = 1, 2, \cdots, n).$$

The sum of these $n+1$ vectors is zero and consequently they form a closed skew $(n+1)$-gon.

If we now choose the starting point of M as our origin, the n vectors to the n remaining vertices of the $(n+1)$-gon are

$$M, \; 2M - \lambda A_1, \; 3M - \lambda(A_1 + A_2), \cdots, \; nM - \lambda(A_1 + A_2 + \cdots + A_{n-1}),$$

where $\lambda = (n+1)/n$. The volume V' of this simplex is given by

$$V'/n! = [M, 2M - \lambda A_1, \cdots, nM - \lambda(A_1 + A_2 + \cdots + A_{n-1})].$$

By elementary operations on the determinant,

$$V'/n! = [M, \lambda A_1, \cdots, \lambda A_{n-1}] = [A_n/n, \lambda A_1, \lambda A_2, \cdots, \lambda A_{n-1}]$$

or

$$V'/n! = \frac{\lambda^{n-1}}{n} [A_1, A_2, \cdots, A_n].$$

It is known [10, p. 139] that in order that two points in the interior or on the boundary of a polygon be farthest apart, they must be two of the vertices that are farthest apart. The polygon need not be convex. An extension of this to n-dimensional polytopes is easily proved by means of the triangle inequality.

THEOREM 17. *If two points in the interior or on the boundary of a polytope are furthest apart, they must be two of the vertices which are furthest apart.*

Proof: We need only consider convex polytopes. For if the result is valid for the convex hull of the polytope, it is also valid for the polytope.

Now let $V_1, V_2, \cdots, V_n$ denote vectors from a common origin to the vertices of the polytope. Then $R = \sum \lambda_i V_i$ and $R' = \sum \lambda_i' V_i$, where $\lambda_i, \lambda_i' \geq 0$, $\sum \lambda_i = \sum \lambda_i' = 1$ will denote two vectors from the common origin to two points within or on the boundary of the polytope. Using the triangle inequality repeatedly and the properties of λ_i and λ_i', we get the following sequence of inequalities:

$$|R' - R| = \left|\sum_i \lambda_i'(V_i - R)\right| \leq \sum_i \lambda_i'|V_i - R|$$

$$\leq \operatorname*{Max}_i |V_i - R| = \operatorname*{Max}_i \left|\sum_j \lambda_j(V_j - V_i)\right|$$

$$\leq \operatorname*{Max}_i \sum_j \lambda_j |V_j - V_i| \leq \operatorname*{Max}_{i,j} |V_j - V_i|.$$

For our next theorem, we consider a special case of "the optimal location of a warehouse" problem [11, p. 394]. Here we have to determine a point R which minimizes $\sum w_i |R - V_i|$, where V_i, are given points and w_i are given positive weights. If we set the appropriate derivatives equal to zero, the resulting equations are given by

$$\sum w_i \frac{R - V_i}{|R - V_i|} = 0$$

(a set of forces in equilibrium). It has been shown that the point R is unique. A minor difficulty arises if R coincides with one of the V_i's.

If the weights w_i are equal and the given points V_i correspond to the vertices of a regular polytope (in any dimension), then one expects intuitively from the symmetry of the configuration that R will correspond to the center of the polytope. We now give an extension of this which is established elementarily using the triangle inequality.

THEOREM 18. *Let V_i denote vectors from a common origin O to n given points on a unit sphere (of any dimension) with center O. If $\sum w_i V_i = 0$, where w_i are arbitrarily given positive weights, then*

$$\operatorname*{Min}_R \sum_i w_i |V_i - R| = \sum_i w_i.$$

Proof: By symmetry (see Fig. 10), $|V_i - R| = |rV_i - R/r|$. Whence,

$$\sum_i w_i |V_i - R| = \sum_i w_i |rV_i - R/r| \geq \left|\sum_i (rw_i V_i - w_i R/r)\right| = \sum_i w_i$$

with equality, if and only if, $r = 0$.

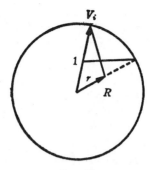

Fig. 10

Our last theorem is an elementary generalization of the convexity of ellipsoids of revolution. The proof via vectors is almost automatic. However, a simpler proof for general ellipsoids follows by transforming the ellipsoid into a sphere by an affine transformation (which preserves convexity).

THEOREM 19. *If $A_1, A_2, \cdots, A_r$ denote arbitrary points, not necessarily distinct, in E_n, then the region spanned by the points P satisfying the inequality $PA_1+PA_2+\cdots+PA_r \leq k$ (constant) is convex.*

Proof: We first show that for any three collinear points P_1, Q, P_2, where Q is strictly between P_1 and P_2, that

$$(11) \quad \sum_{i=1}^{r} |Q - A_i| \leq \max\left\{ \sum_{i=1}^{r} |P_1 - A_i|, \sum_{i=1}^{r} |P_2 - A_i| \right\}.$$

Here A_i denotes the vector from an origin O to the point A_i, etc.

Since Q is between P_1 and P_2, $Q = \lambda P_1 + (1-\lambda) P_2$ with $0 < \lambda < 1$. Then by the triangle inequality,

$$|Q - A_i| = |\lambda(P_1 - A_i) + (1 - \lambda)(P_2 - A_i)|$$
$$\leq \lambda |P_1 - A_i| + (1 - \lambda) |P_2 - A_i|.$$

Whence,

$$\sum_i |Q - A_i| \leq \lambda \sum_i |P_1 - A_i| + (1 - \lambda) \sum_i |P_2 - A_i|$$

with equality if and only if the points are collinear and also (11).

Now if P_1 and P_2 satisfy $\sum_i PA_i \leq k$, so also does Q and thus the region is convex. For a generalization of Theorem 19, see [12].

References

1. M. S. Klamkin and D. J. Newman, The philosophy and applications of transform theory, SIAM Review, January 1961, pp. 10–36.

2. M. S. Klamkin, The teaching of mathematics so as to be useful, Educational Studies in Math., 1 (1968) 126–160.

3. L. Lines, Solid Geometry, Dover, New York, 1965.
4. N. Altschiller-Court, Modern Pure Solid Geometry, Macmillan, New York, 1935.
5. D. Y. Sommerville, Analytical Geometry of Three Dimensions, Cambridge University Press, Cambridge, 1947.
6. T. M. MacRobert and W. Arthur, Trigonometry, Part IV, Methuen, London, 1938.
7. A. J. McConnell, Applications of the Absolute Differential Calculus, Blackie and Son, Glasgow, 1946.
8. L. Brand, Vector and Tensor Analysis, Wiley, New York, 1948.
9. C. V. Durell and A. Robson, Advanced Trigonometry, Bell, London, 1953.
10. H. Rademacher and O. Toeplitz, The Enjoyment of Mathematics, Princeton University Press, N. J., 1957.
11. Problem 62-11, SIAM Review, Oct. 1962.
12. P. Hartman and F. A. Valentine, On general ellipses, Duke Math. J., 26 (1959) 373–385.

(c)
CONSTRUCTIONS WITH STRAIGHTEDGE AND COMPASS

ON THE TRISECTION OF AN ANGLE AND THE CONSTRUCTION OF REGULAR POLYGONS OF 7 AND 9 SIDES*

L. E. DICKSON, University of Chicago

1. Purpose and plan of this note. Frequently a wide-awake student who has learned how to bisect any angle asks if every angle can be trisected and, if not, why not; after learning how to construct regular polygons of 3, 4, 5, 6, 8, and 10 sides, he is apt to ask about the missing ones of 7 and 9 sides. Having several times received a first aid call from teachers of these inquisitive students, the writer would find it convenient to be able to refer to an exposition of these questions which is as elementary as possible. It is the purpose of this note to present such a treatment.

Moreover, it seems necessary that these questions be discussed publicly at regular intervals in order to keep down the number of angle-trisectors, who are partly unable and largely unwilling to understand the standard proofs of the impossibility of these constructions by means of ruler and compasses, but prefer to attempt to make the issue depend upon their own alleged construction involving always a confusing mass of lines and circles and always a child-like error.

With either class of readers, the use of imaginary numbers is not convincing. Hence they are not employed in this note, even though the imaginary roots of unity enter naturally into the questions concerning regular polygons. Moreover, the entire discussion is not beyond a college freshman.

---

* From AMERICAN MATHEMATICAL MONTHLY, vol. 21 (1914), pp. 259–262.

2. The cubic equations. In the problem of the duplication of a cube, we take as the unit of length a side of the given cube, and seek the length x of a side of another cube whose volume is double that of the given cube; thus

(1) $$x^3 = 2.$$

In the problem of the trisection of a given angle A, we are given a line of length $\cos A$ and seek a line of length $\cos(A/3)$. For, if we lay off the unit of length AB on one arm of angle A and draw the perpendicular BC to the other arm, the number of units of length in AC is $\cos A$ or $-\cos A$, according as A is an acute or obtuse angle. We employ the well-known trigonometric identity

$$\cos A = 4\cos^3 \frac{A}{3} - 3\cos \frac{A}{3}.$$

Multiply each member by 2 and set $x = 2\cos(A/3)$. Thus

$$x^3 - 3x = 2\cos A.$$

We are to prove that an arbitrary* angle A cannot be trisected by ruler and compasses. It suffices to prove this for the angle $A = 120°$. Then $\cos A = -1/2$ and the cubic is

(2) $$x^3 - 3x + 1 = 0.$$

After we have proved that angle $120°$ cannot be so trisected and hence that angle $40°$ cannot be constructed by ruler and compasses, it will follow that a regular polygon of nine sides cannot be so constructed, since the angle at the center subtended by one side is $\frac{1}{9}360° = 40°$.

Finally, the problem of the construction of a regular polygon of seven sides by ruler and compasses is equivalent to the construction of angle B containing $\frac{360}{7}$ degrees and hence to the construction of a line of length $x = 2\cos B$. We have

$$\cos 3B = \cos(360° - 3B) = \cos(7B - 3B) = \cos 4B,$$
$$\cos 3B = 4\cos^3 B - 3\cos B,$$
$$\cos 4B = 2\cos^2 2B - 1 = 2(2\cos^2 B - 1)^2 - 1.$$

After multiplication by 2 and setting $x = 2\cos B$, we get

$$x^3 - 3x = 4(\tfrac{1}{2}x^2 - 1)^2 - 2,$$
$$0 = x^4 - 4x^2 + 2 - x^3 + 3x = (x-2)(x^3 + x^2 - 2x - 1).$$

But $x = 2$ would give $\cos B = 1$, whereas B is acute. Hence†

(3) $$x^3 + x^2 - 2x - 1 = 0.$$

---

* Certain angles, like $A = 180°$, can be trisected. Since $\cos 60° = \tfrac{1}{2}$, the cubic then has the root $x = 1$. Hence this case does not invalidate our general theorem in §4.

† We can derive this equation and the corresponding ones for other regular polygons without the use of trigonometry, making use only of a theorem on chords in a circle. Cf. Dickson, *Annals of Mathematics*, 1894, p. 73.

3. Our cubic equations have no rational roots. Suppose for example that equation (2) has the root a/b, where a and b are integers with no common (integral) divisor greater than unity. Then

$$\frac{a^3}{b^3} - 3\frac{a}{b} + 1 = 0, \qquad \frac{a^3}{b} = 3ab - b^2 = \text{integer}.$$

Thus if $b \neq \pm 1$, b has a divisor greater than unity in common with a, contrary to hypothesis. Hence $b = \pm 1$ and the root is an integer.

If a root x of (2) is an integer, it divides x^3 and $3x$ and hence also the constant term 1, so that $x = \pm 1$. By trial, neither $+1$ nor -1 is a root. Hence (2) has no rational root.

The same discussion applies step by step to equation (3). In the case of (1), we must try also the divisors ± 2.

Hence each of our problems has led us to a cubic equation with rational coefficients having no rational root. Each problem is therefore impossible in view of the next theorem.

4. General theorem. *It is not possible to construct by ruler and compasses a line whose length is a root or the negative of a root of a cubic equation with rational coefficients having no rational root.*

We begin by investigating the nature of a positive number p such that a line of length p can be constructed by ruler and compasses. The ends of this line as well as other points found in the course of the construction are located as the intersections of straight lines and circles. Consider the equations of these lines and circles referred to a fixed pair of rectangular axes, the y-axis not being parallel to any of our straight lines. The equation of any one of our lines is

(1) $$y = mx + b.$$

Another line intersecting this has an equation

$$y = m'x + b'$$

and the coördinates of their point of intersection,

$$x = \frac{b' - b}{m - m'}, \quad y = \frac{mb' - m'b}{m - m'},$$

are rational functions of the coefficients of the lines.

To find the coördinates of the intersections of (1) with the circle

$$(x - e)^2 + (y - f)^2 = r^2,$$

we eliminate y and obtain a quadratic equation for x. Thus x, and hence also y, involves no irrationality (besides irrationalities already appearing in m, b, e, f, r) other than a square root.

Finally, the intersections of two circles are given by the intersections of one of them with their common chord, so that this case reduces to the preceding.

Hence the coördinates of the various points located by the construction, and therefore also the length p of the segment joining two of them, are found by a finite number of rational operations and extractions of real square roots, performed upon rational numbers or numbers obtained by such operations. By way of example, note that the side of a regular pentagon inscribed in a circle of radius unity is

$$\tfrac{1}{2}\sqrt{10-2\sqrt{5}}\ .$$

This point settled, consider a cubic equation with rational coefficients and having a constructible root r. Either r is rational or else it involves a real square root. In the latter case, we obtain a second root of the cubic by changing the sign of this square root in the expression for r. Then the third root of the cubic must be rational, since otherwise there would be, as before, a pair of roots in addition to the first pair. Hence in every case the cubic has a rational root, so that the denial of the general theorem stated at the beginning of this section leads to a contradiction. We have merely outlined in a rough way the final step of the proof. The argument in detail is accessible in books by Klein* and the writer;† it is based upon a systematic classification of the square roots involved in r, but employs only elementary algebraic principles.

The final step in the proof can be made in a few lines by means of the Galois theory of equations, which is based upon the theory of groups.

For a more elaborate elementary discussion of these special problems and the general problem relating to regular polygons, the reader may consult the eighth article in *Monographs on Modern Mathematics*, Longmans, Green and Co., 1911, where further references are given on page 386.

ON WHO FIRST PROVED THE IMPOSSIBILITY OF CONSTRUCTING CERTAIN REGULAR POLYGONS WITH RULER AND COMPASS ALONE‡

N. D. KAZARINOFF, University of Michigan

Carl Friedrich Gauss published in 1801 in his *Disquisitiones Arithmeticae* exquisite arithmetic yielding the constructibility of regular p-gons for p a prime of the form $1+2^k$. Closing his discussion, he asserted he had proved that no other regular p-gons were constructible. Gauss never published a proof of this assertion, nor did he ever outline one in his correspondence or notes. Yet Felix Klein in his *Famous Problems of Elementary Geometry* defers to Gauss. Practically all those who have written since on this subject follow Klein in stating that Gauss proved the impossibility theorem in the *Disquisitiones Arithmeticae*. R. C. Archibald partially corrected the record in an article in this MONTHLY in 1914. But he incorrectly asserted that James Pierpont gave the first proof.

In my opinion, Professors Gauss, Klein, Pierpont, and Archibald should each

---

* *Elementarmathematik vom höheren Standpunkte aus*, Leipzig, 1908, vol. 1, p. 125, and 2d ed., 1911.

† *Elementary Theory of Equations*, Wiley and Sons, 1914, p. 90.

‡ From AMERICAN MATHEMATICAL MONTHLY, vol. 75 (1968), p. 647.

have credited Pierre L. Wantzel (1814–1848), who proved the impossibility of constructing non-Gaussian regular n-gons with ruler and compass alone in 1837 in the very article where he solved the problems of angle trisection and cube duplication. Interestingly, Wantzel is cited by many authors, but only for solving the latter two problems.

Should one believe Gauss? There are aspects of this question to be considered aside from Gauss apparently never writing so much as an outline of a proof anywhere. Abel wrote to Holmboe from Paris in 1826 and referred to the "mystery that has reigned over the theory of Mr. Gauss on the division of the circle into equal parts." Gauss in 1796 did not have the bright light of Ruffini's, Abel's, and Galois' researches to guide him. Lastly, it was the general belief in 1796 that the classical ruler and compass construction problems were soluble in the negative, so that Gauss should have been more timid about publishing proofs of constructibility than proofs of nonconstructibility.

Mail Address: Dept. of Math., Angell Hall, The University of Michigan, Ann Arbor, MI 48104.

BIBLIOGRAPHIC ENTRIES: CONSTRUCTIONS WITH STRAIGHTEDGE AND COMPASS

The references below are to the AMERICAN MATHEMATICAL MONTHLY.

1. W. H. Bussey, Geometric constructions without the classical restriction to ruler and compasses, vol. 43, p. 265.

The angle trisection problem and others.

2. M. E. Stark, Constructions with limited means, vol. 48, p. 475.

Performing constructions with or without straight edge or compass.

(d)

MISCELLANEOUS

RATIONALIZING FACTORS AND THE METHOD OF UNDETERMINED COEFFICIENTS*

L. J. PARADISO, Cornell University

In elementary algebra the question is sometimes raised by a student as to how a rationalizing factor can be obtained for an expression which contains a root higher than a square root. For example, to rationalize the denominator of $2/((2)^{1/3}+$

---

* From AMERICAN MATHEMATICAL MONTHLY, vol. 36 (1929), pp. 87–89.

$(3)^{1/2}$). In many cases a rationalizing factor is easily obtained by the use of the conjugate, as in the example $(a^{1/2}+b^{1/2})(a^{1/2}-b^{1/2})=a-b$ or by using a factor of an expression such as x^n+y^n. For example, a rationalizing factor of $2(3)^{1/3}-5(5)^{1/3}$ is $(2(3)^{1/3})^2+10(15)^{1/3}+(5(5)^{1/3})^2$; it is obtained by using $x=2(3)^{1/3}$, $y=5(5)^{1/3}$ and $x^3-y^3=(x-y)(x^2+xy+y^2)$. But in rationalizing expressions of the form

$$a_0+a_1(p)^{1/n}+a_2(p)^{2/n}+\cdots+a_{n-1}(p)^{(n-1)/n},$$

where the a_i are rational coefficients, the process usually given is not readily remembered nor easy to teach without a background of the general theory.*

It is the object of this note to show that theoretically in all cases and practically in many cases a rationalizing factor may be found by the method of undetermined coefficients. Although this method is taught to freshmen in college in the case of partial fractions, it is not commonly employed to find rationalizing factors.

The theorem which we shall use to justify the method of undetermined coefficients is: *If ϕ is an algebraic number of degree m, that is, the root of a uniquely determined equation $f(x)=0$ of degree m with rational coefficients, and if $R(\phi)$ is a rational function of the algebraic number that is, $R(\phi)=g(\phi)/h(\phi)$, then*

$$R(\phi)=r_0+r_1\phi+r_2\phi^2+\cdots+r_{m-1}\phi^{m-1},$$

where the r_i are rational coefficients.

Suppose we wish to find a rationalizing factor for

$$a_0+a_1\phi+\cdots+a_{m-1}\phi^{m-1},$$

where the a_i are rational numbers. The problem amounts to finding rational numbers A_i such that

$$(A_0+A_1\phi+A_2\phi^2+\cdots+A_{m-1}\phi^{m-1})(a_0+a_1\phi+\cdots+a_{m-1}\phi^{m-1})=1.$$

Multiplying the factors in the left hand member of the above and arranging the terms, we can equate the constant term and the coefficients of $\phi,\phi^2,\ldots,\phi^{m-1}$ respectively to $1,0,0,\ldots$. In this way we obtain m simultaneous equations in the A_i from which we can find the required coefficients.

Example 1: To find the rationalizing factor of $2(2)^{1/3}-(4)^{1/3}-2$, we let $\phi=(2)^{1/3}$ so that $\phi^3=2$. We then have $(A_0+A_1\phi+A_2\phi^2)(2\phi-\phi^2-2)=1$. We group the terms

$$(4A_2-2A_1-2A_0)+(-2A_2-2A_1+2A_0)\phi+(2A_1-2A_2-A_0)\phi^2=1,$$

from which we get

$$4A_2-2A_1-2A_0=1,\ 2A_0-2A_1-2A_0=0,\ 2A_1-2A_2-A_0=0,$$

from which we find the rationalizing factor to be

$$-(2/5)-(3/10)\cdot 2^{1/3}-(1/10)\cdot 4^{1/3}.$$

---

* See G. Chrystal, Algebra, Part 1, (1878), p. 197.

By this same method we can of course rationalize numerators as well as denominators.

We can generalize this method to cases where more than one algebraic number is involved in the expression to be rationalized; say ϕ of degree m and θ of degree n. In this case in forming the rationalizing factor with undetermined coefficients one must not only put undetermined coefficients with powers of ϕ up to $m-1$ and powers of θ up to $n-1$, but also one must attach undetermined coefficients to all possible different cross products which can be formed from the various powers of ϕ and θ using the restriction,* of course, that $\phi^m = \phi$ and $\theta^n = \theta$. This can be further generalized to any number of algebraic numbers, but in most cases where the index of the root is greater than 3 and the number of different algebraic numbers is more than three it becomes impractical to solve the simultaneous equations obtained.

Example 2, illustrating a more general case: Rationalize the denominator of $1/((a)^{1/2} + (b)^{1/3})$ where a and b are rational. Let $x = (a)^{1/2}$, $x^2 = a$, and $y = (b)^{1/3}$, $y^3 = b$. Then
$$1/(x+y) = A + Bx + Cy + Dy^2 + Exy + Fxy^2.$$
Expanding, arranging the terms, and equating the coefficients we have
$$aB + bD = 1,\ B + C = 0,\ D + E = 0,\ C + aF = 0,\ A + aE = 0,\ A + bF = 0$$
from which we get
$$1/(a^{1/2} + b^{1/3}) = \frac{ab + a^2 \cdot a^{1/2} - a^2 \cdot b^{1/3} + b \cdot b^{2/3} - b \cdot a^{1/2} \cdot b^{1/3} - a \cdot a^{1/2} \cdot b^{2/3}}{a^3 + b^2}.$$

A NOTE ON PARTIAL FRACTIONS[†]

RAYMOND GARVER, University of Rochester

In any first or second course in the calculus, a topic of some importance is the reduction of a rational function of x, say $n(x)/d(x)$, to the sum of partial fractions whose denominators are factors of $d(x)$. If $d(x)$ can be factored completely into real, linear factors, the work can be carried through quite easily by any one of several familiar methods. But if there are quadratic factors of the form $x^2 + px + q\ (p^2 - 4q < 0)$, the work is much longer.

Thus to reduce the fraction
$$\frac{3x^3 - 4x^2 - 3x + 5}{(x^2 + x + 1)(x^2 - 2x + 3)} \quad \text{to the form} \quad \frac{Ax + B}{x^2 + x + 1} + \frac{Cx + D}{x^2 - 2x + 3},$$
we equate the two, since we have presumably proved that such a reduction is possible. Clearing of fractions, we have
$$(1) \quad 3x^3 - 4x^2 - 3x + 5 = (Ax + B)(x^2 - 2x + 3) + (Cx + D)(x^2 + x + 1),$$

---

* Loc. cit., p. 192, II.
[†] From AMERICAN MATHEMATICAL MONTHLY, vol. 34 (1927), pp. 319–320.

which is an identity. There are then two usual ways of computing A, B, C, D. We may assign four convenient integral values to x, or we may equate corresponding coefficients. Either process leads to four equations in four unknowns, a fairly long problem in algebra.

It seems possible to simplify the work by giving x an imaginary value which makes one of the quadratic factors vanish. The work can be arranged so that we never have to find this root explicitly at all. Thus, if we put $x = x_1$, where x_1 is a root of $x^2 + x + 1 = 0$, we have $x_1^2 = -x_1 - 1$, $x_1^3 = -x_1^2 - x_1 = 1$; and equation (1) reduces to

$$(2) \quad x_1 + 12 = (Ax_1 + B)(-3x_1 + 2) = -3Ax_1^2 + (2A - 3B)x_1 + 2B,$$
$$= (5A - 3B)x_1 + (3A + 2B).$$

Equating real and imaginary parts, $1 = 5A - 3B$, and $12 = 3A + 2B$, which gives $A = 2$, $B = 3$. In this example we can obtain $C = 1$, $D = -4$, by inspection.

This method also has the advantage that it can be used to derive a formula for the reduction of any fraction similar to our example. Thus if we have

$$(3) \quad \frac{ax^3 + bx^2 + cx + d}{(x^2 + p_1 x + q_1)(x^2 + p_2 x + q_2)} = \frac{Ax + B}{x^2 + p_1 x + q_1} + \frac{Cx + D}{x^2 + p_2 x + q_2}$$

where $p_i^2 - 4q_i < 0$ and A, B, C, D are to be determined; and if we go through the above process we obtain

$$(4) \quad A = \frac{d(p_1 - p_2) + (aq_1 - c)(q_1 - q_2) + (ap_1 - b)(p_1 q_2 - p_2 q_1)}{(q_1 - q_2)^2 + (p_1 - p_2)(p_1 q_2 - p_2 q_1)},$$
$$B = \frac{[q_1(aq_1 - c) + dp_1](p_1 - p_2) - [q_1(ap_1 - b) + d](q_1 - q_2)}{(q_1 - q_2)^2 + (p_1 - p_2)(p_1 q_2 - p_2 q_1)},$$

where the denominator does not vanish if the factors in the denominator of (3) are distinct, as we have tacitly assumed. The substitution in the formulas is easier than at first appears, since many of the terms are repetitions. If necessary, C and D can be found by interchanging subscripts, though in our example this was hardly necessary.

A NOTE ON PARTIAL FRACTIONS*

L. S. JOHNSTON, University of Detroit

In manuscript notes left by the late Rear Admiral John P. Merrell, United States Navy, Head of the Department of Mathematics, United States Naval Academy in the 1890's and later (about 1905–1908) President of the Naval War College, the writer discovers the following method of resolving into its partial fractions the proper fraction

$$\frac{f(x)}{(x^2 + ax + b)(x^2 + cx + d)}$$

where the denominator is not separated into linear factors. While it does not appear that the method possesses any advantage on the score of brevity, it is at least somewhat different from the more conventional methods. The method will be illustrated by a particular example rather than proved, though the proof is not difficult.

Consider the equation

(1) $$\frac{x^3 - 8x^2 - 10x - 30}{(x^2 + x + 3)(x^2 + 2x + 5)} = \frac{Ax + B}{x^2 + x + 3} + \frac{Cx + D}{x^2 + 2x + 5}$$

or

(2) $$x^2(x - 8) - 10x - 30 = (Ax + B)(x^2 + 2x + 5) + (Cx + D)(x^2 + x + 3).$$

Replacing x^2 by $-2x-5$, but not disturbing x itself, we have

$$(-2x - 5)(x - 8) - 10x - 30 = (Cx + D)(-x - 2),$$

or

$$-2x^2 + x + 10 = -Cx^2 - Dx - 2Cx - 2D.$$

Again replacing x^2 by $-2x-5$ without disturbing x itself, we have

$$5x + 20 = -Dx + 5C - 2D.$$

Equating coefficients of like powers of x, we have $D = -5$, $C = 2$. Similarly we might have replaced, in (2), x^2 by $-x-3$ and carried through the operation in exactly the same manner, finding $A = -1$, $B = -3$.

In every case, then, the method consists in writing the analogue of (2) in such a way as to display the left member as the sum of linear functions of x multiplied by integral powers of x^2, and then replacing x^2 by the linear function of x which will make any given factor of the denominator vanish. This replacement of x^2 by the proper linear function of x continues so long as the resulting

---

* From AMERICAN MATHEMATICAL MONTHLY, vol. 43 (1936), pp. 413–414.

equation contains powers of x higher than the first. Eventually such substitutions will reduce the equation to linear form, at which time we equate coefficients of like powers of x.

The method is perfectly general for any proper fraction the denominator of which consists of quadratic factors none of which are repeated.

As Admiral Merrell remarks in his note, the reader will readily note that the method is equivalent to resolving the denominator into complex linear factors and expanding in the usual manner, and then equating coefficients of real and imaginary terms in the resulting equation.

A NOTE ON JOINT VARIATION*

R. A. ROSENBAUM, Reed College

There is a small point in connection with the topic of variation and proportionality which frequently disturbs the beginner, for many texts do not treat it adequately. Some students are still bothered by it when they take a course in thermodynamics and have to deal with relations of p, v, and T.

The trouble seems to arise from text-book statements similar to the following: "z is said to vary jointly as x and y if $z = kxy$. It is evident that, if y is held constant, z varies as x, and that, if x is held constant, z varies as y." So far everything is clear, but, when the student comes to work problems, he is expected to assume the converse of the above, that is, to use the theorem: "If z varies as x, and z varies as y, then $z = kxy$." That this theorem is indeed true may be made to seem reasonable to the beginner by replacing the elliptical statement "z varies as x and z varies as y" by the complete statement "z depends on both x and y. When y is held constant, z varies as x, and, when x is held constant, z varies as y." A proof of the theorem which may be given to college students is the following:

If z varies as x when y is held constant, then $z = Rx$. For a fixed y, R is a constant, but for a different y, R may have a different value. In other words, R is a function of y; $R = f_1(y)$, say. Then

(1) $$z = f_1(y) \cdot x.$$

Similarly, if z varies as y when x is held constant,

(2) $$z = f_2(x) \cdot y.$$

Dividing (1) by (2), we obtain:

$$\frac{f_1(y) \cdot x}{f_2(x) \cdot y} = 1$$

---

* From AMERICAN MATHEMATICAL MONTHLY, vol. 49 (1942), pp. 537–538.

or

(3) $$\frac{f_1(y)}{y} = \frac{f_2(x)}{x}.$$

Since the left-hand side of (3) is a function of y and the right-hand side is a function of x, and since the equality holds for all x and y, each side must be equal to some constant, k.
i.e.

$$\frac{f_1(y)}{y} = k = \frac{f_2(x)}{x}$$

or

(4) $$f_1(y) = ky$$

and

(5) $$f_2(x) = kx.$$

Substituting from (4) in (1) or from (5) in (2), we have

$$z = kxy.$$

A similar type of proof may be used when more than three variables are involved, or when not all the variation is direct.

NEW PROOF OF A CLASSIC COMBINATORIAL THEOREM*

S. W. GOLOMB, University of Southern California

THEOREM. *The number of samples of r objects from a set of k objects, allowing repetition but disregarding order, is*

$$\binom{k+r-1}{r}.$$

Proof. Consider the k objects to be cards numbered from 1 to k, and adjoin $r-1$ extra cards numbered from $k+1$ to $k+r-1$, and bearing the respective instructions "repeat lowest numbered card," "repeat 2nd-lowest numbered card," $\cdots$, "repeat $(r-1)^{st}$ lowest numbered card." Then a sample of size r without replacement from this enlarged ($k+r-1$ card) deck corresponds uniquely to a sample of size r from the original deck allowing replacement. The number of such samples is accordingly

$$\binom{k+r-1}{r}.$$

---

* From AMERICAN MATHEMATICAL MONTHLY, vol. 75 (1968), pp. 530–531.

Examples:

1. To form a 5-card poker hand allowing repetition, it suffices to adjoin four "jokers" with the respective instructions: (a) repeat lowest card, (b) repeat 2nd-lowest card, (c) repeat 3rd-lowest card, and (d) repeat 4th-lowest card. We regard these four jokers as being the highest cards in the enlarged deck, with $a<b<c<d$.

A hand with no jokers is an ordinary poker hand (no repetition).

A hand with one joker has any one of its four ordinary cards repeated, depending upon which joker is held.

A hand with two jokers either has two ordinary cards duplicated (in the cases $a-b$, $a-c$, and $b-c$), or one ordinary card triplicated (in the cases $a-d$, $b-d$, $c-d$).

A hand with three jokers becomes one of the following: $xxyyy$, $xxxyy$, $xyyyy$, or $xxxxy$, where x and y are the two ordinary cards, depending on which three jokers are held.

A hand with all four jokers becomes a five-fold repetition of the ordinary card it contains.

2. If we form a 17-card deck with cards labeled A, 2, 3, 4, 5, 6, 7, 8, 9, 10, J, Q, K, a, b, c, d, where a, b, c, d are as in the previous example, then 5-card hands have all the following possibilities: 1 pair, 2 pair, 3 of a kind, full house, 4 of a kind, 5 of a kind, straight, and "bust."

ANOTHER GENERALIZATION OF THE BIRTHDAY PROBLEM*

J. E. NYMANN, The University of Texas at El Paso

While discussing the birthday problem with a class of 100 students I decided to run an experiment with the class. Knowing that the probability that two of them would have the same birthday was very close to one, I decided on the following alternative. I asked each student, in succession, to call out his birthday and if anyone in the room had the same birthday we would stop. I offered to bet that the procedure would stop on or before the tenth student. Unfortunately (since it turns out the probability of my winning is .928) no one was willing to bet with me. The object of this note is to report on this type of generalization of the birthday problem.

For n and k positive integers with $k \leq n$, $P(n,k)$ will denote the probability that in a group of n people, at least one pair have the same birthday with at least one such pair among the first k people. Determining $P(n,n)$ is, of course, the ordinary birthday problem and the assertion made in the preceding paragraph is that $P(100, 10)$ = .928. In determining $P(n,k)$ we will make the standard assumptions that each person in the group can have his birthday on any one of the 365 days in a year (ignoring leap years) and that each day of the year is equally likely to be the person's birthday.

---

* From MATHEMATICS MAGAZINE, vol. 48 (1975), pp. 46–47.

The sample space can be taken to be the 365^n n-tuples with integer components between 1 and 365 inclusive. The event in which we are interested is the set of n-tuples for which at least one of the first k components is repeated. By elementary counting techniques the number of elements in the complementary event (i.e., the set of n-tuples for which none of the first k components is repeated) is easily seen to be $365 \cdot 364 \cdots (365 - k + 1)(365 - k)^{n-k}$. Hence

$$P(n, k) = 1 - \frac{365 \cdot 364 \cdots (365 - k + 1)(365 - k)^{n-k}}{365^n}.$$

From knowing the solution to the ordinary birthday problem we see that if $Q_k = 1 - P(k, k)$, then

$$P(n, k) = 1 - Q_k \left(1 - \frac{k}{365}\right)^{n-k}.$$

In the table below $k(n)$ will denote the smallest value of k for which $P(n, k) \geq 1/2$. From the solution to the ordinary birthday problem it is clear that n must be at least 23 for $k(n)$ to exist. It is also clear that $k(n + 1) \leq k(n)$. Consequently the table begins with $n = 23$ and only those n for which $k(n)$ changes are entered. This table is constructed from a computer printout which gave $P(n, k)$ for n between 23 and 100 and $1 \leq k \leq n$. The entries for $n > 100$ were done separately in order to complete this table.

n	23	24	25	26	27	28	29	31	33	36	40	46	54	66	86	128	254
$k(n)$	20	17	15	14	13	12	11	10	9	8	7	6	5	4	3	2	1

The last entry in this table is at first glance rather surprising. It says that if you were to bet that in a group of n people someone else has the same birthday as you, n must be at least 254 for this to be a favorable bet. One would perhaps think, initially, that this entry should be $1 + [365/2] = 183$. However, by the solution to the ordinary birthday problem, in a group of 183 people there would be many repeated birthdays; hence, not nearly half the birthdays would be represented. Clearly, the bet described above will have favorable odds when at least half of the possible birthdays are represented among the n people. Therefore, the last entry in the table seems to show that it takes at least 254 people before the expected number of birthdays represented is 183.

BIBLIOGRAPHIC ENTRIES: MISCELLANEOUS

Except for the entries labeled MATHEMATICS MAGAZINE, the references below are to the AMERICAN MATHEMATICAL MONTHLY.

1. S. Gandz, The origin of the term "algebra", vol. 33, p. 437.
2. E. B. Stouffer, Expressions for the general determinant in terms of its principal minors, vol. 35, p. 18.
3. J. Williamson, Determinants whose elements are 0 or 1, vol. 53, p. 427.

Proof that when $n=7$, the maximum value for a determinant whose elements are ± 1 is $2^6 9$ and that there is essentially only one type of determinant with this value.

4. O. Taussky, A recurring theorem on determinants, vol. 56, p. 672.

Discussion of a theorem which deals with determinants of matrices with a "dominant" main diagonal.

5. A. W. Goodman, On sets of acquaintances and strangers at any party, vol. 66, p. 778.

Relationships between N persons at a gathering.

6. D. J. Newman, How to play baseball, vol. 67, p. 865.

The use of probabilities to determine playing strategies.

7. T. L. Bartlow, An historical note on the parity of permutations, vol. 79, p. 766.
8. T. A. Brown, A note on "Instant Insanity", MATHEMATICS MAGAZINE, vol. 41, p. 167.

Analysis of the popular puzzle known as Instant Insanity.

9. A. P. Grecos, A diagrammatic solution to "Instant Insanity" problem, MATHEMATICS MAGAZINE, vol. 44, p. 119.
10. M. S. Klamkin, A probability of more heads, MATHEMATICS MAGAZINE, vol. 44, p. 146.

Finds the probability that if A and B toss $n+m$ and n coins respectively that A gets more heads than B.

6

SOLUTIONS OF EQUATIONS

RELATING TO SOLUTIONS OF QUADRATIC EQUATIONS*

GEO. R. DEAN, Missouri School of Mines

I. Solution of the quadratic without factoring or completing the square. Let the equation be

$$ax^2 + bx + c = 0.$$

Put

$$x = u + iv, \quad \text{where} \quad i = \sqrt{-1}.$$

Then

$$a(u^2 - v^2 + 2uvi) + b(u + iv) + c = 0,$$
$$a(u^2 - v^2) + bu + c + i(2auv + bv) = 0.$$

Since the real and imaginary parts vanish separately,

$$a(u^2 - v^2) + bu + c = 0, \quad \text{and} \quad (2au + b)v = 0.$$

And since v is not, in general, equal to zero, we get

$$u = -\frac{b}{2a},$$

from which

$$au^2 + bu + c = c - \frac{b^2}{4a}.$$

Hence,

$$av^2 = c - \frac{b^2}{4a}, \quad v = \frac{\pm\sqrt{4ac - b^2}}{2a};$$

$$u + iv = \frac{-b \pm i\sqrt{4ac - b^2}}{2a} = \frac{-b \pm \sqrt{b^2 - 4ac}}{2a}.$$

When v, that is, $\dfrac{\sqrt{4ac - b^2}}{2a}$ is imaginary the equation has real roots; and when $v = 0$, equal roots.

---

* From AMERICAN MATHEMATICAL MONTHLY, vol. 22 (1915), pp. 243–244.

There is probably nothing new about this solution, but it affords a good example of the part played by the imaginary unit in higher mathematics, and would not be out of place in our elementary text-books on algebra.

II. Solution of a pair of simultaneous equations which occurs in the theory of cables and transmission lines. In the following equations the unknown quantities are α and β:

(1) $$\alpha^2 - \beta^2 = RS - LCp^2$$
(2) $$2\alpha\beta = (RC + LS)p.$$

The solution by the regular algebraic process gives

$$\alpha = \sqrt{\tfrac{1}{2}\left\{\sqrt{(R^2+p^2L^2)(S^2+p^2C^2)} + (RS - LCp^2)\right\}}$$
$$\beta = \sqrt{\tfrac{1}{2}\left\{\sqrt{(R^2+p^2L^2)(S^2+p^2C^2)} - (RS - LCp^2)\right\}}.$$

A more elegant solution, from the mathematician's point of view, more convenient for the computer, and furnishing at the same time the value of three other quantities that are needed in other computations, is obtained by using trigonometric functions.

Let

$$\alpha = N\cos\xi, \beta = N\sin\xi, \text{ then } \alpha^2 - \beta^2 = N^2 \cos 2\xi;$$
$$R = Z\cos\delta, pL = Z\sin\delta, \text{ then } \tan\delta = \frac{pL}{R}, Z = \frac{R}{\cos\delta} = \frac{pL}{\sin\delta};$$
$$S = Y\cos\gamma, pC = Y\sin\gamma, \text{ then } \tan\gamma = \frac{pC}{S}, Y = \frac{S}{\cos\gamma} = \frac{pC}{\sin\gamma};$$
$$RS - LCp^2 = ZY\cos(\gamma + \delta), (RC + LS)p = ZY\sin(\gamma + \delta).$$

Therefore

$$N^2 \cos 2\xi = ZY\cos(\gamma + \delta), N^2 \sin 2\xi = ZY\sin(\gamma + \delta),$$

and it is easy to see that

$$N = \sqrt{ZY}, \quad \text{and} \quad \xi = \tfrac{1}{2}(\gamma + \delta).$$

As a numerical illustration, take $R = 0.30$, $L = 0.00196$, $C = 0.0153 \times 10^{-6}$, $S = 0$, $p = 377$. Then, using a slide rule,

$$\tan\delta = \frac{pL}{R} = 2.4600, \delta = 67°53', \cos\delta = 0.3765, Z = 0.795;$$
$$\tan\gamma = \frac{pC}{S} = \infty, \gamma = 90°, \xi = \tfrac{1}{2}(\gamma + \delta) = 78°56'30'', Y = 5.77 \times 10^{-6};$$
$$\alpha = \sqrt{ZY}\cos\xi = 0.000412, \beta = \sqrt{ZY}\sin\xi = 0.002100.$$

The quantities γ, δ, ξ are useful in other computations.

NOTE ON THE ALGEBRAIC SOLUTION OF THE CUBIC*

E. J. OGLESBY, New York University

Take the general cubic equation

(1) $$a_0 x^3 + 3a_1 x^2 + 3a_2 x + a_3 = 0$$

and assume that the roots x_1, x_2, x_3 have the form†

$$x_1 = a + b + c,$$
$$x_2 = a + \omega b + \omega^2 c,$$
$$x_3 = a + \omega^2 b + \omega c,$$

where $\omega^2 + \omega + 1 = 0$ and a, b, and c are to be determined.

We then have the following identities:

(2) $$(a+b+c)+(a+\omega b+\omega^2 c)+(a+\omega^2 b+\omega c)=3a,$$

(3) $$(a+b+c)(a+\omega b+\omega^2 c)+(a+\omega b+\omega^2 c)(a+\omega^2 b+\omega c) \\ +(a+\omega^2 b+\omega c)(a+b+c)=3(a^2-bc),$$

(4) $$(a+b+c)(a+\omega b+\omega^2 c)(a+\omega^2 b+\omega c)=a^3+b^3+c^3-3abc.$$

From (2), (3), and (4)

$$x_1 + x_2 + x_3 = 3a,$$
$$x_1 x_2 + x_2 x_3 + x_3 x_1 = 3(a^2 - bc),$$
$$x_1 x_2 x_3 = a^3 + b^3 + c^3 - 3abc.$$

From (1)

$$x_1 + x_2 + x_3 = -\frac{3a_1}{a_0},$$
$$x_1 x_2 + x_2 x_3 + x_3 x_1 = \frac{3a_2}{a_0},$$
$$x_1 x_2 x_3 = -\frac{a_3}{a_0};$$

whence,

(5) $$a = -\frac{a_1}{a_0},$$

(6) $$a^2 - bc = \frac{a_2}{a_0},$$

(7) $$a^3 + b^3 + c^3 - 3abc = -\frac{a_3}{a_0}.$$

---

* From AMERICAN MATHEMATICAL MONTHLY, vol. 30 (1923), pp. 321–323.

† This assumption may be justified in advance, since a, b, and c can always be found to satis these equations, for any x_1, x_2, x_3.—EDITOR.

From (5) and (6) we get

(8) $$bc = \frac{a_1^2 - a_0 a_2}{a_0^2}$$

and substituting the values of a and of bc into (7), we have

(9) $$b^3 + c^3 = \frac{3a_0 a_1 a_2 - a_0^2 a_3 - 2a_1^3}{a_0^3}.$$

Put $a_0 a_2 - a_1^2 = H$ and $a_0^2 a_3 - 3a_0 a_1 a_2 + 2a_1^3 = G$. Then $bc = -(H/a_0^2)$ and $b^3 + c^3 = (-G/a_0^3)$, from which we get, by the solution of a quadratic,

$$b^3 = \frac{-G + \sqrt{G^2 + 4H^3}}{2a_0^3}, \qquad c^3 = \frac{-G - \sqrt{G^2 + 4H^3}}{2a_0^3};$$

whence the roots of the cubic are

$$x_1 = \frac{1}{a_0}\left[-a_1 + \sqrt[3]{\frac{-G + \sqrt{G^2 + 4H^3}}{2}} + \sqrt[3]{\frac{-G - \sqrt{G^2 + 4H^3}}{2}} \right],$$

$$x_2 = \frac{1}{a_0}\left[-a_1 + \omega\sqrt[3]{\frac{-G + \sqrt{G^2 + 4H^3}}{2}} + \omega^2\sqrt[3]{\frac{-G - \sqrt{G^2 + 4H^3}}{2}} \right],$$

$$x_3 = \frac{1}{a_0}\left[-a_1 + \omega^2\sqrt[3]{\frac{-G + \sqrt{G^2 + 4H^3}}{2}} + \omega\sqrt[3]{\frac{-G - \sqrt{G^2 + 4H^3}}{2}} \right].$$

It should be noted that this is substantially the same as Cardan's solution but from an entirely different method of attack.

As an example, consider the equation

$$x^3 + 2x^2 + 3x + 4 = 0.$$

Here

$$a = -\frac{a_1}{a_0} = -\frac{2}{3},$$

$$a^2 - bc = \frac{a_2}{a_0} = 1,$$

whence

$$bc = -\frac{5}{9};$$

$$a^3 + b^3 + c^3 - 3abc = -\frac{a_3}{a_0} = -4,$$

whence

$$b^3 + c^3 = -\frac{70}{27}.$$

Therefore
$$b^3 - \frac{125}{729b^3} = -\frac{70}{27},$$

so that
$$b^3 = \frac{-35+15\sqrt{6}}{27}, \quad c^3 = \frac{-35-15\sqrt{6}}{27},$$
$$b = -\tfrac{1}{3}\sqrt[3]{35-15\sqrt{6}}, \quad c = -\tfrac{1}{3}\sqrt[3]{35+15\sqrt{6}},$$

whence
$$x_1 = a+b+c = -\tfrac{1}{3}\left[2 + \sqrt[3]{35-15\sqrt{6}} + \sqrt[3]{35+15\sqrt{6}}\right],$$
$$x_2 = a+\omega b+\omega^2 c = -\tfrac{1}{3}\left[2 + \omega\sqrt[3]{35-15\sqrt{6}} + \omega^2\sqrt[3]{35+15\sqrt{6}}\right],$$
$$x_3 = a+\omega^2 b+\omega c = -\tfrac{1}{3}\left[2 + \omega^2\sqrt[3]{35-15\sqrt{6}} + \omega\sqrt[3]{35+15\sqrt{6}}\right].$$

A GRAPHICAL METHOD OF SOLVING SIMULTANEOUS LINEAR EQUATIONS*

J. P. BALLANTINE, Columbia University

Suppose it is desired to find a solution of the following simultaneous linear equations:
$$3x - 2y = 4,$$
$$2x + y = 5.$$

It is desired to find a number x which multiplied by the vector $(3,2)$ and a number y which multiplied by the vector $(-2,1)$ will be so chosen that the sum of the two resulting vectors is the vector $(4,5)$. This gives rise to the following geometrical representation of the problem. The vectors $(3,2)$ and $(-2,1)$ are laid off, and it is geometrically apparent that, in order to obtain $(4,5)$, it is necessary to take the first vector twice and the second vector once. Hence the solution is obtained by taking $x=2$ and $y=1$. It is also clear that if the solution were to come out fractional, as good estimates of these fractions could be obtained by the above geometrical interpretation as from the usual method of plotting the two loci, and estimating the coördinates of the point of intersection.

Each of the two geometrical methods has its own advantages. For instance, if the problem were varied by altering one of the two equations, the method which plots the loci of the two equations would be preferable. On the other hand, if the problem were varied by replacing the right-hand members, 4 and 5, by different pairs of numbers, then the method suggested in this paper would be preferable. In

---

* From AMERICAN MATHEMATICAL MONTHLY, vol. 30 (1923), pp. 442–443.

practical work this situation often arises. Suppose, for instance, one is using the so-called method of diminishing the constant terms. In this method one assumes an approximate solution X and Y, and finds that the differences $x - X$ and $y - Y$ satisfy simultaneous equations differing from the original ones only in the right-hand members. The same figure which suggested that X and Y were approximations can be used with the new right-hand members to suggest second approximations. Thus a process similar to Horner's method may be carried out.

No novelty is claimed for the above method. It is one in common use in general analysis. It is laid before the readers of the MONTHLY as a method of familiarizing the freshman with the notion of a vector as a number pair.

NOTE ON THE SOLUTION OF A SET OF LINEAR EQUATIONS*

J. P. BALLANTINE, Columbia University

Let x_1, x_2, x_3, x_4 satisfy the set of equations

$$\begin{aligned} a_1x_1 + a_2x_2 + a_3x_3 + a_4x_4 &= a_5, \\ b_1x_1 + \cdots &= b_5, \\ c_1x_1 + \cdots &= c_5, \\ d_1x_1 + \cdots &= d_5, \end{aligned}$$

supposed independent.

Then[†]

$$\begin{vmatrix} a_1x_1 - a_5 & a_2 & a_3 & a_4 \\ b_1x_1 - b_5 & b_2 & b_3 & b_4 \\ c_1x_1 - c_5 & c_2 & c_3 & c_4 \\ d_1x_1 - d_5 & d_2 & d_3 & d_4 \end{vmatrix} = 0;$$

for, on multiplying the respective columns by $1, x_2, x_3,$ and x_4, and adding, a column of zeros is obtained. From the vanishing of the above determinant may be obtained the usual formula for the value of x_1.

Suppose it is desired to solve the set of equations for only one of the unknowns, and to check the result. The usual method affords no check on the value of x_1 until all of the other unknowns have been evaluated; except, perhaps, the consistency of the four equations in the other three unknowns when the value of x_1 is substituted. This consistency condition is precisely the vanishing of the above determinant, a fact which is not usually brought out in an elementary treatment.

---

* From AMERICAN MATHEMATICAL MONTHLY, vol. 31 (1924), p. 341.
† Dickson, *Elementary Theory of Equations*, 1914, p. 145.

A NOTE ON THE ROOTS OF A CUBIC*

E. C. KENNEDY, College of Mines, El Paso, Texas

If the cubic

$$X^3 + AX^2 + BX + K = 0$$

(A, B, and K rational) has one or more rational roots it may often be solved very quickly by the scheme illustrated below. If c is a rational root and $a \pm \sqrt{b}$ are roots, a and b rational, it can be shown that

(1) $$2a + c = -A$$
(2) $$b = K/c + a^2$$
(3) $$2ac^2 - Bc - K = 0.$$

From (3), $D = B^2 + 8Ka$ must be made a perfect square by some rational value of a. This almost instantly gives us the three roots of our cubic. To illustrate the brevity of our method let us examine, for integral roots,

(4) $$X^3 - 304X - 1920 = 0.$$

Here,

(5) $$2a + c = 0$$
(6) $$b = -1920/c + a^2$$
(7) $$D/4 = 23104 - 3840a.$$

To find a we arrange the work as follows:

$$\begin{array}{r} 23104 \\ 3840 \\ \hline 19264 \\ 3840 \\ \hline 15424 \\ 3840 \\ \hline 11584 \\ 3840 \\ \hline 7744 \end{array} = \text{a perfect square.}$$

Thus $a = 4$. From (5) $c = -8$ and from (6) $b = 256$, and the roots of the cubic are $X_1 = 4 + 16 = 20, X_2 = 4 - 16 = -12, X_3 = -8$.

This scheme is often much quicker than finding the roots by trial, especially if the constant term has a large number of factors as does 1920. The method will always work if the roots are integers—though a may be negative or zero (the latter if $A = K/B$).

---

* From AMERICAN MATHEMATICAL MONTHLY, vol. 40 (1933), pp. 411–412.

Rational roots present no difficulty. For example, to solve

(8) $\qquad 6X^3 - 21X^2 + 2X - 56 = 0,$
$\qquad\qquad 2a + c = 21/6$
$\qquad\qquad\; b = -56/c + a^2$
$$D = (2/6)^2 - \frac{8(56)a}{6} = \frac{4 - 224a'}{36}. \qquad (a' = 12a)$$

Evidently $a' < 0$ (otherwise c would be complex since a' is an integer) so we write

$$\begin{array}{r} 4 \\ 224 \\ \hline 228 \\ 224 \\ \hline 452 \\ 224 \\ \hline 676 \end{array} = \text{a perfect square.}$$

Thus $a' = -3$, $a = -3/12$, $c = 4$, $b = -109/48$. This gives us immediately all three of the roots. We set a' equal to $12a$ instead of equal to $6a$ because the equation might conceivably be resolved into two factors one of which is a quadratic with the coefficient of X (in the quadratic) odd.

An interesting fact may be observed at this stage. By Descartes' Rule there are either three or one positive real root of (8). There are no negative real roots. Now if there are three real roots, it follows that $a > 0$. But we must have $a < 0$ in order to make $4 - 224a > 0$ which it must be in order for c to be real. Therefore, (8) has exactly one real root. This is quicker than computing the discriminant and determining the number of real roots in that way.

As a final illustration let us test

(9) $\qquad\qquad\qquad X^3 - 14X^2 + 9X - 12 = 0.$

Here $a > 0$ and we write mechanically $D = 81 - 48a$, $81 - 48 = 33$, not a perfect square. Therefore no rational roots.

Note: The scheme described above will always work and is often quite short. It seems that every method for finding rational roots has some drawbacks. In our method we cannot always determine the sign of a immediately, though we may often find very restricted limits for it. Again it sometimes happens that we obtain a value of a that will not serve. In such a case we quickly eliminate that possibility by considering (1), (2), or (3). If the roots are all rational then there are three acceptable values of a, any one of which will serve. The scheme has one big advantage over the usual method in that the amount of labor involved is wholly independent of the number of factors in the constant term. It is not expected to displace other methods of determining rational roots, but it is often useful.

THE GRAPHICAL INTERPRETATION OF THE COMPLEX ROOTS OF CUBIC EQUATIONS*

GARCIA HENRIQUEZ, Santo Domingo, Rep. Dominicana

The following method for interpreting graphically the complex roots of cubic equations, while possibly not new, may be of interest.

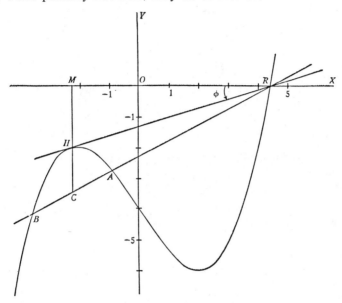

The figure represents a cubic with one real root. (The curve is drawn for the equation $8y = x^3 - 12x - 32$.) From the intersection, R, of the curve with the X-axis, a tangent RH is drawn to the cubic. The abscissa, OM, of H is the real part of the complex roots. The absolute value of the imaginary parts is the square root of $\tan \phi = \tan ORH$. The tangent required may be easily constructed by drawing any secant, RAB, through R, intersecting the curve in A and B. An ordinate through C, the midpoint of AB, cuts the curve in H, the point of tangency.

Proof. The cubic with roots r, $a \pm ib$, where r, a and b are real, has for its equation

$$f(x) = (x - r)(x^2 - 2ax + a^2 + b^2) = 0.$$

Any secant through R is

$$y = m(x - r),$$

which intersects the curve in two additional points given by

$$x^2 - 2ax + a^2 + b^2 - m = 0,$$

---

* From AMERICAN MATHEMATICAL MONTHLY, vol. 42 (1935), pp. 383–384.

i.e., in the points whose abscissas are

$$a \pm \sqrt{m - b^2}.$$

The secant will be a tangent if $m = b^2$. In this case the abscissa of H is a, as required. The slope of RH is b^2, which is the square of the imaginary coefficient of the complex roots.

Editorial Note. As this issue is about to go to press, Professor C. F. Barr of the University of Wyoming calls our attention to the fact that this same construction was given in this MONTHLY, vol. 25 (1918), p. 268, and in the Annals of Mathematics, vol. 19 (1917), p. 157. Professor Barr points out the further interesting fact that if the slope of the line RAB is twice that of the line RH, then the distance from A (or B) to the line MC will be b.

COMPLEX ROOTS OF A POLYNOMIAL EQUATION*†

H. M. GEHMAN, University of Buffalo

1. Introduction. Although a graphic interpretation of the complex roots of a cubic polynomial has been given previously by various authors,‡ it does not seem to have been noticed by any of them that the interpretation given in the case of the cubic is capable of being generalized to the case of a polynomial of any degree. Accordingly the theorems of this paper are believed to be new, and the description given in section 5 of the complex roots of a quartic polynomial with two real roots is also believed to be new.

2. A graphic interpretation of a pair of complex roots. It is well known that if we wish to represent graphically the real roots of a polynomial equation with real coefficients, $F(x) \equiv a_0 x^n + \cdots + a_n = 0$, we may plot the graph of $y = F(x)$ on a pair of cartesian axes. If this is done, there corresponds to each real root x_0 of $F(x) = 0$, a point $(x_0, 0)$ in which the curve $y = F(x)$ intersects the x-axis. We shall show that if $F(x) = 0$ has complex roots, this same graph has certain properties which are dependent upon the values of these complex roots.

Suppose then that $n \geq 2$, and that the complex numbers $a \pm bi$ are roots of $F(x) = 0$.§ For simplicity we shall let $F(x) = (x^2 - 2ax + a^2 + b^2)f(x)$, where $f(x)$ is

---

* From AMERICAN MATHEMATICAL MONTHLY, vol. 48 (1941), pp. 237–239.

† Presented to the Upper New York State Section of the Mathematical Association of America at Hamilton, N. Y., May 11, 1940.

‡ R. E. Gleason, Popular Astronomy, vol. 17, 1909, p. 119. R. E. Gleason, Annals of Mathematics, vol. 11, 1909–10, pp. 95–96. Frank Irwin and H. N. Wright, Annals of Mathematics, vol. 19, 1918, pp. 152–158. E. S. Crawley, this MONTHLY, vol. 25, 1918, pp. 268–269. Garcia Henriquez, this MONTHLY, vol. 42, 1935, pp. 383–384. G. A. Yanosik, National Mathematics Magazine, vol. 10, 1935–36, pp. 139–140.

§ The discussion may be made to include the case where these two roots are real. This has been done by Gleason, Annals, *loc. cit.*, in the case of the cubic. When the roots are real, the parameter m is negative, b is a pure imaginary, and hence $a \pm bi$ are real numbers.

a polynomial of degree $n-2$ with real coefficients. We shall next consider some relations between the curve $y = F(x)$ and the curves of the family $y = mf(x)$, where m is a constant.

Each curve of the family $y = mf(x)$ intersects the curve $y = F(x)$ on the x-axis in the $n-2$ real or complex points whose abscissas satisfy the equation $f(x) = 0$, and also in the two real or complex points $S = (x_1, y_1)$, $T = (x_2, y_2)$, whose abscissas are the roots of $x^2 - 2ax + a^2 + b^2 = m$. Let U be the midpoint of the line segment ST. Since $\frac{1}{2}(x_1 + x_2) = a$ is independent of m, it follows that the point U is real and lies on the line $x = a$ for any value of m.

If $m < b^2$, the points S and T are complex. But if $m > b^2$, the points S and T are real and distinct. Furthermore, neither of these points is on the x-axis, except when the parameter m has one of that finite set of values for which either x_1 or x_2 is also a root of $f(x) = 0$.

While the above remarks hold true for general curves of the family $y = mf(x)$, we are particularly interested in the special curve for which $m = b^2$. In this case, and only in this case, are the roots of $x^2 - 2ax + a^2 + b^2 = m$ real and equal. Accordingly the points S and T coincide with the point $R = (a, b^2 f(a))$, and the curve $y = b^2 f(x)$ is tangent to $y = F(x)$ at that point.

These remarks are summarized in the following theorem. It is easily established that the curve $y = b^2 f(x)$ is the only curve of the family $y = mf(x)$ which intersects $y = F(x)$ at a point at which the functions and their derivatives agree in value up to and including the non-vanishing derivative of lowest order, and that the only point of intersection at which $y = b^2 f(x)$ and $y = F(x)$ have this property is the point $R = (a, b^2 f(a))$.

THEOREM 1. *Let $F(x) = (x^2 - 2ax + a^2 + b^2) f(x)$, where $b \neq 0$, and $f(x)$ is a polynomial with real coefficients. The parameter m may be so chosen that the curve $y = mf(x)$ intersects the curve $y = F(x)$ in two points S and T which are real and not on the x-axis. The abscissa of the midpoint of the line segment ST is a.*

If $f(a) \neq 0$, there is a unique curve of the family $y = mf(x)$ which is tangent to the curve $y = F(x)$ at a real point not on the x-axis. The abscissa of the point of tangency is a, and the parameter m for the tangent curve is b^2.

If $f(a) = 0$, but $f(x) = 0$ has no real multiple roots except possibly a, there is a unique curve of the family $y = mf(x)$ which is tangent to $y = F(x)$ at a real point on the x-axis. The abscissa of the point of tangency is a, and the parameter m is b^2.

In general, there is a unique curve of the family $y = mf(x)$ which is tangent to the curve $y = F(x)$ at a real point R, having the property that the derivatives of the two functions at the point R are equal from the first to the kth inclusive, where k is the order of the lowest-ordered non-vanishing derivative of $f(x)$ at the point R. The abscissa of the point of tangency is a, and the parameter m for the tangent curve is b^2.

We next consider a few polynomials of low degree with complex roots, and consider the special form which the theorem takes on in each of these cases.

3. Quadratic with two complex roots. Let $F(x) = x^2 - 2ax + a^2 + b^2$. Here $f(x) = 1$, and the family of curves is the family $y = m$ of lines parallel to the x-axis. Of these, $y = b^2$ is the only line which is tangent to the parabola $y = F(x)$, the point of tangency being its vertex (a, b^2). Hence the complex roots of a quadratic may be described thus:

The real part is the abscissa of the point of tangency of the horizontal tangent to $y = F(x)$; the coefficient of i is the square root of the distance from the x-axis to the tangent line.

This description is in accord with the vocabulary of Theorem 1. A briefer description would be:

*The real part is the abscissa of the vertex of the parabola $y = F(x)$; the coefficient of i is the square root of the ordinate of the vertex.**

4. Cubic with one real and two complex roots. Let $F(x) = (x^2 - 2ax + a^2 + b^2)(x - r)$. Here $f(x) = x - r$, and the family of curves is the family $y = m(x - r)$ of lines passing through the point $(r, 0)$ which corresponds to the real root of $F(x) = 0$. Of this family of lines, just one is tangent to the cubic curve $y = F(x)$. If $r = a$, the point of tangency is $(r, 0)$, which is in this case the inflection point of the cubic. The complex roots of a cubic may be described thus:

The real part is the abscissa of the point of tangency of the tangent line passing through the real intersection of the curve $y = F(x)$ with the x-axis; the coefficient of i is the square root of the slope of the tangent line.†

5. Quartic with two real and two complex roots. Let $F(x) = (x^2 - 2ax + a^2 + b^2)(x - r_1)(x - r_2)$. Here $f(x) = (x - r_1)(x - r_2)$, and the family of curves is the family $y = m(x - r_1)(x - r_2)$ of parabolas passing through the points $(r_1, 0)$, $(r_2, 0)$ which correspond to the real roots of $F(x) = 0$. If $r_1 = r_2$, the parabolas are tangent to the x-axis at the point $(r_1, 0)$. Of this family of parabolas, just one is tangent to the quartic curve $y = F(x)$. It is either (1) an ordinary tangent at a point R not on the x-axis, if $a \neq r_1$, $a \neq r_2$; or (2) an inflection tangent with three-point contact at $(r_1, 0)$, if $a = r_1 \neq r_2$; or (3) a tangent with four-point contact at $(r_1, 0)$, if $a = r_1 = r_2$. The complex roots of a quartic with two real roots may be described thus:

The real part is the abscissa of the point of tangency of the tangent parabola passing through the real intersections of the curve $y = F(x)$ with the x-axis; the coefficient of i is the square root of the reciprocal of the latus rectum of the tangent parabola.‡

6. Note on a curve suggested by C. F. Barr. In Theorem 1, the number a is described as a length, but b is the square root of a parameter. The result given next enables us to describe b also as a length.

Let us consider the relation between the curve $y = F(x)$ and the curve $y = 2b^2 f(x)$ of the family $y = mf(x)$.§ This curve is the curve of the family which passes through the point $(a, 2b^2 f(a))$, whose abscissa is equal to that of R, but whose ordinate is double that of R.

---

* Gleason, Popular Astronomy, *loc. cit.*; Irwin and Wright, *loc. cit.*

† Cf. the six references given in section 1.

‡ An entirely different description is given by Irwin and Wright, *loc. cit.*, in which various lines are drawn in place of the parabola used here.

§ The use of this curve was suggested by a note of C. F. Barr on the paper by Henriquez, *loc. cit.*

THEOREM 2. *The curve $y=2b^2f(x)$ of the family $y=mf(x)$ intersects the curve $y=F(x)$ in points S and T whose abscissas are $a-b$ and $a+b$, respectively.*

This fact may be used in connection with the results of Sections 3, 4, and 5 to give a new description of the coefficient of i in each case. This new description is particularly useful in case $f(a)\neq 0$.

COWS AND COSINES*

L. R. FORD, Armour Institute of Technology

The problem solved here was brought in by a resident of a Chicago suburb. The formula embodying the solution caused him no little astonishment. It happens rather often, I think, that connections which are all a part of the day's work in some field seem far-fetched and unexpected to the uninitiated.

One is reminded of DeMorgan's actuarial friend.[†] While explaining certain probabilities of survival, DeMorgan used a formula involving π, explaining that it was the ratio of the circumference of a circle to its diameter. At this the friend exclaimed, "That must be a delusion. What can the circle have to do with the numbers alive at a given time?" Our suburban friend might have exclaimed in like vein, "What have *angles* to do with the question, either?"

The Problem.—A boy is given a young heifer calf. On its third birthday and each year thereafter, it gives birth to a heifer calf. Each of its descendants likewise produces a heifer calf on each birthday, beginning with the third. Assuming no deaths, find a formula for the number of animals in existence n years hence.

Let u_n be the required number. We have

$$n = 0, 1, 2, 3, 4, 5, 6, \cdots,$$
$$u_n = 1, 1, 1, 2, 3, 4, 6, \cdots.$$

The number for any year is equal to the number the year before plus the number of births which have just occurred, and the latter is equal to the number of animals three years earlier. This gives the difference equation

(1) $$u_{n+3} = u_{n+2} + u_n.$$

To solve this, set $u_n = r^n$. We have $r^{n+3} = r^{n+2} + r^n$, and cancelling r^n, we have a solution provided

(2) $$r^3 - r^2 - 1 = 0.$$

The general solution of (1) is then

(3) $$u_n = ar_1^n + br_2^n + cr_3^n,$$

where r_1, r_2, r_3, are the roots of (2) and a, b, c are constants.

The constants in (3) are to be determined so that u_n shall have proper values

---

* From AMERICAN MATHEMATICAL MONTHLY, vol. 46 (1939), pp. 586–587.
† A. DeMorgan, A Budget of Paradoxes, London, 1872, p. 172.

for $n = 0, 1, 2$, whence
$$a + b + c = 1,$$
$$ar_1 + br_2 + cr_3 = 1,$$
$$ar_1^2 + br_2^2 + cr_3^2 = 1.$$

Solving these and simplifying by the use of the relations between the roots of (2),
$$r_1 + r_2 + r_3 = 1, \qquad r_1r_2 + r_1r_3 + r_2r_3 = 0, \qquad r_1r_2r_3 = 1,$$
we have the solution in the form

(4) $$u_n = \frac{r_1^2 + 1}{r_1^2 + 3} r_1^n + \frac{r_2^2 + 1}{r_2^2 + 3} r_2^n + \frac{r_3^2 + 1}{r_3^2 + 3} r_3^n.$$

Now (2) has one real positive root and two imaginary roots. We find
$$r_1 = 1.465576, \qquad r_2 = -0.232788 + 0.792557i$$
and r_3 is the conjugate imaginary of r_2. We write r_2 in the polar form
$$r_2 = 0.826030(\cos 106.369° + i \sin 106.369°).$$

Using these values to determine the coefficients in the solution, we find, after some calculation,
$$a = 0.61149, \qquad b = 0.229679(\cos 32.248° - i \sin 32.248°),$$
c being the conjugate of the latter.

We can now write the solution. Noting that the second and third terms of (4) are conjugates, their sum is twice the real part of br_2^n. We have

(5) $$u_n = 0.61149(1.465576)^n$$
$$+ 0.45936(0.826030)^n \cos (n \cdot 106.369° - 32.248°).$$

This is the required formula. As a matter of fact, the second term of the formula damps out rather rapidly. Thus for $n = 10$, the two terms are as follows:
$$u_{10} = 27.956 + 0.045 = 28.001.$$

From the very first ($n = 0$) the second term is numerically less than one-half; so u_n is the nearest integer to the first term in all cases.

A RULE FOR COMPUTING THE INVERSE OF A MATRIX*

A. A. ALBERT, University of Chicago

If one uses the usual formula to compute the inverse of an n-rowed nonsingular matrix A, one must compute the determinant of A as well as all n^2 of its $(n-1)$-rowed minors. This is actually the way inverses of numerical matrices are customarily computed, and it is regrettable that a really simple method for this computation seems to have been overlooked† in the literature. This latter method may not be new but it is certainly known to very few mathematicians and, since it involves very little more than the computation of the determinant of A alone, it deserves some publicity.

The method for constructing A^{-1} depends upon the fact that if A is nonsingular it is possible to carry A into the identity matrix I by elementary *row* transformations alone. We may now state the method as the following rule:

Apply elementary row transformations to A which carry it into I, noting each transformation as used. Apply the same transformations to I and thereby obtain A^{-1}.

The rule is a consequence of the following rather evident

LEMMA. *Let $C = BA$ and apply an elementary row transformation to B resulting in what we shall designate by B_0, and then the same transformation to C resulting in C_0. Then $C_0 = B_0 A$.*

The lemma evidently generalizes to the case where C_0 and B_0 are obtained respectively from C and B by a finite sequence of elementary row transformations. We now take $B = I$, $C = A = IA$, so that $A_0 = I = I_0 A$. Thus $I_0 = A^{-1}$ and we have our rule. To illustrate the rule we may, for example, take the simple case

$$A = \begin{pmatrix} -2 & 1 & 0 & 1 \\ 1 & 0 & 2 & -1 \\ -4 & 1 & -3 & 1 \\ -1 & 0 & -2 & 2 \end{pmatrix}.$$

We add the second row to the fourth row of A, subtract twice the first row from the third, add twice the second row to the first, and interchange the first and second rows obtaining as the result

$$A_1 = \begin{pmatrix} 1 & 0 & 2 & -1 \\ 0 & 1 & 4 & -1 \\ 0 & -1 & -3 & -1 \\ 0 & 0 & 0 & 1 \end{pmatrix}.$$

---

* From AMERICAN MATHEMATICAL MONTHLY, vol. 48 (1941), pp. 198–199.

† The rule was unknown to me until I observed it recently, while engaged in constructing numerical exercises for my Introduction to Algebraic Theories. I give the rule in that text only in exercises, and was persuaded by W. D. Cairns to present it to the readers of this MONTHLY.

We next add the fourth row of A_1 to each of its other rows, add the second row of the result to this third row to obtain

$$A_2 = \begin{pmatrix} 1 & 0 & 2 & 0 \\ 0 & 1 & 4 & 0 \\ 0 & 0 & 1 & 0 \\ 0 & 0 & 0 & 1 \end{pmatrix},$$

and then subtract twice the third row from the first, four times the third row from the second to obtain I. Applying the same transformations we obtain

$$I = \begin{pmatrix} 1 & 0 & 0 & 0 \\ 0 & 1 & 0 & 0 \\ 0 & 0 & 1 & 0 \\ 0 & 0 & 0 & 1 \end{pmatrix}, \quad I_1 = \begin{pmatrix} 0 & 1 & 0 & 0 \\ 1 & 2 & 0 & 0 \\ -2 & 0 & 1 & 0 \\ 0 & 1 & 0 & 1 \end{pmatrix}, \quad I_2 = \begin{pmatrix} 0 & 2 & 0 & 1 \\ 1 & 3 & 0 & 1 \\ -1 & 4 & 1 & 2 \\ 0 & 1 & 0 & 1 \end{pmatrix},$$

and finally

$$A^{-1} = \begin{pmatrix} 2 & -6 & -2 & -3 \\ 5 & -13 & -4 & -7 \\ -1 & 4 & 1 & 2 \\ 0 & 1 & 0 & 1 \end{pmatrix}.$$

A MODERN TRICK*

CLAIRE ADLER, New York University

There are old methods for showing that

$$\sqrt[3]{2 + \sqrt{5}} + \sqrt[3]{2 - \sqrt{5}} = 1,$$

but this one seems extremely simple:
Let

$$\sqrt[3]{2 + \sqrt{5}} = a$$

$$\sqrt[3]{2 - \sqrt{5}} = b$$

and

$$a + b = x.$$

Then

(1) $\qquad x^3 = a^3 + b^3 + 3abx = -3x + 4$

---

* From AMERICAN MATHEMATICAL MONTHLY, vol. 59 (1952), p. 328.

or

(2) $$x^3 + 3x - 4 = 0$$

and the only real root of (2) is

$$x = 1.$$

ON AN ELEMENTARY DERIVATION OF CRAMER'S RULE*

D. E. WHITFORD and M. S. KLAMKIN, Polytechnic Institute of Brooklyn

The purpose of this note is to point out an elementary derivation of Cramer's rule which should be easily understood by freshman students.

Consider the simultaneous set of equations

(1)
$$a_1 x + b_1 y + c_1 z = d_1$$
$$a_2 x + b_2 y + c_2 z = d_2$$
$$a_3 x + b_3 y + c_3 z = d_3.$$

Now

$$x \begin{vmatrix} a_1 & b_1 & c_1 \\ a_2 & b_2 & c_2 \\ a_3 & b_3 & c_3 \end{vmatrix} = \begin{vmatrix} a_1 x & b_1 & c_1 \\ a_2 x & b_2 & c_2 \\ a_3 x & b_3 & c_3 \end{vmatrix} = \begin{vmatrix} a_1 x + b_1 y + c_1 z & b_1 & c_1 \\ a_2 x + b_2 y + c_2 z & b_2 & c_2 \\ a_3 x + b_3 y + c_3 z & b_3 & c_3 \end{vmatrix}$$

by the elementary transformations of a determinant. Hence if x is to satisfy equations (1) it is necessary that

$$x \begin{vmatrix} a_1 & b_1 & c_1 \\ a_2 & b_2 & c_2 \\ a_3 & b_3 & c_3 \end{vmatrix} = \begin{vmatrix} d_1 & b_1 & c_1 \\ d_2 & b_2 & c_2 \\ d_3 & b_3 & c_3 \end{vmatrix}$$

or

$$x = \begin{vmatrix} d_1 & b_1 & c_1 \\ d_2 & b_2 & c_2 \\ d_3 & b_3 & c_3 \end{vmatrix} \div \Delta, \quad \text{provided} \quad \Delta \neq 0,$$

where Δ is the determinant of the coefficient matrix of the system (1). Similarly

$$y = \begin{vmatrix} a_1 & d_1 & c_1 \\ a_2 & d_2 & c_2 \\ a_3 & d_3 & c_3 \end{vmatrix} \div \Delta, \quad z = \begin{vmatrix} a_1 & b_1 & d_1 \\ a_2 & b_2 & d_2 \\ a_3 & b_3 & d_3 \end{vmatrix} \div \Delta.$$

---

* From AMERICAN MATHEMATICAL MONTHLY, vol. 60 (1953), pp. 186–187.

That these conditions are sufficient, when $\Delta \neq 0$, can be established by substituting back into (1), which gives

$$a_r \begin{vmatrix} d_1 & b_1 & c_1 \\ d_2 & b_2 & c_2 \\ d_3 & b_3 & c_3 \end{vmatrix} + b_r \begin{vmatrix} a_1 & d_1 & c_1 \\ a_2 & d_2 & c_2 \\ a_3 & d_3 & c_3 \end{vmatrix} + c_r \begin{vmatrix} a_1 & b_1 & d_1 \\ a_2 & b_2 & d_2 \\ a_3 & b_3 & d_3 \end{vmatrix} = d_r \begin{vmatrix} a_1 & b_1 & c_1 \\ a_2 & b_2 & c_2 \\ a_3 & b_3 & c_3 \end{vmatrix}.$$

That this is true follows from

$$\begin{vmatrix} a_r & b_r & c_r & d_r \\ a_1 & b_1 & c_1 & d_1 \\ a_2 & b_2 & c_2 & d_2 \\ a_3 & b_3 & c_3 & d_3 \end{vmatrix} = 0$$

since the top row is equivalent to one of the other rows.

The method can be extended immediately to n linear equations in n unknowns.

A UNIFYING TECHNIQUE FOR THE SOLUTION OF THE QUADRATIC, CUBIC, AND QUARTIC*

MORTON J. HELLMAN, Rutgers University

In this paper, starting with Gauss's fundamental theorem of algebra and its immediate extension that every algebraic equation of degree n has n roots, the solutions of the quadratic, cubic, and quartic are derived. The only other results used are the relations between the roots and coefficients of the equations. The solution in the cubic case is identical with Cardan's and, in the quartic case, the result is the same as that obtained by the method of Descartes. In each case, the motivation for the solution is clearer than in conventional methods. The quadratic case, which is very simple, is included to show the unifying technique and, in order to obtain the familiar formula, the leading coefficient will be taken to be a, but in the case of the cubic and the quartic, the equations will be in reduced form with leading coefficients unity.

Case I ($n=2$). The general quadratic equation is $ax^2+bx+c=0$, $a \neq 0$. Let r_1 and r_2 be its two roots so that

(1) $\qquad r_1 + r_2 = -b/a, \qquad r_1 r_2 = c/a.$

Then

(2) $\qquad (r_1 - r_2)^2 = \dfrac{b^2}{a^2} - \dfrac{4c}{a}.$

---

* From AMERICAN MATHEMATICAL MONTHLY, vol. 65 (1958), pp. 274–276.

SOLUTION OF THE QUADRATIC, CUBIC, AND QUARTIC

Extracting the square root of (2) and solving with the first equation of (1) results in

$$r_1 = (-b \pm \sqrt{b^2 - 4ac})/(2a), \qquad r_2 = (-b \mp \sqrt{b^2 - 4ac})/(2a).$$

These give permuted values for r_1 and r_2, and the two roots are given by $x = (-b \pm \sqrt{b^2 - 4ac})/(2a)$.

Case II ($n = 3$). It is sufficient to consider the reduced cubic

(3) $$x^3 + ax + b = 0.$$

Let r_1, r_2, and r_3 be its three roots so that

(4) $$r_1 + r_2 + r_3 = 0,$$
(5) $$r_1 r_2 + r_1 r_3 + r_2 r_3 = a,$$
(6) $$r_1 r_2 r_3 = -b.$$

Substituting r_1, r_2, and r_3 successively in (3), adding the resulting equations, and using (4) and (6) yields $r_1^3 + r_2^3 + r_3^3 - 3r_1 r_2 r_3 = 0$, which factors into

(7) $$(r_1 + r_2 + r_3)(r_1 + \omega r_2 + \omega^2 r_3)(r_1 + \omega^2 r_2 + \omega r_3) = 0,$$

where ω and ω^2 are the imaginary cube roots of unity. From (7) we have the three possibilities $r_1 = -(r_2 + r_3)$, $r_1 = -(\omega r_2 + \omega^2 r_3)$, $r_1 = -(\omega^2 r_2 + \omega r_3)$. These results suggest that the three roots of (3) can be found by setting

(8) $$r_1 = -(s + t), \qquad r_2 = -(\omega s + \omega^2 t), \qquad r_3 = -(\omega^2 s + \omega t)$$

and determining s and t so that (8) satisfies (4), (5), and (6).

Now (4) is satisfied for all s and t since $\omega^2 + \omega + 1 = 0$; therefore, (5) and (6) become, respectively,

(9) $$st = -a/3,$$
(10) $$s^3 + t^3 = b.$$

Raising (9) to the third power and solving simultaneously with (10), using the method of Case I, yields

$$s = \left\{ \frac{b}{2} + \sqrt{\frac{b^2}{4} + \frac{a^3}{27}} \right\}^{1/3}, \qquad t = \left\{ \frac{b}{2} - \sqrt{\frac{b^2}{4} + \frac{a^3}{27}} \right\}^{1/3}.$$

The first equation of (8) then becomes

$$r_1 = \left\{ -\frac{b}{2} + \sqrt{\frac{b^2}{4} + \frac{a^3}{27}} \right\}^{1/3} + \left\{ -\frac{b}{2} - \sqrt{\frac{b^2}{4} + \frac{a^3}{27}} \right\}^{1/3}.$$

which is Cardan's result. The other two roots are given by the second and third equations of (8).

Case III ($n=4$). Here it is sufficient to consider $x^4+ax^2+bx+c=0$ whose roots will be denoted by r_1, r_2, r_3, and r_4. Then

(11)
$$r_1 + r_2 + r_3 + r_4 = 0,$$
$$r_1r_2 + r_1r_3 + r_1r_4 + r_2r_3 + r_2r_4 + r_3r_4 = a,$$
$$r_1r_2r_3 + r_1r_2r_4 + r_2r_3r_4 + r_1r_3r_4 = -b,$$
$$r_1r_2r_3r_4 = c.$$

Making the substitutions

(12) $$u = r_1 + r_2, \quad v = r_3 + r_4, \quad s = r_1r_2, \quad t = r_3r_4$$

in equations (11) yields

$$u + v = 0, \quad s + t + uv = a, \quad sv + tu = -b, \quad st = c.$$

These reduce to

(13) $$s + t = a + u^2,$$

(14) $$u(s - t) = b,$$

(15) $$st = c.$$

Solving (13) and (14) for s, t in terms of u, a, and b results in

(16) $$s = \frac{1}{2}\left\{a + u^2 + \frac{b}{u}\right\}, \quad t = \frac{1}{2}\left\{a + u^2 - \frac{b}{u}\right\}.$$

Substituting s, t from (16) in (15) and simplifying leads to

(17) $$u^6 + 2au^4 + (a^2 - 4c)u^2 - b^2 = 0.$$

This is a cubic equation in u^2 and can be solved by the method of Case II; from this u is found and hence v. Using the values of u, s and v, t in pairs according to equation (12) and the technique of Case I determines r_1, r_2, r_3, and r_4.

Equation (17) is identical with that obtained by the method of Descartes for the solution of the quartic.

THE INSOLVABILITY OF THE QUINTIC RE-EXAMINED*

MORTON J. HELLMAN, Rutgers University

In the April 1958 issue of this MONTHLY a unifying technique for the solution of the quadratic, cubic, and quartic was presented using the relations between the roots and coefficients. It is interesting to pursue the editor's suggestion to see where this method breaks down in the case of the quintic. That is the purpose of this note.

It is sufficient to consider the reduced quintic

(1) $$x^5 + ax^3 + bx^2 + cx + d = 0.$$

Denote the five roots by r_1, r_2, r_3, r_4, r_5 and set

$$r_1 + r_2 + r_3 = s, \quad r_1r_2 + r_1r_3 + r_2r_3 = t,$$
$$r_1r_2r_3 = v, \quad r_4 + r_5 = -s.$$
$$r_4r_5 = w,$$

Then the remaining four equations connecting the roots and coefficients are

(2) $$-s^2 + t + w = a,$$
(3) $$-ts + sw + v = -b,$$
(4) $$tw - vs = c,$$
(5) $$vw = -d.$$

From (5) $v = -d/w$ and from (4) $t = (c+vs)/w = (wc-ds)/w^2$ and substituting these relations in (2) and (3) we obtain, after simplification,

$$-w^2s^2 + wc - sd + w^3 = aw^2,$$
$$s^2d - wsc + sw^3 - dw = -bw^2$$

which upon elimination of either w or s yields an equation of higher degree than the fifth. Hence, as expected, this technique fails in the case of the quintic. It is easy to show, as one might expect, that equations (2) to (5) are equivalent to the equations which result when one attempts to factor (1) into cubic and quadratic factors.

---

* From AMERICAN MATHEMATICAL MONTHLY, vol. 66 (1959), p. 410.

BIBLIOGRAPHIC ENTRIES: SOLUTIONS OF EQUATIONS

The references below are to the AMERICAN MATHEMATICAL MONTHLY.

1. H. Heaton, A method of solving quadratic equations, vol. 3, p. 236.

Rewrite $ax^2+bx+c=0$ by transposing middle term, then square both sides, subtract $4acx^2$, extract square roots and combine with original equation to get usual formula.

2. G. James, On the solution of algebraic equations with rational coefficients, vol. 31, p. 283.

3. N. Anning, A cubic equation of Newton's, vol. 33, p. 211.

Relation of a geometry problem to roots of a cubic equation.

4. W. C. Graustein, A geometrical method for solving the biquadratic equation, vol. 35, p. 236.

5. A. Henderson, Observations on simultaneous quadratic equations, vol. 35, p. 337.

6. T. H. Cronwall, The number of arithmetical operations involved in the solution of a system of linear equations, vol. 36, p. 325.

7. J. P. Ballantine, A graphical derivation of Cramer's rule, vol. 36, p. 439.

8. J. P. Ballantine, Numerical solution of linear equations by vectors, vol. 38, p. 275.

9. R. P. Boas, Jr., A proof of the fundamental theorem of algebra, vol. 42, p. 501.

10. L. C. Karpinski, Simultaneous quadratics solvable in quadratic irrationalities, vol. 43, p. 362.

11. E. Beaman, The moduli of the roots of an algebraic equation, vol. 53, p. 506.

12. R. M. Robinson, A note on linear equations, vol. 56, p. 251.

Method of being sure that a solution to a linear system is actually a solution by forming linear combinations of the given equations.

13. M. R. Spiegel, Reciprocal quadratic equations, vol. 59, p. 175.

Solution of the equation $x^2-rx+1=0$ using the methods of reciprocal equations.

14. G. B. Huff and D. F. Barrow, A minute theory of radical equations, vol. 59, p. 320.

A short development of theory for radical equations which supports why students of elementary algebra test for "extraneous roots".

15. D. E. Whitford and M. S. Klamkin, On an elementary derivation of Cramer's Rule, vol. 60, p. 186.

7
SYNTHETIC GEOMETRY

(a)

TRIANGLES

HISTORICAL NOTE*

FLORIAN CAJORI, Colorado College

If the lengths of the sides of a triangle are, respectively, 3, 4, 5 units, then the figure is a right triangle. This fact was known to the early Egyptians, who, it appears, based upon it a method of laying out their temples. They determined a N. and S. line by accurate astronomical observation, then ran a line at right angles to this by means of a rope stretched around three pegs in such a way that the three sides of a triangle thus formed were to each other as $3:4:5$, one of the legs of the right triangle being made to coincide with the N. and S. line.† Essentially the same process was described later by Heron of Alexandria, by the Hindu astronomers, and by Chinese writers. The Hindus took for the lengths of the sides 15, 36, 39, respectively. There is reason to believe that the Egyptian "rope-stretchers" existed as early as the time of King Amenemhat I, about 2300 B.C. If this date is correct, then this method of laying out right angles in the field by rope-stretching was in vogue fully 3000 years!

The discovery of the well-known property of the right triangle is ascribed by Greek writers to Pythagoras. The truth of the theorem for the special case when the sides are 3, 4, 5, respectively, he may have learned from the Egyptians. That the importance and beauty of this theorem of three squares was thoroughly appreciated by the Greeks is evident from the legend to which its discovery gave rise. Pythagoras is said to have been so jubilant over his great achievement, that he offered a hecatomb to the muses who inspired him. As the Pythagoreans believed in the transmigration of the soul and, for that reason, opposed the shedding of blood, the sacrifice was replaced in the traditions of the Neo-Pythagoreans by that of "an ox made of flour"! The proof given by Pythagoras for this theorem has not

---

* From AMERICAN MATHEMATICAL MONTHLY, vol. 6 (1899), pp. 72–73.
† M. Cantor, *Vorlesungen ueber Geschichte der Mathematik*, Vol. I, 1894, page 64.

been handed down to us. That in *Euclid* I, 47 is due to Euclid himself. Much ingenuity has been expended in conjecture as to the nature of the proof given by Pythagoras. Some critics believe that the proof involved the consideration of special cases; that it was essentially that for the isosceles right triangle outlined by Plato in *Meno*,* in which a square is divided into isosceles right triangles. Other critics surmise that the Pythagorean proof was substantially the same as that given by the Hindu astronomer Bhaskara (about 1150 A. D.), who draws the right triangle four times in the square upon its hypotenuse, so that in the middle there remains a square whose side equals the difference between the two sides of the right triangle. Arranging the small square and the four triangles in a different way, they can be shown, together, to make up the sum of the squares of the two sides. In another place Bhaskara gives a second demonstration of this theorem by drawing from the vertex of the right triangle a perpendicular to the hypotenuse and then suitably manipulating the proportions yielded by the similar triangles. This proof was unknown in Europe until it was rediscovered by the English mathematician, John Wallis.

Among Arabic authors the earliest proof, for the case of the isosceles right triangle, was given by Alchwarizmî, who lived in the early part of the 9th century. It is the same as that in Plato's *Meno*. The Persian mathematician, Nasîr Eddîn, who flourished during the early part of the 13th century, gave a new proof, which required the consideration of eight special cases.† Until six years ago this proof was attributed to more recent writers.

The theorem of Pythagoras has received several nicknames. In European universities of the Middle Ages it was called "magister matheseos," because examinations for the degree of A.M. (when held at all) appear usually not to have extended beyond this theorem, which, with its converse, is the last in the first book of Euclid. The name, "pons asinorum," has sometimes been applied to it, though usually this is the sobriquet for *Euclid*, I., 5. Some Arabic writers, Behâ Eddîn for instance, call the Pythagorean theorem, "figure of the bride." Curiously enough, this romantic appellation appears to have originated from a mistranslation of the Greek word νύμφη, applied to the theorem by a Byzantine writer of the 13th century. This Greek word admits of two meanings, "bride" and "winged insect." The figure of the right triangle with the three squares suggests an insect, but Behâ Eddîn apparently translated the word as "bride."‡

---

* Cantor, *op. cit.* page 205.
† See H. Suter in *Bibliotheca Mathematica*, 1892, pages 3 and 4.
‡ See P. Tannery in *L'Intermédiaire des Mathématiciens*, 1894, Vol. I, page 254.

A MODIFICATION OF A PROOF BY STEINER*

OTTO DUNKEL, Washington University

Introduction. An elegant and elementary proof was given by Steiner of the theorem that the equilateral triangle has the greatest area of all triangles having the same perimeter.† This proof is interesting in that no use is made of either parallels or metrical expressions for the area; it applies therefore whether the sum of the angles of a triangle is supposed to be less than, equal to, or more than 180°, and Steiner showed that his proof applied to spherical triangles without essential change. His proof consists of two parts of which the first part is essentially the proof under Theorem I below, while the second part has been altered to the form of proof under Theorem II. This modified form of proof is applicable to other similar geometrical theorems, and two such theorems are proved in this way without the use of parallels or metrical expressions for the area. The following proofs are worded for spherical triangles since in a few places restrictions are required peculiar to this form of geometry. For the cases where the sum of the angles of the triangle is less than or equal to 180° the proofs are essentially the same but simpler. In conclusion two theorems are given which result from the consideration of a metrical expression for the area. In the discussions below when one side of a triangle is designated as a base the term side will be considered to apply only to the two remaining sides.

THEOREM I. *Two triangles which have equal perimeters and bases of equal lengths have unequal areas if they are neither congruent nor symmetric. The triangle having the smaller area has the smallest base angle, the greatest base angle, the shortest side and the longest side.*

Proof. Let ABC and $A'B'C'$ be two triangles which are neither congruent nor symmetric, but are such that $AB = A'B'$, $AC + BC = A'C' + B'C'$, $A \leq B$, $A' \leq B'$, where A denotes the angle BAC etc. The equality signs in the last relation are assumed to hold for only one triangle, for otherwise the two triangles would be congruent. Let the bases be made to coincide so that A' falls at A and B' at B. If then C and C' fall on opposite sides of the common base, we shall replace one triangle by its symmetric triangle and we shall suppose that the lettering of the vertices of the new triangle is the same as that for the old. The new triangle has the same area and parts as the old. With this understanding C and C' lie on the same side of the base, and C' cannot fall within ABC or upon a side, for then we should have $AC' + BC' < AC + BC$; for the same reason C cannot fall within ABC' or upon a side. It follows then that a longer side of one triangle, say AC', must cut in a point M a shorter side, BC, of the other triangle, and the point M must lie within each of the segments AC' and BC. Hence $\angle BAC' < \angle BAC \leq \angle ABC < \angle ABC'$,

---

* From AMERICAN MATHEMATICAL MONTHLY, vol. 36 (1929), pp. 418–421.

† Steiner, *Sur le maximum et le minimum des figures dans le plan, sur la sphère et dans l'espace en général*, Crelle's Journal, vol. 24 (1842), pp. 96–99.

and therefore $BM < AM$. On MA lay off $MD = MB$, and on MC lay off $ME = MC'$, and draw DE. The two triangles MBC' and MDE have equal areas and $DE = BC'$. It will be shown that E lies within the segment MC. We have $AC + CM + MB = AD + DM + MC' + C'B$, and, since $MB = DM$, $MC' = ME$, $C'B = DE$, this equality reduces to $AC = AD + DE + (ME - CM)$. If $ME \geqq CM$ we would have the length of the broken line $ADEC$ equal to the length of the unbroken line AC, which is impossible. Hence $CM > ME$. It now follows that the area of ABC exceeds the area of ABC' by the area of the quadrilateral $ADEC$. Moreover, $BC' < BM + MC' = BM + ME < BC$, and from this inequality follows that $AC' > AC$. The theorem will be used in the following form:

If two triangles have bases equal to c and sides $a < b$, $a' \leqq b'$, respectively, such that $a + b = a' + b'$; then, if $a < a'$, the triangle with the side, a, has a smaller area than the triangle with side a'.

THEOREM II. *Of all triangles having the same length of perimeter the equilateral triangle has the greatest area.*

Proof. Let a, b, c be the lengths of the sides of a triangle which is not equilateral, and let S be its area; let e be the side of an equilateral triangle having the area E, and such that $3e = a + b + c$. Let a and c be the shortest and longest side, respectively; then $a < e < c$. Consider an isosceles triangle with the base c and the equal sides $a' = \frac{1}{2}(a + b)$, and let its area be S'. If $a = b = a'$, then $S' = S$; but if a and b are unequal, then $a < a' < b \leqq c$. Hence $S' > S$. Since $2a' + c = 3e$ and $c > e$, we have $a' < e < c$. Consider now a triangle with base a', one side of length e and the other side of length b' so that $a' + b' = 2e$, and let its area be S''. Since $a' < e$, it follows that $e < b'$. Hence in the two triangles S' and S'' having bases of length a' the shorter side of the first, a', is less than the shorter side, e, of the second, and therefore $S'' > S'$. We compare finally E and S'' considered as having the base e. From the inequalities above $a' < b'$, $a' < e$. Hence $E > S''$, and $E > S'' > S' \geqq S$, and the theorem is proved.

THEOREM III. *If two triangles which circumscribe the same small circle have an angle of one equal to an angle of the other, the one having the shortest adjacent side has the longest adjacent side and the greater area.*

Proof. Let ABC and $A'B'C'$ be two triangles circumscribing the same small circle with center I, and such that $C = C'$, $CB < CA$, $C'B' \leqq C'A'$, $CB < C'B'$. Let the triangles be placed so that C and C' coincide, and if in this position $C'B'$ does not fall along CB we shall replace one triangle by its symmetric triangle so that this will be the case. Let L and M be the points of contact of CB and CA, and N and N' the points of contact of AB and $A'B'$. Let $\angle LIM = \gamma$, $\angle MIN = \alpha$, $\angle NIL = \beta$, $\angle MIN' = \alpha'$, $\angle N'IL = \beta'$. Then $\alpha + \beta = \alpha' + \beta'$. In the right triangles LIB and LIB', the common side IL is less than 90° and $180° > LB' > LB$; hence $\frac{1}{2}\beta < \frac{1}{2}\beta'$ and therefore $\alpha' < \alpha$. In the right triangles MIA and MIA' we have again the common side $MI < 90°$, $\angle MIA > \angle MIA'$; therefore $MA > MA'$ or $CA > C'A'$. Hence AB and $A'B'$ meet in a point P which lies within each segment, and it is easily shown that $PN = PN'$. In the triangles BIN and BIP with the common side BI and the angle at B in common, $\angle BIN = \frac{1}{2}\beta < \frac{1}{2}\beta' = \angle BIP$, and hence P lies within the segment NA. We show in a similar manner that P lies within the segment $N'B'$. We have then $AN = AM > A'M = A'N' \geq B'L = B'N'$. Hence $AP = AN - NP > B'N' - PN' = B'P$. Also $A'N' = A'M \geq B'L > BL = BN$, and hence $A'P = A'N' + N'P > BN + NP = BP$. Thus in the two triangles $PA'A$ and PBB' with equal angles at P, $A'P > BP$ and $AP > B'P$. Hence the area of the first triangle is greater than the area of the second, and from this follows at once that the area of ABC is greater than the area of $A'B'C$ or of $A'B'C'$. The angles α, β, γ for the triangle ABC will be designated as its central angles.

COROLLARY. *Of two triangles circumscribing the same circle and having a central angle of one equal to a central angle of the other, the one with the smallest of the remaining four central angles has the greatest central angle and the greater area.*

This follows at once from the above since $\beta < \beta' \leq \alpha' < \alpha$ and $\gamma = \gamma'$.

THEOREM IV. *If two triangles circumscribe the same small circle and have an angle of one equal to an angle of the other, the one having the smallest adjacent side has the longest adjacent side and the greater perimeter.*

Proof. The assumptions here and the proof of the first part are the same as in the proof of III, and it remains to prove that the perimeter of ABC is greater than the perimeter of $A'B'C'$. We have $\angle A'IA = \frac{1}{2}(\alpha - \alpha')$, $\angle B'IB = \frac{1}{2}(\beta' - \beta)$, and therefore $\angle A'IA = \angle B'IB$, since $\alpha + \beta = \alpha' + \beta'$. In the triangle AIB, $IN < 90°$ and $AB < 180°$. Hence AI produced meets ANB produced in A_1 so that $AIA_1 = ANA_1 = 180°$, and since $NB < NA_1 < 180°$, we have $IB < IA_1$, or $AI + IB < AI + IA_1 = 180°$. Lay off on MA the lengths $MB_1 = LB$ and $MB_1' = LB'$, and draw IB_1 and IB_1'. Then in the triangle AB_1I, $B_1I + IA < 180°$, and $\angle AIA' = \angle B_1'IB_1 \leq \frac{1}{2}\angle AIB_1 < 90°$. Therefore $AA' > B_1'B_1 = BB'$.* In the two triangles ABC and $A'B'C$ the part $CM + CL$ is common and the corresponding remaining parts of the perimeters are $2(MA + LB)$ and $2(MA' + LB')$. But $MA + LB = MA' + A'A + LB' - BB' > MA' + LB'$. Hence the perimeter of ABC is greater than the perimeter of $A'B'C'$.

COROLLARY. *Of two triangles circumscribing the small circle and having a central angle of one equal to a central angle of the other, the one with the smallest of the four remaining central angles has the greatest of the four and the greater perimeter.*

THEOREM V. *Of all triangles circumscribing the same small circle, the equilateral triangle has the smallest area and the shortest perimeter.*

Proof. Let ABC be a triangle which is not equilateral and which circumscribes a given small circle, and let its area be S and its perimeter, p. Then the central

---

* This theorem has been proposed as problem 3331 [1928, 321] in the Monthly, and a proof will appear later. The theorem is true whether the sum of the angles of the triangle is more than, equal to, or less than 180°; but in the last two cases the condition on the sides is of course omitted and the proof is much simpler.

angles α, β, γ are not all equal and we may suppose that the smallest is α and that the largest is γ. Hence $\alpha < 120° < \gamma$. Consider the circumscribing triangle with a central angle α and $\beta' = \gamma' = 180° - \frac{1}{2}\alpha$. Then $\beta' = \gamma' > 120° > \alpha$. If $\beta = \gamma$, the two triangles are congruent and the area S' and the perimeter p' of the second are the same as for the first. If $\beta \neq \gamma$, $\beta < \gamma$ and since $\beta + \gamma = 2\beta' = 2\gamma'$, we have $\beta < \beta'$ and $\gamma > \gamma' = \beta' > \beta$. Hence $S > S'$ and $p > p'$. Now consider the triangle with central angle $\beta'' = \beta' > 120°$, $\gamma'' = 120°$, $\alpha'' = 240° - \beta'' < 120°$, and let S'' and p'' denote its area and perimeter. The two triangles S' and S'' have the central angle β' and of the remaining four angles $\alpha < \gamma'$, $\alpha'' < \gamma''$, and $\alpha < \alpha''$. The last inequality follows form $\alpha'' + \gamma'' = \alpha + \gamma'$, $\gamma'' = 120° < \gamma'$. Hence $S' > S''$ and $p' > p''$. Now the circumscribing equilateral triangle has $\alpha''' = \beta''' = \gamma''' = 120°$. Let its area be E and its perimeter $3e$. Then S'' and E have $\gamma'' = \gamma''' = 120°$, and of the four remaining angles $\alpha'' < 120° < \beta''$; hence $S'' > E$ and $p'' > 3e$ and $S \stackrel{\geq}{=} S' > S'' > E$ and $p \stackrel{\geq}{=} p' > p'' > 3e$, and the theorem follows.

Some special forms of proof: In spherical geometry the properties of polar triangles and the measure of the area of a triangle enable us to prove the following theorem:

THEOREM VI. *Of all triangles inscribed in a given small circle on the surface of a sphere the equilateral triangle has the greatest area and the greatest perimeter.*

Proof. Let S be the area of a triangle with the angles A, B, C and the opposite sides a, b, c, inscribed in a circle with the center I and of radius $R < 90°$. Let E be the area of the inscribed equilateral triangle with the equal angles ϕ and the equal sides e. The polar triangle of S has the angles $180° - a$, $180° - b$, $180° - c$; the sides $180° - A$, $180° - B$, $180° - C$; and the area S'. The polar of E has the equal angles $180° - e$, the equal sides $180° - \phi$, the area E'. The polar triangle of S circumscribes the circle with center I and with the radius $90° - R$ and similarly for the polar of E. Hence by V we have $S' > E'$ or $540° - (a + b + c) > 540° - 3e$. Therefore $3e > a + b + c$. Also by the same theorem $540° - (A + B + C) > 540° - 3\phi$ or $3\phi > A + B + C$. Hence $E > S$.

In Euclidean geometry of the plane the area of a triangle is equal to the product of the radius of the inscribed circle and one half of the perimeter. Hence in this geometry we can infer* II from V or V from II.

---

* This is shown in the article, *Maximum and minimum areas in geometry*, School Science and Mathematics, vol. 28 (1928), pp. 710–716.

A SIMPLE PROOF OF THE THEOREM OF MORLEY*

JACOB O. ENGELHARDT, Brooklyn, N. Y.

THEOREM†. *If the three angles of a triangle be trisected, the triangle whose vertices are each the intersection of a pair of trisectors adjacent to a side, is equilateral.*

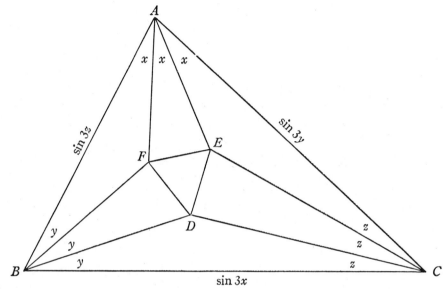

$2R = 1$, where R is the radius of the circumscribed circle.

LEMMA 1. $\sin 3a = 4 \sin a \sin(60° + a) \sin(60° - a)$.

LEMMA 2. If $a + b + c = 180°$, $\cos^2 a + \cos^2 b + \cos^2 c = 1 - 2\cos(a)\cos(b)\cos(c)$.

LEMMA 3. $x + y + z = 60°$.

Proof.

$$BD = \frac{\sin 3x \sin z}{\sin(y+z)} = 4 \sin x \sin z \sin(60 + x).$$

Similarly $BF = 4 \sin x \sin z \sin(60 + z)$.

---

* From AMERICAN MATHEMATICAL MONTHLY, vol. 37 (1930), pp. 493–494.

† For other proofs of this theorem see R. A. Johnson, *Modern Geometry*, p. 253; and Philip Franklin, in Contributions of the Mathematics Department of the Massachusetts Institute of Technology, Second Series, No. 117 (Nov., 1926), p. 57.

Therefore

$$(FD)^2 = 16\sin^2 x \sin^2 z \big[\sin^2(60+x) + \sin^2(60+z)$$
$$- 2\sin(60+x)\sin(60+z)\cos y\big] = 16\sin^2 x \sin^2 z \big[\cos^2(30-x)$$
$$+ \cos^2(30-z) - 2\cos(30-x)\cos(30-z)\cos y\big].$$

But, by Lemma 2,

$$\cos^2(30-x) + \cos^2(30-z) = 1 - \cos^2(180-y) - 2\cos(30-x)\cos(30-z)\cos(180-y)$$
$$= \sin^2(180-y) + 2\cos(30-x)\cos(30-z)\cos y.$$

Therefore $(FD)^2 = 16 \sin^2 x \sin^2 z \sin^2(180-y)$ and $FD = 4\sin x \sin y \sin z$.

Similarly $DE = 4\sin x \sin y \sin z$, $EF = 4\sin x \sin y \sin z$, and therefore $FD = DE = EF$.

GENERALIZATION, SPECIALIZATION, ANALOGY*†

GEORGE PÓLYA, Stanford University

My personal opinion is that the choice of problems and their discussion in class must be, first and foremost, *instructive*. I shall be in a better position to explain the meaning of the word "instructive" after an example. I take as an example the proof of the best known theorem of elementary geometry, the theorem of Pythagoras. The proof on which I shall comment is not new; it is due to Euclid himself (*Elements* VI, 31).

1. We consider a right triangle with sides a, b and c, of which the first, a, is the hypotenuse. We wish to show that

(1) $$a^2 = b^2 + c^2.$$

This aim suggests that we describe squares on the three sides of our right triangle. And so we arrive at the not unfamiliar part I of our compound figure. (The reader should draw the parts of this figure as they arise, in order to see it in the making.)

2. Discoveries, even very modest discoveries, need some remark, the recognition of some relation. We can discover the following proof by observing the *analogy* between the familiar part I of our compound figure and the scarcely less familiar part II: the same right triangle that arises in I is divided in II into two parts by the altitude perpendicular to the hypotenuse.

3. Perhaps, you fail to perceive the analogy between Figures I and II. This analogy, however, can be made quite explicit by a common *generalization* of I and II which is expressed by Figure III. There we find again the same right

---

* From AMERICAN MATHEMATICAL MONTHLY, vol. 55 (1948), pp. 241–243.

† Presented at the summer meeting of the Mathematical Association of America, New Haven, Conn., September 1, 1947.

triangle, and on its three sides three polygons are described which are similar to each other but arbitrary otherwise.

4. The area of the square described on the hypotenuse in Figure I is a^2. The area of the irregular polygon described on the hypotenuse in Figure III can be put equal to λa^2; the factor λ is determined as the ratio of two given areas. Yet then, it follows from the similarity of the three polygons described on the sides a, b and c of the triangle in Figure III that their areas are equal to λa^2, λb^2 and λc^2, respectively.

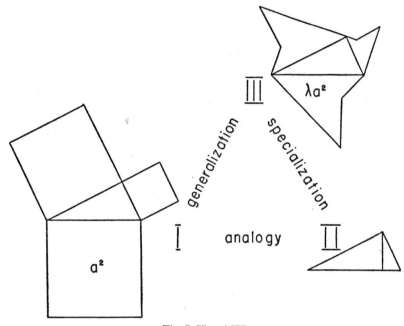

Fig. I, II and III

Now, if the equation (1) should be true (as stated by the theorem that we wish to prove), then also the following would be true:

(2) $$\lambda a^2 = \lambda b^2 + \lambda c^2.$$

In fact, very little algebra is needed to derive (2) from (1). Now, (2) represents a *generalization* of the original theorem of Pythagoras: *If three similar polygons are described on the three sides of a right triangle, the one described on the hypotenuse is equal in area to the sum of the two others.*

It is instructive to observe that this generalization is *equivalent* to the special case from which we started. In fact, we can derive the equations (1) and (2) from each other, by multiplying or dividing by λ (which is, as the ratio of two areas, different from 0).

5. The general theorem expressed by (2) is equivalent not only to the special case (1), but to any other special case. Therefore, if any such special case should turn out to be obvious, the general case would be demonstrated.

Now, trying to *specialize* usefully, we look around for a suitable special case. Indeed Figure II represents such a case. In fact, the right triangle described on its own hypotenuse is similar to the two other triangles described on the two legs, as is well known and easy to see. And, obviously, the area of the whole triangle is equal to the sum of its two parts. And so, the theorem of Pythagoras has been proved.

6. I took the liberty of presenting the foregoing reasoning so broadly because, in almost all its phases, it is so eminently instructive. A case is instructive if we can learn from its something applicable to other cases, and the more instructive the wider the range of possible applications. Now, from the foregoing example we can learn the use of such fundamental mental operations as generalization, specialization and the perception of analogies. There is perhaps no discovery either in elementary or in advanced mathematics or, for that matter, in any other subject that could do without these operations, especially without analogy.

The foregoing example shows how we can ascend by generalization from a special case, as from the one represented by Figure I, to a more general situation as to that of Figure III, and redescend hence by specialization to an analogous case, as to that of Figure II. It shows also the fact, so usual in mathematics and still so surprising to the beginner, or to the philosopher who takes himself for advanced, that the general case can be logically equivalent to a special case. Our example shows, naively and suggestively, how generalization, specialization and analogy are naturally combined in the effort to attain the desired solution. Observe that only a minimum of preliminary knowledge is needed to understand fully the foregoing reasoning. And then we can really regret that mathematics teachers usually do not emphasize such things and neglect such excellent opportunities to teach their students to think.*

---

* The author's views are presented more fully in his booklet, *How to Solve It* (Princeton, 5th enlarged printing 1948). For more about generalization, specialization and analogy see the sections starting on pp. 97, 164 and 37.

ANGLE BISECTORS OF AN ISOSCELES TRIANGLE*

W. E. BLEICK, U.S. Naval Postgraduate School

The following is an indirect proof that a triangle ABC is isosceles if the angle bisectors AD and BE are equal. If the triangle is not isosceles, one of the bisected angles must be greater than the other. Assume that angle EAB is greater than angle ABD. Construct a line through E parallel to BC and intersecting AB at G. Construct a line through D parallel to AC and intersecting AB at F. Then $EG>EA$ since $BC>AC$ by hypothesis. Also $DF>EG$ from triangles DFA and EGB. Now $AG=AB-EG$ and $BF=AB-DF$. Hence $BF<AG$ since $DF>EG$. The triangles ABC, AGE and FBD are similar. Hence

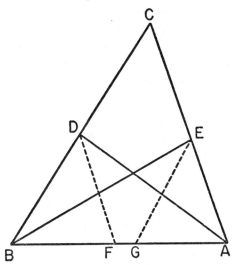

$BF/AG=DF/EA$, from which it follows that $DF<EA$ since $BF<AG$. But $EA<EG$ from above. Hence $DF<EG$ which contradicts the inequality $DF>EG$ above. Hence the bisected angles must be equal. This proof was obtained by Maria Goeppert Mayer in 1932.

---

* From AMERICAN MATHEMATICAL MONTHLY, vol. 55 (1948), p. 495.

THE STEINER-LEHMUS THEOREM*

G. GILBERT, B.I.C.C. Research Organization, London, and
D. MACDONNELL, Fenlow Electronics, Weybridge, England

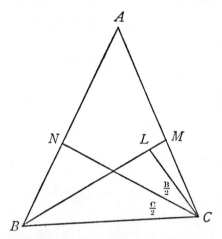

THEOREM. *Any triangle having two equal internal angle bisectors (each measured from a vertex to the opposite side) is isosceles.*

Proof. Let ABC be the triangle with equal angle bisectors BM and CN, as in the figure. If the angles B and C are not equal, one must be less, say $B<C$. Take L on BM so that $\angle LCN = \frac{1}{2}B$. Since this is equal to $\angle LBN$, the four points L, N, B, C are concyclic (lie on a circle). Since

$$B < \tfrac{1}{2}(B+C) < \tfrac{1}{2}(A+B+C),$$

$\angle CBN < \angle LCB < 90°$. Since smaller chords of a circle subtend smaller acute angles, and $BL < CN$,

$$\angle LCB < \angle CBN.$$

We thus have a contradiction.

Editorial note. Martin Gardner, in his review of Coxeter's *Introduction to geometry* (Scientific American, 204 (1961) 166–168) described this famous theorem in such an interesting manner that hundreds of readers sent him their own proofs. He took the trouble to refine this massive lump of material until only the above gem remained. This theorem was proposed in 1840 by C. L. Lehmus, and proved by Jacob Steiner. For its history until 1940 see J. A. McBride, Edinburgh Math. Notes, 33 (1943) 1–13.

---

* From AMERICAN MATHEMATICAL MONTHLY, vol. 70 (1963), pp. 79–80.

ON (WHAT SHOULD BE) A WELL-KNOWN THEOREM IN GEOMETRY*

DANIEL PEDOE, University of Minnesota

The following theorem is not as well known as it should be:

(a) *If ABC is a given triangle, V a point in the plane of ABC which does not lie on a side of the triangle, and $A'B'C'$ is a triangle which has $B'C'$ parallel to VA, $C'A'$ parallel to VB and $A'B'$ parallel to VC, then lines through A' parallel to BC, through B' parallel to CA and through C' parallel to AB will be concurrent at a point V'.*

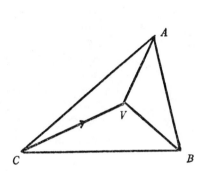 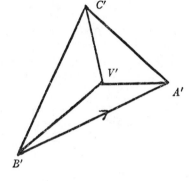

Fig. 1

The theorem is, however, more familiar to geometers in the form:

(b) *If ABC and $A'B'C'$ are two triangles which are such that lines through A parallel to $B'C'$, through B parallel to $C'A'$ and through C parallel to $A'B'$ meet in a point V, then the lines through A', B' and C' parallel to BC, CA and AB respectively will also meet in a point V'.*

It is in this second form that this theorem appears as an exercise in the older books on coordinate geometry. There is a tendency for this theorem to be rediscovered from time to time. It has a respectable history, and is at least a hundred years old. It should be better known, perhaps.

In the form (a) it can be found in books on statics. The quadrangles $ABCV$, $A'B'C'V'$ are said to form *reciprocal figures*, and the interesting thing about the two figures is that in any one figure three lines forming the sides of a triangle are parallel to three concurrent lines in the other.

The connection with graphical statics is evident, to anyone who has studied force-diagrams. To quote Clerk Maxwell (Philosophical Magazine, (1864), p. 258): "If forces represented in magnitude by the lines of the first figure be made to act between the extremities of the corresponding lines of the reciprocal

---

* From AMERICAN MATHEMATICAL MONTHLY, vol. 74 (1967), pp. 839–841.

figure, then the points of the reciprocal figure will all be in equilibrium under the action of these forces."

This statement refers to general reciprocal figures. Not every figure of straight lines and points in the plane admits a reciprocal figure. Maxwell obtained his general reciprocal figures by the orthogonal projection of polyhedra reciprocal with respect to a paraboloid of revolution. After the projection, one figure has to be rotated through a right angle. The great Italian geometer Cremona, who published a book called *Reciprocal figures in graphical statics* (Oxford, 1890), preferred to use polyhedra reciprocal with respect to a null system, on the grounds that after a suitable projection the reciprocal figures were produced without one having to be rotated through a right angle!

Clerk Maxwell's early papers were on geometry, and a knowledge of elementary geometry does not seem to have hurt him! It is interesting to see how he proves Theorem (a). He considers circles with centers at the points A, B, C and V and notes that the chord of intersection of two circles is perpendicular to the line joining the centers and the chords of intersection of three circles, taken in pairs, all pass through a point. Hence the common chords of the four circles are six lines which pass in threes through four points A', B', C' and V', where $B'C' \perp VA$, $C'A' \perp VB$, $A'B' \perp VC$, $V'A' \perp BC$, $V'B' \perp CA$ and $V'C' \perp AB$. All we have to do now is to rotate the figure $A'B'C'V'$ through a right angle, and the figure reciprocal to $ABCV$ is in existence!

This involvement with right angles suggests another theorem, which also appears in books on coordinate geometry:

(c) *If the perpendiculars from the vertices A, B and C of a triangle ABC onto the sides $B'C'$, $C'A'$ and $A'B'$ of a triangle $A'B'C'$ are concurrent, then so are the perpendiculars from the vertices A', B' and C' of $A'B'C'$ onto the sides BC, CA and AB of triangle ABC.*

This theorem can be found in Steiner (Gesammelte Werke, 1 (1881) p. 157, G. Reimer, Berlin), but it is immediately derived from Theorem (b). Rotate the triangle $A'B'C'$ through a right angle. If it be called $A^*B^*C^*$ in its new position, we are told that lines through A, B and C parallel to B^*C^*, C^*A^* and A^*B^* respectively meet in a point V^*. Hence lines through A^*, B^* and C^* respectively parallel to BC, CA and AB will be concurrent in a point V. Rotate $A^*B^*C^*$ back to $A'B'C'$, and let V' be the new position of V. Then lines through A', B' and C' perpendicular to BC, CA and AB respectively intersect at V'.

To make the theorem apparently harder, we can rotate the triangle $A'B'C'$ through an angle α. The formulation of the theorem can be left to the student. Rotation brings relief, in any case.

There are a number of different proofs of (a), not as ingenious as Maxwell's proof. That version of Ceva's theorem which introduces the ratio of the sines of the angles which VA, VB and VC make with the sides of triangle ABC gives an immediate proof on rearrangement, since the Ceva condition for concurrency is also a sufficient condition. But a more fundamental proof can be given if we use projective geometry.

The six sides of the quadrangle $ABCV$ intersect any line in the plane in three pairs of points in involution, and this is also true if the line be the line at infinity. Let the intersections with the line at infinity of BC and AV be x and x', of CA and BV be y and y' and of AB and CV be z and z'. Then since $B'C'$ is parallel to VA, $C'A'$ to VB and $A'B'$ to VC we know that $B'C'$ passes through x', that $C'A'$ passes through y' and that $A'B'$ passes through z'. The fact that $A'x$, $B'y$ and $C'z$ are concurrent is immediate. For if $A'x$ meets $B'y$ in V', then $C'V'$ meets the line at infinity in the mate of z' in the involution which has (x, x') and (y, y') as pairs, and this involution is the one we are considering already. Hence $C'V'$ passes through z, and so $A'x$, $B'y$ and $C'z$ are concurrent.

MORLEY'S TRIANGLE THEOREM*

R. J. WEBSTER, University of Sheffield

Let ABC be a triangle with inradius r and circumradius R, and let the adjacent trisectors of angles A, B, C meet in A', B', C' as illustrated in Figure 1. Then Morley's theorem states that the triangle A', B', C' is equilateral. The proofs of this theorem, which are usually given, do not include a calculation of

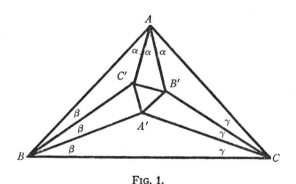

FIG. 1.

the side of the equilateral triangle $A'B'C'$. In this note we prove Morley's theorem by showing that the side of this triangle is $8R \sin A/3 \sin B/3 \sin C/3$. This is a particularly interesting result, when one recalls the formula $r = 4R \sin A/2 \sin B/2 \sin C/2$.

In any triangle ABC, we have $a = 2R \sin A$, etc. Let $A = 3\alpha$, $B = 3\beta$, $C = 3\gamma$.

---

* From MATHEMATICS MAGAZINE, vol. 43 (1970), pp. 209–210.

Then applying the sine rule to $A'BC$ we obtain

$$A'B = 2R \sin 3\alpha \sin \gamma / \sin (\beta + \gamma),$$
$$= 2R(4 \sin \alpha \sin (60 + \alpha) \sin (60 - \alpha) \sin \gamma / \sin (60 - \alpha),$$
$$= 8R \sin \alpha \sin \gamma \sin (60 + \alpha).$$

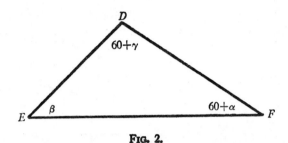

FIG. 2.

Similarly $BC' = 8R \sin \alpha \sin \gamma \sin (60+\gamma)$. Consider now the triangle DEF shown in Figure 2, where $DE = 8R \sin \alpha \sin \gamma \sin (60+\alpha)$. It follows, using the sine rule on DEF, that $EF = 8R \sin \alpha \sin \gamma \sin (60+\gamma)$ and $DF = 8R \sin \alpha \sin \beta \sin \gamma$. However, the triangles $A'BC'$ and DEF are congruent side-angle-side, so $A'C' = DF = 8R \sin \alpha \sin \beta \sin \gamma$. Thus, by the symmetry of this expression, the triangle $A'B'C'$ is equilateral and Morley's theorem is proved.

Editorial Note: Professor C. N. Mills of Illinois State University at Normal as a *tour de force*. found the above expression for a side of the Morley triangle by a straightforward use of elementary Cartesian analysis! His complete proof required some twenty $8\frac{1}{2} \times 11$ sheets of paper.

BIBLIOGRAPHIC ENTRIES: TRIANGLES

Except for the entries labeled MATHEMATICS MAGAZINE, the references below are to the AMERICAN MATHEMATICAL MONTHLY.

1. A. Ingraham, A proof of the Pythagorean proposition, vol. 1, p. 223.

Rotates the right triangle in its own plane about a vertex and calculates in two ways the area of the disk so generated.

2. W. J. Hazard, Generalizations of the theorems of Pythagoras and Euclid's theorem of the gnomon, vol. 36, p. 32.

Extensions to parallelograms and parallelepipeds.

3. H. D. Grossman, The Morley triangle: a new geometric proof, vol. 50, p. 552.
4. C. Lubin, A proof of Morley's theorem, vol. 62, p. 110.

Proof using complex numbers.

5. K. Venkatachaliengar, An elementary proof of Morley's theorem, vol. 65, p. 612.

6. L. Bankoff, A simple proof of Morley's theorem, MATHEMATICS MAGAZINE, vol. 35, p. 223.

7. G. L. Neidhard and V. Milenkovic, Morley's triangle, MATHEMATICS MAGAZINE, vol. 42, p. 87.

8. J. C. Burns, Morley's triangle, MATHEMATICS MAGAZINE, vol. 43, p. 210.

9. W. R. Spickerman, An extension of Morley's theorem, MATHEMATICS MAGAZINE, vol. 44, p. 191.

Extends Morley's theorem to the trisectors of the exterior angles of a triangle.

10. Z. Usiskin and S. G. Wayment, Partitioning a triangle into 5 triangles similar to it, MATHEMATICS MAGAZINE, vol. 45, p. 37.

11. F. Abeles, The affine theorems of Pasch, Menelaus, and Ceva, MATHEMATICS MAGAZINE, vol. 45, p. 78.

Proves Pasch's theorem in various ways in a convex plane.

12. K. R. S. Sastry, Constellation Morley, MATHEMATICS MAGAZINE, vol. 47, p. 15.

Considers numerous properties of Morley's triangle. Proofs are not given, in order to encourage use of the article for undergraduate research projects.

13. L. Bankoff and C. W. Trigg, The ubiquitous 3:4:5 triangle, MATHEMATICS MAGAZINE, vol. 47, p. 61.

A collection of constructions involving a 3:4:5 triangle.

14. M. Lewin, On the Steiner–Lehmus theorem, MATHEMATICS MAGAZINE, vol. 47, p. 87.

An interesting history of the theorem.

15. R. T. Jones and B. B. Peterson, Almost congruent triangles, MATHEMATICS MAGAZINE, vol. 47, p. 180.

An analysis of the extent to which a triangle is determined by knowledge of various combinations of its parts.

(b)

OTHER CONFIGURATIONS

A NOTE ON KNOTS*

F. V. MORLEY, New College, Oxford, England

1. The construction of a regular pentagon by tying a simple knot in a strip of paper leads directly to a generalization for the construction of regular polygons of any odd number of sides.

---

* From AMERICAN MATHEMATICAL MONTHLY, vol. 31 (1924), pp. 237–239.

300   SYNTHETIC GEOMETRY

The construction of the pentagon came to me by oral tradition, and I am at a loss for a reference to it.* Though no doubt familiar to many, it may be recapitulated here. To perform the operation, take a strip of smooth, pliable paper, sides cut parallel and even, say half an inch wide and ten inches long. Visualization may be aided by Fig. 1. The process is simpler in action than in words.

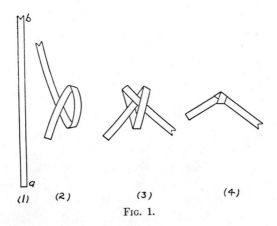

FIG. 1.

1. Hold the strip vertical, the lower end, a, between thumb and forefinger of the left hand; the upper end, b, in the right hand.

2. Carry the end b forward and pass it, from the right, behind a—catching the double thickness between the left thumb and forefinger. We have thus a simple loop, with long end b projecting to the left.

3. Carry b forward again, and pass it from the left axially through the loop.

4. Now pull ends a and b delicately. Folding the paper neatly flat when the knot is tight, we have the pentagon.

If, with a somewhat thinner strip of the same length, say a quarter of an inch wide, we start from stage 3 in the above process, and, instead of pulling the ends, pass b again back and behind the double thickness—so as now to catch three thicknesses between the left thumb and forefinger—we have a double loop, with long end b projecting to the left. This is shown in Fig. 2 (5). Carry b forward and pass it from the left axially through the double loop (Fig. 2 (6)). Pulling both ends

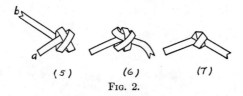

FIG. 2.

---

* Mr. H. W. Richmond tells me that the construction is mentioned in *Scientific Amusements*, by Tom Tit; translated from the French by C. G. Knott, and published by Nelson (no date mentioned).

of the strip until the knot is tight, and flattening, we have the regular heptagon (Fig. 2 (7)).

This construction applies to any regular polygon of $2n+3$ sides, where n is the number of loops in the knot. Thus from Fig. 2 (6) we might, instead of pulling the ends, form a triple loop by passing b back and behind. Tying the knot with this triple loop, the result will be a regular nonagon; and so on.

2. The question rises of constructing the even regular polygons by knots. I do not think this can be done with a single strip of paper. Whenever we tie a "four-in-hand" tie, we construct a hexagonal knot; but it does not flatten into a regular hexagon.

However, we may take two strips, of equal width, and make a simple loop in each, as in Fig. 1 (2). Call the ends of the strips a, b and α, β respectively. Turn the second loop over, and join it with the first as shown in Fig. 3 (8); that is, with ends α, β through the loop of a, b, and ends a, b through the loop of α, β. Now pulling the ends, and flattening, we get the regular hexagon (Fig. 3 (9)).

If, instead of using two simple loops, we had in this way combined a double loop and a single loop, the result would have been the regular octagon. Combining a triple loop and a single, we get the regular decagon. And so on.

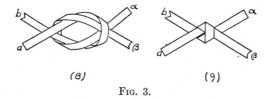

Fig. 3.

3. The above constructions give any regular polygon of five or more sides. Physically, the strips are not easy to manipulate when the loopage is high. Theoretically the construction may be thought of as that of tying knots in parallel

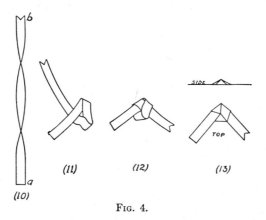

Fig. 4.

lines, and may so continue *ad infinitum*. If an analysis can be developed to handle such processes of knotting, we shall have, in a treatment of these constructions, a method for solving particular equations of any degree.

There are plenty of possible complications to be studied. One interesting development is to twist the strip. Here it is well to have one face colored, to aid in seeing what happens. If a is held and b twisted through two right angles perpendicularly to the length of the strip—so that b presents the same face as a—we say the strip has one complete twist (Fig. 4 (10)). Tying the pentagonal knot in a strip once twisted, we have in Fig. 4 (12) a knot which will not flatten into a plane polygon; but it "flattens" naturally on a pentagonal pyramid (Fig. 4 (13)). In other words, a simple knot tied in a helical strip may give a pentagonal pyramid. In the strict sense I have no proof, other than having seen it happen.

ON THE DIVISION OF A CIRCUMFERENCE INTO FIVE EQUAL PARTS*

H. C. BRADLEY, Massachusetts Institute of Technology

The following simple construction for dividing a circumference into 5 or 10 parts is, in the opinion of the author, new and differs considerably from methods commonly described in text-books of plane geometry. Instead of constructing the sides of regular inscribed decagon and pentagon, this method constructs $\operatorname{arcsec}(\sqrt{5}-1)$ and $\operatorname{arcsec}(\sqrt{5}+1)$ and assumes that it has been proved by trigonometry that these are, respectively, 36° and 72°. The details follow:

At a point A on the circumference of a circle of radius R and center O, draw the tangent AH. On AH lay off $AC = 2R$. On OC, towards O, lay off $CD = R$. With center O and radius OD draw an arc intersecting AH at F. Then $\angle FOA = \operatorname{arcsec}(\sqrt{5}-1) = 36°$.

On OC, away from O, lay off $CE = R$. With center O and radius OE draw an arc intersecting AH at G. Then $\angle GOA = \operatorname{arcsec}(\sqrt{5}+1) = 72°$.

---

* From AMERICAN MATHEMATICAL MONTHLY, vol. 31 (1924), p. 342.

ON TWO INTERSECTING SPHERES*

N. A. COURT, University of Oklahoma

I. Introduction. The object of this paper is to extend to two intersecting spheres some properties of two intersecting circles.† For the convenience of the reader I shall reproduce here those properties of the circles which will be referred to in what follows.

(a) *The two lines joining the two (real) points of intersection of two circles to a variable point of one of these circles determine in the other circle a chord of constant length.*

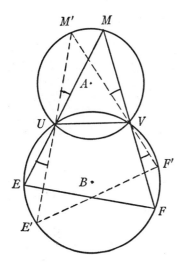 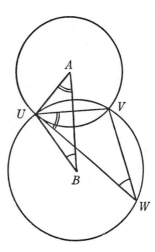

Let U, V, be the points common to the two circles (A), (B) (Fig. 1), and let E, F, be the traces on (B) of the lines MU, MV, joining U, V, to any point M of (A). If M' is any other position of M, and E', F', the corresponding pair of points on (B), it is readily seen that the arcs EE', FF', are equal, and that EF, $E'F'$, are the diagonals of an isosceles trapezoid; hence the proposition.

(b) When the point M coincides with U (Fig. 2), the point F also coincides with U, the line MUE becomes the tangent to (A) at U, and the chord EF is the segment UW intercepted on this tangent by the circle (B).

(c) Let A, B, denote the centers of (A), (B), (Fig. 2). The two triangles AUB, UVW, are equiangular, hence $UW:AB = UV:AU$; or, if $AU=a$, $AB=d$, $UV=c$, and $UW=y$, then $ya=cd$. This relation gives the length of the constant chord $EF=y$ of the above theorem (a).

---

* From AMERICAN MATHEMATICAL MONTHLY, vol. 40 (1933), pp. 265–269.
† Nathan Altshiller-Court. *Sur deux cercles secants.* Mathesis, vol. 39 (1925), p. 453.

II. Theorem. *The variable cone having for base the (real) circle of intersection of two spheres and for vertex a variable point of one of these spheres cuts the other sphere again along a circle whose radius is constant.*

The cone Σ having for base the common (real) circle (S) of the two given spheres (A), (B), and for vertex any point M of (A) penetrates into (B) along (S), hence Σ emerges from (B) along another circle (T).

The principal plane of Σ is determined by the line MS joining M to the center S of (S) and by the perpendicular at S to the plane of (S); hence the principal plane of Σ contains the centers A, B, of (A), (B), and cuts these spheres along great circles. This principal plane cuts (S) along the principal diameter UV, and the principal elements MU, MV, of Σ meet (B) again in the ends E, F, of a diameter of (T).* The proposition thus follows immediately from the case of the plane (see Ia).

III. If $AU = a$, $AB = d$, $UV = c$, and $EF = y$, we have (see Ic)

(1) $$ya = cd.$$

We have thus a simple expression for the length y of the diameter EF of (T) in terms of the fixed elements of the two given spheres.

When the vertex M of the cone Σ coincides with a point L of the circle (S), the cone degenerates, and (T) becomes the circle of intersection of the sphere (B) with the tangent plane to (A) at L; the formula (1) gives the diameter y of this circle.

The segment BT joining B to the center T of the circle (T) is equal to the distance h of B from the plane of (T). If we denote by b the radius of (B) we obtain from the right triangle BTE and the formula (1),

(2) $$h^2 = (4a^2b^2 - d^2c^2) : 4a^2.$$

If the sphere (A) is bisected by the sphere (B), then $c = 2a$, and $d^2 = b^2 - a^2$, and the formulas (1), (2), give

$$y = 2d, \quad h = a.$$

The reader may formulate verbal statements expressing these results.

IV. Consider the cone Σ' having (S) for its base and its vertex on the sphere (B). The formula (1) gives for the diameter x of the circle of intersection of Σ' with (A),

$$xb = cd;$$

hence

$$cd = yb = xa, \text{ or } x : y = a : b.$$

Thus: *The two cones having for base the common circle of two spheres and for vertices any two points on the two spheres cut these spheres again along two circles whose radii are proportional to the radii of the spheres on which the circles are situated.*

---

* Rouché et Comberousse. *Traité de géométrie*, vol. II, pp. 223–224, Gauthier-Villars. Paris, 1900.

As a limiting case we have: *The two tangent planes to two spheres at a point common to these spheres cut the spheres along two circles whose radii are proportional to the radii of the spheres on which the circles are situated.*

V. If the sphere (A) and the circle (S) remain fixed while the sphere (B) varies describing the coaxal pencil having (S) for basic circle, we have, by (1),

$$y:d=c:a,$$

hence: *The cone passing through the basic circle of a coaxal pencil of spheres and having its vertex on a given sphere of the pencil cuts the spheres of the pencil again along circles whose radii are proportional to the distances of the centers of the respective spheres from the center of the given sphere.*

VI. When the vertex M of the cone Σ lies on (S) the preceding proposition takes the following form: *The tangent plane to a sphere of a coaxal pencil at a point of the basic circle of the pencil cuts the other spheres along circles whose radii are proportional to the distances of the centers of the respective spheres from the center of the first sphere.*

The special case of the last two propositions when (S) is a great circle on (A) is left for the consideration of the reader.

VII. *If p, q, are the squares of the radii of two spheres, t the square of their line of centers, and s the square of the radius of their common circle, we have*

(3) $$4st = 4pq - (p+q-t)^2.$$

If u is the power, with respect to either of the two given spheres (P), (Q), of the trace X of the radical plane of (P), (Q), on their line of centers PQ, we have

$$u = XP^2 - p = XQ^2 - q,$$

and hence

(4) $$XP^2 + XQ^2 = 2u + p + q.$$

On the other hand,

$$XP^2 - XQ^2 = p - q,$$

and we have, both in magnitude and in sign,

$$(XP + XQ)(XP - XQ) = p - q,$$

or

$$(XP + XP + PQ)(QX + XP) = p - q.$$

Hence we have, both in magnitude and in sign,

$$2PX \cdot PQ = PQ^2 + (p - q);$$

and similarly,

$$2QX \cdot QP = QP^2 + (q - p).$$

Multiplying these two equalities and putting $PQ^2 = t$, we have, both in magnitude and in sign,

(5) $$4PX \cdot XQ \cdot t = t^2 - (p-q)^2.$$

Multiplying (4) by $2t$ and adding to (5) we have, after simplification,

$$4ut = (p+q-t)^2 - 4pq.$$

Now u and s are equal in magnitude and opposite in sign, hence the announced relation (3).

In the above proof no use has been made of points on either sphere, hence the formula (3) is valid for any two spheres whose centers are real and the squares of whose radii, as well as the square of the radius of their common circle, are either positive, or negative.

If the spheres (P), (Q), are orthogonal, we have $t = p+q$, and (3) becomes $4(p+q)s = 4pq$ or*

$$\frac{1}{s} = \frac{1}{p} + \frac{1}{q}.$$

VIII. The formula (3) applied to the spheres (A), (B), (see II, III) becomes

(6) $$c^2 d^2 = 4a^2 b^2 - (a^2 + b^2 - d^2)^2;$$

hence, using the formula (1),

(7) $$y^2 a^2 = 4a^2 b^2 - (a^2 + b^2 - d^2)^2.$$

We have thus the value of the diameter $EF = y$ in terms of the radii of the given spheres and of the length of their line of centers.

From the formulas (6), (2), we have

(8) $$h = (a^2 + b^2 - d^2) : 2a.$$

IX. If (A), (B), are orthogonal, we have $a^2 + b^2 = d^2$, and (7) gives $y = 2b$; hence:

The cone passing through the common circle of two orthogonal spheres and having its vertex on one of these spheres cuts the second sphere along a great circle.

This result also follows from (8), since h in this case becomes zero.

The same formula (8) shows that the tangent plane to one of two orthogonal spheres at a point common to the two spheres passes through the center of the second sphere, as was to be expected.

X. As the vertex M of Σ varies on (A) the plane of the circle (T) remains at a fixed distance from the center B of (B), hence the plane of (T) envelopes a sphere

---

* Nathan Altshiller-Court. *On five mutually orthogonal spheres*. Annals of Mathematics, vol. 30 (1929), p. 614.

(C) concentric with (B). The sphere (C) is tangent to the planes which touch (A) at the points along (S) (see III), i.e., (C) is inscribed in the cone which circumscribes (A) along (S). Stated in this form the proposition admits of the following projective generalization:

Given two quadric surfaces (A), (B), intersecting along two conics (S), (S'), the cone projecting one of these conics, say (S), from a variable point of one of the surfaces, say (A), cuts the second surface (B) along a conic whose plane envelopes a third quadric (C) tangent to (B) along the conic (S'). Moreover (C) is inscribed in the cone circumscribing (A) along the conic (S).

TWO ISOPERIMETRIC PROBLEMS*

M. S. KNEBELMAN, State College of Washington

In a paper entitled *Ein isoperimetrisches Problem*,[†] G. Bol proved that a curve of given perimeter and prescribed "corners" enclosing the greatest area is obtained from a circle by replacing an arc by a pair of tangents forming each corner, his method of proof being based on Minkowski's inequality for mixed areas. In this note we show that this result is half of a pair of dual theorems, and our proof involves only freshman mathematics.

1. Polygons with prescribed sides. In this section we shall prove the companion to Bol's theorem. Suppose we have given n (straight) rods of lengths $a_1, a_2, \cdots, a_n$; let $\sum_1^n a_i = 2p$ and impose the condition that $a_i < p$, that is, that each rod is shorter then the sum of the remaining ones.

THEOREM 1. *The polygon of greatest area with prescribed sides is cyclic. The area and the radius of the circumscribing circle are independent of the order of the sides.*

Proof. If $n=3$ there is no question of maximum, and since any triangle is cyclic the theorem is obviously true. For $n=4$ let $0 < \omega < 2\pi$ be the sum of either pair of opposite angles; the area S of the quadrilateral is given by[‡]

$$S^2 = (p - a_1)(p - a_2)(p - a_3)(p - a_4) - a_1 a_2 a_3 a_4 \cos^2 \omega/2.$$

Obviously for maximum S, $\cos \omega/2 = 0$ and $\omega = \pi$, from which it follows that the quadrilateral is inscribable in a circle.

The theorem for any value of n may now be proved by induction; assume it to be true for polygons of n sides and consider a polygon of $n+1$ given sides. Let $P_1, P_2, \cdots, P_{n+1}$ be its vertices; if this polygon is not cyclic its area may be increased by keeping, say, $P_1P_2P_3$ fixed and making the polygon of n sides $P_1P_3 \cdots P_{n+1}$ cyclic by our assumption. If P_2 is not on this circle we keep,

---

* From AMERICAN MATHEMATICAL MONTHLY, vol. 48 (1941), pp. 623–627.
† Niew Arch. Wiskde., vol. 20, 1940, pp. 171–175.
‡ *Cf.* any old-fashioned trigonometry book.

say, $P_1P_{n+1}P_n$ fixed and make the polygon cyclic, thus increasing the area further. This may be kept up as long as the polygon of $n+1$ sides is not cyclic, and consequently our proposition is true for $n+1$ sides if it is true for n. Being true for $n=4$, it is true for $n=5$, *etc.*, and so it is true in general, which proves the first part of Theorem 1.

To prove the second part of the theorem we need only observe that the interchange of any two sides may be accomplished by successive interchanges of consecutive sides, and since the area and circumcircle of a triangle are unaltered when two sides are interchanged, the polygon and its circumcircle are also unaltered. This completes the proof of our theorem.

This result has an interesting immediate generalization; given a number of rods and a number of strings, the greatest area is enclosed by them when the strings are arcs of a circle and the rods are chords of the circle. The order in which the rods and strings are joined is immaterial. It may also be observed that the rods need not be straight; the same result holds if the rods are simple arcs of plane curves.

2. Polygons with prescribed angles. We suppose now that we are to construct a polygon of n vertices with angles $\alpha_1, \alpha_2, \cdots, \alpha_n$, where $\sum_1^n \alpha_i = (n-2)\pi$, $0 < \alpha_i < \pi$, and with prescribed perimeter $2p$ enclosing the greatest area.

THEOREM 2. *The angles and perimeter of a convex polygon being given, the greatest area is enclosed by it when the polygon circumscribes a circle. The order in which the angles are placed is immaterial.*

Proof. If $n=3$ there is no question of maximum, since a circle may be inscribed in any triangle. For $n=4$ we first consider the case when the two pairs of opposite sides are parallel. If α is one of the angles, a and c are a pair of adjacent sides, and S is its area, we have $2S = ac \sin \alpha$, $a+c=p$, so that

$$2S = a(p - a) \sin \alpha = - a^2 \sin \alpha + ap \sin \alpha.$$

This quadratic function has a maximum for $a = p/2 = c$, and the parallelogram is a rhombus and therefore circumscribes a circle. Hence, we consider the case when opposite sides are not parallel for at least one pair (Fig. 1). By the law of sines we have

(1) $$2S \sin \theta = a^2 \sin \alpha \sin \beta - b^2 \sin \gamma \sin \delta,$$

$$2p = a + b + a \frac{\sin \alpha + \sin \beta}{\sin \theta} - b \frac{\sin \gamma + \sin \delta}{\sin \theta},$$

while

$$\theta = \pi - (\alpha + \beta) = (\gamma + \delta) - \pi.$$

A little trigonometric manipulation gives

(2) $$p \sin \frac{\theta}{2} = a \cos \frac{\alpha}{2} \cos \frac{\beta}{2} - b \cos \frac{\gamma}{2} \cos \frac{\delta}{2}.$$

Substituting the value of b from (2) into (1) we find

(3)
$$\frac{8S \sin \theta \cos^2 \frac{\gamma}{2} \cos^2 \frac{\delta}{2}}{\prod \sin \alpha} = -a^2 \cot \frac{\alpha+\beta}{2} \sum \cot \frac{\alpha}{2} + 2ap \frac{\cos \frac{\alpha+\beta}{2}}{\sin \frac{\alpha}{2} \sin \frac{\beta}{2}} - 4p^2 \frac{\sin^2 \frac{\theta}{2}}{\sin \alpha \sin \beta},$$

where $\prod \sin \alpha$ is the product of the sines of the four angles and $\sum \cot \alpha/2$ is the

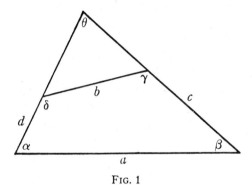

Fig. 1

sum of the cotangents of the four half-angles. From (3) we see that S has its maximum value for

(4)
$$a = p \frac{\sin \frac{\alpha+\beta}{2}}{\sin \frac{\alpha}{2} \sin \frac{\beta}{2} \sum \cot \frac{\alpha}{2}} = p \frac{\cot \frac{\alpha}{2} + \cot \frac{\beta}{2}}{\sum \cot \frac{\alpha}{2}}.$$

Changing the notation so that the angle at vertex i is α_i and the side joining the vertices i and j is a_{ij}, a little calculation shows that

(5)
$$a_{ij} = p \frac{\cot \frac{\alpha_i}{2} + \cot \frac{\alpha_j}{2}}{\sum \cot \frac{\alpha}{2}}.$$

Now if we construct a circle tangent to d, b, and c, we find its radius r to be $p/\sum \cot \alpha/2$, and the circle tangent to d, a, c, has the same radius. Therefore, since the sides d and c are not parallel, it must be the same circle and our quadrangle circumscribes a circle. This proves Theorem 2 for $n=4$.

For general n we use induction; we suppose the theorem true for a given n and consider a polygon of $n+1$ angles $\alpha_1, \alpha_2, \cdots, \alpha_{n+1}$. Let $p_1, p_2, \cdots, p_{n+1}$

be its sides; we keep the sides, say, p_1, p_2, p_3 fixed and consider the polygon*
of n angles formed by p_1, p_3, $\cdots$, p_{n+1} (Fig. 2). The area of this polygon will

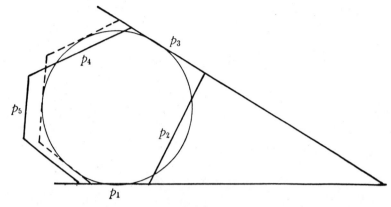

Fig. 2

be increased by making it circumscribe a circle, and therefore the area of the original polygon will also be increased. If p_2 is not tangent to this circle we keep, say, $p_1 p_{n+1} p_n$ fixed and again increase the area of our polygon by making $p_1 p_2 \cdots p_n$ circumscribe a circle so that our theorem is true for $n+1$ if it is true for n. But being true for $n=4$, it is true for $n=5$, *etc.*, and so it is true for any n.

The second part of Theorem 2 is proved by the fact that the interchange of any two angles may be accomplished by the interchange of consecutive angles, and the area of a triangle (of fixed perimeter) is not altered by the interchange of two angles. This completes the proof of Theorem 2.

In this case we can give an explicit expression for the maximum area, radius of the circle, and each side:

$$(6) \quad a_{ij} = p \frac{\cot \frac{\alpha_i}{2} + \cot \frac{\alpha_j}{2}}{\sum \cot \frac{\alpha}{2}}, \quad r = \frac{p}{\sum \cot \frac{\alpha}{2}}, \quad S = \frac{p^2}{\sum \cot \frac{\alpha}{2}}.$$

An interesting consequence of this theorem may be obtained by letting the number of angles tend to infinity, and therefore letting each of a sub-set of the

---

* If p_1 is parallel to p_3 we choose another set of three sides for $p_1 p_2 p_3$ so that p_1 and p_3 are not parallel. This is always possible for $n>4$, for the quadrangle is the only convex polygon which may have more than one set of adjacent sides $p_1 p_2 p_3$ such that p_1 is parallel to p_3. This interesting property of the quadrangle follows from the fact that $\sum \alpha = (n-2)\pi$, and if there were two sets of sides of the required sort, four of the angles would add up to 2π and the remaining $n-4$ would then have to add up to $(n-4)\pi$, so that at least one of them would have to be $\geq \pi$ and the polygon would not be convex.

angles tend to a straight angle. The greatest area is then given by a convex figure consisting of arcs of a circle and tangents to this circle. The explicit expression for each arc and each tangent is given by (6). Thus, in case the figure is to have n angles α_i, we have

$$r = \frac{p}{\sum \cot \frac{\alpha}{2} + \frac{1}{2} \sum \alpha - \frac{n-2}{2}\pi}, \quad S = pr,$$

and each of the tangents forming the angle α_i is of length $r \cot \alpha_i/2$. The only additional fact needed to establish this last result is that $\lim_{\theta \to 0} \sin \theta/\theta = 1$.

In conclusion we may state an interesting problem in elimination. According to Theorem 1, the area of the greatest polygon with n given sides depends only on their magnitude but not on their order. Hence S must be a root of an algebraic equation whose coefficients are functions of the elementary symmetric functions of the sides a_i. The problem is to determine this equation. Thus in case $n=4$ and

$$x^4 - 2px^3 + qx^2 - rx + s = 0$$

is the equation whose roots are the four sides, we have

$$S_4^2 = -p^4 + qp^2 - rp + s,$$

and for $n=3$ we need only make the length of one side zero, which makes $s=0$, and obtain

$$S_3^2 = -p^4 + qp^2 - rp.$$

A GENERALIZATION OF THE MEDIAN THEOREM FOR TRIANGLES*

ROGER BURR KIRCHNER, Harvard University

Let $p_1, p_2, \cdots, p_n$ be points in three-space, and let $P_n = [p_1, p_2, \cdots, p_n]$ be the n-gon obtained by connecting p_1 to p_2, p_2 to p_3, $\cdots$, and p_n to p_1. Regarding the points as triples of numbers (their coordinates with respect to some basis) we can operate with them as vectors.

DEFINITION 1. $p = (p_1 + p_2 + \cdots + p_n)/n$ is defined to be the center of $P_n = [p_1, p_2, \cdots, p_n]$.

The center of a 1-gon (a point) is the point itself, and the center of a 2-gon (line segment) is its midpoint. It is the purpose of this note to give a geometrical characterization of the center of an n-gon and thereby provide examples of applications of vector algebra. We first make the following definition.

---

* From AMERICAN MATHEMATICAL MONTHLY, vol. 69 (1962), p. 650.

DEFINITION 2. *If $Q_k = [q_1, \cdots, q_k]$ and $Q'_{n-k} = [q'_1, \cdots, q'_{n-k}]$ are such that $P_n = [q_1, \cdots, q_k, q'_1, \cdots, q'_{n-k}]$, then Q_k and Q'_{n-k} are called opposite (k and $n-k$) subgons of P_n for $0 < k < n$.*

Our main result is the following theorem.

THEOREM. *The line segments joining the centers of opposite subgons of an n-gon P_n meet in a point, the center of P_n. If Q_k and Q'_{n-k} are opposite subgons of P_n, this point is $(n-k)/n$ of the way from the center of Q_k to the center of Q'_{n-k}.*

Proof. Let $q = (q_1 + \cdots + q_k)/k$ and $q' = (q'_1 + \cdots + q'_{n-k})/(n-k)$ be the centers of Q_k and Q'_{n-k} respectively. We observe that $p = kq/n + (n-k)q'/n$ is the center of P_n. Since $k/n + (n-k)/n = 1$, it follows that p is on the line segment joining q and q'. The first statement of the theorem follows from the symmetry of p in the vertices of P_n. The second statement is obvious if we write $p = q + (n-k)(q'-q)/n$.

The case $n = 3$ is a familiar application of vector algebra, and motivated the present note. We state it as

COROLLARY 1. *The medians of a triangle meet in a point two-thirds of the way from a vertex to the midpoint of its opposite side.*

Not so well known is

COROLLARY 2. *The segments joining the centers of opposite n-gons of a $2n$-gon bisect each other.*

This has the special case,

COROLLARY 3. *The lines joining the midpoints of opposite sides of a quadrilateral bisect each other.*

These theorems are especially interesting as we do not need to assume that the vertices lie on a plane.

SQUARING RECTANGLES AND SQUARES*

N. D. KAZARINOFF, State University of New York at Buffalo, and
ROGER WEITZENKAMP, The University of Michigan

1. Introduction. A **squared rectangle** is a closed rectangular region subdivided into a finite number of square regions that intersect only at their boundaries. The **order** of a squared rectangle is the number of its component squares. A squaring (of a rectangle) is **perfect** if no two component squares are congruent; otherwise it is **imperfect**. A **simple squared rectangle** properly contains no squared rectangle of order more than one. All other squared rectangles are **compound**.

Perfect squared rectangles of low order are easy to find, once one knows how to generate them. Perfect squared squares of low order are exceedingly rare at best. The perfect squared square of least order known has order 24 and is compound. It was found by T. H. Willcocks in 1948 [23, 24]. In 1965, W. T. Tutte [21] reported in this MONTHLY that A. J. W. Duijvestijn [9] had shown no perfect squared squares of order less than 20 exist. But, in fact, Duijvestijn only resolved the problem of determining all *simple* squared rectangles of order less than 20. We have recently [12] proved that there does not exist a *compound* perfect squared square of order less than 22. Thus there exists no perfect squared square of order less than 20, and Tutte's generous restatement of Duijvestijn's result is true.

Study of squarings of rectangles involves some graph theory, topology, combinatorics, number theory, and computer programming, which makes it an attractive subject. In this article we introduce the reader to the theory of squared rectangles, and we give an account of both recent and past results. The prerequisites we require are an elementary knowledge of topology, of how to solve a system of simultaneous linear equations, and of Kirchhoff's Laws. For an exposition less technical than ours, we refer the reader to an article by Tutte [19].

---

* From AMERICAN MATHEMATICAL MONTHLY, vol. 80 (1973), pp. 877–888.

N. D. Kazarinoff did his University of Wisconsin Ph. D. under R. E. Langer. He held positions at Purdue University and the University of Michigan before joining SUNY at Buffalo as Chairman of the Mathematics Department, and now also Martin Professor of Mathematics. He spent a year leave at the University of Wisconsin, was an exchange professor at the Steklov Institute of Mathematics, Moscow in 1960–61, and again in the spring semester of 1965.

He served as managing editor of the Michigan Mathematical Journal, as the consulting editor of Mathematical Reviews, as Chairman of the MAA Putnam Examination Committee, and on numerous MAA and AMS Committees. He is an elected member of CBMS, and in 1968 he received an award for Distinguished Undergraduate Teaching from the University of Michigan. His main research is in differential equations, and he is the author of *Geometric Inequalities* (1961), *Analytic Inequalities* (1961), and *Ruler and the Round* (1970).

Roger Weitzenkamp did his undergraduate and master's degrees at the University of Nebraska. He is a graduate student at the University of Michigan, and his main interest is combinatorics. *Editor.*

Although the study of squaring rectangles is old (see Section 6 for an historical account) a mathematical theory for squared rectangles is much younger. In 1940 Brooks, Smith, Stone, and Tutte [6] constructed an elegant, and indeed, definitive theory of squared rectangles. They related squaring rectangles to determining current distributions in certain electrical networks (planar graphs). These networks are composed of wires of one ohm resistance, except for one wire that contains a battery whose potential produces the current distribution. The central results of Brooks, Smith, Stone, and Tutte are: (1) *there exists a one-to-one correspondence between squared rectangles and certain equivalence classes of planar graphs,* and (2) *each simple perfect squared rectangle corresponds to an electrical network of unit resistances and battery that is equivalent to a 3-connected planar graph.* (See Section 3 for definitions of the terms **3-connected** and **planar graph**.) Our article is based on the theory of Brooks *et al.*

Somewhat after 1940, Tutte found, and later published [20], an algorithm for creating a complete list of 3-connected, finite, planar graphs. Duijvestijn [9] used Tutte's algorithm to search with an electronic computer for simple perfect squared squares by inductively creating a first portion of the list of all 3-connected planar graphs, an induction beginning with the 3-connected planar graphs of six and eight edges (Fig. 1).

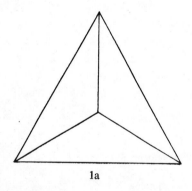
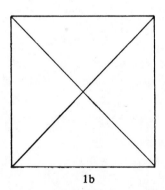

1a   1b

Fig. 1.

The theory of Brooks *et al*, Tutte's theorem [20], and our extension of it (Theorems 2 and 4 below) provide an algorithm for generating all perfect squared rectangles — simple and compound. But almost nothing is known of the obvious problem: *given a closed rectangular region how can it be subdivided to yield a perfect squared rectangle?* Max Dehn [8] proved that *a rectangle can be squared if and only if its sides are commensurable,* and first R. Sprague [16] and, independently, Brooks *et al* proved that *each rectangle with commensurable sides can be squared perfectly in infinitely many totally distinct ways.* But no one has found any algorithm for determining the perfect squaring of least order or even a reasonable estimate of that minimal order. For example, I. M. Yaglom [25] has shown that an *a by b rectangle*

(a/b rational) *always can be subdivided to yield a perfect squared rectangle of order at most* $13a^2b^2 - 11ab - 1$, which for a 32 by 33 rectangle yields the upper bound 14,485,151. But in actuality the perfect squaring of a 32 by 33 rectangle of minimal order has order 9.

2. Generation of squared rectangles using Kirchhoff's Laws. The following example illustrates the method of Brooks *et al* for constructing squared rectangles from planar graphs. Tutte [**19**] has also written an elementary account of the relationship between planar graphs and squared rectangles. We consider the electrical network S^+ illustrated in Fig. 2.

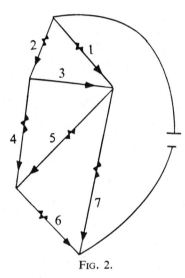

FIG. 2.

Suppose each resistance R_k is 1 ohm and that the current i_k in the resistance R_k is positive if it flows in the direction indicated and negative if it does not. Kirchhoff's Laws applied to a network are expressed as mesh equations (the change in potential around any closed path in the network is zero) and vertex equations (the flow of current into a vertex equals the flow out). For the illustrated network the vertex equations are:

$$i_1 + i_3 = i_5 + i_7$$
$$i_2 = i_3 + i_4$$
$$i_4 + i_5 = i_6.$$

The mesh equations are:

$$1 \cdot i_1 - 1 \cdot i_3 - 1 \cdot i_2 = 0$$
$$1 \cdot i_3 + 1 \cdot i_5 - 1 \cdot i_4 = 0$$
$$1 \cdot i_7 - 1 \cdot i_6 - 1 \cdot i_5 = 0.$$

These six equations are solvable for the seven currents $(i_1, \cdots, i_7)$ up to a constant factor of the unknowns, which we may choose so as to obtain a least solution in integers. This solution is (4,3,1,2,1,3,4). We now imagine that in the network S^+ each wire containing a resistance corresponds to a rectangle of width equal to the absolute value of current in the wire and height equal to the absolute value of the drop in potential over the wire. Since each R_k equals 1 ohm, the drop in potential numerically equals the current; and the associated rectangle is a square. The imperfect squared rectangle that is thus obtained from the network S^+ of Fig. 2 is illustrated in Fig. 3.

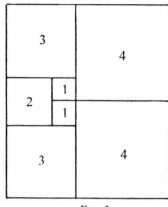

Fig. 3

Note that each horizontal line segment in this figure corresponds to a vertex of S^+ and that each square corresponds to an edge of S^+. Given any planar electrical network of 1 ohm resistances and a battery, a squared rectangle can be derived from it in this way.

Let us vary this example. Suppose now $R_1 = b$ and $R_2 = R_3 = \cdots = R_7 = 1$, that is, suppose that not all resistances in S^+ are 1 ohm. This adds one unknown, namely b, to the system of equations we obtain from S^+ and changes the first mesh equation to

$$b \cdot i_1 - 1 \cdot i_3 - 1 \cdot i_2 = 0.$$

To solve for $(i_1, \cdots, i_7, b)$ we need another equation. To provide it we add a "fixed ratio" condition, one that introduces the width-to-length ratio c of the "rectangled" rectangle to be derived from S^+. This condition is:

$$c(b \cdot i_1 + 1 \cdot i_7) = i_1 + i_2.$$

For each c, the seven equations in the eight unknowns are again solvable up to a constant factor of the currents. For $c = 1$, a solution is $(8, 4, 1, 3, 2, 5, 7, 5/8)$. Again we imagine that in S^+ each wire containing a resistance corresponds to a rectangle. Only this time one rectangle, the one corresponding to R_1, is not a square. Because of the choice $c = 1$, the "rectangled" rectangle corresponding to S^+ is a square; see Fig. 4.

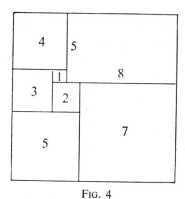

Fig. 4

The point of this variation of the example is that if we knew a squaring of a 5 by 8 rectangle, then it would yield a compound squaring of a square. We shall seek to make clear a process of exhaustively producing such compound squarings of squares.

There is, however, a difficulty arising from the first example. The resulting squared rectangle is compound, but the graph S^+ is 3-connected. We should like **simple** squared rectangles to correspond to 3-connected planar graphs, and **compound** squared rectangles to correspond to 2-connected planar graphs. A more careful scrutiny in the next section will allow us to make such a classification.

3. Graphs corresponding to perfect squared rectangles. A **finite planar graph** is a finite, planar collection of points called **vertices** and closed connected arcs called **edges,** together with a correspondence associating edges with vertices, namely, the vertices are the endpoints of the appropriate edges. A **loop** is an edge whose endpoints coincide. The **order** of a graph is the number of its edges. Throughout this article we deal only with connected, finite, planar graphs of positive order which we shall simply call **nets**, and we consider them as point sets.

A simple closed curve contained in a net that either contains no edges in its interior or all edges in the closure of its interior is a **mesh** of that net. If two vertices on the same mesh of a net are designated as poles, the net is a **polar** net. If A and B are polar nets, $A \subset B$, and A meets $B - A$ only at the poles of A, then A is a **polar subnet** of B. If S is a polar net with poles V and W, the **completion** S^+ of S is the net formed by joining the poles of S with one additional edge joining v and w.

Let S be a net. If there exists a vertex v of S such that $S - v$ is not connected, then S is **1-connected**. If S is not 1-connected, H and K are subsets of S, each containing at least two edges, and v and w are vertices of S, such that $S = H \cup K$ and $H \cap K = v \cup w$, then S is **2-connected**. If S is neither 1-connected nor 2-connected, then S is **3-connected**.

Our first theorem provides the means for defining a polar net corresponding to a squared rectangle. This theorem is a rewording of a theorem formulated by Tutte [17, §2.2] for triangles.

THEOREM 1. *For each squaring of a rectangle R with component squares S_j, there exists an orientation of the rectangle and a set of closed line segments p_i^σ ($\sigma = h, v;\ i = 1, 2, \cdots, m_\sigma$), where m_h and m_v are positive integers, such that:*

(a) *The union of the p_i^σ is the union of the sides of the component squares S_j, each side of each S_j being contained in some p_i^σ.*

(b) *Each p_i^σ is horizontal or vertical as $\sigma = h$ or $\sigma = v$.*

(c) *Two distinct segments have at most one point in common.*

(d) *If w is a vertex of a square S_j but not a vertex of R, then w is an interior point of just one of the segments p_i^σ. If such a vertex w is common to four of the squares S_j, then w is an interior point of some p_i^v.*

Given a squared rectangle R, let $P = P(R)$ denote that polar net whose vertices correspond to the segments p_i^h and whose edges correspond to the squares S_j. If we consider P as an electrical network with unit resistance in each edge, a voltage applied to the poles of P induces currents in the edges which are proportional to the sides of the squares they represent. Brooks et al [6, p. 324] show that if R is simple, then P^+ is 3-connected, while Theorem 3 (below) shows that if R is compound, then P^+ is 2-connected. Since R can also be determined from P^+, we have the desired correspondence between nets and rectangles which was mentioned at the end of the last section. Part (d) of the conclusion of Theorem 1 plays a key rôle in this correspondence.

DEFINITION. For $n \geq 5$, let $\mathscr{L}_n$ denote the set of all finite planar graphs S such that:

(a) S is a polar net of order n.

(b) S^+ is 2-connected or 3-connected.

(c) No two edges of S have the same pair of endpoints.

(d) Each vertex of S that is not a pole is an endpoint of at least 3 edges.

It is easy to construct a family of nets which shows that $\mathscr{L}_n$ is nonempty for each $n \geq 5$.

LEMMA. *Let R be a perfect squared rectangle of order n. Then $P = P(R) \in \mathscr{L}_n$.*

This lemma shows that the class $\mathscr{L}_n$ was well chosen.

Proof of the Lemma. The net P is polar by construction, its poles corresponding to the p_i^h at the top and bottom of R. It has more than 5 edges because a perfect squared rectangle must contain at least nine component squares [6, p. 324]. To establish the second property of membership in $\mathscr{L}_n$ it is sufficient to show that for any vertex v of P^+ there exists a circuit containing v and the poles of P. Such a circuit may be found by tracing a path from v to each pole via the corresponding squares in R, and including the edge $P^+ - P$. Finally, if either of the last two properties of membership in $\mathscr{L}_n$ were violated by P, then two edges of P would carry the same (nonzero) current in the electrical model of P. This is not possible because R is perfect.

The set $\mathscr{L}_n$ can be broken conveniently into four parts by the following theorem.

THEOREM 2. *Each element S of $\mathscr{L}_n$ satisfies exactly one of the following*:

(1) S^+ *is 3-connected*.

(2) $S = X \cup x$, *where* $X \in \mathscr{L}_{n-1}$ *and x is an edge added to a pole p of X in such a way that x connects p to one pole of S and* $X \cap x = p$. *The second pole of X is the second pole of S*.

(3) $S = Y \cup y$, *where* $Y \in \mathscr{L}_{n-1}$, *the poles of Y are the poles of S, and y is an edge joining the poles of Y*.

(4) *There exist integers m and k with* $m, k \geq 5$ *and polar nets A in* $\mathscr{L}_m$ *and B in* $\mathscr{L}_k$ *such that* A^+ *is 3-connected, and* S^+ *is formed by joining A and B at their poles*.

We omit the proof of this theorem. It is long and somewhat tedious, involving counting and connectivity arguments.

The following theorem is implicit in [6, p. 323].

THEOREM 3. *Let R be a compound perfect squared rectangle, and let* $P = P(R)$. *Then* P^+ *is 2-connected*.

Proof. By the lemma, $P \in \mathscr{L}_n$ for some n. Therefore P^+ is 2-connected or 3-connected. Since R is compound, it properly contains a perfect squared subrectangle R_1. Let $P_1 = P(R_1)$. From Theorem 1(d), we conclude that a vertex of P_1 that is not a pole of P_1 is incident only with edges corresponding to squares of R_1. Thus P_1 is a polar subnet of P, and P^+ is 2-connected.

We shall call a net S in $\mathscr{L}_n$ a T_i net ($i = 1,2,3,4$) if S satisfies the ith conclusion of Theorem 2. To discover all compound perfect squarings of rectangles one need not consider T_1 nets because of the above theorem. Also, perfect rectangles corresponding to T_2 and T_3 nets consist of a square adjoined to one side of a smaller perfect squared rectangle, so that they are easy to find inductively. We are left with T_4 nets.

THEOREM 4. *If Q is a compound perfect squared square of order n, then* $P = P(Q)$ *is a* T_4 *net of order n*.

4. Gnomons. Defining a **gnomon** as the completion of a T_4 net, we know that to determine all compound perfect squared rectangles it is sufficient to create a hierarchal list of gnomons. Theorem 2 provides the means for doing this. Conclusion (4) of Theorem 2 describes the compound structure of gnomons, and all the conclusions describe the basic parts of gnomons. In this section we present an algorithm for creating a complete, hierarchal list of gnomons.

It is possible with forethought to eliminate certain portions of this list from consideration. For example if C is a polar net that corresponds to an imperfect squared rectangle, no gnomon G containing C can yield a perfect squared rectangle so long as the "battery" edge is an edge of $G - C$. After eliminating as many gnomons from the list as we easily could through mathematical analysis, we generated the remainder of the gnomons in the list having 22 or fewer edges by IBM-360 computer and dissected them one by one, also by computer. Over 17,000 gnomons were dissected by the computer. The program we used to perform the dissections is a modification of Duijvestijn's program [9] that was written by James Reeds III.

We emphasize that when we dissected a gnomon of order n, we did so in all possible ways; that is, we solved the electrical networks determined by placing a battery in turn in each edge of the gnomon and unit resistances in each of the other $n - 1$ edges. The n squared rectangles resulting from these n dissections may be all different or there may be several alike. Each may be perfect or imperfect. The final result of this analysis of gnomons was the following theorem.

THEOREM 5. *There exists no compound perfect squared square of order* 21 *or less.*

We describe one method of constructing gnomons. Consider a T_4 net $S = S_0$ in $\mathscr{L}_n$, and let A, B, m, and k be as in conclusion (4) of Theorem 2. Since B belongs to $\mathscr{L}_k$, it is a T_i net for some i. If B is a T_2 or a T_3 net, remove the edge that corresponds to the edge x or y of Theorem 2 to obtain a net B_1 in $\mathscr{L}_{k-1}$. Repeat the procedure, if possible, to obtain $B_0 = B \in \mathscr{L}_k$, $B_1 \in \mathscr{L}_{k-1}, \cdots, B_j \in \mathscr{L}_{k-j}, \cdots$. The procedure terminates at some $S_1 = B_l$, where S_1 is either a T_1 or a T_4 net. Then the gnomon S^+ can be realized as the union of S_1, a T_1 net $A_0 = A$, and $i_1 = l$ edges of the types of x and y in Theorem 2. If S_1 is a T_1 net, no further decompositions are needed. Otherwise, S_1 is a T_4 net, and S_1 is decomposed in the way S was decomposed. Repeat the procedure for each S_j ($j = 1, 2, \cdots$) until a T_1 net, say S_r, is reached. At this final stage the original T_4 net S is described by:

(1) a sequence of T_1 nets $A_0, \cdots, A_{r-1}, A_r = S_r$,
(2) a sequence of T_4 nets $S_0, \cdots, S_{r-1}$, and
(3) $q \equiv i_1 + \cdots + i_r$ edges of the types of x and y in Theorem 2,

in such a way that for each $j < r$, the gnomon S_j^+ is the union of A_j, S_{j+1}, and i_{j+1} edges of the types of x and y; see Fig. 5 for an example.

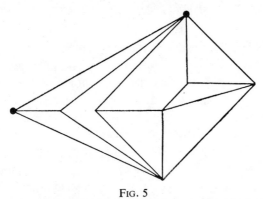

FIG. 5

By reversing the steps in the above procedure we reconstruct S^+ from the A_j's and the extra edges. Indeed, beginning with arbitrary T_1 nets $\{A_j\}$ and as many extra edges as needed, we (theoretically) can construct all gnomons of a given order m.

Another procedure for searching for compound squared squares is illustrated by the second example in Section 2. One can substitute an unknown resistance in one or more wires in the electrical analogue of a net and solve that network by Kirchhoff's laws subject to the constraint that the resultant squared rectangle be a square. This method depends upon a knowledge of squared rectangles and, ultimately, simple perfect squared rectangles. In the example of Section 2 knowledge of a 5 by 8 rectangle is required. If there were one of order 17 or less, then it would yield a compound squared square of order less than 24, which could perhaps be perfect. (No such 5 by 8 rectangle does exist.) This method of search has been used by P. J. Federico [10].

5. 3-connected nets and simple squared rectangles. The fundamental data for generating compound squarings of rectangles are the 3-connected nets whose dissections yield simple squared rectangles. The basic theorem for generating 3-connected nets is Tutte's [20].

THEOREM 6. (Tutte). *Let G be a 3-connected planar graph with no loops, at least 4 vertices and such that no two edges have the same endpoints. Suppose further that G is not a wheel. Then either G or its dual graph can be derived from a simple 3-connected planar graph H by adjoining a new edge e to H whose ends are vertices of the same mesh of H and are not joined by an edge of H.*

Duijvestijn used this theorem to write a program for computer. Using the computer, he generated the 3-connected nets of orders less than 21 and showed there exists no simple perfect squared square of order less than 20.

We repeated some of his work. By computer we found and printed all dissections of rectangles corresponding to 3-connected nets with 17 or fewer edges. We also generated all such nets of orders 18 and 19, dissected them, and printed perfect dissections yielding ratios p/q with $p + q < 300$. (Duijvestijn [9] counts eight 3-connected nets of 12 edges. We found nine. All other counts agree.) We present some statistics arising from these data in Table I.

In Table II we give the Bouwkamp codes of the simple perfect squared rectangles of orders 16, 17 and 18 having sides with reduced ratios p/q such that $p + q < 30$. The Bouwkamp code of a 32 by 33 simple perfect squared rectangle of order 9, the least order possible, is: (18,15) (7,8) (14,4) (10,1) (9). The edge lengths of squares whose upper edges are segments of the same horizontal dissector are grouped in parentheses; the groups are listed in order of decreasing levels of the horizontal dissectors. These levels correspond to the levels of the potential in the corresponding electrical network.

In all perfect squared rectangles of small order (9 or 10 or 11) the largest subsquare appears in a corner. This phenomenon tends to persist, although occasionally the largest square appears at the "middle" of one of the sides. An example of a simple perfect squared rectangle of order 22 in which the largest subsquare appears in the "center" of the dissection is: (419,366,174,156,255) (18,138) (192) (39,216) (177) (53,505) (472) (393) (7,386) (133,379) (359,113) (246). This is a 1370 by 1250 rectangle.

Table I

No. of edges	Total No. of 3-connected nets	% of nets yielding at least one squared rectangle with sides having reduced ratio p/q such that					
		$p+q<300$ (perfect)	$p+q<30$ (perfect)	$p+q<30$ (imperfect)	p or $q<40$ (perfect)	$p/q=1$ (imperfect)	$p/q<1/2$ (perfect)
10	2	50	0	100	50	0	0
11	2	100	0	100	0	0	0
12	9	22.2	0	66.7	0	11.1	0
13	11	54.5	0	63.6	9.1	9.1	0
14	37	32.4	0	48.6	0	10.8	2.7
15	79	22.8	0	35.4	0	3.8	3.8
16	249	14.1	0	20.5	1.2	4.0	6.8
17	671	14.0	.15	18.2	2.4	1.8	6.9
18	2182	11.5	.23	—	—	2.1	—
19	6692	8.7	.09	—	—	1	—
20	12,123*	5.5*	.1*	—	—	.8*	—
21	5,998*	3.0*	.1*	—	—	.5*	—
22	4,949*	2.2*	.04*	—	—	.3*	—

* only the number of nets sampled and percentages thereof

Table II

Order	p/q	Bouwkamp code
16	14/15	(87, 95) (39, 48) (40, 55) (27, 12) (3, 60, 25) (15) (10, 45) (42) (35)
16	11/18	(70, 73) (67, 3) (76) (39, 9, 7, 12) (2, 5) (11) (8, 85) (19) (58)
17	13/14	(51, 30, 88) (13, 17) (8, 5) (1, 16) (6) (56, 3) (9) (25) (19, 94) (75)
17	5/7	(17, 19, 27, 21) (15, 2) (13, 8) (5, 16) (1, 4) (33, 3) (7) (28) (23)
17	3/5	(40, 29) (13, 16) (36, 4) (2, 8, 3) (6) (19) (14) (12, 21) (39, 9) (30)
17	14/15	(145, 93) (45, 48) (42, 3) (23, 28) (110, 35) (18, 5) (33) (75,2) (20) (53)
17	11/15	(67, 52, 46) (6, 19, 21) (28, 30) (17, 2) (54, 13) (23) (41) (39, 8) (31)
18	9/10	(163, 98) (44, 54) (21, 23) (13, 41) (127, 57) (36) (8, 33) (19, 25) (70, 6) (64)
18	13/14	(123, 150) (91, 32) (6, 144) (23, 8, 1) (7) (15) (38) (80, 11) (9, 29) (20) (49)
18	14/15	(95, 61, 30, 54) (17, 13) (7, 6) (14, 3) (1, 5) (11) (59) (34, 52) (129) (111)
18	11/15	(76, 69, 119) (19, 50) (64, 12) (31) (21, 67, 112) (85) (39, 28) (11, 17) (135) (129)
18	11/15	(60, 29, 43) (15, 14) (1, 56) (16) (37, 39) (32, 5) (3, 11, 81) (8) (19) (51)
18	10/11	(129, 61, 70) (23, 29, 9) (20, 59) (17, 6) (11, 44) (28) (157) (5, 54) (49) (103)

An analogous phenomenon is a compound perfect squared rectangle composed of a perfect squared subrectangle surrounded by squares. Here is an example, first found by Federico (private communication). The elements of the squared subrectangle are set off by square brackets:

(390,389) ([3·14, 3·18], 293) (292,98) [3·10, 3·4] [3·7, 3·15] [3·9, 3·1] [3·8] (194), a 779 by 682 rectangle of order 15.

We have no statistics corresponding to those in Table I relative to compound squared rectangles because of the duplications that may occur among the various collections of gnomons of a given order that we separately constructed. We did observe, however, that compound perfect rectangles with $p + q < 30$ and $(p, q) = 1$ first occur with higher orders than in the case of simple perfect rectangles.

6. Historical notes. Max Dehn [8] proved in 1903 that a rectangle can be squared if and only if its sides are commensurable (see [25, §3] for a modern version of Dehn's proof). Yet no example of a perfect squaring was discovered until 1925 when Z. Moroń [14] found the 32 by 33 simple perfect rectangle of order 9. In 1930, M. Kraitchik [13, p. 272] quoted the famous N. Lusin to the effect that there exists no perfect squared square. Kraitchik only knew nontrivial examples of imperfect squared rectangles as well. Apparently Moroń's example was published in too obscure a journal (see also [7]). Finally, in 1939, the German geometer R. Sprague [15] discovered a compound perfect squared square of order 55 and side $5 \cdot 16 \cdot 29$, after he had attempted to prove Lusin's conjecture. Sprague built his example from two different 13 by 16 compound perfect squared rectangles and two squares. The next year Sprague [16] proved that each rectangle with commensurable sides has a perfect, perhaps compound, squaring, and, indeed, that each such rectangle has infinitely many totally distinct perfect squarings.

Almost simultaneously with Sprague and independently the paper by Brooks, Smith, Stone, and Tutte [6] appeared. Bouwkamp [1,2] also found all the low order squarings of rectangles, but constructed no developed theory. Brooks *et al* did. They obtained Sprague's result, developed the analogy with electrical networks (which shows that each squared rectangle has commensurable sides and subsquares), and proved much more: there exists no perfect squared rectangle of order 8 or less; there exist two simple perfect squared rectangles of order 9; there exists a simple perfect squared square of order 55 and a compound perfect squared square of order 26—they gave examples [6, p. 333 and p. 334]. Using the electrical network analogy, C.J. Bouwkamp *et al* [3] gave a catalogue of all simple perfect squared rectangles of orders less than sixteen. In 1948 T. H. Willcocks, then a clerk for the Bank of England in Bristol, discovered a compound perfect squared square of order 24 and side 175 [18, 23, 24]. He holds the record still.

Duijvestijn [9] extended Bouwkamp's work. Federico [10, 11], Brooks [5], Bouwkamp [4], and John C. Wilson [21, 22] found interesting examples of squared squares and rectangles, mostly by computer. I. M. Yaglom [25] finally published the first book on squared rectangles. It contains much original material.

The outstanding open question remains: how to find the perfect squaring of least order of a given rectangle with commensurable sides. Perhaps it will prove easier to estimate closely this minimal order. The perfect squared square of least order will soon be found if computational difficulties are overcome or if much faster computers are built that will compute much more per dollar. It may well turn out to be Willcocks's gem: (64,56,55) (16,39) (38,18) (33,31) (3,4,9) (20,1) (5) (14) (30,81) (2,29) (35) (8,51) (43).

References

1. C. J. Bouwkamp, On the dissection of rectangles into squares, I–III, Nederl. Akad. Wetensch. Proc., 49 (1946) 1176–1188; 50 (1947) 58–71 and 72–78.
2. ———, On the construction of simple perfect squared rectangles, Nederl. Akad. Wetensch. Proc., 50 (1947) 1296–1299.
3. ———, A. J. W. Duijvestijn and P. Medema, Catalogue of simple squared rectangles of orders nine through fifteen, Department of Math. and Mech., Technische Hogeschool, Eindhoven 1960.
4. ———, On some special squared rectangles, J. Combinatorial Theory, 10 (1971) 206–211.
5. R. L. Brooks, A procedure for dissecting a rectangle into squares, and an example for the rectangle whose sides are in the ratio 2:1, J. Combinatorial Theory, 8 (1970) 232–243.
6. R. L. Brooks, C. A. B. Smith, A. H. Stone, and W. T. Tutte, The dissection of rectangles into squares, Duke Math. J., 7 (1940) 312–340.
7. S. Chowla, The division of a rectangle into unequal squares, Math. Student, 7 (1939) 69–70.
8. Max Dehn, Über die Zerlegung von Rechtecken in Rechtecke, Math. Ann., 57 (1903) 314–332.
9. A. J. W. Duijvestijn, Electronic computation of squared rectangles, Thesis, Technische Wetenschap aan de Tech. Hogeschool te Eindhoven, 1962.
10. P. J. Federico, Note on some low-order perfect squared squares, Canad. J. Math., 15 (1963) 350–362.
11. ———, Some simple perfect 2 × 1 rectangles, J. Combinatorial Theory, 8 (1970) 244–246.
12. N. D. Kazarinoff and Roger Weitzenkamp, On existence of compound perfect squared squares of small order, J. Combinatorial Theory, B 14 (1973) 163–179.
13. Maurice Kraitchik, La mathématique des jeux ou Récréations Mathématiques, Stevens Frères, Bruxelles, 1930.
14. Z. Moroń, O rozkladach prostokatów na kwadraty, Przeglad. Matem. — Fizyczny, 3 (1925) 152–153.
15. R. Sprague, Beispiel einer Zerlegung des Quadrats in lauter verschiedene Quadrate, Math. Z., 45 (1939) 607–608.
16. ———, Über die Zerlegung von Rechtecken in lauter verschiedene Quadrate, J. Reine Angew. Math., 182 (1940) 60–64.
17. W. T. Tutte, The dissection of equilateral triangles into equilateral triangles, Proc. Cambridge Philos. Soc., 44 (1948) 463–482.
18. ———, Squaring the square, Canad. J. Math., 2 (1950) 197–209.
19. ———, Squaring the square, "Second Scientific American Book of Mathematical Puzzles and Diversions," by Martin Gardner, Simon and Schuster, New York, 1961, 186–209. Reprinted from *Scientific American* November, 1958, 136–142.
20. ———, A theory of 3-connected graphs, Indag. Math., 23 (1961) 441–455.
21. ———, The quest of the perfect square, this MONTHLY, No. 2, Part II, 72 (1965) 29–35.
22. ———, Squared rectangles, Proc. I. B. M. Scientific Computing Symposium on Combinatorial Problems (March, 1964) 3–9, I. B. M. Data Processing Div., White Plains, N. Y. 1966.
23. T. H. Willcocks, Problem 7795 and Solution, Fairy Chess Review, 7 (1948) 97, 106.
24. ———, A note on some perfect squared squares, Canad. J. Math., 3 (1951) 304–308.
25. I. M. Yaglom, How to cut up a square? (Russian) Nauka, Moskva 1968.

PACKING CYLINDERS INTO CYLINDRICAL CONTAINERS*

SIDNEY KRAVITZ, Dover, New Jersey

The problem of determining the minimum radius of the cylindrical container which can contain N equal cylinders is important in packaging and in the design of rope and conductor cables. A few general results and a few special cases have been proved [1, 2]. The rest is empirical.

Figures 2 to 17 inclusive show good ways to pack cylinders into cylindrical containers for $N \leq 19$. Readers are invited to find smaller containers than those given here.

To assist those who would like to try their hand at this problem, a do-it-yourself kit of formulas and tables is presented. First it is noted that if (ρ_1, θ_1) and (ρ_2, θ_2) are the polar coordinates of the centers of two cylinders of unit radius then they can be packed without interference if

$$\rho_1^2 + \rho_2^2 - 2\rho_1\rho_2 \cos(\theta_1 - \theta_2) \geq 4.$$

The equality holds when the cylinders touch. Second, if we pack N cylinders into an annular ring as shown in Figure 1, then

$$\frac{R}{r} = \frac{2}{1 - \tan^2\left(\frac{\pi}{4} - \frac{\pi}{2N}\right)}$$

where R is the outer radius of the ring and r is the radius of each equal cylinder inside the annulus. (R/r) is given in Table I, column 2, as a function of N. If we place an additional cylinder, A, external to the ring, but touching two of the cylinders inside the annulus, then the furthermost point, R_0, of the external cylinder is given by

$$\frac{R_0}{r} = \frac{1}{\tan \frac{\pi}{N}} + (\sqrt{3} + 1).$$

For an internal cylinder, B, the innermost point, R_1, is given by

$$\frac{R_1}{r} = \frac{1}{\tan \frac{\pi}{N}} - (\sqrt{3} + 1).$$

(R_0/r) and (R_1/r) are given in Table I, columns 3 and 4, as functions of N.

Imagine hexagons which circumscribe the cylinders packed within the container. The packing efficiency is defined as the area of these hexagons divided

---

* From MATHEMATICS MAGAZINE, vol. 40 (1967), pp. 65–71.

by the area of the container:

$$\phi = \frac{2\sqrt{3}}{\pi} \cdot \frac{N}{(R/r)^2} = 1.10266 \frac{N}{(R/r)^2},$$

where R is the radius of the container. The efficiency cannot exceed one. Efficiencies are presented in column 6, Table I, and in Figures 2 to 17 inclusive. $N=7$ and $N=19$ give high efficiencies.

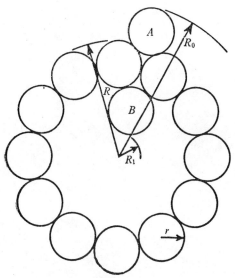

FIG. 1.

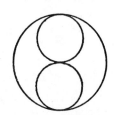

FIG. 2. $N=2$, $R/r=2$, $\phi=.5513$.

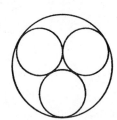

FIG. 3. $N = 3$, $R/r = 1 + \dfrac{2}{\sqrt{3}} = 2.1547, \phi = .7125$.

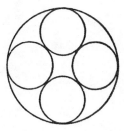

FIG. 4. $N=4$, $R/r=1+\sqrt{2}=2.4142$, $\phi=.7568$.

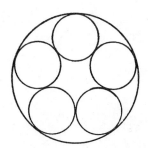

FIG. 5. $N = 5, R/r = 1 + \sqrt{\dfrac{(10 + 2\sqrt{5})}{5}} = 2.7013, \phi = .7555$.

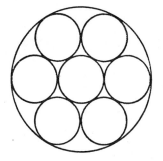

Fig. 6. $R/r=3$ for $N=6$, $\phi=.7351$
for $N=7$, $\phi=.8576$.

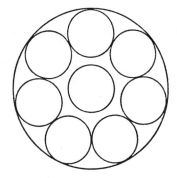

Fig. 7. $N=8$, $R/r=1+\csc(\pi/7)=3.3046$
$\phi=.8078$.

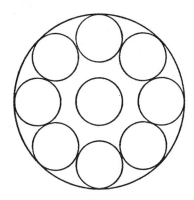

Fig. 8. $N=9$, $R/r=1+\sqrt{(4+2\sqrt{2})}=3.6131, \phi=.7602$.

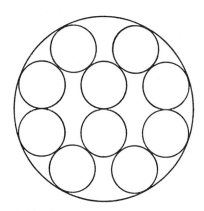

Fig. 9. $N=10$, $R/r=1+2\sqrt{2}=3.8284$, $\phi=.7523$.

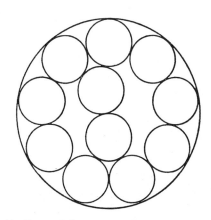

Fig. 10. $N=11$, $R/r=1+\csc 20°=3.9238$, $\phi=.7879$.

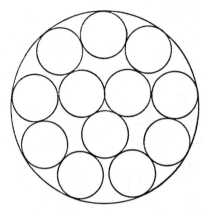

FIG. 11. $N=12$, $R/r=4.0294$, $\phi=.8150$.

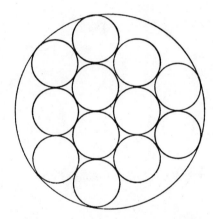

FIG. 12. $N=12$, $R/r=4.0550$. This is *not* the minimum solution for $N=12$. See Figure 11.

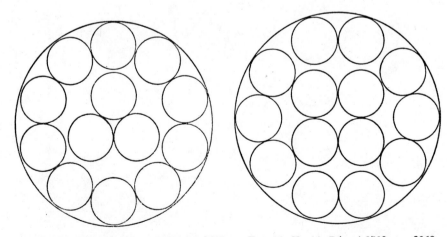

FIG. 13. $N=13$, $R/r=2+\sqrt{5}=4.2361$, $\phi=.7989$.   FIG. 14. $N=14$, $R/r=4.3738$, $\phi=.8069$.

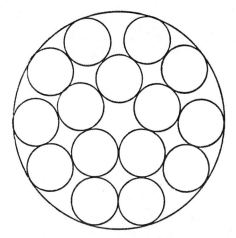

Fig. 15. $N = 15$, $R/r = 1 + \sqrt{1 + \left\{2 + \left(\sqrt{\dfrac{5+1}{4}}\right)\left(\sqrt{\dfrac{10+2\sqrt{5}}{5}}\right)\right\}^2}$
$= 4.5213$, $\phi = .8091$.

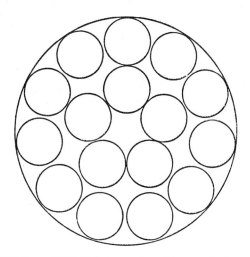

Fig. 16. $N = 16$, $R/r = 4.7013$, $\phi = .7982$. This is not the minimum solution. It is possible to squeeze the outer ring in a bit.

TABLE I.

N	Annular Ring (Figure 1)			Designs Presented in Figures 2 to 17 Inclusive	
	R/r	R_0/r	R_1/r	R/r	Efficiency, ϕ
2				2.0000	.5513
3	2.15470	3.30940		2.1547	.7125
4	2.41421	3.73205		2.4142	.7568
5	2.70130	4.10843		2.7013	.7555
6	3.00000	4.46410		3.0000	.7351
7	3.30477	4.80857		3.0000	.8576
8	3.61313	5.14626		3.3046	.8078
9	3.92381	5.47953	0.01543	3.6131	.7602
10	4.23607	5.80973	0.34563	3.8284	.7523
11	4.54947	6.13774	0.67364	3.9238	.7879
12	4.86370	6.46410	1.00000	4.0294	.8150
13	5.17858	6.78921	1.32511	4.2361	.7989
14	5.49396	7.11334	1.64923	4.3738	.8069
15	5.80974	7.43668	1.97258	4.5213	.8091
16	6.12583	7.75939	2.29529	4.7013	.7982
17	6.44219	8.08158	2.61748	4.8637	.7924
18	6.75877	8.40333	2.93923	4.8637	.8390
19	7.07554	8.72472	3.26062	4.8637	.8857
20	7.39246	9.04580	3.58170	Column 5	Column 6
21	7.70951	9.36662	3.90252		
22	8.02668	9.68720	4.22310		
23	8.34395	10.00759	4.54349		
24	8.66130	10.32780	4.86370		
25	8.97874	10.64787	5.18376		
26	9.29624	10.96779	5.50369		
27	9.61380	11.28760	5.82350		
28	9.93141	11.60730	6.14319		
29	10.24908	11.92690	6.46280		
30	10.56678	12.24641	6.78231		
31	10.88453	12.56585	7.10175		
32	11.20231	12.88522	7.42112		
33	11.52012	13.20452	7.74042		
34	11.83797	13.52377	8.05967		
35	12.15583	13.84296	8.37886		
36	12.47373	14.16210	8.69800		
37	12.79164	14.48120	9.01710		
38	13.10958	14.80025	9.33615		
39	13.42754	15.11927	9.65517		
40	13.74551	15.43825	9.97415		
41	14.06350	15.75720	10.29310		
42	14.38151	16.07612	10.61202		
43	14.69953	16.39501	10.93091		
44	15.01756	16.71388	11.24977		
45	15.33561	17.03272	11.56861		
46	15.65367	17.35153	11.88743		
47	15.97174	17.67033	12.20623		
48	16.28982	17.98910	12.52500		
49	16.60790	18.30786	12.84375		
50	16.92600	18.62659	13.16249		
Column 1	Column 2	Column 3	Column 4		

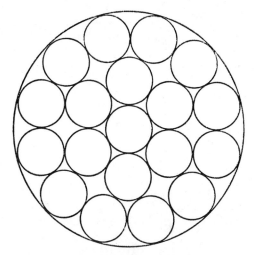

Fig. 17. $R/r = 1 + \sqrt{2} + \sqrt{6} = 4.8637$
for $N=17$, $\phi = .7924$
$N=18$, $\phi = .8390$
$N=19$, $\phi = .8857$.

References

1. L. Fejes Toth, Lagerungen in der Ebene, auf der Kugel und im Raum, Springer, Berlin, 1953.
2. C. A. Rogers, Packing and Covering, Cambridge Press, New York, 1964.
3. H. Pender, Handbook for Electrical Engineers, W. A. Del Mar, 1958, pp. 14–164.

PACKING EQUAL CIRCLES IN A SQUARE*

MICHAEL GOLDBERG, Washington, D.C.

1. Introduction. There are many interesting extremal problems associated with the packing of circles and spheres. An excellent summary of these problems is contained in the book by Fejes Tóth [1]. These include the packing of circles on the surface of the sphere which has been further investigated by the present author [2].

One of the oldest of packing problems is the packing of equal circles in a rectangle or a square. Its most common application today is the packing of bottles or cans in a box. In spite of its antiquity, and its common utility, very little has been done on the problem from an analytical standpoint. The best recent references are the papers by Schaer and Meir [3, 4]. They derived the "best" arrangements for the packing of n circles in a square for $n \leq 9$. The term "best" implies the smallest square for the given n circles, or the largest diameter of the circles for a given square.

---

* From MATHEMATICS MAGAZINE, vol. 43 (1970), pp. 24–30.

Some of the techniques employed by the author in finding solutions for the sphere are here used to find solution in a square. The sphere had no boundary conditions. The boundary of the square introduces further complications.

The results presented here should be compared with the results of Kravitz [5] on the packing of equal circles in a larger circle.

2. Notation. The problem is the same as maximizing the minimum distance between the pairs of n points in a unit square. If this distance is designated by m, then the edge e of the square enclosing n circles of diameter m, centered on these points, is given by $e = 1 + m$. Hence $d(n)$, the density of coverage of the latter square by the circles, is given by $d(n) = n(\pi m^2)/4(1+m)^2 = (\pi/4)nm^2/(1+m)^2$.

Many efficient arrangements are obtained by packing in rows of equally spaced circles. The symbol $n(p, q, r, \cdots)$ indicates n circles arranged in rows so that p equally spaced circles are in the first row, q circles in the second row, etc. One may pack them closely in interlaced rows parallel to the edge of the square, as shown in the figures for $n = 10, 12, 13$, etc.; or they may be packed parallel to the diagonal of the square, as shown in the figures for $n = 9, 16, 25$, etc.

3. Stability of packing. For extremal solutions, it is necessary that a structure of circles must connect all the sides of the square. Each circle in the structure must make at least three contacts with other circles or with the sides of the square. These contacts cannot be limited to a semicircle, for then the circle could be moved to separate it from the structure. There may be, however, other circles which are not part of the structure. Such solutions appear for $n = 7, 19, 21, 22$ and 26. The foregoing conditions are necessary for static stability and hence, for an extremal solution. However, for a given n, there may be several such stable arrangements with different packing densities. They correspond to local extrema. As an example, the arrangement of 20 circles in Table 2 and its figure, seems to be stable. Yet, 22 circles of the same size can be arranged in the square, as shown in Table 1 and its figure.

4. Description of the special arrangements. Some of the best results are not obtained as regular arrangements in equally spaced rows. Instead, some may be considered as combinations of several sets of the best arrangements for smaller numbers of circles. For example, for $n = 13$, the arrangement may be considered as the combination of four sets of five circles in which some of the circles have been superimposed. For $n = 18$, the arrangement may be considered as the combination of four sets of the six circles shown for $n = 6$. Also, the efficient cluster of eight circles is visible in the upper right corner of the figures for $n = 14$ and $n = 15$, and in the upper right and lower left corners of the figure for $n = 23$.

A dense arrangement of 20 circles consisting of five rows of four circles in each row will make a rectangle which is slightly longer than its height. Therefore, a lateral compression must be applied to shorten and heighten the rectangle to make it into a square. Two of the rows are shifted laterally, while the circles in each row remain in contact. The spacing between adjacent rows is increased. An interior circle, which had made six contacts with its neighboring circles in the dense arrangement, now makes only four contacts in the square arrangement.

On the other hand, the arrangement for $n=11$ has no symmetry and only a trace of regularity. It is obtained by the removal of a circle from the arrangement for 12 circles, and a readjustment of the remaining circles until stability is restored. Similarly, the arrangements for $n=15$ and $n=24$ are obtained by removal of a circle from $n=16$ and $n=25$, and collapsing the corresponding rows and columns. For $n=14$, 17 and 21, more complicated adjustments were required.

5. Tabulation of results. For the regular arrangements, it is possible to obtain exact solutions since they involve only linear or quadratic equations. In the cases of irregular arrangements, equations of higher degree are involved. For example, for $n=19$, the distance m is found from the equation $5\sqrt{m^2-1/16}+\sqrt{m^2-(1/4-m/2)^2}=1$. In these cases, the numerical results were obtained by successive approximation.

In many cases, there are several promising arrangements, and each must be investigated to determine the best. An analytical attack is lacking, and a rigorous demonstration of the attainment of the best result is still needed. The best results obtained are summarized in Table 1 and shown in the corresponding figures. Less efficient arrangements which have been tried are given in the upper part of Table 2 and its corresponding figures.

6. Larger values of n. As n increases without limit, the arrangement approaches *regular dense packing* in the plane. Then $d(n)$ approaches $\pi/\sqrt{12}=0.9069$ as a limit [1, p. 58]. Hence, $4d(n)/\pi=1.1547$ for $n\to\infty$. In this arrangement, each interior circle is surrounded and in contact with six other circles.

For *rectangular lattice packing*, each interior circle is surrounded and in contact with four other circles. Then, $4d(n)/\pi=1.000$ and this value is independent of n. This is shown in Table 1 for $n=1, 4, 9, 16$ and 25. For $n25$, $\leq$ these are seen to be the most efficient packings.

For $n=30$, it seems that $4d(n)/\pi$ may always be greater than unity. See the lower part of Table 2. As a particularly interesting example, a square of edge 14 can enclose 196 circles of unit diameter in rectangular lattice packing. Yet, a dense array of 16 rows, each row containing 13 circles (a total of 208 circles), can be enclosed in a rectangle of dimensions 13.50 by 13.98. Thus, a square of edge 14, which contains this rectangle, can enclose at least 208 circles, instead of only 196.

The shift method, used for $n=20$, was used also for the determination of the data for $n=30, 42, 143, 168$ and 340 in Table 2. It is illustrated in the figure for $n=42$.

If a dense collection is enclosed in a rectangle which approximated a square, but in which the height is greater than the length, then a vertical compression will produce a square by reducing the height and spreading the circles in the rows. This method was used for obtaining the data for $n=39, 52, 80, 99, 120, 161, 188, 270$ and 304. This method was used also for $n=24(4,4,4,4,4,4)$ and is illustrated among the figures for Table 2.

If the rectangle enclosing the dense array is a close approximation of a square, then only a small distortion is needed to reshape it into a square. The greater the distortion, which may be a shift or a compression, the greater is the loss in efficiency. Hence, the function which expresses the density in terms of n, as seen in Tables 1 and 2, is not monotonic. The arrangements listed at the bottom of Table 2 were selected from the more efficient arrangements.

TABLE 1

n	Arrangement	m		$4d(n)/\pi$
2		$\sqrt{2}$	$=1.414$	.6850
3	1,2	$\sqrt{6}-\sqrt{2}$	$=1.035$	.7755
4	1,2,1	1	$=1.000$	1.0000
5	2,1,2	$\sqrt{2}/2$	$=.707$	.8580
6	2,2,2	$\sqrt{13}/6$	$=.601$	.8460
7	Irregular	$2(2-\sqrt{3})$	$=.536$	.8533
8	Based on 3 circles	$(\sqrt{6}-\sqrt{2})/2$	$=.518$	.9320
9	1,2,3,2,1	$1/2$	$=.500$	1.0000
10	2,3,2,3	$5/12$	$=.4167$	.8650
11	Irregular		$=.398$	.8921
12	3,3,3,3	$\sqrt{34}/15$	$=.389$	.9420
13	3,2,3,2,3	$\sqrt{2}/4$	$=.353$	.885
14	Modified 15	$(\sqrt{6}-\sqrt{2})/3$	$=.3451$	.921
15	Modified 16	$4/(8+\sqrt{6}+\sqrt{2})$	$=.3372$	.954
16	1,2,3,4,3,2,1	$1/3$	$=.3333$	1.000
17	Modified 18		$=.3045$	.926
18	4,3,4,3,4	$\sqrt{13}/12$	$=.3005$	.961
19	Modified 20		$=.290$	.960
20	4,4,4,4,4, shift	$3/8-\sqrt{2}/16$	$=.2866$	.992
21	Modified 22		$=.2704$	.952
22	Modified 25	$2-\sqrt{3}$	$=.2680$	.981
23	Based on 8 circles	$(\sqrt{6}-\sqrt{2})/4$	$=.2588$	.972
24	Modified 25	$1/(2+\sqrt{3/2}+1/\sqrt{2})$	$=.2543$	.987
25	1,2,3,4,5,4,3,2,1	$1/4$	$=.2500$	1.000
26	Modified 27		$=.2373$	.957
27	5,4,5,4,5,4	$\sqrt{89}/40$	$=.2358$	.983

Figures For Table 1

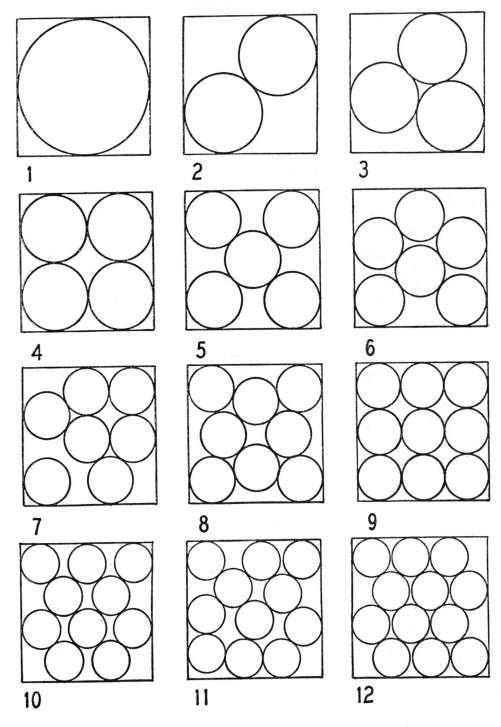

336　　　　　　　　　SYNTHETIC GEOMETRY

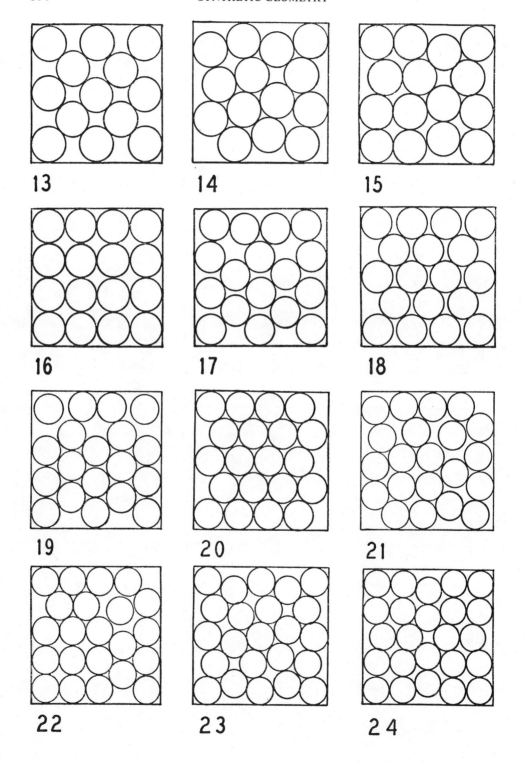

13　　　　　　14　　　　　　15

16　　　　　　17　　　　　　18

19　　　　　　20　　　　　　21

22　　　　　　23　　　　　　24

PACKING EQUAL CIRCLES IN A SQUARE

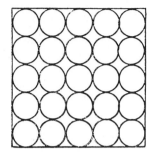

25

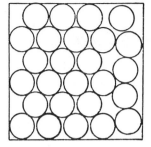

26

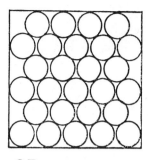

27

TABLE 2

n	Arrangement	m		$4d(n)/\pi$
14	Modified 16	$1/(1+\sqrt{3/2}+1/\sqrt{2})$	= .3413	.908
15	3,3,3,3,3	$\sqrt{41}/20$	= .3202	.882
17	21 circles minus 4	$1/\sqrt{2}-1/\sqrt{6}$	= .2988	.899
20	Irregular	$2-\sqrt{3}$	= .2680	.900
21	4,3,4,3,4,3	$\sqrt{61}/30$	= .2603	.896
23	Modified 25	$2/(4+\sqrt{6}+\sqrt{2})$	= .2542	.948
24	4,4,4,4,4,4	$\sqrt{74}/35$	= .2458	.933
24	Remove corner from 25	$4/(12+\sqrt{6}+\sqrt{2})$	= .2522	.974
30	5,5,5,5,5,5, shift	$(20-\sqrt{10})/75$	= .2245	1.008
39	6,5,6,5,6,5,6	$\sqrt{34}/30$	= .1944	1.033
42	6,6,6,6,6,6,6 shift	$(15-\sqrt{3})/72$	= .1887	1.017
52	7,6,7,6,7,6,7,6	$\sqrt{193}/84$	= .1654	1.047
80	8,8,8, $\cdots$	$\sqrt{34}/45$	= .1296	1.053
99	9,9,9, $\cdots$	$\sqrt{389}/170$	= .1160	1.070
120	10,10,10, $\cdots$	$\sqrt{482}/209$	= .1050	1.0844
143	11,11,11, $\cdots$, shift	$(40-\sqrt{5})/396$	= .0954	1.0839
161	12,11,12,11, $\cdots$	$\sqrt{653}/282$	= .0906	1.114
168	12,12,12, $\cdots$, shift	$\sqrt{17}/195$	= .0871	1.0804
188	13,12,13,12, $\cdots$	$\sqrt{193}/168$	= .0827	1.0967
270	15,15,15, $\cdots$	$\sqrt{1073}/476$	= .0688	1.1193
304	16,16,16, $\cdots$	$\sqrt{34}/90$	= .0648	1.1255
340	17,17,17, $\cdots$, shift	$(304-\sqrt{106}/4845$	= .0606	1.1107
∞	Regular dense packing			1.1547

FIGURES FOR TABLE 2

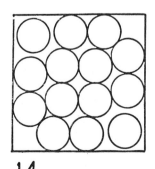

14

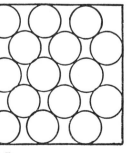

15

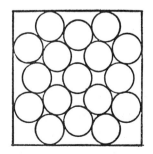

17

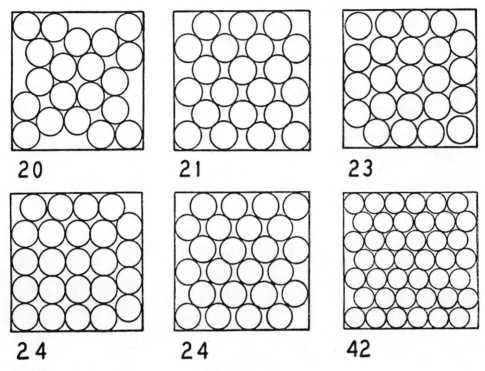

References

1. L. Fejes Toth, Lagerungen in der Ebene, auf der Kugel und im Raum, Springer, Berlin 1953.
2. M. Goldberg, Axially symmetric packing of equal circles on a sphere, Ann. Univ. Sci. Budapest. Sect. Math., 10 (1967) 37–48.
3. J. Schaer and A. Meir, On a geometric extremum problem, Canad. Math. Bull., 8 (1965) 21–27.
4. J. Schaer, The densest packing of 9 circles in a square, Canad. Math. Bull., 8 (1965) 273–277.
5. S. Kravitz, Packing cylinders into cylindrical containers, this MAGAZINE, 40 (1967) 65–71.

A GEOMETRIC APPLICATION OF $f(n) = n/(n+1)$*

FRANCINE ABELES, Newark State College

Introduction. The elementary geometric theorem which states that the diagonals of a parallelogram bisect each other is well known. It can, however, be placed in a much broader framework providing an interesting example of the limit of the real function $f(n)_{n\to\infty} = n/(n+1)$. The purpose of this paper is to develop the essentials utilizing the notion of points of division.

Preliminaries. In the figure below, the ratio in which C divides $\overline{DC}$ is undefined (∞), and F divides diagonal $\overline{DB}$ in the ratio $1/1$. This is the familiar theorem on the diagonals of a parallelogram. It can easily be proved that if E' is the midpoint of $\overline{DC}$, i.e., E' divides $\overline{DC}$ in the ratio $1/1$), then F' divides diagonal $\overline{DB}$ in the ratio $1/2$ (i.e., F' is a point of trisection). Similarly, if E'' divides $\overline{DC}$ in the ratio $-3/1$, then F'' divides $\overline{DB}$ in the ratio $3/2$. And if E''' divides $\overline{DC}$ in the ratio $-1/5$, then F''' divides $\overline{DB}$ (externally) in the ratio $-1/4$.

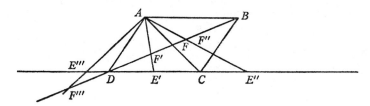

THEOREM. *In parallelogram ABCD, if a line through A intersects the base DC or its prolongation in a point E such that $(\overline{DE}/\overline{EC}) = n$, then the distinct point F in which AE intersects diagonal DB effects the division, $(\overline{DF}/\overline{FB}) = n/(n+1)$.*

Proof. Let E be any point on the base DC or its extension, $E \neq D$, such that AE meets diagonal DB or its extension in point F. (If $E = D = F$, there is nothing to prove.)

(1) $$\vec{DF} = x\vec{DB}, \quad x \neq 0, 1; \quad \vec{DB} = \vec{DC} + \vec{CB},$$

(2) $$\vec{DE} = y\vec{DC}, \quad y \neq 0, -1.$$

So,

(3) $$\vec{DF} = x/y\vec{DE} + x\vec{CB}.$$

$$\vec{DF} = \vec{DA} + \vec{AF}; \quad \vec{AF} = z\vec{AE}, \; z \neq 0,1; \quad \vec{AE} = \vec{DE} - \vec{DA}, \; \vec{DA} = \vec{CB}.$$

So,

(4) $$\vec{DF} = z\vec{DE} + (1-z)\vec{CB}.$$

---

* From MATHEMATICS MAGAZINE, vol. 41 (1968), pp. 259–260.

Equating coefficients in (3) and (4), we obtain

(5) $$x/y = z \quad \text{and} \quad x = 1 - z.$$

From (1),

(6) $$\overrightarrow{DF} = x(\overrightarrow{DF} + \overrightarrow{FB}) \quad \text{and so} \quad \frac{\overline{DF}}{\overline{FB}} = \frac{x}{1-x}.$$

From (2),

(7) $$\overrightarrow{DE} = y(\overrightarrow{DE} + \overrightarrow{EC}) \quad \text{and so} \quad \frac{\overline{DE}}{\overline{EC}} = \frac{y}{1-y}.$$

Using (5) to solve (7), the ratio of the base, we have $(y/1-y) = x/(1-2x)$. Letting $n = x/(1-2x)$, it is easily shown that (6), the ratio of the diagonal, is $n/(n+1)$.

Some Remarks. If E is strictly between D and C, we have the following specific cases: if the base is bisected, the diagonal is trisected; if the base is trisected, the diagonal is quadrisected, etc. Note that when E is a point of internal division, then $0 < n < \infty$. If E is coincident with D or C, then $n = 0$ or is undefined (∞), respectively. The latter situation, when the diagonal is bisected, illustrates $\lim_{n \to \infty} (n/n+1) = 1$. If D is strictly between E and C, E and F points of external division, then $-1 < n < 0$. As E moves further to the left of D, n approaches -1. $f(n)$ is undefined at this point, corresponding to the statement that it is impossible to divide a segment in the ratio -1. When $n = -1/2$, $f(n) = -1$, i.e., $AE \| DB$.

If C is strictly between D and E, E a point of external division, then $-\infty < n < -1$. As E moves to the right of C, n approaches -1. Geometrically, $n = -1$ when $AE \| DC$.

The entire argument obviously applies to the consideration of diagonal CA with suitable adjustments. For example, the ratio of the base would be $(\overline{CE}/\overline{ED})$ and $(\overline{CF}/\overline{FA})$ for the diagonal. This indicates that the ratios are independent of any orientation assigned to the base or diagonal.

BIBLIOGRAPHIC ENTRIES: OTHER CONFIGURATIONS

Except for the entries labeled MATHEMATICS MAGAZINE, the references below are to the AMERICAN MATHEMATICAL MONTHLY.

1. R. A. Johnson, On the approximate division of the circle, vol. 34, p. 429.

    Improving accuracy in division of circle into equal parts.

2. E. Kasner, A projective theorem on the plane pentagon, vol. 35, p. 352.
3. L. Hoffman, Synthetic proof of Professor Kasner's pentagon theorem, vol. 35, p. 356.

    Simplification of Kasner, vol. 35, p. 352.

4. H. Grossman, On the perpendiculars from a point on a circle to the sides of a regular circumscribed n-gon, vol. 39, p. 226.

    Application of binomial coefficients to geometry.

5. C. W. Williams, An elementary proof of the problem of Poncelet pentagons, vol. 45, p. 677.
6. H. Eves, Feuerbach's Theorem by "Mean Position", vol. 52, p. 35.

    Application of the "theory of mean position" employed in proof of Feuerbach's theorem.

7. H. F. Sandham, A simple proof of Feuerbach's theorem, vol. 52, p. 571.
8. H. F. Sandham, A generalization of Feuerbach's theorem, vol. 56, p. 620.

    A generalization that the pedal circle of two isogonal conjugates which are collinear with the circumcenter touches the nine-point circle.

9. S. S. Cairns, Peculiarities of polyhedra, vol. 58, p. 684.

    Facts about polyhedra including Euler's theorem,

10. O. L. Lacey, A note on Bertrand's problem, vol. 65, p. 279.

    Explanation of a paradox arising from a seemingly unambiguous probability problem involving the length of a random chord of a circle.

11. I. Niven and H. S. Zuckerman, Lattice points and polygonal area, vol. 74, p. 1195.
12. H. S. M. Coxeter, The problem of Apollonius, vol. 75, p. 5.
13. R. A. Jacobson and K. L. Yocom, Shortest paths within polygons, MATHEMATICS MAGAZINE, vol. 39, p. 290.
14. T. M. Apostol, Ptolemy's inequality and the chordal metric, MATHEMATICS MAGAZINE, vol. 40, p. 233.
15. M. Goldberg, On the original Malfatti problem, MATHEMATICS MAGAZINE, vol. 40, p. 241.

    A discussion of the problem of packing three circles of maximum total area in a triangle.

16. R. F. Demar, The problem of the shortest network joining n points, MATHEMATICS MAGAZINE, vol. 41, p. 225.
17. M. Goldberg, The converse Malfatti problem, MATHEMATICS MAGAZINE, VOL. 41, p. 262.

    The problem of enclosing three nonoverlapping circles in a triangle of least area.

18. W. I. Jacobson, The butterfly problem—extensions, generalizations, MATHEMATICS MAGAZINE, vol. 42, p. 17.
19. G. D. Chakerian, M. S. Klamkin and G. T. Sallee, On the butterfly property, MATHEMATICS MAGAZINE, vol. 42, p. 21.

    Proof that the butterfly property characterizes the ellipse.

20. M. Goldberg, Packing of 14, 16, 17, and 20 circles in a circle, MATHEMATICS MAGAZINE, vol. 44, p. 134.

21. J. Shaer, On the packing of ten equal circles in a square, MATHEMATICS MAGAZINE, vol. 44, p. 139.

22. L. Bankoff and J. Garfunkel, The heptagonal triangle, MATHEMATICS MAGAZINE, vol. 46, p. 7.

Derives numerous properties of the regular heptagon and triangles associated with it.

23. F. Pavlick, If n lines in the Euclidean plane meet in 2 points then they meet in at least $n-1$ points, MATHEMATICS MAGAZINE, vol. 46, p. 221.

24. S. R. Conrad, Another simple solution of the Butterfly Problem, MATHEMATICS MAGAZINE, vol. 46, p. 278.

Provides an excellent set of references on the Butterfly Problem.

(c)

MISCELLANEOUS

HOW TO TRISECT AN ANGLE WITH A CARPENTER'S SQUARE*†

HENRY T. SCUDDER, Marinette, Wisconsin

Let ABC be the given angle. By means of the larger arm of a carpenter's square, which is two inches wide, draw the line DE parallel to and two inches distant from the line CB. Adjust the square (as indicated in the diagram) so that the inner edge of its larger arm is on the point B, its outer corner is on the line DE (at F), and the mark for four inches on the shorter arm is on the line AB at G. Make dots at F and G and half way between them (at the mark for 2 inches) at H. Remove the square. Draw lines from the dots at F and H to the point B. These lines, FB and HB, trisect the given angle ABC: because if we draw the line FG and the line FJ perpendicular to CB, we shall have three congruent triangles with their equal angles at B. (See Figure 1.)

---

* From AMERICAN MATHEMATICAL MONTHLY, vol. 35 (1928), pp. 250–251.

† NOTE BY THE EDITOR: This construction for the trisection of an angle is of the familiar type involving three-point contact of an instrument with a drawing. For other constructions of that type, the reader is referred to August Adler's *Theorie der geometrischen Konstruktionen*; to F. Enriques' *Fragen der Elementargeometrie*, vol. 2; and to K. Th. Vahlen's *Konstruktionen und Approximationen*.

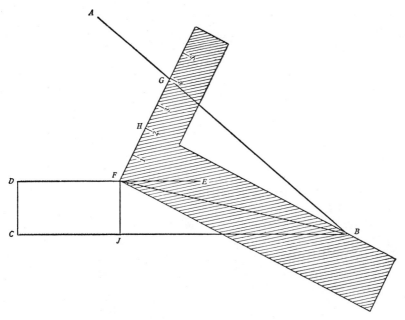

Fig. 1.

HILBERT'S AXIOMS OF PLANE ORDER*

C. R. WYLIE, JR., Ohio State University

1. Introduction. Beyond the bare facts of the courses they will be called upon to teach, there are probably few things which can contribute more to the training of teachers of secondary school geometry than participation in a critical examination of the definitions and axioms of Euclid, and a comparison of them with a carefully developed modern equivalent, for instance the axiomatic system of David Hilbert.[†]

Of the various sets of axioms included in Hilbert's system, the axiom of parallels is in some ways the most interesting, opening up as it does the spectacular fields of non-euclidean geometry through its denial. However in another sense a careful scrutiny of the axioms of order affords a more profitable investment of time for those who have no lasting interest in any geometry but euclidean.

In the first place, the very existence of these axioms, devoted to such an obvious and familiar notion as *betweenness* serves to emphasize the fundamental modern attitude that geometry is an abstract science concerned with undefined elements whose properties are to be inferred from a set of consistent but otherwise essentially arbitrary assumptions, and not from any pseudo-isomorphism

---

* From AMERICAN MATHEMATICAL MONTHLY, vol. 51 (1944) pp. 371–376.
† The Foundations of Geometry, David Hilbert; La Salle, Ill., 1938.

between the geometry elements and objects in the universe of experience. Secondly, it is in connection with the axioms of order that the logical structure of Euclid is weakest. In fact most of the fallacies of elementary geometry, such as the well-known "proofs" that all angles are right angles, and that all triangles are isosceles, spring immediately from the lack of clear-cut order relations.

The present note is essentially an expository account of the independence of these axioms, which may be of interest to those engaged in the training of secondary school teachers, and to those interested in the foundations of geometry, for some of the results obtained here are new, so far as the author is aware.

2. Hilbert's axioms. We shall consider exclusively the order axioms of Hilbert as they appear in the Carus Monograph, The Foundations of Geometry. (*loc. cit.*) They are

1. If A, B, C, are points of a straight line, and B lies between A and C, then B lies between C and A.
2. If A and C are two distinct points of a straight line, then there exists at least one point B lying between A and C, and at least one point D, so situated that C lies between A and D.
3. Of any three points situated on a straight line, there is always one and only one which is between the other two.
4. Any four points A, B, C, D of a straight line can always be so arranged [*i.e.*, named] that B shall lie between A and C and also between A and D, and furthermore that C shall lie between A and D, and also between B and D.

If we now define the segment AB to be the set of all points which are between A and B, we can add to the above axioms which define the notion of *betweenness* for points on a single line, the plane order axiom of Pasch.

5. Let A, B, C be three points not lying in the same straight line and let a be a straight line lying in the plane of A, B, C and not passing through any of the points A, B, C. Then if the straight line a passes through a point of the segment AB it will also pass through either a point of the segment BC or a point of the segment AC.

3. Independence of the linear axioms. Considering the linear axioms 1–4 by themselves it is easy to establish their independence by concrete representations. For instance the following example shows that axiom 1 is independent of axioms 2–4.

Let the system consist of all the points on a line, and define "between A and B" to mean *actually between* if B is to the right of A, and *between but not the midpoint of* if B is to the left of A.

Similarly axioms 2, 3, 4 are shown to be independent by the following examples respectively. The system of points with coordinates 0, ± 1, ± 2, $\cdots$ on a line with "between" defined to mean *actually between*. The system of all points on a line with "between" defined to mean *not between* [!]. The system of all points on a line with "between A and B" defined to mean *to the right of both A and B*.

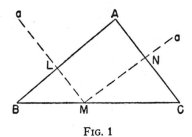

Fig. 1

4. Precise meaning of axiom 5. When axiom 5 is added to the four linear axioms certain dependencies are introduced. Before proceeding to a discussion of these it is desirable, however, to clarify a possible ambiguity in the meaning of this axiom. As stated, it is not entirely clear whether its alternatives are exclusive or not. That they are is an easy matter to demonstrate, and we now show that by appealing only to axiom 3 and the weak form of axiom 5, we may conclude that the line a which by hypothesis passes through a point of the segment AB can pass through a point in *only one* of the remaining segments BC or AC. In fact let the line a in Fig. 1, not passing through any of the points A, B, C, pass through points L, M, N of the segments AB, BC, and AC respectively. By axiom 3 one of the points L, M, N, say M, lies between the other two.* Now consider the three non-collinear points L, N, A. The straight line BC passes through the point M of the segment LN. Therefore it must pass through a point in *at least* one of the segments NA or LA. But this is impossible, for by axiom 3 since N lies between A and C, C cannot simultaneously be between N and A, and for the same reason B cannot lie between L and A. This contradiction of the weak form of axiom 5 establishes the result.

5. Dependence of axiom 1. We now prove that axiom 1 is a consequence of axioms 2, 3, and 5. Let (AXB) be true on a line l, Fig. 2. Choose C not on l and Y on the line AC so that (ACY), (axiom 2). Now consider the line XY in relation to the three non-collinear points A, B, C. It passes through the point X, of the segment AB. It must therefore pass through a point of the segment BC

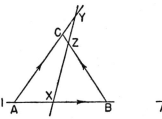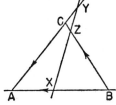

Fig. 2

or a point of the segment AC, by axiom 5. By axiom 3 it cannot pass through a point of AC, therefore it must pass through a point, say Z, of the segment BC.

---

* Henceforth we shall indicate such a "betweenness" relationship by writing simply (LMN).

Now consider this same line in relation to the three points B, C, A. It surely passes through a point, namely Z, of the segment BC. It cannot pass through a point of CA, hence it must pass through a point of the segment BA. This is possible if and only if X is between B and A.

6. Dependence of axiom 4. Axiom 4 is also a consequence of axioms 2, 3, and 5, as was first pointed out by E. H. Moore* in 1902. His proof though entirely straightforward was somewhat lengthy, and a shorter proof may be of interest here. This seems doubly appropriate since in subsequent editions of Hilbert's *Foundations* this result is still not noted.†

We base our proof upon the following series of observations.

LEMMA: *Of four points, A, B, C, D, situated on a line l, no one can be "between" only once.*

We assume the contrary and suppose that of the four points A, B, C, D on l, Fig. 3, the point B lies between A and C and between no other pair. Choose E not on l, and on CE choose F so that (CEF), axiom 2. Now consider the line BF in relation to the three points A, C, E. It passes through the point B of the segment AC. Since by axiom 3 it cannot pass through a point of the segment CE, it must pass through a point, say G, of AE. Now consider this same line in relation to the three points A, E, D. It passes through the point G of the segment AE. It cannot pass through a point of the segment AD, since by hypothesis B is only between A and C on l. Therefore it must pass through a point of ED, say H. Now finally consider this same line in relation to the points E, D, C. It passes through a point of ED as we have just seen. However it cannot pass through a point of DC by hypothesis, and by axiom 3 it cannot pass through a point of EC. This contradicts axiom 5 and establishes our result.

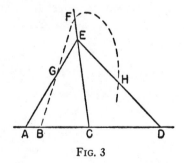

FIG. 3

As a first corollary of this lemma we conclude that of our points on a line, no one can be "between" three times. For there are only four combinations of three points each, existing in the set of points A, B, C, D, and by axiom 3 only one

---
* E. H. Moore, The projective axioms of geometry, Trans. Amer. Math. Soc., 1902.

† It is, of course, to be found in various places outside of Moore's paper. For instance J. W. Young makes explicit mention of this dependence in his book, Fundamental Concepts of Algebra and Geometry.

point in each triad can be "between." Hence there are only four possible "betweeness" relations insofar as the central term is concerned. If then, one point is the middle point in three of these, some other point must be "between" only once, which contradicts the lemma.

As a second corollary we observe that of four points on a line, two are necessarily "between" exactly twice, and two are "between" not at all.

The proof of axiom 4 as a consequence of axioms 2, 3, 5, and 1 (now known to be a theorem independent of axiom 4) follows at once from the second corollary. Of the four points A, B, C, D on a line l let the points which are never "between" be called A and D, the others B and C. Then of the respective sets of relations

$$(ABC) \quad (ABD) \quad (CBD)$$

and

$$(ACB) \quad (ACD) \quad (BCD)$$

two and only two of each set must be true. If first, (ABC) is true, then (ACB) is necessarily false by axiom 3. Hence (ACD) and (BCD) are true, and since the last is true, (CBD) is false, leaving (ABD) true. The points are then named according to axiom 4. If on the contrary (ABC) is false, then inevitably (ABD), (CBD), (ACB), and (ACD) are true, and if the names B and C are interchanged, the ordering of the axiom is again accomplished.

7. Independence of axioms 2, 3, and 5. The three axioms that remain may now be shown to be independent by the following concrete representations.

For the system consisting of three non-collinear points and the three lines through them, with "between" defined to mean *actually between*, axiom 2 is false and axioms 3 and 5 are (vacuously) true.

In the ordinary euclidean plane with "between" defined to mean *collinear with* axioms 2 and 5 are true but axiom 3 is false.

In the ordinary euclidean plane with "between" defined to mean *above both* on all vertical lines, and *to the right of both* on all other lines, axioms 2 and 3 are valid but axiom 5 is false.

8. A new form of axiom 5. We have now completed our exposition of the dependence and independence of the axioms of betweenness in the specific form in which Hilbert presented them. Numerous other sets of axioms have been devised to define this same notion. In general these differ substantially from Hilbert's system, and we do not propose to consider them here. There is however one other system almost identical with that of Hilbert which we shall touch on in conclusion.

Suppose we keep the axioms as quoted in paragraph I intact except that in axiom 5 we change the last phrase to read " . . . it will also pass through either a point of the segment *BC or a point of the segment CA.*" The nature of this new set of axioms must be regarded as a new problem, entirely independent of those observations which led us to our conclusions concerning the dependence of Hilbert's own set. It may well be that axioms dependent in the first formulation

348   SYNTHETIC GEOMETRY

will now be independent or vice-versa. And the proofs of such relations as are true in both systems may differ widely in their character.

9. Dependence of the new axioms. We shall not explore this new system in detail. The final results are curiously enough the same here as before: axioms 1 and 4 are consequences of axioms 2, 3, and 5. The proof of this follows roughly the outline used above. For preliminary clarification we can prove a strong form of axiom 5 precisely as we did in treating Hilbert's version of the axioms.

When we attempt to establish the dependence of axiom 1, however, we must devise an entirely new form of proof. In fact it must be based upon the lemma of paragraph 6 and its corollaries, instead of being independent of them. We omit the new proof of this lemma which follows *mutatis mutandis* from the earlier proof, and give only the new proof of the dependence of axiom 1.

Suppose that on a line l, (AXC) is true while (CXA) is false. Choose B not on l, Fig. 4, and choose Y so that (BYA), axiom 2. Now consider the line XY in relation to the points B, A, C. It passes through the point Y of the segment BA, and the point X of the segment AC. Therefore by the strong form of axiom 5* it cannot pass through a point of CB. Now choose D so that (AYD) and consider the same line in relation to the points A, D, C. It passes through a point of AD, namely Y, does not pass through a point of CA, since by hypothesis (CXA) is false, hence must pass through a point, say Z, of DC. Finally consider this line in relation to the points D, C, B. It passes through the point Z of DC, does not pass through a point of CB, as pointed out above, and therefore must pass through a point of BD. This is possible if and only if Y is between B and D. But then in the range A, Y, B, D, the point Y is between three times which is impossible by the first corollary. This contradiction establishes the result.

The proof that axiom 4 is dependent is literally identical in each system and we pass over it without mention.

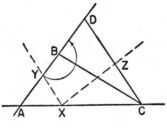

Fig. 4

10. Conclusion. We have considered as a separate problem the modified system that we have just discussed, not only as a simple experiment designed to make clear how a very slight change in an axiom may completely alter the character of a proof, but also to illumine a careless treatment of these axioms

---

* Note that in all our discussion of Hilbert's axioms we did not need to use the strong form of axiom 5, and only called attention to it in order to secure a clearer understanding of the system.

that is sometimes encountered. Axiom 5 which introduced the dependencies we have been discussing is sometimes loosely stated "A straight line which meets one side of a triangle and lies in the plane of the triangle meets at least one other side of the triangle."

Until a triangle is defined, this statement is meaningless. If the word triangle is taken to mean the configuration shown in Fig. 5a, associated as it were with vector addition, the axiom is simply an ill-stated version of Hilbert's axiom 5.

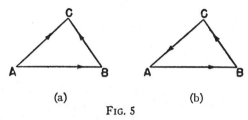

Fig. 5

If, equally naturally, the word triangle is taken to imply the cyclic configuration shown in Fig. 5b we have the second version of axiom 5. If finally the word triangle is defined in the sense of elementary high school geometry, we pemit either of the above definitions, and commit the paralogism of incorporating axiom 1 in the definition of a term employed in another axiom. Small wonder that we can then prove it dependent!

AFTER THE DELUGE*

D. A. MORAN, Michigan State University

The purpose of this note is to provide a somewhat simpler proof than that given by Professor Marston Morse of his elementary theorem about pits, peaks, and passes on the sphere [1]. It should be recalled that Professor Morse views a positive, real-valued, bounded, differentiable function on the sphere as an altitude, measured from the center of some hypothetical spherical planet. Critical points of the function then correspond to pits (=minima), peaks (=maxima), and passes (=saddle points of index -1) on the planet. It is assumed that no critical points more complicated than these three types ever occur, and that no two of these singularities occur at precisely the same altitude. Our viewpoint is essentially the same as this, but we start with the following slightly different

ADDITIONAL HYPOTHESIS: *No pass is as high as a peak, or as low as a pit.*

This hypothesis is easily fulfilled, if we agree to drill deep holes at the bottom of each pit, and raise tall flagpoles atop each peak. Let N_0, N_1, and N_2 be the number of pits, passes, and peaks.

Now let rain begin to fall on the planet which represents the sphere. Immediately N_0 lakes are created. As the water level rises to the altitude of a pass,

---

* From AMERICAN MATHEMATICAL MONTHLY, vol. 77 (1970), p. 1096.

a lake can merge with itself, creating an island, or else a lake can merge with another lake, resulting in a net decrease by 1 in the number of lakes. When the water level has risen to inundate every pass, but is not yet as high as any peak, the number of islands will be

$$1 + \text{number of island-increasing passes,}$$

and the number of lakes will be

$$N_0 - \text{number of lake-decreasing passes.}$$

The number of lakes less the number of islands is therefore

$$N_0 - N_1 - 1.$$

On the other hand, at this point in time there is clearly one lake and N_2 islands, so

$$N_0 - N_1 - 1 = 1 - N_2,$$

proving Morse's generalization of the theorem of Euler.

<small>Written while the author was partially supported by NSF grant GP 8962.</small>

References

1. Marston Morse, *Pits, peaks, and passes* (motion picture film), Modern Learning Aids #3462, New York.
2. George Polya, Induction and Analogy in Mathematics, Princeton Univ. Press, 1954, pp. 163-165.

GEOMETRIC INTERPRETATIONS OF SOME CLASSICAL INEQUALITIES*

JOSEPH L. ERCOLANO, Baruch College, CUNY

For a, b any positive real numbers, consider the following classical means relating them:

$$\text{A.M.} = \tfrac{1}{2}(a+b) \qquad \text{(arithmetic mean)}$$

$$\text{G.M.} = \sqrt{ab} \qquad \text{(geometric mean)}$$

$$\text{H.M.} = 2ab/(a+b) \qquad \text{(harmonic mean)}$$

$$\text{R.M.} = \sqrt{\tfrac{1}{2}(a^2+b^2)} \qquad \text{(root-mean square)}$$

Let $\max(a,b)$ and $\min(a,b)$ denote respectively the maximum and minimum values of a and b. Then it is well known [1, 2] that the following set of inequalities relating

---

<small>* From MATHEMATICS MAGAZINE, vol. 45 (1972), p. 226.</small>

these six quantities is true:

$$\max(a,b) \geqq \text{R.M.} \geqq \text{A.M.} \geqq \text{G.M.} \geqq \text{H.M.} \geqq \min(a,b).$$

The purpose of this note is to present a geometric interpretation of these inequalities. We will do this, with straightedge-and-compass constructions, in two different settings.

Suppose $a > b > 0$. In Figure 1, the lengths of AC and BC are a and b, respectively, while in Figure 2, these are the lengths of AB and BC. Then $2ab/(a+b)$, $\sqrt{ab}$, $\frac{1}{2}(a+b)$, and $\sqrt{\frac{1}{2}(a^2+b^2)}$ are the lengths of EC, DC, OC, and FC respectively in Figure 1, and of ED, BD, OC, and BF respectively in Figure 2. In Figure 2, $a = b$ is permissible.

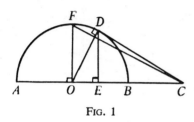

Fig. 1

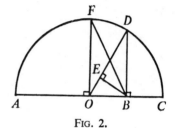

Fig. 2.

References

1. E. Beckenbach and R. Bellman, An Introduction to Inequalities, Random House, New York, 1961.

2. D. S. Mitrinovic, Elementary Inequalities, Noordhoff, Groningen, The Netherlands, 1964.

TWO MATHEMATICAL PAPERS WITHOUT WORDS*

RUFUS ISAACS, The Johns Hopkins University

ON TRISECTING AN ANGLE

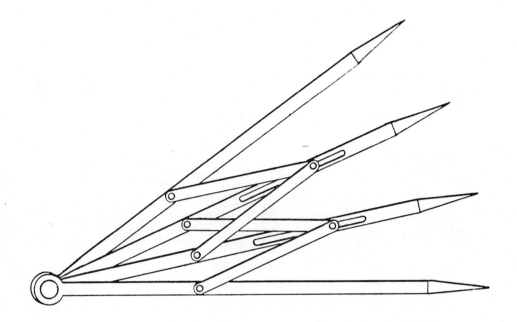

A PROOF OF THE PYTHAGOREAN THEOREM

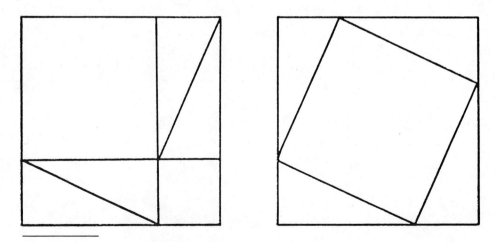

* From MATHEMATICS MAGAZINE, vol. 48 (1975), p. 198.

BIBLIOGRAPHIC ENTRIES: MISCELLANEOUS

Except for the entries labeled MATHEMATICS MAGAZINE, the references below are to the AMERICAN MATHEMATICAL MONTHLY.

1. G. Gibbens, Some constructions for the classical problems of geometry, vol. 37, p. 343.

> Outline of one person's efforts to solve some classical unsolved problems.

2. P. B. Johnson, The geometric significance of the standard deviation and coefficient of variation, vol. 62, p. 38.

3. M. C. Ehrmann, Finite geometries on a torus, vol. 79, p. 279.

4. C. S. Ogilvy, A proof that would please N. D. Kazarinoff, MATHEMATICS MAGAZINE, vol. 38, p. 110.

> A geometric proof of the inequality of the arithmetic and geometric means.

5. O. Shisha, Geometrical interpretations of the inequalities between the arithmetic, geometric, and harmonic means, MATHEMATICS MAGAZINE, vol. 39, p. 268.

8

CONIC SECTIONS

(a)

EQUATIONS

A METHOD OF DEFINING THE ELLIPSE, HYPERBOLA AND PARABOLA AS CONIC SECTIONS*

W. W. LANDIS, Dickinson College

ABC is a right circular cone, the angle at the vertex being 2α. DFG is a plane section making an angle θ with the axis of the cone. Take D as the origin, DH as the axis of x, and a perpendicular through D as the axis of y. We seek to find a relation between x and y, the parameters being $g(=AD)$ and α, and the variable parameter θ. In the circle BGC, $y^2 = (a \pm b)c = ac \pm bc$ (1), where $a = BL$, $b = LH$, and $c = HC$. In the isosceles triangle DLC, $x^2 = d^2 \mp bc$ (2), where $d = DC = DL$. Adding (1) and (2), $x^2 + y^2 = ac + d^2$ (3).

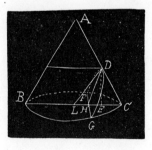

Now

$$c = x\sin\theta + d\sin\alpha,$$
$$d = x\cos\theta\sec\alpha = x\cos\theta/\cos\alpha, \text{ and}$$
$$a = 2g\sin\alpha.$$

---

* From AMERICAN MATHEMATICAL MONTHLY, vol. 5 (1898), pp. 72–73.

Making this substitution we get

$$x^2 + y^2 = x^2 \frac{\cos^2\theta}{\cos^2\alpha} + 2g\sin\alpha\left[x\sin\theta + x\frac{\cos\theta\sin\alpha}{\cos\alpha}\right]$$

or

(4) $$x^2\left[1 - \frac{\cos^2\theta}{\cos^2\alpha}\right] + y^2 - 2gx\sin\alpha[\sin\theta + \cos\theta\tan\alpha] = 0,$$

which we may write

(5) $$x^2\left[\frac{\sin^2\theta - \sin^2\alpha}{1 - \sin^2\alpha}\right] + y^2 - 2gx\sin\alpha[\sin\theta + \cos\theta\tan\alpha] = 0.$$

Now if $\theta = \alpha$, the plane then being parallel to one and only one element, (5) reduces to $y^2 - 4gx\sin^2\alpha = 0$, a parabola of latus rectum $= 4g\sin^2\alpha$.

If $\theta > \alpha$ the section cuts all elements and the coefficients of x^2 and y^2 are both positive, and we have an ellipse whose center, axes, and eccentricity are readily found; and in particular if $\theta = 90°$, the section is parallel to the base, the coefficients of x^2 and y^2 are unity and we have a circle, whose center is $(g\sin\alpha, 0)$. If $\theta < \alpha$, the section cuts both nappes, the coefficients of x^2 and y^2 are of unlike sign and we have a hyperbola.

If $g = 0$, (5) becomes

(6) $$y = \pm x\sqrt{\frac{\sin^2\alpha - \sin^2\theta}{\cos^2\alpha}}.$$

Now if $\theta = \alpha$, (6) becomes $y = 0$, a straight line, the limit of the parabola. If $\theta < \alpha$, (6) represents two real straight lines, the limiting case of the hyperbola. And if $\theta > \alpha$, (6) represents two imaginary lines intersecting in the real point $(0,0)$, which is the limiting form of the ellipse.

The equations (5) and (6) show the dependence of the nature of the conic sections upon the angle which the cutting plane makes with the axis, and the dependence of their shape upon the angle of the cone and the distance from the vertex to the first element cut.

REMARK 1. If the section be a parabola, the foot of the perpendicular from the middle point of the line through D parallel to BC, upon DH, is the focus.

REMARK 2. The eccentricity of any conic section is $\epsilon = [\cos\theta/\cos\alpha]$.

A BIQUADRATIC EQUATION CONNECTED WITH THE REDUCTION OF A QUADRATIC LOCUS*

ARTHUR C. LUNN, University of Chicago

If the equation of a conic section be written in the form

$$Ax^2 + By^2 + 2Cxy + 2Dx + 2Ey + F = 0,$$

then it is known that a rotation of the coordinate axes through an angle a will bring them into parallelism with the axes of symmetry of the curve, provided this angle is determined by

$$\tan 2a = \frac{2C}{A-B}.$$

This rotation corresponds to the substitution

(1) $$\begin{aligned} x &= x'\cos a - y'\sin a, \\ y &= y'\cos a + x'\sin a, \end{aligned}$$

with a so chosen as to eliminate the term in $x'y'$. But the sine and cosine may be expressed in terms of the tangent of the half-angle, thus:

(2) $$t = \tan\frac{a}{2}, \quad \cos a = \frac{1-t^2}{1+t^2}, \quad \sin a = \frac{2t}{1+t^2},$$

and the use of these in (1) gives the substitution expressed rationally in terms of the parameter t. Without reference to its trigonometric source, the substitution in that form is seen to be orthogonal or rotational for all values of t, since the equation of constancy of distances:

$$(x_1' - x_2')^2 + (y_1' - y_2')^2 = (x_1 - x_2)^2 + (y_1 - y_2)^2,$$

is directly verifiable as an identity in t.

The use of this parameter makes it possible to effect the reduction of the conic by purely algebraic processes, independently of the trigonometric formulae. For the term in $x'y'$ will have as coefficient

$$\frac{2}{(1+t^2)^2}\left\{2(B-A)\cdot t(1-t^2) + C\left[(1-t^2)^2 - (2t)^2\right]\right\},$$

which will vanish if t satisfy the biquadratic equation:

(3) $$t^4 + 4mt^3 - 6t^2 - 4mt + 1 = 0,$$

in which is put $m = \dfrac{A-B}{2C}$, the trigonometric value of which is $\cot 2a$. The four roots of this equation must therefore be real for all real values of m, and correspond to the four semi-axes of the conic.

---

* From AMERICAN MATHEMATICAL MONTHLY, vol. 15 (1908), pp. 5–6.

But this equation must be solvable by quadratics. For, by a familiar formula of trigonometry

$$\tan 2a = \frac{2\tan a}{1 - \tan^2 a}$$

which, regarded as a quadratic equation in $\tan a$ gives

$$\tan a = -\cot 2a \pm \sqrt{[1 + \cot^2 2a]},$$

and a repetition of such solution gives $\tan\frac{1}{2}a$ in terms of $\cot a$. This suggests at once the four roots of the biquadratic in t, which after a little reduction prove to be:

$$\begin{aligned}
t_1 &= -m + r + R_1 & &\text{where} \\
t_2 &= -m - r + R_2 & r &= \sqrt{[1 + m^2]}, \\
t_3 &= -m + r - R_1 & R_1 &= \sqrt{[2(r^2 - rm)]}, \\
t_4 &= -m - r - R_2 & R_2 &= \sqrt{[2(r^2 + rm)]}.
\end{aligned}$$

These roots are obviously all real, and are easily shown to be distinct. Direct computation shows that the product $(t - t_1)(t - t_2)(t - t_3)(t - t_4)$ gives the biquadratic polynomial on the left of the equation in t.

The reducibility of the biquadratic equation by quadratics is connected intimately with the existence of rational relations among the roots, which in the present case are the following:

$$t_1 t_3 = -1, \quad t_2 t_4 = -1, \quad \frac{t_2 - t_1}{1 + t_1 t_2} = \frac{t_3 - t_2}{1 + t_3 t_2},$$

and others (not independent) similar to the last. These correspond to the fact that, since the axes of the conic are mutually perpendicular, the various values of $\frac{1}{2}a$ must be spaced at intervals of $45°$.

AN ELEMENTARY ANALYSIS OF THE GENERAL EQUATION OF SECOND DEGREE*†

E. S. ALLEN, Iowa State College

1. Introduction. Almost all elementary books on analytic geometry analyze the general quadratic equation in x and y by means of two changes of axes. It is, so far as I know, only in Smith's *Conic Sections* (p. 207) that a method of finding the foci of a conic section without change of coördinates is described. This method makes use of the polar properties of the foci and directrices. While making a class acquainted with conics recently, I was asked how it was possible to find the eccentricity, directrices, and foci of such a curve directly from its equation. The

---

* From AMERICAN MATHEMATICAL MONTHLY, vol. 31 (1924), pp. 479–481.

† Read at the meeting of the Iowa Section of the Mathematical Association of America, May 2, 1924.

following answer to the question, which uses only "elementary" mathematical ideas, is, I believe, new.

2. General considerations. If a conic has the focus (α, β), the directrix
$$x\cos\theta + y\sin\theta - p = 0$$
and the eccentricity e; its equation is
$$\sqrt{(x-\alpha)^2 + (y-\beta)^2} = e(x\cos\theta + y\sin\theta - p);$$
that is,

(1) $$\begin{aligned}&x^2(1-e^2\cos^2\theta) - 2xy(e^2\sin\theta\cos\theta) + y^2(1-e^2\sin^2\theta)\\&-2x(\alpha - e^2 p\cos\theta) - 2y(\beta - e^2 p\sin\theta) + (\alpha^2 + \beta^2 - e^2 p^2) = 0.\end{aligned}$$

In order that the equation

(2) $$ax^2 + 2hxy + by^2 + 2gx + 2fy + c = 0$$

may have the same coefficients as an equation of the form (1), it is necessary that
$$(a-1)(b-1) = h^2.$$
This results from the first three of the following equations:

(3) $$\begin{aligned}a &= 1 - e^2\cos^2\theta, & g &= -(\alpha - e^2 p\cos\theta),\\h &= -e^2\sin\theta\cos\theta, & f &= -(\beta - e^2 p\sin\theta),\\b &= 1 - e^2\sin^2\theta, & c &= \alpha^2 + \beta^2 - e^2 p^2.\end{aligned}$$

Consequently, if we have any equation of second degree, with real coefficients,

(4) $$a'x^2 + 2h'xy + b'y^2 + 2g'x + 2f'y + c' = 0,$$

we must multiply the first member by such a number k that

(5) $$(ka' - 1)(kb' - 1) = (kh')^2.$$

3. Parabola. In the case of the parabola, $a'b' - h'^2 = 0$, equation (5) is linear, and the values of $a, b, \ldots, c$ are uniquely determined. If we demand that $0 \leqq \theta < \pi$, all the numbers $e, \alpha, \beta, \theta, p$ are found without ambiguity.

4. Central conic. If we have a central conic, then $a'b' - h'^2 \neq 0$, and the roots of (5) are

(6) $$\begin{aligned}k_1 &= \frac{(a'+b') + \sqrt{(a'+b')^2 - 4(a'b' - h'^2)}}{2(a'b' - h'^2)},\\k_2 &= \frac{(a'+b') - \sqrt{(a'+b')^2 - 4(a'b' - h'^2)}}{2(a'b' - h'^2)}.\end{aligned}$$

Thus, k_1 and k_2 are evidently real.

If follows from (3) that

$$e^2 = 2 - (a+b) = 2 - k(a'+b')$$

(7)
$$= \frac{-(a'+b')^2 + 4(a'b' - h'^2) \mp (a'+b')\sqrt{(a'+b')^2 - 4(a'b' - h'^2)}}{2(a'b' - h'^2)}.$$

5. Ellipse. In the case of the ellipse (where $a'b' - h'^2 > 0$) the absolute value of the last term of the numerator is greater than that of the other two. Accordingly, one value of e is real, the other imaginary. To obtain the real foci and directrices of a real ellipse, it is necessary to take the last term of (7) positive.

From the known values of a, h, b, that of θ is then determined. Equations (3) for g, f, c give us a quadratic equation for the determination of p; its roots are the values of p for the two real directrices. The values of α and β are then found immediately.

6. Hyperbola. If $a'b' - h'^2 < 0$, the curve is a hyperbola. Equation (7) shows that e^2 is necessarily positive; for the absolute value of the last term of the numerator is in this case less than that of the other terms. The latter are negative, as is the denominator. That eccentricity which corresponds to the real foci has, then, to be chosen. Having e, we are now able to find p.

The equation which p must satisfy is

$$(e^4 - e^2)p^2 - 2e^2(g\cos\theta + f\sin\theta)p + g^2 + f^2 - c = 0.$$

The quotient of the discriminant of this equation by the positive number e^2 is

$$B = g^2(1-a) - 2fgh + f^2(1-b) - (e^2 - 1)(g^2 + f^2 - c)$$
$$= bg^2 - 2fgh + af^2 + c(1 - a - b)$$
$$= k^3(b'g'^2 - 2f'g'h' + a'f'^2) - k^2c'(a' + b') + kc'.$$

Now, from (5),

$$k(a'+b') - 1 = k^2(a'b' - h'^2);$$

consequently

$$B = -k^3 \begin{vmatrix} a' & h' & g' \\ h' & b' & f' \\ g' & f' & c' \end{vmatrix}.$$

In view of the fact that $k_1 k_2 = \dfrac{1}{a'b' - h'^2} < 0$, the two values of B will have opposite signs, and that one of the two k's must be chosen whose sign is opposite that of the determinant. k having been found, the solution is identical with that for the ellipse.

7. Perpendicularity of axes. Since

$$\tan\theta_1 \tan\theta_2 = \frac{k_1 h'}{k_1 a' - 1} \cdot \frac{k_2 h'}{k_2 a' - 1} = \frac{\dfrac{h'^2}{a'b' - h'^2}}{\dfrac{a'^2}{a'b' - h'^2} - \dfrac{(a'+b')a'}{a'b' - h'^2} + 1} = -1,$$

we find, of course, that the two values of θ differ by $\dfrac{\pi}{2}$.

SIMPLIFICATION OF THE EQUATIONS OF CONICS*

H. B. THORNTON, Sumner Junior College, Kansas City, Kansas

The process given in most textbooks on analytic geometry for simplifying the equation of a conic by means of substitutions from trigonometric formulae is usually long and laborious. By means of the formulae derived below (which as far as I know are not given in any text) the simplification can be accomplished easily and quickly. Let

(1)
$$x = x_0 + \lambda t$$
$$y = y_0 + \mu t$$

be the parametric equations of a line intersecting the conic

$$ax^2 + by^2 + 2hxy + 2gx + 2fy + c = 0.$$

Then the roots of the equation

$$t^2(a\lambda^2 + b\mu^2 + 2h\lambda\mu) + t(2a\lambda x_0 + 2b\mu y_0 + 2h\lambda y_0 + 2h\mu x_0 + 2g\lambda + 2f\mu)$$
$$+ ax_0^2 + by_0^2 + 2hx_0y_0 + 2gx_0 + 2fy_0 + c = 0$$

are, except for the factor $(\lambda^2+\mu^2)^{1/2}$, the distances from the point (x_0, y_0) on the line to the points of intersection of the line with the conic. The condition that the roots of this equation be numerically equal but opposite in sign is

$$\lambda(ax_0 + hy_0 + g) + \mu(hx_0 + by_0 + f) = 0.$$

If x_0, y_0 are taken as the variables this is the equation of a system of diameters. The condition that any line of this system be perpendicular to the system of chords (1) is

$$\frac{a\lambda + h\mu}{\lambda} = \frac{h\lambda + b\mu}{\mu}.$$

Set each of these ratios equal to β. Then the condition that the equations

$$a\lambda + h\mu = \lambda\beta$$
$$h\lambda + b\mu = \mu\beta$$

---

* From AMERICAN MATHEMATICAL MONTHLY, vol. 41 (1934), pp. 36–37.

have a common solution other than 0, 0 is

$$\begin{vmatrix} a - \beta & h \\ h & b - \beta \end{vmatrix} = 0,$$

or

$$\beta^2 - (a + b)\beta - h^2 + ab = 0,$$

whence

$$\beta_1, \beta_2 = \frac{a + b \pm \sqrt{(a - b)^2 + 4h^2}}{2}.$$

Then, upon making the suitable transformation, the simplified equation of the conic will take the form

(2) $$\beta_1 x^2 + \beta_2 y^2 - \frac{D}{d} = 0,$$

where

$$D = \begin{vmatrix} a & h & g \\ h & b & f \\ g & f & c \end{vmatrix}$$

and $d = h^2 - ab$. If $d = 0$ (a parabola), the simplified form becomes

(3) $$\beta_1 x^2 + 2\sqrt{-D/\beta_1}\, y = 0.$$

CONIC SECTIONS FROM WHOSE EQUATIONS THE XY-TERM MAY BE ELIMINATED BY A ROTATION OF AXES INVOLVING NO SURD NUMBERS*

D. C. DUNCAN, University of California

The removal of the xy-term from the equation of the real conic, $Ax^2 + Bxy + Cy^2 + Dx + Ex + F = 0$, by rotating the axes through an angle θ determined by the usual relation, $\tan 2\theta = B/(A - C)$, in general involves "two-story" radicals in the formulas of transformation. In introducing the process to the student, specially devised equations are invariably used which do not involve additional algebraic difficulties in irrationalities. The standard procedure of finding 2θ by the relation $\tan 2\theta = B/(A - C)$, then $\cos 2\theta$, from which $\sin \theta$ and $\cos \theta$ are available by $\sqrt{(1 - \cos 2\theta)/2}$ and $\sqrt{(1 + \cos 2\theta)/2}$, when applied, for example, to $12x^2 + 24xy + 5y^2 + Dx + Ey + F = 0$, leads most agreeably to $\sin \theta = 3/5$ and $\cos \theta = 4/5$. In fact, the illustrative examples in texts usually involve the ratio

---

* From AMERICAN MATHEMATICAL MONTHLY, vol. 41 (1934), pp. 441–442.

±24/7 for $B/(A-C)$ for the very reason that no surds are involved in the reductions. In this note I wish to call attention to the comparatively rare occurrence of absolute values of the ratio $B/(A-C)$ which involve only rational operations in the subsequent reduction, and to indicate a means of finding all values of the ratio $B/(A-C)$ from which these desirable equations may be formed "for purposes of illustration."

Suppose that $\sin \theta$ and $\cos \theta$ are to be rational; then they must be of the form $(m^2-n^2)/(m^2+n^2)$ and $2mn/(m^2+n^2)$, or vice versa. We then have

$$2mn/(m^2 + n^2) = \sqrt{(1 + \cos 2\theta)/2}, \quad (m^2 - n^2)/(m^2 + n^2) = \sqrt{(1 - \cos 2\theta)/2},$$

from which one obtains

$$\cos 2\theta = [8m^2n^2 - (m^2 + n^2)^2]/(m^2 + n^2)^2,$$

and hence

$$\tan 2\theta = 4mn(m^2 - n^2)/[8m^2n^2 - (m^2 + n^2)^2] = \pm B/(A - C),$$

a necessary condition that no surds appear in the calculations. This condition is readily noted to be also sufficient in that these steps are immediately reversible and lead back to $\sin \theta$ and $\cos \theta$.

Accordingly one may build special equations at pleasure whose reduction to standard forms by rotation and translation involve only rational numbers[1] by choosing B to be of the form $\pm 4mn(m^2-n^2)$ and $A-C$ of the form $\pm [8m^2n^2 - (m^2+n^2)^2]$. There are only 7 absolute values of these ratios in which the members are less than 2,000, namely, 24/7, 120/119, 240/161, 336/527, 840/41, 840/1081, and 720/1519.

SIMPLIFICATION OF EQUATIONS OF CONICS*

L. S. JOHNSTON, University of Detroit

Mr. H. B. Thornton has pointed out in a recent note† that the process given in most texts on analytic geometry for simplifying the equation of a conic by means of substitutions from trigonometric formulas is usually long and laborious, and he exhibits a simpler method of accomplishing such simplifications. I offer a method which uses those same trigonometric formulas, but uses them in such manner as materially to shorten the actual work of simplification, accomplishing, I believe, the desired results with much less difficulty than in even Mr. Thornton's method. This method is not to my knowledge mentioned in any text, but is mentioned rather sketchily in my note "Additional Marginal Notes," this MONTHLY, October 1932.

---

[1] All the different ratios will be obtained without duplication by letting m and n range over all pairs of positive relatively prime odd integers. (EDITOR.)

* From AMERICAN MATHEMATICAL MONTHLY, vol. 44 (1937), pp. 30–31.

† Simplification of the equations of conics, this MONTHLY, January 1934, p. 36.

SIMPLIFICATION OF EQUATIONS OF CONICS

Consider the general equation of the conic

(1) $$Ax^2 + Bxy + Cy^2 + Dx + Ey + F = 0.$$

Under the usual transformation rotating axes this becomes

(2) $$A'x'^2 + C'y'^2 + D'x' + E'y' + F' = 0$$

the coefficients being well known functions of the original coefficients and of the angle θ defined by the equation

(3) $$\tan 2\theta = \frac{B}{A - C}, \text{ with } 2\theta \text{ less than } \pi.$$

Most texts in analytics either prove or set as exercises the relations

(4) $$A' + C' = A + C$$
(5) $$A' - C' = \pm \sqrt{B^2 + (A - C)^2},$$

but make little or no mention of their consequences. Now it is evident that if the proper sign to use before the radical in (5) can be definitely determined, then from (4) and (5) we can at once determine A' and C' without ambiguity, and without the explicit use of θ at all.

It is easily shown that if $\tan 2\alpha = m/n$, then

$$\tan \alpha = \frac{m}{n \pm \sqrt{m^2 + n^2}}.$$

(Incidentally, this formula, or the equivalent identity

$$\tan \alpha = \frac{\tan 2\alpha}{1 + \sec 2\alpha} = \frac{\tan 2\alpha}{1 \pm \sqrt{1 + \tan^2 2\alpha}},$$

is not to my knowledge mentioned in any text in trigonometry, though it is a much simpler method than the one usually used for finding $\tan \alpha$ when $\tan 2\alpha$ is given). Using this formula, we have

$$\tan \theta = \frac{B}{A - C \pm \sqrt{B^2 + (A - C)^2}}.$$

Now since this fraction must always be positive, we must choose the sign before the radical to be the same as that of B. Furthermore, the calculations of $\sin \theta$ and $\cos \theta$ which enter into the constants D', E', and F' of (2), are much simpler by this method than by the usual formulas $\sin \theta = \sqrt{(1 - \cos 2\theta)/2}$, and $\cos \theta = \sqrt{(1 + \cos 2\theta)/2}$.

For the central conics it is usually the practice to translate axes first, removing the linear terms from (1). Since this does not affect in any way the coefficients of the second degree terms, it is immaterial whether the angle of rotation be determined before or after the translation, for either the central conics or the parabola. Hence all the geometric properties of any conic except the position of the center (for the central conics) or the vertex (for the parabola) are very rapidly and conveniently calculated from A, B, and C without recourse to the angle θ.

The increasing tendency to omit simplification of conics by rotation from elementary courses is no doubt traceable in large part to the difficulties inherent in the computations involved. The method shown here seems to eliminate most of the difficulties, and I have found the method so satisfactory in my own classes that several years ago I abandoned the usual methods entirely, changing what had formerly been to most students a very distasteful section of the course to one which has been one of the most interesting.

THE REDUCED EQUATION OF THE GENERAL CONIC*

A. E. JOHNS, McMaster University

EDITOR'S NOTE. The subject matter of this note is familiar to everyone, and has received a satisfactory treatment in a number of standard texts. Among such expositions, that of Professor Johns is outstanding for its elegance and precision, and it is being published for this reason.

1. Notation. The general equation of a conic may be written:

$$aX^2 + 2hXY + bY^2 + 2gX + 2fY + c = 0.$$

Let

$$D \equiv \begin{vmatrix} a & h & g \\ h & b & f \\ g & f & c \end{vmatrix} = abc + 2fgh - af^2 - bg^2 - ch^2$$

and let A, B, C, F, G, H be the cofactors of the corresponding small letters in relation D. The fundamental invariants under rotation and translation of rectangular axes are: $a+b$; $C=ab-h^2$; and D. In addition, when $C=0$ and $D=0$, $A+B=bc+ca-f^2-g^2$ is also invariant. The discriminating quadratic is:

$$\lambda^2 - (a+b)\lambda + ab - h^2 = 0$$

with roots λ_1 and λ_2.

2. Rotation of axes. The equations of transformation for a rotation through an angle θ are:

$$X = lx - my; \quad Y = mx + ly$$

where $l=\cos\theta$ and $m=\sin\theta$; and the new equation of the conic takes the form:

$$a'x^2 + 2h'xy + b'y^2 + 2g'x + 2f'y + c' = 0$$

---

* From AMERICAN MATHEMATICAL MONTHLY, vol. 54 (1947), pp. 100–104.

where:

$$a' = al^2 + 2hlm + bm^2 \qquad f' = fl - gm$$
$$b' = am^2 - 2hlm + bl^2 \qquad g' = gl + fm$$
$$c' = c \qquad h' = (b-a)lm + h(l^2 - m^2).$$

It is always possible to choose a positive acute angle θ to satisfy the equation: $\tan 2\theta = 2h/(a-b)$ and so make $h' = 0$. For this θ, l and m are both positive. After this rotation through the positive acute angle θ, and a suitable translation if necessary, the equation of the conic takes on one of the following types of reduced forms.

Case 1. When $C \neq 0$, $\lambda_1 x^2 + \lambda_2 y^2 + D/C = 0$.

These are the central conics, including two intersecting lines.

Case 2. When $C = 0$ and $D \neq 0$,

(i) $$\lambda_1 x^2 + 2f'y = 0, \quad \text{where} \quad f' = \pm \left(\frac{-D}{\lambda_1}\right)^{1/2},$$

or

(ii) $$\lambda_2 y^2 + 2g'x = 0, \quad \text{where} \quad g' = \pm \left(\frac{-D}{\lambda_2}\right)^{1/2}.$$

These are true parabolas.

Case 3. When $C = 0$ and $D = 0$,

(i) $$\lambda_1 x^2 + \frac{A+B}{\lambda_1} = 0,$$

or

(ii) $$\lambda_2 y^2 + \frac{A+B}{\lambda_2} = 0.$$

These are parallel lines.

The above forms follow directly from the fact that the expressions listed in §1 are invariant.

3. Discussion. When a student asks the instructor which root of the discriminating quadratic is λ_1, or which of the alternative forms in Cases 2 and 3 is correct, or which sign for f' and g' should be chosen, the answer sometimes given is that it does not matter and that either is correct depending on how the new axes are named. However, if θ is to be a positive acute angle this reply is not correct. The following is an attempt to determine the proper alternatives on the above assumption.

From the equations of §2 above we have when $h' = 0$

$$\lambda_1 - \lambda_2 = a' - b' = (a-b)(l^2 - m^2) + 4hlm,$$

and
$$h' = (b - a)lm + h(l^2 - m^2) = 0.$$
So
$$\lambda_1 - \lambda_2 = \left\{\frac{(a-b)^2}{h} + 4h\right\}lm = \{(a-b)^2 + 4h^2\}\frac{lm}{h}.$$

Since lm and $(a-b)^2+4h^2$ are always positive, λ_1 and λ_2 must be so chosen that $\lambda_1-\lambda_2$ has the same sign as h.

For the central conics of Case 1 where $C \neq 0$, this rule is sufficient to determine which root is λ_1, and which is λ_2 in the reduced form
$$\lambda_1 x^2 + \lambda_2 y^2 + \frac{D}{C} = 0.$$

For Cases 2 and 3 where $C=0=ab-h^2$, a and b must always have the same sign. For simplicity, let us suppose that our original equation is written with these coefficients both positive. Then $a+b$ is also positive. When $C=0$, one root of the discriminating quadratic is zero and the other is $a+b$. Two possibilities now arise depending on the sign of h. If h is positive, $\lambda_1=a+b$ and $\lambda_2=0$, since $\lambda_1-\lambda_2$ must have the same sign as h. So the reduced forms are, respectively,
$$\lambda_1 x^2 + 2f'y = 0 \quad \text{and} \quad \lambda_1 x^2 + \frac{A+B}{\lambda_1} = 0.$$

If h is negative, $\lambda_1=0$ and $\lambda_2=a+b$, since $\lambda_1-\lambda_2$ must now be negative to match h in sign. The reduced forms are now, respectively,
$$\lambda_2 y^2 + 2g'x = 0 \quad \text{and} \quad \lambda_2 y^2 + \frac{A+B}{\lambda_2} = 0.$$

It remains to determine the signs of f' and g' on the understanding that θ is a positive acute angle.

When $ab=h^2$ and $h'=hl^2+(b-a)lm-hm^2=0$, we have
$$h^2l^2 + (b-a)hlm - abm^2 = 0$$
or
$$(hl - am)(hl + bm) = 0,$$
whence
$$\frac{l}{a} = \frac{m}{h} \quad \text{or} \quad \frac{l}{b} = \frac{m}{-h}.$$

When h is positive,
$$\frac{l}{a} = \frac{m}{h} = \frac{1}{(a^2+h^2)^{\frac{1}{2}}} = \frac{1}{(a^2+ab)^{\frac{1}{2}}}.$$

From §2

$$f' = fl - gm = \frac{af - gh}{(a^2 + ab)^{\frac{1}{2}}}.$$

This form, which is equivalent to $\pm\left(\frac{-D}{a+b}\right)^{\frac{1}{2}}$ with the proper sign, shows that the sign of f' is the same as the sign of $af - gh$, or opposite to that of F, the co-factor of f in D.

When h is negative,

$$\frac{l}{b} = \frac{m}{-h} = \frac{1}{(b^2 + h^2)^{\frac{1}{2}}} = \frac{1}{(ab + b^2)^{\frac{1}{2}}}.$$

From §2

$$g' = gl + fm = \frac{bg - fh}{(ab + b^2)^{\frac{1}{2}}}.$$

This form, which is equivalent to $\pm\left(\frac{-D}{a+b}\right)^{\frac{1}{2}}$ with the proper sign, shows that the sign of g' is the same as the sign of $bg - fh$, or opposite to that of G, the co-factor of g in D.

4. Examples:

EXAMPLE 1. $8x^2 - 4xy + 5y^2 - 36x + 18y + 9 = 0$.

$a = 8 \quad f = 9 \quad a + b = 13$
$b = 5 \quad g = -18 \quad C = ab - h^2 = 36$
$c = 9 \quad h = -2 \quad D \equiv abc + 2fgh - af^2 - bg^2 - ch^2 = -36^2$.

$\tan 2\theta = -4/3$, $\tan \theta = 2$. Choose θ a positive acute angle.

$$\lambda^2 - 13\lambda + 36 = 0 \quad \text{or} \quad (\lambda - 4)(\lambda - 9) = 0.$$

Since h is negative, choose $\lambda_1 = 4$, $\lambda_2 = 9$, to make $\lambda_1 - \lambda_2$ negative. The reduced equation $\lambda_1 x^2 + \lambda_2 y^2 + D/C = 0$ becomes $4x^2 + 9y^2 - 36 = 0$.

EXAMPLE 2. $x^2 - 4xy + 4y^2 + 5y - 6 = 0$.

$a = 1 \quad f = \dfrac{5}{2} \quad a + b = 5$
$b = 4 \quad g = 0 \quad ab - h^2 = 0$
$c = -6 \quad h = -2 \quad D = \dfrac{-25}{4}$

$\tan 2\theta = 4/3$, $\tan \theta = 1/2$, $\lambda^2 - 5\lambda = 0$.
Since h is negative, choose $\lambda_1 = 0$ and $\lambda_2 = 5$ to make $\lambda_1 - \lambda_2$ negative. $g' = +5^{\frac{1}{2}}/2$

and the positive sign must be chosen since G is negative. The reduced equation is $5y^2+5^{\frac{1}{2}}x=0$.

EXAMPLE 3. $4x^2+12xy+9y^2+2x+3y-42=0$.

$$a = 4 \qquad f = \frac{3}{2} \qquad a + b = 13$$
$$b = 9 \qquad g = 1 \qquad ab - h^2 = 0$$
$$c = -42 \qquad h = 6 \qquad D = 0 \qquad A + B = -\frac{13^3}{4}$$

$\tan 2\theta = -12/5$, $\tan \theta = 3/2$, $\lambda^2 - 13\lambda = 0$.
Since h is positive, $\lambda_1 = 13$, $\lambda_2 = 0$ and the reduced form is $13x^2 - 13^2/4 = 0$ or $x = \pm 13^{\frac{1}{2}}/2$.

SIMPLIFICATION OF EQUATIONS OF CONICS*

M. T. BIRD, Allegheny College

I should like to extend the discussion of the simplification of equations of conics initiated by Thornton[†] and continued by Johnston.[‡]

Consider the general equation of the conic in the form

(1) $$Ax^2 + Bxy + Cy^2 + Dx + Ey + F = 0.$$

It is assumed that B is not zero. The equation of the conic may be written in the alternative form

(2) $$A(x^2 + y^2) + y[Bx - (A - C)y] + Dx + Ey + F = 0.$$

The angles between the pair of intersecting lines

(3) $$y[Bx - (A - C)y] = 0$$

are bisected by the mutually perpendicular lines

(4) $$\pm y = \frac{Bx - (A - C)y}{\sqrt{B^2 + (A - C)^2}}.$$

It follows that the lines (4) are axes of symmetry for the degenerate conic (3). This fact may be used to simplify equation (2).

Let the mutually perpendicular lines (4) be written in the normal form

(5) $$0 = mx + ny, \quad 0 = -nx + my; \quad m > 0, \; n \geq 0, \; m^2 + n^2 = 1.$$

The quantities x', y' defined by the equations,

(6) $$x' = mx + ny, \quad y' = -nx + my; \quad m > 0, \; n \geq 0, \; m^2 + n^2 = 1,$$

---

* From AMERICAN MATHEMATICAL MONTHLY, vol. 54 (1947), pp. 104–106.
† Simplification of the equations of conics, this MONTHLY, vol. 41, 1934, p. 36.
† Simplification of equations of conics, this MONTHLY, vol. 44, 1937, p. 30.

measure the distances of the point $P(x, y)$ from the mutually perpendicular lines (5). The quantities x', y' constitute coördinates of the point $P(x, y)$ after a rotation of axes, and the equations (6) define the coördinate transformation whose inverse is seen to be

(7) $\quad x = mx' - ny', \ y = nx' + my'; \quad m > 0, \ n \geq 0, \ m^2 + n^2 = 1.$

The symmetry of the degenerate conic (3) about the lines (5) leads to the conclusion that the equation of the degenerate conic (3) will assume the form

$$y[Bx - (A - C)y] \equiv Mx'^2 + Ny'^2 = 0,$$

under the transformation (7). It is readily verified that the expression $x^2 + y^2$ is invariant under the transformation (7). Consequently the transformation (7) carries the general equation of the conic (2) into the form

$$A'x'^2 + C'y'^2 + D'x' + E'y' + F' = 0.$$

Consider the example

$$9x^2 - 24xy + 16y^2 - 186x - 252y - 63 = 0.$$

Following the method outlined above we have

$$9(x^2 + y^2) - y(24x - 7y) - 186x - 252y - 63 = 0.$$

The new coördinate axes are defined by the equations

$$\pm y = (24x - 7y)/25,$$

which may be written

$$24x + 18y = 0, \quad 24x - 32y = 0.$$

The new coördinates are defined by the equations

$$x' = (4x + 3y)/5, \quad y' = (-3x + 4y)/5.$$

The transformation

$$x = (4x' - 3y')/5, \quad y = (3x' + 4y')/5$$

leads to the equation of the conic in the form

$$25y'^2 - 300x' - 90y' - 63 = 0.$$

In this example the given equation may be written

$$(3x - 4y)^2 - 186x - 252y - 63 = 0.$$

The parenthetical expression suggests at once the introduction of the coördinates defined above in order to obtain the equation

$$y'^2 + ax' + by' + c = 0.$$

DERIVATION OF THE EQUATION OF CONICS*

F. HAWTHORNE, Hofstra College

In most analytic geometry texts the equations of conics are derived by use of the distance formula. Alternative derivations which do not involve radicals seem to provoke some student interest. The procedure can be illustrated by the following derivation of the equation of the ellipse.

Let the foci be the points $(\pm c, 0)$; and let R be the distance from a point on the curve to the left-hand focus and r be the corresponding distance to the right hand focus. Then by definition:

(1) $$R^2 = y^2 + (x + c)^2, \quad \text{and}$$
(2) $$r^2 = y^2 + (x - c)^2.$$

Adding and subtracting we have:

(3) $$R^2 + r^2 = 2(y^2 + x^2 + c^2), \quad \text{and}$$
(4) $$R^2 - r^2 = 4cx = (R + r)(R - r).$$

But by definition of the ellipse

(5) $$R + r = 2a.$$

Hence from (4) and (5) we have that

(6) $$R - r = 2cx/a.$$

Adding and subtracting (5) and (6) gives:

(7) $$R = a + cx/a, \quad \text{and}$$
(8) $$r = a - cx/a.$$

Putting the results of (7) and (8) into (3) gives:

(9) $$a^2 + \frac{c^2 x^2}{a^2} = c^2 + x^2 + y^2,$$

from which the standard equation is found by putting $b^2 = a^2 - c^2$. The expressions for the focal radii in terms of the eccentricity may be pointed out in equations (7) and (8).

---

* From AMERICAN MATHEMATICAL MONTHLY, vol. 54 (1947), pp. 219–220.

ON INTEGRAL COORDINATES*

NORMAN ANNING, University of Michigan

In the classroom integral coordinates have some obvious uses. Obvious is a teacher's word for "as plain as the truck you overtake on the road (*ob viam*)." The teacher who wants a circle or a parabola with integral points has no trouble in making the arrangements. Here are some handy ellipses.

Obvious solutions of the equation $x^2+3y^2=4(a^2+ab+b^2)$ are

$$x^2 = (a-b)^2, \quad (2b+a)^2, \quad (2a+b)^2,$$
$$y^2 = (a+b)^2, \quad a^2, \quad b^2.$$

Choose for a and b any positive integers and the stage is set. If at any time such a conic with its dozen integral points begins to seem too pat and regular, variety can be introduced by the transformations $x=Y+X$, $y=Y-X$. Thus $x^2+3y^2=52$ becomes $X^2-XY+Y^2=13$ and it will be found that the twelve integral points have gone over unharmed. Anyone who feels the need of more than a dozen integral points on one ellipse should try replacing 52 by 196, or 13 by 91, or 13 by 1729.

Only one application will be suggested: the verification of the Pascal Theorem can be brought within the range of the freshman.

Editorial Note to Students. Professor Anning's note suggests avenues of further research. Can you discover other "handy" curves, in particular conic sections? We shall be glad to receive any that you find.

BIBLIOGRAPHIC ENTRIES: EQUATIONS OF CONIC SECTIONS

The references below are to the AMERICAN MATHEMATICAL MONTHLY.

1. G. R. Dean, Note on the general equation of the second degree, vol. 7, p. 217.
   Directrix, focus and eccentricity expressed in terms of the coefficients.
2. H. W. Bailey, On the general equation of the parabola, vol. 41, p. 316.
   Properties of parabola in terms of coefficients in equations.
3. N. Anning, On integral coordinates, vol. 60, p. 199.
   A scheme for generating integral coordinates for ellipses.

---

* From AMERICAN MATHEMATICAL MONTHLY, vol. 60 (1953), pp. 199–200.

(b)

CLASSIFICATION AND CONSTRUCTION

SIMPLE CONSTRUCTIONS FOR THE CONICS*

L. S. JOHNSTON, University of Detroit

Let the ordered points $O(0, 0)$, $F(c, 0)$, and $A(a, 0)$ be respectively the center, focus, and vertex of an ellipse, and let FR and FR' be perpendicular lines through F, FR intersecting $x=a$ at $T[a, (a-c)\tan\theta]$ and FR' intersecting $x=-a$ at $T'[-a, (a+c)\cot\theta]$, where $\theta = $ angle AFR. It is well known that TT' is tangent to the ellipse $b^2x^2+a^2y^2=a^2b^2$, but it is not so generally known that the point of tangency may be located by a very simple construction.

The line TT' is given by the equation

$$x(a\cos 2\theta + c) + ay\sin 2\theta = a(a + c\cos 2\theta)$$

and the point of tangency is given parametrically by

$$x = a\frac{a\cos 2\theta + c}{a + c\cos 2\theta}, \qquad y = \frac{b^2\sin 2\theta}{a + c\cos 2\theta}.$$

These coördinates satisfy the equation $y=(x-c)\tan 2\theta$, and hence the focal radius through the point of tangency makes the angle 2θ with the x axis. From this fact we derive a simple construction for locating the point at which any arbitrary line through F intersects the ellipse, without drawing the ellipse itself. Let FM be such a line, making with the x axis the angles AFM and MFO. Let the bisector of AFM intersect the line $x=a$ at T and let the bisector of MFO intersect $x=-a$ at T'. Then TT' intersects FM on the ellipse, and TT' is tangent to the ellipse at this point. In particular, if FM be the line $x=c$, the intersection of TT' with $x=c$ is $(c, b^2/a)$, the intersection of the latus rectum with the ellipse. This particular case is especially useful for rapid sketching of the ellipse for blackboard demonstration, as well as for cutting templates for elliptical masonry arches.

In particular cases the scale of the figure may be too large to use both $x=a$ and $x=-a$ conveniently. For such cases we note that the line TT' intersects $x=0$ at $(0, a+c\cos 2\theta/\sin 2\theta)$ and $x=c$ at $(c, b^2/a\sin 2\theta)$. For $\theta=45°$ the intersection with $x=0$, viz. $(0, a)$, is particularly convenient; for other values of θ the intersection with $x=c$ is more convenient, and the construction is easily performed.

---

* From AMERICAN MATHEMATICAL MONTHLY, vol. 44 (1937), pp. 524–525.

The theory and constructions just explained are quite easily adapted to the other conics, and need not be discussed in detail here. The writer has found that his students use these constructions habitually in their board work once they have become familiar with them.

CLASSIFICATION OF THE CONICS*

R. C. YATES, U.S. Military Academy

The universal method of obtaining the discriminant relation for the conics by a transformation of axes is tedious, involved, and often omitted in freshman courses. As a first introduction to geometrical invariants, its value is unquestionable. But if the primary object is the identification of the conics, the following approach has several desirable features that warrant its notice.

The three types of conics represented by the equation

(1) $$Ax^2 + Bxy + Cy^2 + Dx + Ey + F = 0, \qquad (D, E \neq 0)$$

are characterized† as follows.

The *parabola* is such that there is *one and only one* line of the pencil $y = mx$ that cuts the curve in only one point (*i.e.*, the line parallel to the axis of the curve).‡ The *hyperbola* is such that there are *two and only two* distinct lines of the pencil $y = mx$ that cut the curve each in only one point (*i.e.*, the lines parallel to the asymptotes of the curve). The *ellipse* is such that there is *no* line of the pencil $y = mx$ that cuts the curve in only one point.

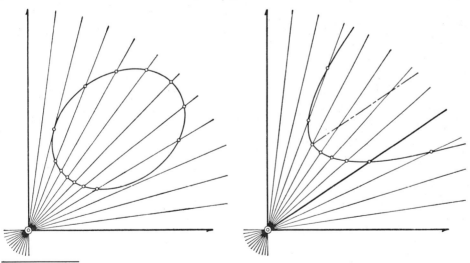

---

* From AMERICAN MATHEMATICAL MONTHLY, vol. 50 (1943), pp. 112–115.

† If thought desirable, these characterizations may be justified by means of the "standard" forms with which the student is already familiar.

‡ We consider here "algebraic" points and real lines $y = mx$. A point of tangency, for instance, is counted twice.

374   CONIC SECTIONS

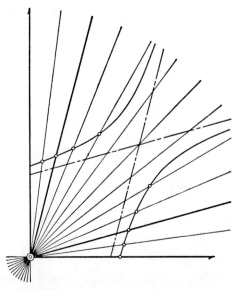

The pencil $y=mx$ meets (1) in points whose abscissas are given by
(2) $$(A + Bm + Cm^2)x^2 + (D + Em)x + F = 0.$$
If some lines of the pencil are to yield but one point of intersection, then
$$A + Bm + Cm^2 = 0,$$
or,*
$$\boxed{m = \frac{-B \pm \sqrt{(B^2 - 4AC)}}{2C}}$$

For the parabola, only one such line exists. Thus†
$$\boxed{B^2 - 4AC = 0}.$$

For the hyperbola, there are two such lines. Thus‡
$$\boxed{B^2 - 4AC > 0}.$$

---

* If $C=0$, $A \neq 0$, we establish, instead of (2), the corresponding equation for the ordinates of points of intersection of the pencil $x=ny$ and the conic.

† Here, obviously, $m = -B/2C$ is the slope of the axis of the curve. Rotation of axes through the angle arc tan $(-B/2C)$, accordingly, removes the xy term.

‡ The bisector of an angle formed by the two lines here is parallel to an axis of the curve. The angle of rotation to remove the xy term is thus easily obtained.

For the ellipse, there is no such line. Thus

$$B^2 - 4AC < 0.$$

The characterization fails when $D=E=0$. However, the abscissas of the points of the curve cut out by the pencil are given by

$$(A + Bm + Cm^2)x^2 + F = 0.$$

Since there are always two values of x, real or imaginary, equal and opposite in sign, it is evident that such conics are central.

If $B^2-4AC>0$, the quantity $A+Bm+Cm^2$ changes sign as m ranges through all real values. Thus some lines of $y=mx$ cut the curve and some do not. The conic is thus a *hyperbola*.

If $B^2-4AC<0$, the quantity $A+Bm+Cm^2$ has the same sign for all real values of m. If this sign is opposite that of F, *all* lines $y=mx$ cut the curve. The conic is therefore an *ellipse*. If the sign is the same as that of F, the curve is imaginary.

FOLDING THE CONICS*

R. C. YATES, U.S. Military Academy

The idea of applying the process of paper folding to the construction of the conics originated with Row.† The methods he gives, however, are quite involved in a structural sense. Improvements (at least for the central conics) are offered by Lotka‡ but the usefulness and charm of his methods are somewhat obscured by an accompanying analytical justification.

The purpose of this note is to offer the best of the methods of Lotka and Row and to present simple proofs to establish the processes involved. It is hoped that this formation of the conics by paper folding will somehow find its way into the clasroom where it will undoubtedly be received with enthusiasm.

---

* From AMERICAN MATHEMATICAL MONTHLY, vol. 50 (1943), pp. 228–230.

† T. Sundara Row, Geometric Exercises in Paper Folding (trans. by Beman and Smith) Chicago, 1901.

‡ A. J. Lotka, School Science and Mathematics, VII, 1907, 595–597; Scientific American Supplement, 1912, 112.

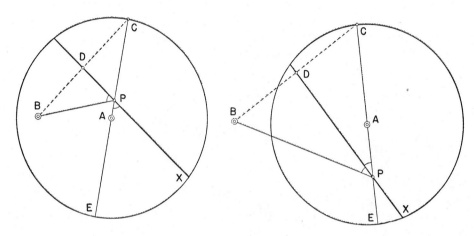

Fig. 1

A fixed point B is selected within a circle of radius r and center A. The point B is folded over upon the circle, as at C, forming the crease DX (the perpendicular bisector of BC) which meets the diameter EC in the point P. As B takes positions along the circle, the path of P is the *ellipse* having foci at A and B, major axis r, and the creases as tangents. For, since P lies on the perpendicular bisector of BC,

$$AP + BP = AP + PC = r$$

and

$$\angle BPD = \angle DPC = \angle APX.$$

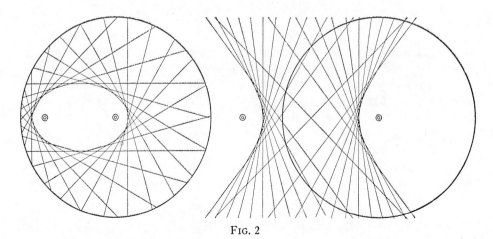

Fig. 2

(If B is taken at A, the locus of P is the circle with center at A and radius $r/2$.)

If B is selected outside of the circle, the locus of P (the intersection of crease and corresponding diameter) is the *hyperbola* having A and B as foci, real axis r, and the creases as tangents. For, since P lies on the perpendicular bisector of BC,

$$BP - AP = CP - AP = r$$

and

$$\angle BPD = \angle DPA.$$

The rectangular hyperbola is formed if $AB = r\sqrt{2}$. If B is taken on the circle, the locus of P is the point A.

The *parabola* (see Figure 3) presents a special case in which a line L replaces the circle of the central conics. The point B is folded over upon L, as at C, producing the crease DX. The perpendicular to L at C meets the crease in P. Points P form the parabola having B as focus, L as directrix, and the creases as tangents. For, since P lies on the perpendicular bisector of BC,

$$BP = PC$$

and

$$\angle BPD = \angle DPC.$$

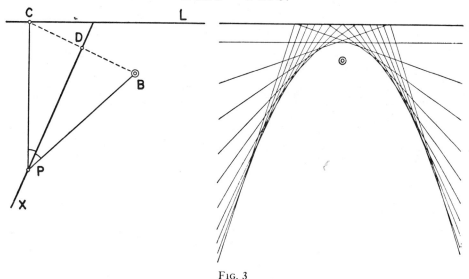

Fig. 3

This fascinating art of paper folding, which seems unfortunately relegated to the limbo, may be brought to full expression through the medium of ordinary wax paper found in every kitchen cabinet.

THE ELLIPSE AS A CIRCLE WITH A MOVING CENTER*

F. H. YOUNG, Oregon State College

Our purpose is to examine the possibility of exhibiting the ellipse as the locus of a point on a rotating circle with a center moving on the focal axis. This condition may be satisfied if we can express the ellipse in the form $(x-x')^2+y^2=b^2$, for the symmetry of the figure requires that the radius be equal in length to b, half the minor axis. Throughout this discussion we shall assume that $a>b$.

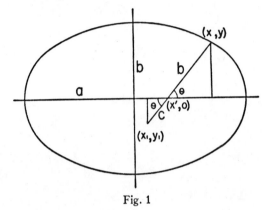

Fig. 1

In order to have x' move in such a way as to fit the ellipse $x^2/a^2+y^2/b^2=1$, we must let $x'=kx$ so that $x^2(1-k)^2/b^2+y^2/b^2=1$ where $b^2/(1-k)^2=a^2$, whence $k=(a-b)/a$.

Further, $x'=x-b\cos\theta$, but since $x'=[(a-b)/a]x$, then $x=a\cos\theta$. Also, $y=b\sin\theta$. Hence $x'=kx=(a-b)\cos\theta$. Therefore, $(x', 0)$, the center of the moving circle moves on the focal axis with simple harmonic motion over the range $-(a-b)\leq x'\leq(a-b)$ as θ increases constantly. Conversely, any such moving circle generates an unique ellipse, for the radius is of fixed length, b.

Let us now examine some of the properties revealed by this method of analysis of the ellipse. First, observe that the parametric equations $x=a\cos\theta$; $y=b\sin\theta$ can refer to the parameter θ, the angle the moving radius makes with the focal axis.

Consider next the situation arising when the radius is extended in the direction $\theta+\pi$ to give a point (x_1, y_1) lying the distance $b+c$ from (x, y) on the given ellipse. The point (x_1, y_1) then has the coordinates $y_1=-c\sin\theta$ and $x_1=x'-c\cos\theta=(a-b)\cos\theta-c\cos\theta=(a-b-c)\cos\theta$. Eliminating θ by squaring and adding, we obtain $y^2/c^2+x^2/(a-b-c)^2=1$, again an ellipse, provided $b+c\neq a$. In the particular case in which $b+c=a$, we find that $x=(a-b-c)\cos\theta=0$ and $y=(b-a)\sin\theta$, and as θ goes from 0 to 2π, we generate that section of the y-axis such that $-(a-b)\leq y\leq(a-b)$. If $b+c>a$, and we continue our assumption that $a>b$, then $c>a-b>0$: hence $|a-b-c|>c$ and our resulting locus is an

---

* From AMERICAN MATHEMATICAL MONTHLY, vol. 55 (1948), pp. 156–158.

ellipse whose major axis lies along the y-axis. It is interesting to note that if $c=a$, we have our original ellipse with the axes interchanged.

Notice now that the slope of the generating radius is ay/bx. Hence, the radius is normal to the ellipse only at the endpoints of the axes. This result gives the answer to the problem of the type of curve generated by a normal of fixed length N moving about the ellipse. For a fixed N, $b-a$ remains unchanged. Thus, x' moves in the same manner as for the original ellipse, and the radius is now $b+N$. If (x_1, y_1) lies on an ellipse, its distance from $(x', 0)$ is $b+N$, and the line

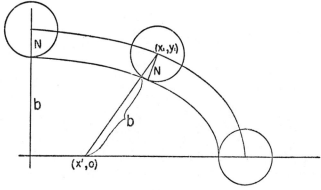

Fig. 2

joining these points is distinct from the normal. This requires the circle of radius N with center at (x_1, y_1) to be tangent to the ellipse externally in one place and to cut it in another, a contradiction. Hence, the curve generated by the moving normal is definitely not an ellipse. This can similarly be shown if N is measured in from the ellipse.

So far we have constrained b to be smaller than a. What results if we have our circle still moving on the x-axis, but $b>a$? This offers no difficulty, for from the relationship that $x'=(a-b)\cos\theta$, we can see that the center of the moving circle now must move from $[-(b-a), 0]$ to $[(b-a), 0]$, as θ increases positively, in describing the upper half of the ellipse whose equation is $x^2/a^2+y^2/b^2=1$, $b>a$.

ON DEFINING CONIC SECTIONS*

G. B. HUFF, The University of Georgia

The conventional American text in analytic geometry begins the study of conic sections with a paragraph describing these curves as sections of a right circular cone by suitably chosen planes. It is customary to remark that this definition of a conic section is not adapted to the methods and objectives of the course, and each one of the parabola, the ellipse, and the hyperbola is then given its own definition as a locus in a later paragraph. Ordinarily there is no attempt to give even an intuitive proof of the equivalence of the two definitions.

Apparently it is not well known that this may be done quickly and simply in the case of the ellipse and the hyperbola. In *Anschauliche Geometrie* (Hilbert, Cohn-Vossen; Dover; 1944; pp. 7, 8), a figure is given from which a simple

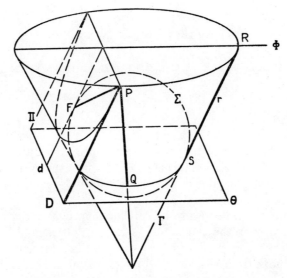

intuitive argument may be made that the two definitions for the ellipse (or the hyperbola) are equivalent. This note is written to call these figures to the attention of analytic geometry teachers and to give the construction of a third figure which leads to a demonstration of the equivalence of the two definitions of the parabola.

Let the parabola be defined as the intersection of a right circular cone Γ and a plane Π parallel to a ruling r of Γ and perpendicular to the plane on r and the axis of Γ. Construct a sphere Σ internally tangent to Γ and tangent to the plane Π. (The existence and unicity of Σ is made clear by thinking of fitting a small balloon in Γ and inflating it until it meets Π.) Let Θ be the plane determined by the intersection of Σ and Γ. Let F be the common point of Π and Σ,

---

* From AMERICAN MATHEMATICAL MONTHLY, vol. 62 (1955), pp. 250–251.

and let d be the common line of Π and Θ. I claim that F is the focus and d is the directrix of the parabola defined by Γ and Π.

Let P be any point of the parabola and let Φ be the plane through P and perpendicular to the axis of the cone Γ. Let D be the foot of the perpendicular from P to d; let Q be the intersection of the ruling on P with the plane Θ, and let R, S be the points on r cut out by Φ and Θ. A simple argument shows that RS is parallel to PD.

Since PF and PQ are tangents to the sphere, then $PF = PQ$.

Since PQ and RS are segments of rulings of Γ cut out by the parallel planes Θ and Φ, then $PQ = RS$.

Since PD is parallel to RS, these segments are cut out on parallel lines by parallel planes and $RS = PD$. It follows that $PF = PD$, as claimed.

It is easy to generalize the figure and the argument to definition of a conic of eccentricity e in terms of its directrix and focus. (Zwikker, *Advanced Plane Geometry*, Amsterdam, 1950, p. 108.)

ON DEFINING THE CONIC SECTIONS*

MICHAEL PASCUAL, Siena College

A merging of the analytic and the geometric definitions of the conic sections could be done neatly when three dimensional analysis is being covered. By obtaining the equation of a right circular cone and considering the equation of the curve of intersection by a cutting plane, we arrive at the equations of the various conic sections. Now to get the different sections we may either vary the cutting plane or the vertex angle of the cone. We use the latter method, since it lends itself more readily to analysis. For convenience, we place the vertex of the cone at $(0, 1, 1)$ and let the axis be the line $z = y$, $x = 0$. Letting α designate the angle formed by the axis and any ruling, we find the equation of the cone to be

$$2x^2 + (y - z)^2 = (y + z - 2)^2 \tan^2 \alpha.$$

By setting $z = 0$, we obtain the equation of the curve of intersection of the cone with the xy plane:

$$2x^2 + (1 - \tan^2 \alpha)y^2 + 4y \tan^2 \alpha - 4 \tan^2 \alpha = 0.$$

1. For $0 < \alpha < \pi/4$, an ellipse is the intersection; its equation is

$$\frac{x^2}{a^2} + \frac{(y + a^2)^2}{b^2} = 1 \quad \text{where} \quad a = \frac{\sqrt{2} \tan \alpha}{\sqrt{1 - \tan^2 \alpha}} \quad \text{and} \quad b = \frac{2 \tan \alpha}{1 - \tan^2 \alpha}.$$

2. For $\alpha = \pi/4$, a parabola is the intersection; it's equation is

$$x^2 = -2(y - 1).$$

---

* From AMERICAN MATHEMATICAL MONTHLY, vol. 63 (1956), pp. 719–720.

3. For $\pi/4 < \alpha < \pi/2$, a hyperbola is the intersection; it's equation is

$$\frac{x^2}{a^2} - \frac{(y+a^2)^2}{b^2} = 1.$$

If we wish to continue further and obtain a focal point and directrix, we need only get the equation of a sphere tangent to the cone and the xy plane, and the equation of the plane determined by the circle of tangency. By showing that the line of intersection of this plane and the xy plane is such that the ratio of the distance from any point on the section to the point of tangency of the sphere and the xy plane to the distance from that same point to the line of intersection is a constant, we fulfill the requirements of the focal point and directrix.

A HISTORICAL NOTE ON A PROBLEM IN THIS MONTHLY*

JOY B. EASTON, West Virginia University

Problem E 1560, January 1963, asks for the locus of the third vertex P of an equilateral triangle with the other two vertices Q and R remaining on two mutually perpendicular lines. It is easy to set up a rectangular or polar coordinate system and find the equations of the ellipses in the given problem. It is a generalization of the well-known locus of a point *on* a fixed line segment whose endpoints remain on two perpendicular axes. The general problem is also frequently referred to: if the two axes are perpendicular, the locus of any point fixed with relation to the segment will be an ellipse. In modern times this seems to have been given first by Francis van Schooten (1615–1660) in his *Exercitationum Mathematicarum*, Leiden, 1657, and probably dates back to Nasir Eddin (13th century).

A less familiar problem is the generalization of the eccentric angle construction of the ellipse, i.e., when the given segment rotates about a fixed point rather than slides along two fixed lines.

First, let C be a point on AB, or AB extended. As AB rotates about point B, the ellipse is generated in the following manner: A lies on a circle with diameter DD' and C on a circle with diameter EE' (Fig. 1). Draw parallels from A and C to a pair of fixed perpendicular diameters. The locus P, of the intersections of these lines will be an ellipse.

Proof. $\triangle BCM \sim \triangle ABN$ giving

$$\frac{BN^2}{AB^2} = \frac{CM^2}{BC^2}$$

---

* From AMERICAN MATHEMATICAL MONTHLY, vol. 72 (1965), pp. 53–56.

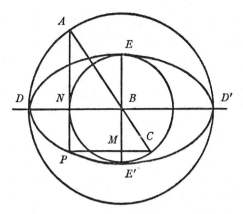

Fig. 1

or, choosing a coordinate system, $BN=x$, $BM=y$, $AB=a$, $BC=b$:

$$\frac{x^2}{a^2} = \frac{b^2 - y^2}{b^2}$$

$$\frac{x^2}{a^2} + \frac{y^2}{b^2} = 1,$$

an ellipse with center at B, semi-major axis $a=DB$, semi-minor axis $b=BE$. If C is taken between A and B, $\angle ABD = \phi$ is the eccentric angle and the parametric equations of the ellipse follow in the usual manner.

Now let ABC form a constant angle, obtuse or acute (Fig. 2). Draw $DE \perp AB$ at A and let BC lie on DF, with $DF = 2DB$. A perpendicular to DF at C intersects AB at G. DE and DF determine an oblique coordinate system with origin at D.

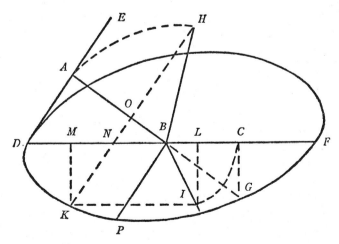

Fig. 2

As ABC rotates about B, A and C describe circles, and, for any position of ABC, say HBI, $HK \parallel DE$ and $IK \parallel DF$ determine a point K. The locus of K will be an ellipse with semi-major axis $DB = a$ and semi-minor axis BP equal in magnitude to $BG = b$.

Proof. Draw perpendiculars through I and K intersecting DF in L and M respectively. HK intersects AB in O and DF in N.

(1) Right triangles BCG, BAD, BON and KMN are similar. Also, since $\angle ABC = \angle HBI$, $\angle HBO = \angle IBL$ and

(2) right triangles BOH and BLI are similar.

From (1)

$$\frac{DB^2}{NB^2} = \frac{AB^2}{OB^2} = \frac{HB^2}{OB^2}$$

since $AB = HB$. By "conversion" of the proportion (taking the antecedent in relation to the excess by which the antecedent exceeds the consequent)

$$\frac{DB^2}{DB^2 - NB^2} = \frac{HB^2}{HB^2 - OB^2}.$$

But in $\triangle BOH$ $HB^2 - OB^2 = HO^2$ and by Euclid II, 5

$$DB^2 - NB^2 = DN \cdot NF,$$

giving

(3) $$\frac{DB^2}{DN \cdot NF} = \frac{HB^2}{HO^2}.$$

Now by (2) $(HB^2/HO^2) = (BI^2/IL^2)$. But $BI = BC$ and $IL = KM$, so that $(HB^2/HO^2) = (BC^2/KM^2)$. By (1) $(BC^2/KM^2) = (BG^2/KN^2)$ and (3) becomes

$$\frac{DB^2}{DN \cdot NF} = \frac{BG^2}{KN^2}$$

or

(4) $$KN^2 = \frac{DG^2}{DB^2} DN \cdot NF,$$

the Apollonian symptom of an ellipse referred to a diameter and the tangent at the vertex as axes.

With $DB = a$, $BG = b$, $DN = x$, $KN = y$ we have

(5) $$y^2 = \frac{b^2}{a^2} x(2a - x).$$

Note that with the aid of our coordinate system the equality $DB^2 - NB^2 = DN \cdot NF$ becomes $a^2 - (a-x)^2 = x(2a-x)$, an algebraic identity.

During the late 16th and 17th centuries there was an awakening of interest in the Greek theory of the conics. Many plane locus constructions were developed by Kepler, van Schooten, de Witt and others. These differed from the Greek approach in that the curves were defined in the plane rather than as sections of cones.

The particular problems discussed above are taken from Johan de Witt (1625–1672), *Elementa Curvarum Linearum*, pp. 238–241 of the 1683 edition. De Witt was a friend and pupil of Francis van Schooten the younger, and his treatise appears as an appendix to van Schooten's 2nd Latin edition of the *Geometria* of René Descartes, Leiden, 1659–1661. I do not know whether the second problem is original with him. Nearly all of our familiar locus problems involving conics were developed during this period. I should like to point out that the synthetic approach based on the Greek theory of proportions is still the best way to handle them. Analytic geometry is a powerful tool, but, in problems of this type can lead to formidable algebraic manipulations.

BIBLIOGRAPHIC ENTRIES: CLASSIFICATION AND CONSTRUCTION OF CONIC SECTIONS

Except for the entry labeled MATHEMATICS MAGAZINE, the references below are to the AMERICAN MATHEMATICAL MONTHLY.

1. J. R. Musselman, A rectangular hyperbola theorem, vol. 40, p. 480.

Construction involving a rectangular hyperbola and a set of chords.

2. W. R. McEwen, Focal points and focal loci, vol. 48, p. 386.

3. J. H. Weaver, Properties of points, lines, and circles associated with a point on an ellipse, vol. 48, p. 435.

4. E. F. Allen, On a triangle inscribed in a rectangular hyperbola, vol. 48, p. 675.

Proof that the equations in inversive geometry of a circle, line through two points and polar line of a point with respect to a circle can be altered by the introduction of a suitable constant so that they hold for any central conic.

5. J. R. Musselman, On the rectangular hyperbola, vol. 59, p. 11.

Some properties of the rectangular hyperbola.

6. A. Porges, The circumscribed circle, vol. 62, p. 361.

A method of finding the equation of the circle circumscribing a triangle formed by three given straight lines.

7. R. C. Yates, Rolling polygons, vol. 66, p. 130.

Perimeters and areas of members of the family of cycloids formed by rolling polygons upon polygons.

8. D. Pedoe, The ellipse as an hypotrochoid, MATHEMATICS MAGAZINE, vol. 48, p. 228.

(c)

TANGENTS

ON THE CHORD OF CONTACT OF TANGENTS TO A CONIC*

W. D. LAMBERT, Washington, D.C.

If we wish to find the equation of the chord of contact of tangents drawn from a point $P_1 \equiv (x_1, y_1)$, to the conic†

(1) $$Ax^2 + Bxy + Cy^2 + Dx + Ey + F = 0$$

the straightforward way is to denote the unknown points of contact $P_2 \equiv (x_2, y_2)$ and $P_3 \equiv (x_3, y_3)$, and proceed to find the values of those coördinates. For this purpose we get two equations by substituting (x_2, y_2) and (x_3, y_3) in (1), and two more by expressing the fact that P_2 lies on the tangents through P_2 and P_3. These are

(2) $$Ax_1x_2 + \frac{B}{2}(x_1y_2 + x_2y_1) + Cy_1y_2 + \frac{D}{2}(x_1 + x_2) + \frac{E}{2}(y_1 + y_2) + F = 0,$$

(3) $$Ax_1x_3 + \frac{B}{2}(x_1y_3 + x_3y_1) + Cy_1y_3 + \frac{D}{2}(x_1 + x_3) + \frac{E}{2}(y_1 + y_3) + F = 0.$$

These four equations give x_2, y_2, x_3, y_3 in terms of A, B, C, etc., and x_1, y_1. The required chord of contact is found by substituting their values in the formula for the straight line

(4) $$\frac{y - y_2}{x - x_2} = \frac{y_3 - y_2}{x_3 - x_2}.$$

This direct solution is obvious in principle, but very tedious in the analytic work, and does not appear in text books. The favorite method in elementary treatises is to *assume* the result, namely, that the equation of the chord of contact of tangents to the conic (1) is

(5) $$Ax_1x + \frac{B}{2}(xy_1 + x_1y) + Cy_1y + \frac{D}{2}(x + x_1) + \frac{E}{2}(y + y_1) + F = 0,$$

and then prove that (5) is a straight line passing through the points of contact P_2 and P_3. This process is certainly contrary to the spirit of *analytic* geometry, and seems obscure to many beginners.

---

* From AMERICAN MATHEMATICAL MONTHLY, vol. 13 (1906), pp. 159–160.

† I carry the work through for the general case, but as beginners are likely to be appalled by the mere length of an equation that is simple enough in principle, it is advisable in teaching to take at first a simpler special form of the equation of the conic.

The direct attack may be made easier by noting that we require not x_2, y_2, x_3, and y_3 themselves, but only certain combinations of them that are easy to find. Subtract (2) from (3) and factor so as to bring out x_3-x_2 and y_3-y_2. We may write the resulting equation in the form

$$\frac{y_3-y_2}{x_3-x_2} = -\frac{Ax_1 + \frac{B}{2}y_1 + \frac{D}{2}}{\frac{B}{2}x_1 + Cy_1 + \frac{E}{2}}.$$

Substitute this value in (4) and clear of fractions; after transposing the result may be written

$$(6) \quad Ax_1x + \frac{B}{2}(x_1y + xy_1) + Cy_1y + \frac{D}{2}x + \frac{E}{2}y \\ - \left[Ax_1x_2 + \frac{B}{2}(x_1y_2 + x_2y_1) + Cy_1y_2\right] + \frac{D}{2}x_2 + \frac{E}{2}y_2 = 0.$$

Adding the identity (2) to (6), we get the required equation (5).

TANGENT TO A CIRCLE FROM AN EXTERIOR POINT*

F. H. YOUNG, Oregon State College

The following method of finding the equation of a line tangent to a circle and passing through a point outside the circle has proved popular with my class in elementary analytic geometry.

Consider the circle

$$(x - h)^2 + (y - k)^2 = r^2$$

and the exterior point (x_1, y_1). Denote by S the length of tangent to the circle from (x_1, y_1). Denote by m_c the slope of the line joining (h, k) and (x_1, y_1). Denote by t the tangent of the angle between the line from (x_1, y_1) to (h, k) and the tangent line. Obviously, $t = r/S$. Then, by using the formula for $\tan(a \pm b)$, we obtain

$$m = \frac{m_c \pm t}{1 \pm m_c t},$$

the slopes of the tangent lines. Knowing the point and the slopes, we can easily obtain the desired equations.

---

* From AMERICAN MATHEMATICAL MONTHLY, vol. 55 (1948), p. 497.

TANGENT LINES AND PLANES*

F. H. YOUNG and J. L. ERICKSEN, Oregon State College

Although the problem of finding a tangent line to a conic or a tangent plane to a quadric presents no great difficulty, the problem can be solved by a method even simpler than that ordinarily seen.

Consider a given circle and a fixed point on the circle. Write the equation of a circle with its center at the point and of radius less than that of the given circle. If the two equations are normalized, the difference between the equations of the first and second circles is then the radical axis, a straight line passing through the points of intersection of the circles. Now, if the radius of the second circle were decreased, the intersection points would approach the tangent line at the fixed point. This suggests the following method: solve simultaneously the equation of the given circle and the equation of the null circle at the fixed point. The method is so simple that in many cases the answer may be obtained by inspection.

In the case of an ellipse, much the same method may be used. At a point on the given ellipse, consider a null ellipse of the same orientation whose eccentricity is the same as that of the given ellipse. Then the difference between the two equations will represent the tangent line.

In the case of a hyperbola, we can make use of the degenerate hyperbola of the same eccentricity and orientation as the given hyperbola. Geometrically, this means the equation of two straight lines, parallel to the asymptotes, intersecting at the fixed point on the given hyperbola. Again the difference is the required tangent.

For a parabola, we use the equation of a degenerate parabola with vertex at the fixed point and axis parallel to that of the given parabola. The method still works.

Geometrically, our method consists of employing a degenerate form of the conic section of the same eccentricity and orientation as that of the curve on which the fixed point lies. Algebraically, we subtract the equation satisfied by the coördinates of the point from the given equation, thus obtaining a linear equation.

The general equation of the conic, and this includes the degenerate, null, and imaginary cases, is

(1) $$ax^2 + bxy + cy^2 + dx + ey + f = 0.$$

Consider the point (u, v) satisfying the above equation. Then our null or degenerate form of the same eccentricity will be

(2) $$a(x - u)^2 + b(x - u)(y - v) + c(y - v)^2 = 0.$$

Expanding and subtracting, we obtain

$$(2au + bv + d)x + (bu + 2cv + e)y - (au^2 + buv + cv^2 - f) = 0.$$

---

* From AMERICAN MATHEMATICAL MONTHLY, vol. 55 (1948), pp. 573–574.

It is then easily verified that this is the equation of the line tangent to the given curve at the point (u, v). If equation (1) represents a null circle or null ellipse, it will be identical with equation (2). This is the only case in which a tangent line cannot be obtained. It is interesting to note that the method works equally well for points on imaginary conics.

An especially simple example of this method is the case of the tangent line at the origin to a conic passing through the origin. Consider, then, the conic

$$ax^2 + bxy + cy^2 + dx + ey = 0.$$

The desired tangent is merely $dx + ey = 0$.

It is apparent that this method may be extended to the case of the general second degree surface in n dimensions to obtain the equation of the tangent plane.

ONE SIDE TANGENTS*

WILLIAM R. RANSOM, Tufts College

In plane geometry the tangent to a circle touches it at but one point and all its other points lie on one side of the curve. This property may also be used for deriving the equation of a tangent to a conic: the commonly used concept of a limiting position of a tangent is therefore not actually necessary.

Take the point (p,q) on the ellipse or hyperbola, $b^2x^2 \pm a^2y^2 = a^2b^2$, and consider the tangent at that point: its equation is $y = q + M(x - p)$, where M is to be determined so that the line does not cross the curve. Let us take the two points (x, C) on the curve and (x, T) on the tangent. According to the one side definition of the tangent $C - T$ must have the same sign on both sides of (p,q). Transforming, we have

$$C - T = C - q - M(x - p) = \frac{C^2 - q^2}{C + q} - M(x - p)$$

$$= \pm \frac{b^2}{a^2} \cdot \frac{p^2 - x^2}{C + q} - M(x - p) = (x - p)\left[\mp \frac{b^2}{a^2} \cdot \frac{x + p}{C + q} - M\right].$$

As (x, C) passes through (p, q) the factor $(x - p)$ changes sign, and as $C - T$ does not change sign, the other factor must change sign. So it is necessary that M have the value $(\mp b^2/a^2)(p+p)/(q+q)$. It is also sufficient that it have this value, for then the factor in the brackets is

$$\mp \frac{b^2}{a^2} \cdot \frac{x + p}{C + q} - \frac{b^2 p}{a^2 q} = \mp \frac{b^2}{a^2} \cdot \frac{xq + pq - pC - pq}{q(C + q)} = \mp \frac{b^2}{a^2} \cdot \frac{q(x - p) + p(q - C)}{q(C + q)}.$$

Both $(x - p)$ and $(q - C)$ change sign as (x, C) passes through (p, q), hence this factor changes sign there, and $C - T$ has the same sign on both sides of (p, q).

---

* From MATHEMATICS MAGAZINE, vol. 29 (1956), pp. 159–160.

With this value of M, the equation for the tangent becomes
$$y = q \mp (b^2 p / a^2 q)(x - p)$$
which reduces to
$$b^2 px \pm a^2 qy = b^2 p^2 \pm a^2 q^2 = a^2 b^2.$$

For the parabola in the form $y = kx^2$, using the same notation we have $C - T = kx^2 - q - M(x-p) = kx^2 - kp^2 - M(x-p) = (x-p)[k(x+p) - M]$. Here we see that it is necessary that $M = k(p+p)$, and this is sufficient for it makes $C - T = k(x-p)^2$, which is positive on both sides of (p,q).

For the hyperbola in the form $xy = a^2$, we have
$$C - T = a^2/x - a^2/p - M(x-p) = (x-p)[-a^2/px - M]$$
and M must have the value $-a^2/p^2$, which is sufficient for it gives
$$C - T = a^2(x+p)^2/p^2 x.$$

The same method may be applied to many other curves. For example, take $y^2 = x^3$, the Neilian parabola. For $C - T$, this gives
$$x\sqrt{x} - p\sqrt{p} - M(x-p) = (\sqrt{x} - \sqrt{p})[x + \sqrt{px} + p - M(\sqrt{x} + \sqrt{p})].$$

For the factor in brackets to change sign at (p,q) we must have
$$M = (p + \sqrt{pp} + p)/(\sqrt{p} + \sqrt{p}) = 3\sqrt{p}/2.$$

$C - T$ then reduces to
$$\tfrac{1}{2}(\sqrt{x} - \sqrt{p})[2x + 2\sqrt{px} + 2p - 3\sqrt{px} - 3p]$$
$$= \tfrac{1}{2}(\sqrt{x} - \sqrt{p})[2x - 2\sqrt{px} + \sqrt{px} - p]$$
$$= \tfrac{1}{2}(\sqrt{x} - \sqrt{p})[2\sqrt{x}(\sqrt{x} - \sqrt{p}) + \sqrt{p}(\sqrt{x} - \sqrt{p})]$$
$$= \tfrac{1}{2}(\sqrt{x} - \sqrt{p})^2(2\sqrt{x} + \sqrt{p})$$

which does not change sign at (p,q).

Note that in this argument x does not approach p as a limit: we take $x = p$ in determining M, and $x \neq p$ in proving that $C - T$ does not change sign.

(d)

APPLICATIONS

MATHEMATICAL FORMS OF CERTAIN ERODED MOUNTAIN SIDES*

T. M. PUTNAM, University of California

In some of the desert regions of the west, visited by occasional heavy rainstorms, the formation is such that erosion takes place in a way that makes it possible to calculate approximately the form of the curve of intersection of the eroded slope and the alluvial fan of the plain.

This problem has been suggested by Professor A. C. Lawson of the department of geology of the University of California. He has formulated the hypotheses used below and the forms of the curves here obtained have been closely approximated by the actual geological conditions observed by him.

In the figure $S_0OR_0T_0$ is the contour of valley, hillside and plateau at some initial time. The table land R_0T_0 may be taken as inclined at a small angle C to the horizontal direction T_0Y. The valley OS_0 is assumed to be inclined at an angle B. The eroded material OR_0RH comes off in layers and is deposited in the valley in the position S_0SHO. The slope of erosion as well as the floor of the valley maintains, in the observed regions, a fairly constant inclination. Both inclinations are assumed constant in these calculations. The line S_0S, at the low point in the valley, is taken vertical. An alluvial fan, similar to the one under discussion, would be deposited from the opposite slope also terminating at S_0S.

The problem is to find the locus of the point H, or in other words the form of the uneroded portion of a cross-section. The geometrical conditions require that the area OR_0RH be equal to the area S_0SHO.

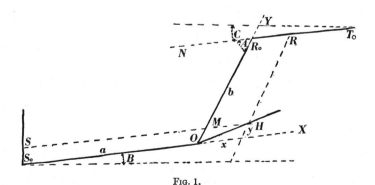

Fig. 1.

---

* From AMERICAN MATHEMATICAL MONTHLY, vol. 24 (1917), pp. 451–453.

Choosing oblique coördinates as indicated in the figure and letting $OS_0 = a$, $OR_0 = b$ and $\angle OR_0N = A$, it follows that

$$MR_0RH = x(b-y)\sin(A+C-B) + \frac{1}{2}\frac{x^2\sin(C-B)\sin(A+C-B)}{\sin A},$$

$$S_0SMO = ay\sin(A+C-B) + \frac{1}{2}\frac{y^2\cos(A+C)\sin(A+C-B)}{\cos B}.$$

Equating these two expressions and letting

$$P = \frac{\sin(C-B)}{\sin A}, \qquad Q = \frac{\cos(A+C)}{\cos B},$$

one obtains the equation of the locus of H in the form

$$Px^2 - 2xy - Qy^2 + 2bx - 2ay = 0.$$

To investigate the form of this conic one considers the sign of $1 + PQ$. Conditions occurring in nature require $A + C > B$ hence $0 < Q < 1$. Since $A > B - C$ then when $B > C$ we have $|P| < 1$. Hence $1 + PQ > 0$. Therefore the conic is always a hyperbola in the cases under consideration.

The discriminant* of the equation is $b^2Q + 2ab - a^2P$ which vanishes when

$$\frac{a}{b} = \frac{1 + \sqrt{1 + PQ}}{P}.$$

Since a/b is positive this requires that $P > 0$ and hence $C > B$. Under this condition there is for every set of angles A, B, C one ratio $a:b$ for which the locus is a straight line.

The tangent to the curve at O is parallel to the line S_0R_0 and at any point H it is parallel to the line SR. The curve is concave downward

$$\text{if} \quad \frac{a}{b} < \frac{1 + \sqrt{1 + PQ}}{P}, \qquad \text{upward if} \quad \frac{a}{b} > \frac{1 + \sqrt{1 + PQ}}{P}.$$

The special case $B = C = 0$ is of interest. The equation now becomes $\cos A\, y^2 + 2xy - 2bx + 2ay = 0$. Using rectangular coördinates, u and v, with the old origin and X-axis,

$$u = x + y\cos A, \qquad v = y\sin A,$$

the equation becomes

$$\cot A \cdot v^2 - 2uv + 2ku - 2hv = 0,$$

where $h = a + b\cos A$, $k = b\sin A$. The center is at the point $(-a, k)$ which is the point where S_0S meets the now horizontal line T_0R_0. One of the asymptotes is the line R_0T_0. The curve is concave downward in this case and the uneroded portion tends to become horizontal as erosion progresses. It might be expected that R_0T_0 is always an asymptote, but this is the case only when $\angle B = \angle C$, i.e., $P = 0$.

---

* C. Smith, *Conic Sections* (1912 ed.), page 43.

THE "REFLECTION PROPERTY" OF THE CONICS*

R. T. COFFMAN, Richland, Washington and C. S. OGILVY, Hamilton College

The ellipse has the property that a tangent at any point P makes equal angles with the two focal radii to P. This can easily be proven by an appeal to the optical principle of reflection of light from a plane mirror.

The parabola has the even more familiar property, exploited in every automobile headlight and radar-scope, that the tangent at P makes equal angles with the focal radius to P and the line through P parallel to the parabola's axis. There are various proofs, some by analytic geometry and some depending on calculus, all slightly involved.

The hyperbola also has a corresponding property, not mentioned by some texts and not even known to many students, although one ought to guess it if he has any feeling for the "unity" of the conics. It is elegantly and immediately evident as a consequence of the orthogonality of confocal central conics; but this, again, is a bit beyond the high school level.

We give here an easy method of demonstrating these properties, requiring only the notion of simple velocity vectors. This approach has the advantage of emphasizing that (1) the three different reflection properties of the ellipse, parabola, and hyperbola are essentially the same property, and that (2) this property is a direct consequence of the definitions of the curves.

First, a review of these definitions. A parabola is the path traced out by a point which moves (in the plane) so that its distance from a fixed point is always equal to its distance from a fixed line. An ellipse (hyperbola) is the path traced out by a point which moves so that the sum (difference) of its distances from two fixed points is constant.

Consider now the point P tracing out an ellipse (see Figure 1). The length of F_1P increases at exactly the rate at which F_2P decreases, in order to maintain $F_1P + F_2P =$ constant. These two equal rates of change are represented by the velocity vectors V_1 and V_2. The resultant velocity V of P itself is in the tangential direction. It can be resolved into any two perpendicular components. Thus V can be resolved into components V_1 and W_1 or it can be resolved into com-

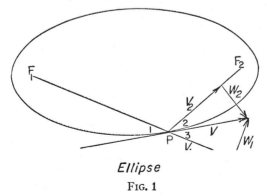

Ellipse

FIG. 1

---

* From MATHEMATICS MAGAZINE, vol. 36 (1963), pp. 11–12.

ponents V_2 and W_2. Here we have two right triangles with a common hypotenuse and two *equal* sides, V_1 and V_2. Hence the right triangles are congruent, $\angle 2 = \angle 3$, and it follows at once that $\angle 1 = \angle 2$, which is what we set out to prove.

For the hyperbola the proof is almost identical, except that F_1P and F_2P are both increasing (or both decreasing) at the same velocity to maintain $F_1P - F_2P = $ constant. The new "reflection" property is that $\angle 1 = \angle 2$.

For the parabola, it is FP and PD, the distances to the focus and to the directrix, which maintain equality and hence change lengths at the same rate. As before, $\angle 1 = \angle 2$.

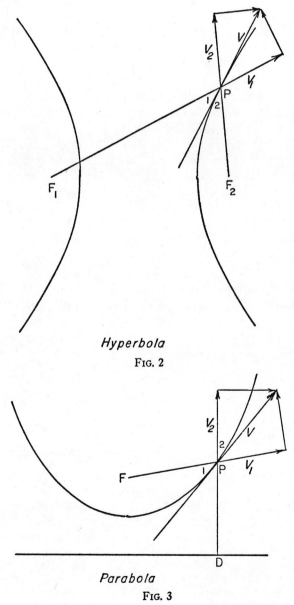

Hyperbola

FIG. 2

Parabola

FIG. 3

A STUDY OF CONIC SECTION ORBITS BY ELEMENTARY MATHEMATICS*

RAPHAEL T. COFFMAN, Richland, Washington

The level of mathematics required for deriving the equations of planetary motion by the usual approach is sufficiently high to place the subject beyond the reach of most people. Because of the importance of the problem and the general interest in the subject, particularly in this era of earth satellites, a treatment is desirable at the lowest level of mathematics which will suffice. In this paper a derivation of the inverse square law and derivations of other equations relating to planetary motion are given, using only elementary mathematics. It is believed that the approach used is to some degree original, particularly in that limit theory is not used in any of the derivations. Also, the representation of acceleration as a vector originating at the tip of the velocity vector on the position-velocity diagram has not been elsewhere observed by the writer; however, it is related to the use of the acceleration vector on the hodograph. As used, the diagram makes it possible to write equations involving simultaneously radius vector, velocity and acceleration. The method used for finding tangents to the conics has been published in this magazine (Reference 1) and is included here because it is essential to the treatment.

Except for the use of the acceleration vector on the velocity vector and the derivation of the required properties of the conics there is little or no mathematics used which should be unfamiliar to a high school student. However, the treatment here given is not intended to be such as could readily be followed at this level. Considerably more exposition than has been given would probably be required in a high school level treatment. The nature of the problem limits the extent to which even an elementary solution can be simplified.

An acceleration which acts in the direction of motion of a point may be represented by a vector attached to the tip of the velocity vector to indicate the rate of change of the length of the latter, Fig. 1-a. If the direction of acceleration does not coincide with the direction of the velocity, Fig. 1-b, the acceleration may be represented by a vector originating at the tip of the velocity vector and parallel to the direction of the acceleration. That this is true may be shown with the aid of Fig. 1-c. Here the velocity, V, has been resolved into two components, V_1 and V_2, parallel and perpendicular, respectively, to the acceleration direction. Hence the acceleration acts only on V_1 and may be represented by a vector attached to the tip of V_1 and having the same direction. The side of the vector parallelogram opposite V_1 is equal to V_1 and its tip coincides with the tip of V. Its acceleration, which is the same as that of V_1, may be represented by a vector, a, with origin at the tip of V and parallel to V_1. This vector represents the acceleration of P and shows graphically the manner in which the acceleration affects V.

If the acceleration vector, hereafter designated by a, and shown as a broken line, is resolved into components parallel and perpendicular to V, the parallel

---

* From MATHEMATICS MAGAZINE, vol. 36 (1963), pp. 271–280.

396   CONIC SECTIONS

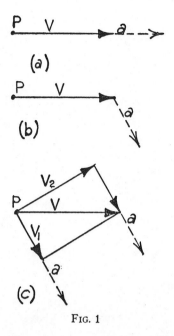

Fig. 1

component will determine the rate of change of magnitude of V and the perpendicular component will determine the rate of change of direction, or angular velocity, of V.

In dealing with orbital motion, in which a body moves under the influence of an acceleration directed toward a fixed point, the approach used in this paper is to determine the acceleration which will keep the velocity vector always tangent to the curve which is the chosen path. The basic diagram is shown in Fig. 2, where S is the fixed point, P the orbiting body, V the velocity and a the acceleration. The acceleration vector is parallel to r, the radius vector. The component of a normal to V changes the direction of V at a rate which keeps V tangent to the curve representing the path of P.

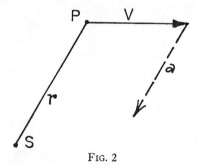

Fig. 2

To illustrate the method, consider the problem of finding an expression for the acceleration which will cause a point to move with constant velocity in a

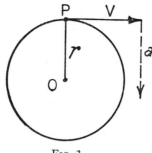

Fig. 3

circular path. See Fig. 3. Since velocity is constant, the acceleration vector must be perpendicular to V. V and r have the same angular velocity since the angle between them remains 90°. Therefore, $V/r = a/V$ or, $a = V^2/r$, and is directed to the center of the circle.

Kepler's second law of planetary motion states that the line joining the sun and a planet sweeps out equal areas in equal times.

Newton proved that this is true for any body moving in any path because of a force directed to a fixed point. Since this theorem is essential in derivations to follow, a proof is given. In Fig. 4, P is a point moving with velocity V, S is a fixed point and r is the line joining S and P. If no acceleration acts on P it

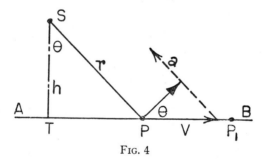

Fig. 4

moves with uniform velocity along line AB. In any equal time intervals the area swept out by r will be the area of a triangle whose base is the distance P moves in the interval and whose altitude is h, both of which are constant. Therefore, under these conditions equal areas are swept out in equal times. The rate of change of area, vA, is given by the equation: $vA = \frac{1}{2} rV \cos \theta$ where $V \cos \theta$ is the component of V perpendicular to r. If an acceleration directed to S acts on P its effect will be to move the tip of V along the line of the acceleration vector, a. This, however, does not produce a change in the vector $V \cos \theta$, hence does not affect the rate of change of area. Hence the rate of change of area is constant for any point in the path. Since $r \cos \theta = h$, $vA = \frac{1}{2} Vh$, or

(1) $\qquad Vh$ is constant,

a corollary to the above theorem in Newton's *Principia*.

In Fig. 5 CDP is a parabola with focus at F and directrix AB. At any point, P, $FP = PE$. If V_1 and V_2 are vectors showing the rate of change of length of FP and PE, they are equal and the line drawn through the intersection of

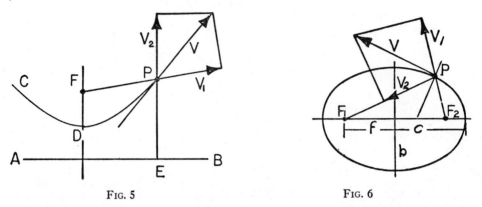

FIG. 5   FIG. 6

perpendiculars to their tips bisects the angle FPE and determines V, the velocity of P.

Fig. 6 is an ellipse with foci F_1 and F_2. P is any point on the ellipse and $PF_1 + PF_2 = 2c$, where c is the major semiaxis. If P is moving to the left PF_2 is increasing in length as shown by vector V_1 and PF_1 is decreasing at the same rate, indicated by vector V_2. The velocity, V, of P also has components perpendicular to both V_1 and V_2. The tip of V is at the point of intersection of perpendiculars drawn from the tips of V_1 and V_2. By congruence of triangles, V bisects the angle formed by the intersection of V_1 and V_2 and it follows that the perpendicular to V at P bisects angle F_1PF_2.

The direction of a point moving on a hyperbola is established in a similar manner, as shown in Fig. 7. Here F_1 and F_2 are the foci, P is any point on the curve and $F_1P - F_2P = 2c$. Since F_1P and F_2P have a constant difference in length, V_1 and V_2 are equal, and the line containing V, the velocity of P, bisects the angle F_1PF_2.

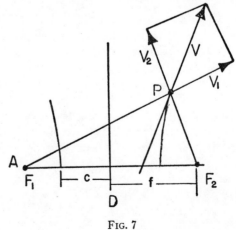

FIG. 7

A STUDY OF CONIC SECTION ORBITS

The preceding material provides the basis for determining how the acceleration toward the focus must vary with distance if a point moves in a conic. The problem will be solved first for the parabola, since the derivation is simpler than for the other two conics.

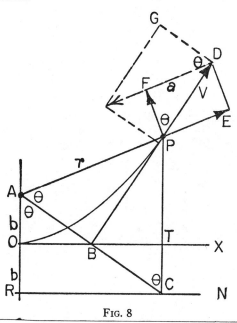

FIG. 8

In Fig. 8, P is a point on the parabola with focus A and directrix RN. OX is parallel to RN and $AO = OR = b$. The velocity vector V lies on PB, the bisector of angle APC. By plane geometry it is readily proved that the angles designated as θ are equal. Using the lower case v to indicate rate of change, or velocity, e.g. vx = rate of change of x, for the angular velocity of AP or r we have $v2\theta = V(\cos \theta)/r$, which, since $v2\theta = 2v\theta$, becomes $v\theta = V(\cos \theta)/2r$. FP remains perpendicular to r, hence has the same angular velocity. The angular velocity of V is DG/V or $a(\cos \theta)/V$ where a is the acceleration vector, which is parallel to r. The angular velocity of angle FPD is the difference between the angular velocities of FP and V or $v\theta = V(\cos \theta)/r - a(\cos \theta)/V$. By giving $v\theta$ the previously obtained value, the above becomes $V(\cos \theta)/2r = V(\cos \theta)/r - a(\cos \theta)/V$, which reduces to:

(2) $$V^2/2r = a.$$

From the law of equal area in equal time, $AB \cdot V = k$, where k is a constant, hence $V^2 = k^2/\overline{AB}^2$. By similar triangles in Fig. 8, $AB : r = b : AB$, or $\overline{AB}^2 = rb$, hence $V^2 = k^2/rb$. Substituting the right hand side of this equation for V^2 in the equation $V^2/2r = a$ we obtain:

$$a = k^2/(2r \cdot rb) = k^2/2br^2.$$

Representing $k^2/2b$, which is constant, by K we have:

(3) $$a = K/r^2$$

which shows that the acceleration toward the focus varies inversely as the square of the distance.

If equation (2) above is solved for V^2 and K/r^2 is substituted for a, the expression:

(4) $$V^2 = K \cdot 2/r$$

is obtained. If the moving point has mass, this equation gives the relation of kinetic energy of the particle to its distance from the focus.

The acceleration to the focus of an ellipse is found with the aid of Fig. 9. Here, P is a point on the ellipse with major axis $2c = r_1 + r_2$ and foci A and B. The velocity of P is V and the acceleration, a, is directed toward A. The perpendicular to V at P bisects angle APB, making angle $APB = 2\theta$. It is apparent that the other angles so designated are equal to θ.

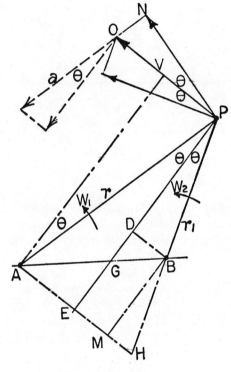

Fig. 9

The angular velocity, W_1, of r is: $W_1 = V(\cos\theta)/r$, and of r_1 is: $W_2 = V(\cos\theta)/r_1 = V(\cos\theta)/(2c-r)$, where $V\cos\theta$ is the component of V perpendicular to r and to r_1. The rate of change of angle APB, designated by $v2\theta$, is $W_2 - W_1$. Since $v2\theta = 2v\theta$ we have: $v\theta = \frac{1}{2}(V(\cos\theta)/2c - r - V(\cos\theta)/r)$ or

$$v\theta = V\cos\theta \frac{(r-c)}{r(2c-r)}.$$

Considering the rate of change of θ, as represented by angle NPO, we have, since the angular velocity of PN is the same as that of r:

$$v\theta = a(\cos \theta)/V - V(\cos \theta)/r.$$

Substituting this expression for $v\theta$ in the above equation:

$$a(\cos \theta)/V - V(\cos \theta)/r = V \cos \theta (r - c))/r(2c - r).$$

When this is reduced and solved for a, we obtain the equation:

$$(5) \qquad a = \frac{V^2}{r}\left(\frac{c}{2c-r}\right).$$

By the law of constant rate of change of area, $Vr \cos \theta = k$, where k is constant, hence, solving for V and squaring: $V^2 = k^2/(r^2 \cos^2 \theta)$. Using this value for V^2 in equation (5)

$$(6) \qquad a = \frac{k^2 c}{r^3 \cos^2 \theta (2c - r)}.$$

This expresses a in terms of two variables, r and θ. The latter is a function of r and may be eliminated by the procedure given below.

In Fig. 9, AH is perpendicular to PE, BM is parallel to PE and BD is parallel to AH. Then, $\overline{AB}^2 = \overline{AM}^2 + \overline{MB}^2$.

$$AE = r \sin \theta$$
$$BD = EM = (2c - r) \sin \theta$$
$$AM = AE + EM = r \sin \theta + (2c - r) \sin \theta = 2c \sin \theta$$
$$MB = DE = r \cos \theta - (2c - r) \cos \theta = 2(r - c) \cos \theta$$
$$\overline{AB}^2 = [2c \sin \theta]^2 + [2(r - c) \cos \theta]^2$$
$$= 4c^2 \sin^2 \theta + 4c^2 \cos^2 \theta + 4r \cos^2 \theta (r - 2c)$$
$$\overline{AB}^2 = 4c^2 + 4r \cos^2 \theta (r - 2c)$$
$$\overline{AB}^2 - 4c^2 = 4r \cos^2 \theta (r - 2c)$$

$$(7) \qquad \frac{4c^2 - \overline{AB}^2}{4} = r \cos^2 \theta (2c - r) = C.$$

The left side of this equation is a constant, hence, $r \cos^2 \theta (2c - r)$ is a constant, which may be called C. Returning to equation (6) above, it is apparent that the denominator is $r^2 C$, hence, $a = k^2 c/r^2 C$, or, calling the three constants K,

$$(8) \qquad a = K/r^2.$$

If $AB = 2f$, the expression $\overline{AB}^2 - 4c^2$ in (7) becomes $4f^2 - 4c^2$ and we have $4f^2 - 4c^2 = 4r \cos^2 \theta (r - 2c)$, or $c^2 - f^2 = r \cos^2 \theta (2c - r)$. Since $c^2 - f^2 = b^2$, where b is the minor semiaxis of the ellipse, we have $C = b^2$ and $a = k^2/r^2 \cdot c/b^2$.

If in the equation, $a = V^2/r \cdot c/(2c-r)$ we substitute K/r^2 for a and solve for V^2 we obtain:

(9) $$V^2 = K(2/r - 1/c)$$

which will be discussed later.

The derivation of the acceleration law for the hyperbola follows, in general, the same method as used for the ellipse.

In Fig. 10, P is a point on the hyperbola with foci A and B. The velocity

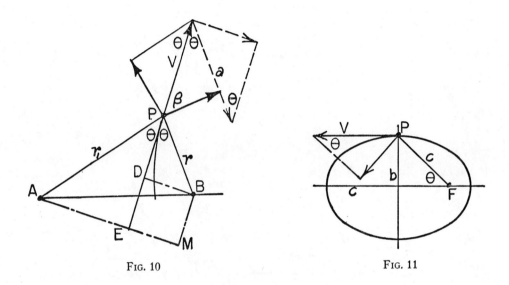

Fig. 10   Fig. 11

of P is V and the acceleration, a, is directed toward B. The constant difference of the distance of P from A and B is:

$$r_1 - r = 2c.$$

If $V \sin \theta$ is the component of V perpendicular to r and to r_1 the rate of change of angle APB is:

$$v2\theta = 2v\theta = -[V(\sin\theta)/r + V(\sin\theta)/r_1]$$
$$= -V\sin\theta[1/r + 1/(2c+r)]$$
$$v\theta = -V\sin\theta \frac{c+r}{r(2c+r)}.$$

Also

$$v\theta = -v\beta = -[V(\sin\theta)/r - a(\sin\theta)/V]$$
$$= -V\sin\theta(1/r - a/V^2).$$

Equating the two expressions for $v\theta$ and solving for a we obtain:

$$(10) \qquad a = \frac{V^2 c}{r(2c + r)}.$$

Using the relation $Vr \sin \theta = k$ to eliminate V^2, the above becomes

$$a = \frac{k^2}{r^3} \cdot \frac{c}{(2c + r) \sin^2 \theta}.$$

By the same steps as used for the ellipse it may be shown that $\frac{1}{4}\overline{AB}^2 - c^2 = r(2c+r) \sin^2 \theta$ and that

$$(11) \qquad a = K/r^2.$$

By substituting K/r^2 for a in the equation $a = V^2 c/r(2c+r)$ and solving for V^2, the following is obtained:

$$(12) \qquad V^2 = K(2/r + 1/c).$$

By the same procedure it may be proved that if the acceleration is directed away from A and the path remains a hyperbola the acceleration varies inversely as r_1^2.

It has now been shown that the law of acceleration, or attraction, to the focus of a conic is that of the inverse square. Also, it has been shown that for the ellipse $V^2 = K(2/r - 1/c)$, for the parabola $V^2 = K \cdot 2/r$ and for the hyperbola $V^2 = K(2/r + 1/c)$. These three equations show that velocity determines which conic is the path. By inspection of the equation $V^2 = K \cdot 2/r$, it is seen that as r increases V decreases, and when r is infinite V is zero. Hence the velocity of the particle in the parabolic path is just sufficient to allow the particle to travel to an infinite distance. In the same equation, if the quantity $2/r$ is decreased by $1/c$ the equation becomes that of an ellipse, $V^2 = K(2/r - 1/c)$. As stated earlier, c is the major semiaxis of the ellipse. If c is infinite the equation is that of a parabola, and as c decreases V decreases. Hence, in an ellipse the velocity is not sufficient to allow the particle to escape. If V is made zero the equation becomes $2/r - 1/c = 0$ or $r = 2c$. This indicates that in an elliptical orbit the velocity at any point is sufficient to carry the particle to a distance $2c$ from the central point.

If $1/c$ is added to $2/r$ the equation becomes that of the hyperbola, $V^2 = K(2/r + 1/c)$. If r becomes infinite this becomes $V^2 = K \cdot 1/c$, which shows the particle reaches infinity with a finite velocity.

Since kinetic energy is proportional to V^2, the three equations above are a measure of the kinetic energy of particles having mass. The difference in kinetic energy for any two distances r_1 and r_2 is the same for all three curves for which K has the same value, for the term $1/c$ is cancelled out:

$$(13) \qquad V_1^2 - V_2^2 = K(1/r_1 - 1/r_2).$$

The term in parentheses is that obtained by integrating dr/r^2 between limits r_1 and r_2 and represents the change in potential energy as the distance changes by the amount $r_1 - r_2$. Here, the expression has been obtained without integration.

Kepler's third law, that T^2 is proportional to r^3, where T is the time for one revolution and r is the mean distance of a planet from the sun, can be shown to be a consequence of the inverse square law. For a circular orbit $T = 2\pi r/V$. Since $a = V^2/r$ and $a = k/r^2$, $V = k^{\frac{1}{2}}/r^{\frac{1}{2}}$ and $T = 2\pi r^{3/2}/k^{\frac{1}{2}}$ or $T^2 = 4\pi^2 r^3/k$. In Fig. 11 the rate of change of area swept out by the radius vector c at the point P, where c equals the major semiaxis, is:

$$vA_1 = \tfrac{1}{2}cV \sin \theta$$

or, since $\sin \theta = b/c$,

$$vA_1 = \tfrac{1}{2}Vb.$$

In a circle with radius c and velocity V the rate of change of area is $vA_2 = \tfrac{1}{2}cV$. Combining the two gives

$$\frac{vA_1}{vA_2} = \frac{b}{c}.$$

The ratio of areas produced is proportional to the ratio of the rate of change of areas. When the point on the circle has made one revolution the area produced is πc^2. Therefore, in the same time:

$$\frac{vA_1}{vA_2} = \frac{A_1}{\pi c^2} = \frac{b}{c},$$

from which $A_1 = \pi bc$. This is the equation of an ellipse with semiaxes b and c. Hence, both the circular orbit and the elliptical orbit in which the major semi-axis is equal to the radius of the circle have the same period, and since Kepler's third law applies to the circle it applies also to the ellipse.

The equations which have been derived in this paper apply to particles. In order to make them applicable to physical bodies, as planets or earth satellites, it is necessary to know that the gravitational effect of a spherical body composed of homogeneous spherical shells is as though all the mass is at the center. An elementary proof of this has been derived by the writer, but because it is long and complex it has not been included here. The equation for the area of an ellipse is not generally derived below the level of integral calculus. It is, however, an easy matter to derive the equation from considerations of the relation of the ellipse to the circle.

Reference

1. The Reflection Property of the Conics. By R. T. Coffman and C. S. Ogilvy, MATHEMATICS MAGAZINE, 36, No. 1, p. 11.

BIBLIOGRAPHIC ENTRY: APPLICATIONS

1. C. T. Ruddick, The circle in Euclid's treatment of optics, AMERICAN MATHEMATICAL MONTHLY, vol. 34, p. 30.

Treatment of geometric properties "as they appear to the viewer" as opposed to "as they are".

9

ANALYTIC GEOMETRY

(a)

LINES AND PLANES

A GENERAL METHOD OF DEDUCING THE EQUATION OF A TANGENT TO A CURVE*

G. W. GREENWOOD, McKendree College

The following method of finding the equation of a tangent to a curve is more general than those usually given in texts in elementary analytical geometry, but can readily be substituted for any of them. At the same time it has many additional advantages.

Through any point $P \equiv (x',y')$ draw a line l making an angle θ with OX. If this line has a point, or points, in common with a locus given by a rational integral equation, denote such a point by Q. If $PQ = r$, the coördinates of Q are

$$x' + r\cos\theta, \qquad y' + r\sin\theta.$$

Since Q is on the locus, its coördinates satisfy the equation of the locus. Substituting, and arranging in ascending powers of r, we have

$$(a) \qquad 0 = u_0 + r u_1 + r^2 u_2 + \cdots .$$

If $u_0 = 0$, one value of r is zero; that is, one position of Q is coincident with P, or, the point P is on the locus.

Another value of r will be zero and the line l is said to have two points in common with the locus at P, if θ be chosen so that $u_1 = 0$. The line l is then said to be a *tangent* at P. [Notice that we do not have Q to coincide with P, and then speak of the tangent as a line through two *coincident* points; neither do we have Q approach P, and define the tangent as the limit of the line PQ. We are not troubled with the idea of limits.] The equation of l is then

$$y - y' = (x - x')\tan\theta,$$

where θ has the value which makes $u_1 = 0$. [If $\theta = 90°$, the equation of l is $x - x' = 0$.]

---

* From AMERICAN MATHEMATICAL MONTHLY, vol. 13 (1906), pp. 54–56.

EXAMPLE I. Find the equation of the tangent at the point $P \equiv (x',y')$ to the locus $y^2 = 4ax$. If a line l through P making an angle θ with OX have a point Q in common with the locus its coördinates

$$x' + r\cos\theta, \qquad y' + r\sin\theta,$$

where $PQ = r$, satisfy the equation of the locus.

Consequently $0 = (y'^2 - 4ax') + 2r(y'\sin\theta - 2a\cos\theta) + r^2\sin^2\theta$.

Since P is on the locus,

(1) $$y'^2 - 4ax' = 0.$$

Hence one value of r is zero for all values of θ. Another value will be zero, that is, the line will be a tangent, if $y'\sin\theta - 2a\cos\theta = 0$. The equation of the tangent is therefore

$$(y - y')y' = (x - x')2a,$$

or, by using (1),

$$yy' = 2a(x + x').$$

If $u_0 = 0$ and $u_1 = 0$ for all values of θ, then two values of r are zero for any position of l through the point P, which is then called a *double point* on the locus. When this is the case, the line l is called a tangent when θ is chosen so that $u_2 = 0$, thus making one more value of r zero.

EXAMPLE II. Find the tangent at the origin to the locus $x^2 - y^2 - 3y^3 = 0$. If a line l through the origin making an angle θ with OX have a point Q in common with the locus, its coördinates,

$$r\cos\theta, \qquad r\sin\theta,$$

where $PQ = r$, satisfy the equation.

Thus $r^2(\cos^2\theta - \sin^2\theta) - 3r^3\sin^3\theta = 0$.

Two values of r are zero for all values of θ. Hence the origin is a double point of the locus. Another value of r will be zero, that is, the line l will be a tangent, if $\cos^2\theta - \sin^2\theta = 0$; that is, if $\tan\theta = \pm 1$. We get therefore two positions of l satisfying the required condition, its equation being $y = x$ or $y = -x$.

EXAMPLE III. If an equation of the second degree represents a locus with a double point, it is the equation of two straight lines. Denote the equation by

$$ax^2 + 2hxy + by^2 + 2gx + 2fy + c = 0$$

and assume that it has a double point $P \equiv (x',y')$. Through P draw a line l making an angle θ with OX, and denote any point common to the line and the locus by Q. The coördinates of Q, viz.,

$$x' + r\cos\theta, \qquad y' + r\sin\theta,$$

where $PQ = r$, satisfy the equation of the locus.

Hence

$$0 = ax'^2 + 2hx'y' + by'^2 + 2gx' + 2fy' + c + 2r\big[(ax' + hy' + g)\cos\theta$$
$$+ (hx' + by' + f)\sin\theta\big] + r^2\big[a\cos^2\theta + 2h\cos\theta\sin\theta + b\sin^2\theta\big].$$

Since at a double point two values of r must be zero, for all values of θ, we must have

(1) $\qquad ax'^2 + 2hx'y' + by'^2 + 2gx' + 2fy' + c = 0,$

(2) $\qquad\qquad\qquad\qquad ax' + hy' + g = 0,$

(3) $\qquad\qquad\qquad\qquad hx' + by' + f = 0$

Multiplying (2) by x', (3) by y', and subtracting their sum from (1), we have

(4) $\qquad\qquad\qquad\qquad gx' + fy' + c = 0.$

Eliminating x', y' from (2), (3), (4), we have

$$\begin{vmatrix} a, & h, & g, \\ h, & b, & f, \\ g, & f, & c, \end{vmatrix} = 0,$$

which is the condition that the given equation represents two straight lines.

EXAMPLE IV. Find the tangent at the origin to the locus $x^2 + 2y^2 + 3y^3 = 0$. Proceeding as in example II, we get

$$r^2(\cos^2\theta + 2\sin^2\theta) + 3r^3\sin^3\theta = 0.$$

Two values of r are zero, so that the origin is a double point on the locus. No value of θ can be found which will make another value of r zero. Hence the locus has no real tangent at the origin. Such a point is called a *conjugate* point, and if we were to trace the curve we would find that we could not find another point satisfying the equation and as near as we pleased to the origin. [While it is true that when there are no tangents at a point on a locus, it is a conjugate point, it does not hold conversely that at a conjugate point tangents are not obtained by the usual methods. For example, we obtain $y = 0$ as the tangent at the origin to the locus $y^2 = 2x^2y + x^4y - 2x^4$, although the origin is a conjugate point.]

If $u_0 = 0$, and the value of θ which makes $u_1 = 0$ also makes $u_2 = 0$, but does not make $u_3 = 0$, the equation gives three zero values of r instead of two for the position of l given by $u_1 = 0$. Such a point is called a *point of inflexion*.

EXAMPLE V. Find the tangent at the origin to the locus $x^3 + x + y = 0$. Proceeding as in example II we get

$$0 = r(\cos\theta + \sin\theta) + r^3\sin^3\theta.$$

If θ be chosen so that $\cos\theta + \sin\theta = 0$, we get *two* additional zero values for r. Hence the origin is a point of inflexion, the tangent at that point being $x + y = 0$.

DERIVATION OF THE NORMAL FORM OF THE EQUATION OF THE STRAIGHT LINE*

KENNETH MAY, Carleton College

The derivations of the normal form of the equation of the straight line which appear most frequently in elementary texts are based on the slope-intercept form, point-slope form, or on projection. The derivation from the intercept form, although less often found, has advantages of greater simplicity, symmetry and easy extension to three dimensions. It requires merely the observation that in all cases where the intercepts exist they are given by

$$a = p/\cos \alpha \quad \text{and} \quad b = p/\sin \alpha,$$

where p is the always positive length of the normal from the origin to the line and α is the positive angle through which the positive x-axis must be rotated to coincide with the positive direction on this normal. Substitution in the intercept form yields the normal form in a single step. Lines parallel to one of the axes or passing through the origin require, as usual, special treatment.

Similarly, the normal form of the equation of the plane in space may be derived from the intercept form by noting that each intercept (when it exists) is equal to the length of the normal from the origin to the plane divided by the corresponding direction cosine. The analogy may be emphasized by defining, for the two dimensional case, a second angle between the y-axis and the normal to the line.

These derivations are easy for the student to understand and remember. They facilitate unified treatment of plane and solid analytics in an elementary course without requiring explicit discussion of direction cosines in the plane. They lend themselves to efforts toward developing the student's appreciation of mathematical form in general and of the similarities between spaces of different dimensions in particular.

ON THE EQUATION OF A LINE†

CAROL S. SCOTT, St. Petersburg Junior College

Given a point (x_1, y_1) and a slope m one usually writes the equation of the corresponding line in the form:

$$y - y_1 = m(x - x_1).$$

Similarly given two points (x_1, y_1) and (x_2, y_2) one finds the slope m and applies the above formula. Simplification of the resulting expression involves several algebraic steps which can be eliminated by the following method of procedure.

---

* From AMERICAN MATHEMATICAL MONTHLY, vol. 55 (1948), pp. 155–156.

† From AMERICAN MATHEMATICAL MONTHLY, vol. 57 (1950), p. 415.

It is well known that for the line

$$Ax + By + C = 0$$

the slope $m = (A/-B)$. So to find the desired equation, write m in fractional form and call its numerator "A" and its denominator "$-B$." The value of C can be determined mentally from:

$$A x_1 + B y_1 + C = 0.$$

Examples: Find the equation of the line through $(4, -2)$ with slope $3/5$. Then $A = 3$, $B = -5$, $C = -22$. The answer is: $3x - 5y - 22 = 0$.

Find the equation of the line through $(2, 6)$ and $(-1, 4)$. Since $m = 2/3$, $A = 2$ and $B = -3$. Using $(2, 6)$ we find $C = 14$. As an easy check we observe that $(-1, 4)$ also satisfies the resulting equation: $2x - 3y + 14 = 0$.

DISTANCE FROM A LINE, OR PLANE, TO A POINT[*][†]

J. P. BALLANTINE and A. R. JERBERT, University of Washington

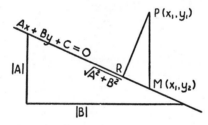

To find the perpendicular distance from the line $Ax + By + C = 0$, to the point $P(x_1, y_1)$, we begin by computing the y-distance,

(1) $$MP = y_1 - y_2 = y_1 - (-A x_1 - C)/B = (A x_1 + B y_1 + C)/B,$$

where y_2 is the y-value obtained by substituting $x = x_1$ in the equation of the line. Since the y and x intercepts of the latter are in the ratio $(-C/B)/(-C/A) = A/B$, the intercept triangle has sides proportional to $|A|$, $|B|$, $\sqrt{A^2 + B^2}$. MRP is evidently a similar triangle so that,

$$RP/MP = |B|/\sqrt{A^2 + B^2},$$

---

[*] From AMERICAN MATHEMATICAL MONTHLY. vol. 59 (1952), pp. 242–243.

[†] This paper is based on Professor J. P. Ballantine's treatment of the distance formula on page 235 of his book Essentials of Engineering Mathematics, published by Prentice Hall, 1938.

whence,

$$d = RP = \frac{|B|}{\sqrt{A^2 + B^2}} \cdot MP$$

$$= \frac{|B|(Ax_1 + By_1 + C)}{B\sqrt{A^2 + B^2}}, \qquad \text{by equation (1)},$$

(2) $$= \frac{Ax_1 + By_1 + C}{(\operatorname{signum} B)\sqrt{A^2 + B^2}} \qquad (\operatorname{signum} B = |B|/B).$$

Any $K \neq 0$, which is multiplied into the coefficients A, B, C of the equation of the line evidently "divides out" of the factors in the right member of equation (2). These factors and their product d are therefore invariant under such a multiplication. The second factor MP, yields $+$ and $-$ values for points above and below the line, respectively, and this remains true for d since the first factor, $|B|/\sqrt{A^2+B^2}$, is always positive.

In three dimensions the x, y, and z distances from the plane $A_1x+A_2y+A_3z+A_4=0$, to the point $P(x_1, y_1, z_1)$ are given by the expressions,

$$(A_1x_1 + A_2y_1 + A_3z_1 + A_4)/A_i, \qquad i = 1, 2, 3,$$

respectively. To obtain the perpendicular distance we multiply by the cosine of the acute angle between the chosen axis and the normal to the plane, *i.e.* by $|A_i|/\sqrt{A_1^2+A_2^2+A_3^2}$. Thus, as before,

$$d = (A_1x_1 + A_2y_1 + A_3z_1 + A_4)/(\operatorname{signum} A_i)\sqrt{A_1^2 + A_2^2 + A_3^2}.$$

In this manner d is $+$ or $-$ according as the point P is right or left, front or back, above or below, the plane, respectively.

DETERMINANTS AND PLANE EQUATIONS*

R. R. STOLL, Oberlin College

In an elementary exposition of solid analytic geometry it is quite common to find the equation of the plane determined by three noncollinear points (x_i, y_i, z_i), $i=1, 2, 3$, written in the form

$$\begin{vmatrix} x & y & z & 1 \\ x_1 & y_1 & z_1 & 1 \\ x_2 & y_2 & z_2 & 1 \\ x_3 & y_3 & z_3 & 1 \end{vmatrix} = 0.$$

That determinants can also be used to advantage in deriving the equations of planes satisfying other standard conditions, seems to have been overlooked.

---

* From AMERICAN MATHEMATICAL MONTHLY, vol. 61 (1954), pp. 255–256.

For example, the equation of the plane through the points (x_i, y_i, z_i), $i=1, 2$, and perpendicular to the plane with equation $ax+by+cz=d$ is

(1)
$$\begin{vmatrix} x & y & z & 1 \\ x_1 & y_1 & z_1 & 1 \\ x_2 & y_2 & z_2 & 1 \\ a & b & c & 0 \end{vmatrix} = 0.$$

Indeed, the only novelty of the statement, *viz.* the perpendicularity of the given plane and (1), follows from the relation

$$\begin{vmatrix} a & b & c & 0 \\ x_1 & y_1 & z_1 & 1 \\ x_2 & y_2 & z_2 & 1 \\ a & b & c & 0 \end{vmatrix} = 0$$

upon expanding the determinant by the elements of the first row, since the coefficients of a, b, c are the coefficients of x, y, z, respectively in (1).

Again, the equation of the plane through (x_1, y_1, z_1) and perpendicular to each of the planes $a_i x + b_i y + c_i z = d_i$, $i=1, 2$ (or in other words, perpendicular to a given line) is

$$\begin{vmatrix} x & y & z & 1 \\ x_1 & y_1 & z_1 & 1 \\ a_1 & b_1 & c_1 & 0 \\ a_2 & b_2 & c_2 & 0 \end{vmatrix} = 0.$$

Finally, it is amusing and sometimes profitable to discuss in a beginning class the equation

$$\begin{vmatrix} x & y & z & 1 \\ a_1 & b_1 & c_1 & 0 \\ a_2 & b_2 & c_2 & 0 \\ a_3 & b_3 & c_3 & 0 \end{vmatrix} = 0$$

which is suggested by the foregoing as a possibility for the equation of a plane which is perpendicular to each of the three planes $a_i x + b_i y + c_i z = d_i$, $i=1, 2, 3$.

ON THE SLOPES OF PERPENDICULAR LINES*

S. LEADER, Rutgers University

The usual proof that the slopes of lines which meet at right angles are negative reciprocals is based upon a trigonometric identity. A geometric proof can be given using the theorem that the interior altitude of a right triangle is the geometric mean of the segments into which it divides the hypotenuse.

Let O be the intersection of the lines l_1 and l_2 which meet at right angles. We exclude the case in which either line is vertical by assuming both lines have slopes. Draw a horizontal segment OA extending one unit to the right of O. Draw a vertical line through A meeting l_1 at B_1 and l_2 at B_2. Then the slopes m_1 and m_2 are given respectively by the directed vertical segments AB_1 and AB_2. Since these segments have opposite directions, the product $m_1 m_2$ is negative. Now OA is the interior altitude of the right triangle $B_1 O B_2$. Therefore, $|OA|^2 = |AB_1||AB_2|$. That is, $1 = |m_1 m_2|$. Finally, since $m_1 m_2$ is negative, $m_1 m_2 = -1$.

THE STRAIGHT LINE TREATED BY TRANSLATION AND ROTATION†

KENNETH MAY, Carleton College

Transformations are usually introduced rather late in the first course in analytical geometry and appear as hardly more than a device to facilitate the graphing of conics. What should be a fundamental notion is merely a footnote. The following remarks indicate one way in which an early introduction of the equations of translation and rotation can be turned to immediate account in developing the basic theorems on the straight line. At the same time it appears that the usual proofs based on the geometrically established constancy of the slope are redundant. The nature of the graph of $Ax + By = C$ follows from the linearity of the axes without reference to the notion of slope, which can then be introduced and proved constant analytically.

Consider the equation $Ax + By = C$, where if $A = 0$, $B \neq 0$; if $B = 0$, $A \neq 0$; or $A \neq 0$, $B \neq 0$. Suppose $A = 0$. Then if $C = 0$, the graph is evidently precisely the x-axis since there and only there is $y = 0$. If $C = 0$, we translate the origin to the point $(0, \frac{C}{B})$, and the graph is now seen to coincide with the x-axis. The case $B = 0$ is treated similarly. If neither A nor B is zero, we translate the origin to the point $(\frac{C}{A}, 0)$, and the equation becomes $Ax + By = 0$. Rotation through an angle θ yields

$$(A \cos\theta + B \sin\theta)x + (B \cos\theta - A \sin\theta)y = 0.$$

Evidently the first term vanishes if $\tan\theta = -\frac{A}{B}$. Moreover, for this θ the coefficient

---

* From AMERICAN MATHEMATICAL MONTHLY, vol. 65 (1958), p. 525.
† From MATHEMATICS MAGAZINE, vol. 22 (1949), p. 211.

of y cannot vanish. Hence the equation becomes $y=0$ and its graph is coincident with the x-axis.

Thus $Ax + By = C$ is shown to be the equation of a straight line since it is in all cases congruent to an x-axis. Moreover, the argument indicates the relation between the coefficients of the equation and the slope and inclination of the line. The special forms of the equation and the constancy of the ratio $\dfrac{(y_2 - y_1)}{(x_2 - x_1)}$ follow immediately.

POINT TO LINE DISTANCE*

WILLIAM R. RANSOM, Tufts University

The derivation of the formula for the distance from a point to a line usually employs a complicated diagram: a simpler diagram will serve.

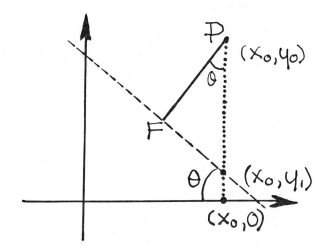

Let (x_0, y_0) be the point and $y = mx + b$ be the line. Then $PF = (y_0 - y_1)\cos\theta$. Since $\tan\theta = m$, $\sec\theta = \sqrt{1 + m^2}$, and $\cos\theta = 1/\sqrt{1 + m^2}$. Since $y_1 = mx_0 + b$, $PF = (y_0 - mx_0 - b)/\sqrt{1 + m^2}$.

An even simpler diagram, which omits the dotted line, will serve. If the point (x_0, y_0) is not on the line we have

$$Ax + By + C = 0,$$
$$Ax_0 + By_0 + C = L \neq 0.$$

Subtraction gives

$$A(x - x_0) + B(y - y_0) = 0$$

---

* From MATHEMATICS MAGAZINE, vol. 33 (1960), p. 218.

while
$$B(x-x_0)-A(y-y_0)=0$$
represents the line on which lies the perpendicular distance PF.

Solving this last pair of equations for the intersection, F, and using the abbreviation $H^2=A^2+B^2$, we get
$$x-x_0=-AL/H^2 \quad \text{and} \quad y-y_0=-BL/H^2$$
and as these are the rise and run of the distance PF, we get
$$(PF)^2=(A^2L^2+B^2L^2)/H^4$$
whose square root gives the distance as L/H, which is
$$(Ax+By+C)/\sqrt{A^2+B^2}\ .$$

THE DISTANCE FORMULA AND CONVENTIONS FOR SIGN*

THOMAS E. MOTT, The Pennsylvania State University

As a mathematician, I have always felt secure in the belief that ours is a very orderly and consistent science. Yet as a teacher faced with the task of converting others to my belief, I have often to be extremely careful lest some inconsistency creep into my own lectures. One of these occasions has been concerned with the distance formula in Analytic Geometry. The troublesome point being the convention to be adopted for the sign of this distance. Therefore I propose here to give an explanation of three such conventions, two of which are common in most texts on the subject.

Since there are a number of satisfactory proofs of the distance formula in Analytic Geometry, we shall merely be concerned here with the result itself. However, a proof which does not require the normal form, such as is to be found in "Analytic Geometry" by John W. Cell, would seem the most suitable; for I do not require the normal form in this paper. If $ax+by+c=0$ is the equation of a line then
$$d=\frac{ax_1+by_1+c}{\pm\sqrt{a^2+b^2}}$$
is the distance from this line to the point $p(x_1,y_1)$ in the plane. Notice that we have yet to adopt a convention for the sign $\pm$, and that we speak of distance from the line to the point. We shall at first be concerned only with oblique lines, hence $a\cdot b\neq 0$.

The first convention under consideration is that the sign of the radical be the same as the sign of b. This will then provide a distance d which is positive when p is "above" the line and negative when p is "below" the line. But what is meant here

---

* From MATHEMATICS MAGAZINE, vol. 35 (1962), pp. 39–42.

THE DISTANCE FORMULA AND CONVENTIONS FOR SIGN

by "above" and "below" the line? Perhaps it would be better to say "above in the y sense" and "below in the y sense." For by p "above the line in the y sense," we mean that the vertical projection of p on the line is below p, while p "below the line in the y sense" means that the vertical projection of p on the line is above p. The proof of this propostion is as follows:

The line $ax+by+c=0$ divides the plane into two half planes, the half plane "above" the line and the half plane "below" the line. These half planes contain respectively all the points "above the line in the y sense" and all the points "below the line in the y sense." Let $ax+by+c_1=0$ be the equation of the line thru $p(x_1,y_1)$ and parallel to the given line $ax+by+c=0$, and assume that p is "above the line in the y sense." Since the line $ax+by+c_1=0$ is "above" the line $ax+by+c=0$, then considering the y intercepts of these lines, we obtain

$$-\frac{c_1}{b} > -\frac{c}{b}.$$

If $b>0$ we have $c_1<c$ and if $b<0$ we have $c_1>c$, hence

$$0=ax_1+by_1+c_1 < ax_1+by_1+c \quad \text{if } b>0$$

and

$$0=ax_1+by_1+c_1 > ax_1+by_1+c \quad \text{if } b<0.$$

But in either case $(ax_1+by_1+c)/(\operatorname{sgn}b)$ is positive. Similarly we find that $(ax_1+by_1+c)/(\operatorname{sgn}b)$ is negative if the point $p(x_1,y_1)$ is "below the line in the y sense." Therefore we now have the desired result that

$$d = \frac{ax_1+by_1+c}{(\operatorname{sgn}b)\cdot\sqrt{a^2+b^2}}$$

is positive if $p(x_1,y_1)$ is "above" the line and negative if $p(x_1,y_1)$ is "below" the line.

By an argument analogous to that given above we treat the case of choosing the sign of the radical the same as the sign of a. We obtain

$$d = \frac{ax_1+by_1+c}{(\operatorname{sgn}a)\cdot\sqrt{a^2+b^2}}$$

positive if $p_1(x_1,y_1)$ is "above the line in the sense of x" and negative if $p_1(x_1,y_1)$ is "below the line in the sense of x."

Next we consider the convention which assigns to the radical the sign of c. This will provide a distance d which is positive when the origin is on the same side of the line as the point $p(x_1,y_1)$ and negative when on the opposite side.

Let the points $p(x_1,y_1)$ and $(0,0)$ be on the same side of the line $ax+by+c=0$ and assume that this is the side "above the line in the y sense." Then on considering the origin we obtain

$$\frac{c}{(\operatorname{sgn}b)\cdot\sqrt{a^2+b^2}} > 0$$

so that $(\operatorname{sgn} c) = (\operatorname{sgn} b)$. Therefore
$$\frac{ax_1 + by_1 + c}{(\operatorname{sgn} c) \cdot \sqrt{a^2 + b^2}} = \frac{ax_1 + by_1 + c}{(\operatorname{sgn} b) \cdot \sqrt{a^2 + b^2}} > 0.$$
On the other hand if the points $p(x_1, y_1)$ and $(0, 0)$ were both on the side "below the line in the y sense" then $(\operatorname{sgn} c) = -(\operatorname{sgn} b)$ and
$$\frac{ax_1 + by_1 + c}{(\operatorname{sgn} c) \cdot \sqrt{a^2 + b^2}} = \frac{-(ax_1 + by_1 + c)}{(\operatorname{sgn} b) \cdot \sqrt{a^2 + b^2}} > 0.$$
Thus in either case we have
$$d = \frac{ax_1 + by_1 + c}{(\operatorname{sgn} c) \cdot \sqrt{a^2 + b^2}} > 0$$
if the points $p(x_1, y_1)$ and $(0, 0)$ are on the same side of the line. By a similar argument one readily verifies that
$$d = \frac{ax_1 + by_1 + c}{(\operatorname{sgn} c) \cdot \sqrt{a^2 + b^2}} < 0$$
if the points $p(x_1, y_1)$ and $(0, 0)$ are on opposite sides of the line.

Finally let us consider the lines which are parallel to the x or y axis, that is lines $ax + by + c = 0$ for which a or b is zero. Let us consider first the line $by + c = 0$ which is parallel to the x axis. The only concept for "above" and "below" which is now admissible being in the sense of y. Whether $p(x_1, y_1)$ is above or below the line the distance with correct sign is $y_1 - y = y_1 + c/b$. From the distance formula we also obtain
$$d = \frac{by_1 + c}{(\operatorname{sgn} b) \cdot \sqrt{b^2}} = \frac{by_1 + c}{b} = y_1 + \frac{c}{b}.$$
Thus the distance formula is valid for lines parallel to the x axis, and by a similar argument one sees that the distance formula is valid for lines parallel to the y axis. It is important to notice however, that the distance formula used with lines parallel to the x axis must be the one with $(\operatorname{sgn} b)$ in the denominator, and that used for lines parallel to the y axis must be the one with $(\operatorname{sgn} a)$ in the denominator.

Since the formula involving $(\operatorname{sgn} b)$ is valid for lines parallel to the x axis then we derive the formula involving $(\operatorname{sgn} c)$ for lines parallel to the x axis just as we did above for oblique lines. But only the formula involving $(\operatorname{sgn} a)$ is valid for lines parallel to the y axis, hence to derive the formula involving $(\operatorname{sgn} c)$ for lines parallel to the y axis, we merely replace the argument given above for oblique lines by one involving $(\operatorname{sgn} a)$ instead of $(\operatorname{sgn} b)$. Thus the formula
$$d = \frac{ax_1 + by_1 + c}{(\operatorname{sgn} c) \cdot \sqrt{a^2 + b^2}}$$
is valid for all lines in the plane, and is positive if the points $p(x_1, y_1)$ and $(0, 0)$ are on the same side of the line $ax + by + c = 0$ and negative if on opposite sides.

The proof may be more complicated, but certainly the results stated above remain valid for the distance formula in Solid Analytic Geometry. The most common form would then be that

$$d = \frac{ax_1 + by_1 + cz_1 + d}{(\operatorname{sgn} c) \cdot \sqrt{a^2 + b^2 + c^2}},$$

with d then positive if the point (x_1, y_1, z_1) is above the plane $ax + by + cz + d = 0$ in the z sense and negative if below it.

AN EASY WAY FROM A POINT TO A LINE*

R. L. EISENMAN, Office of Secretary of Defense, Systems Analysis and Air Force Academy

We should and do publicize the use of coordinate geometry for concise proofs in synthetic geometry. Do we recognize that synthetic geometry can return the favor? This note suggests such an application to derive the distance from a point to a line via similar triangles.

In its fancier dress as "orthogonal projection," distance from a point to a line is a forerunner for orthogonal functions and it is a working tool in such applications as interpreting measures of good fit in statistics.

We usually prove distance from (a, c) to $y = mx + b$ is $(ma + b - c)/\sqrt{1 + m^2}$ by intricate algebra to locate the foot of the perpendicular, and end with a discussion of the significance of the algebraic sign. Consider instead this basic diagram with the numbered observations following in turn:

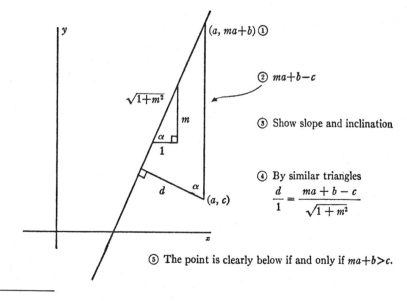

---

* From MATHEMATICS MAGAZINE, vol. 42 (1969), pp. 40–41.

The symmetric version, $a_1x+a_2y+a_3=0$, is useful for extension to more dimensions. Perhaps the students would enjoy formulating an educated guess of the formulas for distances in a general number of dimensions.

BIBLIOGRAPHIC ENTRIES: LINES AND PLANES

Except for the entries labeled MATHEMATICS MAGAZINE, the references below are to the AMERICAN MATHEMATICAL MONTHLY.

1. P. Franklin, The geometric interpretation of some formulas of analytic geometry, vol. 40, p. 143.

Use of determinants in n-space geometric problems.

2. C. B. Boyer, Clairaut and the origin of the distance formula, vol. 55, p. 556.

Gives Clairaut credit for the distance formula.

3. A. Porges, The rotation of axes, vol. 64, p. 37.

Use of DeMoivre's theorem to derive the equations for rotation of axes.

4. F. Scheid, The under-over-under theorem, vol. 68, p. 862.

Fitting a straight line to a set of points in the plane.

5. R. Lariviere, Bisectors of supplementary angles: a new look, MATHEMATICS MAGAZINE, vol. 33, p. 25.

6. C. N. Mills, More about the normal equation of the line $Ax+By+C=0$, MATHEMATICS MAGAZINE, vol. 34, p. 35.

7. T. A. Brown, The distance of a point from a line, MATHEMATICS MAGAZINE, vol. 37, p. 157.

8. D. H. Moore, Distance from line to point, MATHEMATICS MAGAZINE, vol. 38, p. 219.

(b)

CURVES AND SURFACES

THE HYPERBOLOID AS A RULED SURFACE*†

JAMES PERRY WILSON, Columbia University

If a straight line is revolved around another line not in the same plane, as indicated in the figure below, it will generate a hyperboloid of one sheet.

Take the Z-axis as the fixed line. We can then fix the other axes so that the generating line, in its initial position BD, will be parallel to the YZ-plane, and will

---

* From AMERICAN MATHEMATICAL MONTHLY, vol. 18 (1911), pp. 158–159.

† This note was presented by a student in a beginner's class in solid analytical geometry. Its extreme simplicity seems to warrant its publication.

THE HYPERBOLOID AS A RULED SURFACE

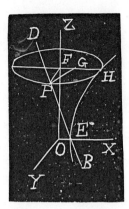

intersect the X-axis at E. The equations of the generating line in this position will then be of the form $x=a, y=mz$.

As the line is revolved around the Z-axis, every point on the line will generate a circle around the Z-axis; and the locus of the intersections of these circles with the XZ-plane will indicate the nature of the surface generated by the line.

Consider any point $P(a, mz, z)$ on the line BD. As the line is revolved around OZ, P will describe a circle around OZ, whose radius $FP = \sqrt{(FG^2 + GP^2)} = \sqrt{(a^2 + m^2z^2)}$. This circle will intersect the XZ-plane at H; then

(1) $$FH = x = \sqrt{(a^2 + m^2z^2)}.$$

Since P is any point on the generating line, in its initial position, equation (1) represents the locus of the intersections with the XZ-plane, of the circles generated by all the points on the moving line.

From (1), by rationalizing, transposing, and dividing by a^2, we obtain

$$\frac{x^2}{a^2} - \frac{m^2z^2}{a^2} = 1.$$

Putting $\frac{a^2}{m^2} = c^2$, we obtain

(2) $$\frac{x^2}{a^2} - \frac{z^2}{c^2} = 1,$$

which is the standard equation of the hyperbola; and the surface generated by revolving BD around OZ is the same as that generated by revolving the hyperbola (2) around OZ, its conjugate axis; that is, the hyperboloid

$$\frac{x^2}{a^2} + \frac{y^2}{a^2} - \frac{z^2}{c^2} = 1.$$

If the moving line is perpendicular to the fixed line, it will lie wholly in the plane $Z = k$, and the hyperboloid will degenerate into the plane figure $Z = k, x^2 + y^2 \stackrel{=}{>} a^2$, a in this case being the shortest distance between the lines OZ and BD.

SIMPLE HINTS ON PLOTTING GRAPHS IN ANALYTIC GEOMETRY*

AUBREY KEMPNER, Urbana, Ill.

In courses in analytic geometry, much importance is generally attached to the problem of plotting curves from given equations. It is therefore surprising that in most textbooks on analytic geometry some very simple and effective hints on plotting should be omitted. The contents of the present note are well known; for §§ III, IV compare the article in the MONTHLY for November, 1916, on "Graphical constructions for a function of a function and for a function given by a pair of

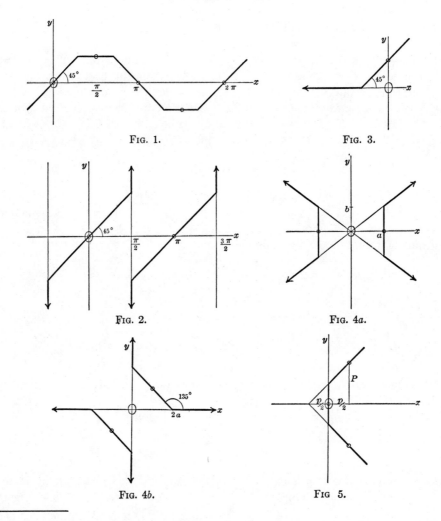

Fig. 1.

Fig. 2.

Fig. 3.

Fig. 4a.

Fig. 4b.

Fig 5.

---

* From AMERICAN MATHEMATICAL MONTHLY, vol. 24 (1917), pp. 17–21.

parametric equations," by Professor W. H. Roever and his reference to Professor E. H. Moore.

    I. In many cases it is very easy to construct a few tangents to the curve with the corresponding points of tangency, and thus to gain a skeleton of the curve. Besides being a help in the actual plotting, such a skeleton is a valuable aid to the memory. In Figs. 1–5 these skeletons are constructed for certain standard curves. The unit of measurement is supposed to be the same on both axes. Asymptotes are marked with an arrow-tip.

Figure 1: $y = \sin x$ (x in radians*), and similarly $y = \cos x$, and $y = \arcsin x$, $y = \arccos x$.
Figure 2: $y = \tan x$, and similarly $y = \cot x$, $y = \arctan x$, and $y = \text{arc}\cot x$.
Figure 3: $y = e^x$, and similarly $y = \log_e x$.
Figure 4a: The well-known figure for the hyperbola $x^2/a^2 - y^2/b^2 = 1$. (The corresponding figure for the ellipse consists simply of a rectangle, and is not here drawn.)
Figure 4b: The equilateral hyperbola $x \cdot y = a^2$.
Figure 5: The parabola $y^2 = 2px$.

    II. To plot the graph of an equation of the type $y^2 = f(x)$ (or, similarly, of $x^2 = \varphi(y)$).

    Rule: Construct the curve whose equation is $Y = f(x)$, using for Y-axis the original y-axis, and change the ordinates,† replacing Y by $y = \sqrt{Y}$.

    The following relations between the curves $y^2 = f(x)$ and $Y = f(x)$ are evident:
    (a) $y^2 = f(x)$ is symmetric with respect to x-axis.
    (b) Corresponding to every loop of $Y = f(x)$ above the x-axis, $y^2 = f(x)$ has an oval.
    (c) The parts of $Y = f(x)$ below the x-axis do not yield real points of $y^2 = f(x)$.
    (d) When $0 < Y < 1$, the curve $y^2 = f(x)$ lies above the curve $Y = f(x)$; and when $Y > 1$ it lies below.
    (e) The curves $Y = f(x)$, $y^2 = f(x)$ cross each other in all points where $Y = 0$ and where $Y = 1$.

    Curves of this type are, for example, the conic sections, the semi-cubical parabola, the Cissoid, the Strophoid, the Conchoid, the Versiera, when the equations are given in the forms commonly used.

Figure 6: The conchoid $y^2 = \dfrac{1}{x^2} \cdot (x+c)^2 \cdot (a^2 - x^2)$, where $a = 1$, $c = \frac{1}{2}$. The dotted curve is $Y = \dfrac{1}{x^2} \cdot (x+c)^2 \cdot (a^2 - x^2)$.

---

    * The fraction $\frac{22}{7}$, as an approximate value for π, is exact to about $1/2000$ of the value of π, and is therefore sufficiently accurate for all graphical purposes.
    † This method of construction is really nothing but a particularly simple case of changing, along the Y-axis, from a uniform scale to a certain non-uniform scale. See, for example, C. RUNGE, *Graphical Methods*, 1912, § 6, for the general method.

Figure 7: The parabola $y^2 = 2px$ (dotted line $Y = 2px$).

Figure 8: The ellipse $\dfrac{x^2}{a^2} + \dfrac{y^2}{b^2} = 1$, or

$$y^2 = b^2 - \frac{b^2}{a^2}x^2 \left(\text{dotted curve } Y = b^2 - \frac{b^2}{a^2}x^2\right),$$

and the hyperbola $\dfrac{x^2}{a^2} - \dfrac{y^2}{b^2} = 1$, or

$$y^2 = -b^2 + \frac{b^2}{a^2}x^2 \left(\text{dotted curve } Y = -b^2 + \frac{b^2}{a^2}x^2\right).$$

The left half of the figure refers to the ellipse, the right half to the hyperbola. The two parabolas are congruent. $b = \tfrac{1}{2}\sqrt{2}$, $a = 1$.

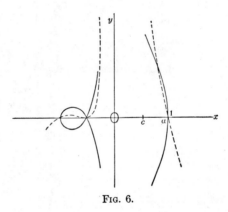

FIG. 6.

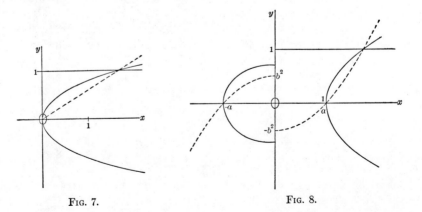

FIG. 7. FIG. 8.

III. To construct the graph of $F(x,y)=0$ when the equation is given in parametric form: $x=\varphi(t)$, $y=\psi(t)$.

*Rule:** In a system of coördinates I, II (see Fig. 9) plot first $y=\psi(t)$, using I for the t-axis and II for the y-axis, and also $x=\varphi(t)$, using II for the t-axis and I for the x-axis. Next, draw the line of slope unity through the origin, assume any point on this line, draw through this point a line parallel to the axis II to its intersection (D in Fig. 9) with the curve $y=\psi(t)$, another line parallel to I to its intersection (E in Fig. 9) with the curve $x=\varphi(t)$, and complete the rectangle. The fourth vertex (G in Fig. 9) is a point $F(x,y)=0$, with I for x-axis and II for y-axis.

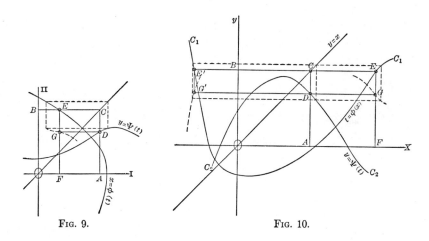

Fig. 9.   Fig. 10.

Proof. $AD=\psi(OA)$, $BE=\varphi(OB)$. (See Fig. 9.) Let $OA=OB=t$, then $BE=OF=\varphi(t)=x$, and $AD=FG=\psi(t)=y$.

IV. By interchanging x and t in $x=\varphi(t)$ we obtain
$$t=\varphi(x), \qquad y=\psi(t)=\psi(\varphi(x)),$$
and have thus the following rule for plotting the graph of a function of a function.

Rule: Plot in the $x-y$ system of coördinates the two curves (see Fig. 10)

C_1: $t=\varphi(x)$, using for t-axis the y-axis,

C_2: $y=\psi(t)$, using for t-axis the x-axis.

Draw the line $y=x$, assume any point C on it, draw through this point a line parallel to the y-axis to its intersection (D) with C_2, and another line parallel to the x-axis to its intersections (E and E' in Fig. 10) with C_1, and complete the rectangle. The fourth vertices (G and G' in Fig. 10) are points of the graph of $y=\psi(\varphi(x))$.†

---

* Compare Professor Roever's article referred to above.

† For another kind of graphical representation of a function of a function compare, for example, C. RUNGE, *Graphical Methods*, 1912, § 6.

Proof. $AD = \psi(OA)$, $OB = \varphi(BE)$. (See Fig. 10.) Let $OA = OB = t$, then $AD = FG = \psi(t)$, $t = \varphi(BE) = \varphi(OF)$, and $FG = \psi(\varphi(OF))$.

In order to reduce the number of cuts, no illustrative examples for Sections III, IV have been given. The reader is referred to the examples worked out in Professor Roever's paper.

From the rule of construction in the present paragraph one can derive in geometrical form the conditions which must be satisfied in order to have $\psi(\varphi(x)) \equiv \varphi(\psi(x))$.

By taking for φ and ψ the same function one finds a simple method for plotting the iterated function $\phi_2(x) = \phi(\phi(x))$; repeating the process for $\phi_2(x)$ instead of for $\phi(x)$, $y = \phi_4(x) = \phi\phi\phi\phi(x)$, and in the same way $y = \phi_k(x)$, may be plotted, when k is any power of 2. When k is not a power of 2, the construction of this paragraph generally leads to complicated figures.

A NEW CURVE CONNECTED WITH TWO CLASSIC PROBLEMS*

G. M. JUREDINI, Syracuse University

The curve is the locus of a point on a defined triangle which revolves about a circle in a particular manner. In Fig. 1, it is the path traced by the point P, which is the foot of the perpendicular let fall onto the hypotenuse of a right-angle triangle, RQX, the triangle revolving about a circle in such a manner that one of its sides, RQ, is always equal to and an extension of a radius of the circle, and the other side, QX, always reaches a fixed diameter of the circle, OX, as initial line, produced.

The equation for the curve may be deduced as follows: In Fig. 1, we have angle ROX = angle $ORX = \phi$. Let angle $POX = \theta$, radius of circle, or $RQ = a$, and line $PO = r$. Then, angle $RXO = 180° - 2\phi$, $OX = a \cdot \sec\phi$, $RP = a \cdot \cos\phi$, $PX = a(\sec\phi - \cos\phi)$, so that, in triangle PXO, we have $\sin(180° - 2\phi)/r = \sin\theta/a(\sec\phi - \cos\phi)$,

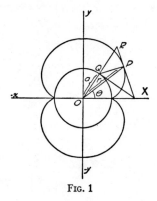

Fig. 1

---

* From AMERICAN MATHEMATICAL MONTHLY, vol. 33 (1926), pp. 377–378.

and from this it follows readily that

(1) $$\sin\theta = 2(a/r)(\sin^3\phi).$$

Also, in triangle PRO, we have $r^2 = 4a^2 + a^2\cos^2\phi - 4a^2\cos^2\phi$, $\cos^2\phi = (4a^2 - r^2)/3a^2$,

(2) $$\sin^2\phi = 1 - (4a^2 - r^2)/3a^2 = (r^2 - a^2)/3a^2.$$

From (1) and (2), we get

(3) $$\sin\theta = 2a/r\left(\frac{r^2-a^2}{3a^2}\right)^{3/2}, \quad \text{or} \quad \sqrt[3]{y^2} = \frac{\sqrt[3]{4a^2}}{3a^2}(x^2+y^2-a^2)$$

which is the equation of the curve.* The graph takes the form indicated in Fig. 1.

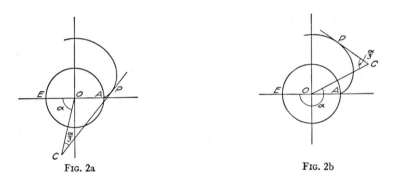

Fig. 2a   Fig. 2b

Application to Trisection of an Angle: In Fig. 2, let A be the starting point of the curve, and let EOC be the angle required to be trisected. Place OE on the side of the initial line that is opposite A, and measure off OC equal to diameter of circle. From C draw tangent to curve at P. Then PCO is one-third of EOC.

Application to Duplication, or Dimidiation of a Cube: In Fig. 1, let $QP = b$, and let P be at a distance $y = a$ from the line OX. From the right triangle QPR we have $\sin\varphi = (b/a)$, and when this value is inserted in (1) there results, after easy simplification,

(4) $$2b^3 = a^3.$$

---

* The curve of this note is a two-cusped epicycloid as may be verified by eliminating the parameter φ from the equations:

$$2x = a(3\cos\varphi - \cos 3\varphi), \quad 2y = a(3\sin\varphi - \sin 3\varphi).$$

For such a curve, $(dy/dx) = \tan 2\varphi$, and the given construction is a consequence.

Thus the author has added a new curve to the galaxy of trisectrices already known. Concerning such curves Henri Brocard (*Notes de Bibliographie des Courbes Géométriques*, 1897, p. 290) says "Trisectrices are infinite in number. One may mention: the quadratrix of Dinostratos, the spiral of Archimedes, the limaçon $r = 1 + 2\cos\theta$, the conchoid of Nicomedes, the equilateral hyperbola, Maclaurin's trisectrix, and a host of others listed by Aubry, *Journal de Mathématiques Spéciales*, 1896, pp. 76–84, 106–112."

N. Anning

AN UNUSUAL SPIRAL*

L. S. JOHNSTON, University of Detroit

We discuss in this note a spiral differing from the more familiar spirals in that its whorls all have common point, and that all its whorls have a common tangent at this common point. The writer encountered the spiral while investigating a problem submitted to him by a member of the designing staff of a company manufacturing reflectors.† This problem may be stated as follows:

Using the convention polar coordinate net, let $(a, \pi/2)$ be the coordinates of a given point P_0. Let the perpendicular to OP_0 at P_0 intersect the line $\theta = \pi/4$ at P_1, the perpendicular to OP_1 at P_1 intersect the line $\theta = \pi/8$ at P_2, etc., thus establishing the sequence $[P_n]$ of points $P_0, P_1, P_2, \ldots$. We wish to find

(a) the limit point P_∞ of $[P_n]$,
(b) a continuous function $\rho = f(\theta)$ which includes $[P_n]$ and P_∞, and
(c) an elementary geometric construction by which other points on $\rho = f(\theta)$ may be located.

We discuss these parts in order.

(a) Let $OP_i = \rho_i$ $(i = 0, 1, 2, \ldots)$ and $\theta_i = $ angle $P_i OX$, OX being the line $\theta = 0$. We then have

$$\rho_0 = a, \rho_1 = a \sec \frac{\pi}{4}, \rho_2 = a \sec \frac{\pi}{4} \sec \frac{\pi}{8}, \ldots$$

(1) $$\rho_n = a \sec \frac{\pi}{4} \sec \frac{\pi}{8} \sec \frac{\pi}{16} \cdots \sec \frac{\pi}{2^{n+1}}, \text{ and}$$

(2) $$\theta_0 = \frac{\pi}{2}, \theta_1 = \frac{\pi}{4}, \theta_2 = \frac{\pi}{8}, \ldots, \theta_n = \frac{\pi}{2^{n+1}}.$$

From the identity $\sec \theta = 2 \sin \theta / \sin 2\theta$ we have

$$\rho_n = 2^n a \sin \frac{\pi}{2^{n+1}} = \frac{\pi a}{2} \left[\frac{\sin \frac{\pi}{2^{n+1}}}{\frac{\pi}{2^{n+1}}} \right].$$

Hence

$$\rho_\infty = \lim_{n \to \infty} \rho_n = \frac{\pi a}{2}.$$

Also, since

$$\lim_{n \to \infty} \theta_n = 0,$$

we have $(\pi a/2, 0)$ as the coordinates of the point P_∞.

It may be remarked here that this furnishes an excellent and convenient method of finding the approximate circumference of a circle of given material

---

* From AMERICAN MATHEMATICAL MONTHLY, vol. 40 (1933), pp. 596–598.

† See "Cochleoid", Encyclopaedia Britannica (14th edition), vol. 6, p. 894. Also solution of problem E41, this issue of the MONTHLY, p. 609.

radius, for OP_4 differs from the true length of the quadrant by less than one six hundredth part of the quadrant.

(b) We note that
$$\rho_n = \frac{\pi a}{2} \frac{\sin \theta_n}{\theta_n}$$
when n is finite, and that
$$\lim_{n \to \infty} \rho_n = \frac{\pi a}{2}.$$
Hence the function
$$\rho = \frac{\pi a}{2} \frac{\sin \theta}{\theta} \qquad (\theta \neq 0)$$
$$\rho = \frac{\pi a}{2} \qquad (\theta = 0)$$
is completely defined and continuous at every point, and is a function of the kind required.

(c) In the function derived above let $\theta = p\pi$, p being any number, integral or fractional, such that $p\pi$ can be constructed by elementary geometry. Then $\rho = (a \sin p\pi)/2p$. Since $\sin p\pi$ can be constructed by elementary geometry, we may find by elementary geometry the value of ρ for any admissible value of p.

It is evident that as θ increases through positive values the point (ρ, θ) traces a spiral in the first and second quadrants, passing through the origin for every value $\theta = n\pi$, n being integral. Furthermore it is easily shown that the horizontal axis is tangent to the curve at the origin. The curve is symmetric with respect to horizontal axis; hence there are two sets of whorls tangent to each other at the origin.

Some interesting properties of this spiral may be mentioned.

(1) the chord joining (ρ_1, α_1) and (ρ_2, α_2), where $\alpha_2 = 2\alpha_1$, is perpendicular to the line $\theta = \alpha_2$. Hence if we locate any point (ρ, α) on the curve, we may establish a sequence of points on the curve in a manner exactly like that by which $[P_n]$ was established. Every such sequence has $(\pi a/2, 0)$ as the limit point.

(2) The tangent to the curve at any point (ρ, α) intersects the line $\theta = 2\alpha$ on the circle $\rho = \pi a/2$. Hence if any line $\theta = \alpha$ be drawn through the origin, all the tangents to the curve at the points of intersection of this line and the curve pass through the same point, $(\pi a/2, 2\alpha)$.

(3) Let any line $\theta = \alpha$ intersect the successive whorls of the spiral, reading inward from the outermost intersection, at $R_0, R_1, R_2, \ldots$. Then $OR_0, OR_1, OR_2, \ldots$, form a harmonic sequence. This property is, of course, the same as that of the reciprocal spiral.

PLOTTING CURVES IN POLAR COORDINATES*

P. C. HAMMER, Oregon State College

To many calculus students the plotting of curves in polar coördinates is a process which is both tedious and uncertain. I have found that the elementary mapping approach given below is appreciated by most calculus students. This "solution" of the problem presumes with some justification that students are more familiar with cartesian than with polar coördinates.

Let $f(\rho, \theta) = 0$ be the equation of a curve to be plotted in the polar coördinate system. Taking ρ as the ordinate and θ (in radians, say) as the abscissa, plot the graph of this function in cartesian coördinates and then map this curve into the polar coördinate system in the obvious manner—*i.e.*, interpret the abscissas as angles and the ordinates as distances from the origin. Using compasses, many points can be plotted in the polar coördinate system with relative ease.

This method is not a panacea for all the ills of plotting curves in polar coördinates. In many cases the customary device of reverting to a cartesian coördinate system superimposed upon the polar coördinate system is better, and in other cases direct plotting is superior. This mapping method has, nonetheless, many advantages in certain types of problems. The graph of any equation of the form $\rho = a + b \cos(c\theta + d)$ is made very simple by mapping, in that the symmetries and required range of θ are immediately clear from the "cartesian" curve and the general shape of the curve is evident at a glance. Thus all the ordinary n-leaf roses, the cardioids, and the limaçons are easily plotted. Other common curves which are easily done are the lemniscates, the conchoid of Nicomedes, and the customary spirals. This method has the advantage of introducing in a natural way a continuous transformation which is not one-to-one. For the student who likes to play around with oddities, the mapping of other plane curves from the cartesian to the polar system will provide some sport. Letting c be an irrational number in the above equation, the range of θ must be infinite and the student is introduced to curves rarely mentioned in the texts.

---

* From AMERICAN MATHEMATICAL MONTHLY, vol. 48 (1941), p. 397.

EQUATIONS AND LOCI IN POLAR COORDINATES*

R. W. WAGNER, Oberlin College

The purpose of this note is to further publicize and to urge a wider application of an idea discussed by C. Fox† to clarify the difficulties which are encountered in seeking the intersections of loci defined by equations in polar coördinates.

Everyone who has worked with polar coördinates is aware of the fact that a point has an infinite number of coördinates. Thus, the point with coördinates (r, θ) has the other coördinates $([-1]^n r, \theta + n\pi)$, $n = 0, \pm 1, \pm 2, \cdots$. Writing the coördinates in this way implies that the common conventions that r may be any real number and that θ is measured in radians are being used.

The locus or graph of an equation is usually taken to mean the set of all points for which some pair of coördinates satisfy the equation.

Fox's idea is this: *The lack of uniqueness in the coördinates of a point leads, in general, to a lack of uniqueness in the equation of a locus.* This principle is overlooked or ignored in the majority of text books which the author has examined. A few books‡ mention it and use it to find the points of intersection of polar loci. However, none of these books use the principle to its full extent, for example, to clarify the tests for symmetry or to expedite the plotting of curves.

To investigate the validity of this principle, consider the locus defined by the equation

(1) $$F(r, \theta) = 0.$$

A point will be on this locus if, and only if, for some integer n $(n = 0, \pm 1, \pm 2, \cdots)$

(2) $$F([-1]^n r, \theta + n\pi) = 0.$$

The locus defined by assigning any permissible value to n in equation (2) will be the same as the locus defined by equation (1). For, any point which has a pair of coördinates which satisfy (1) will have a pair which satisfy (2), and conversely.

The equations obtained from (2) by assigning various values to n may be algebraically equivalent. This cannot occur unless $F(r, \theta)$ is a periodic function of θ with a period which is a rational multiple of π. If this period be $p\pi/q$, each equation obtained from (2) is algebraically equivalent to one of at most $2p$ equations which are algebraically non-equivalent. For example, all alternative forms for the equation

$$r - a \cos \theta - b \sin \theta = 0$$

---

* From AMERICAN MATHEMATICAL MONTHLY, vol. 55 (1948), pp. 360–363.

† C. Fox, The Polar Equations of a Curve, Math. Gazette, vol. 15 (1931), pp. 486–7.

‡ They are the analytic geometry texts by Curtiss and Moulton (1930), by Middlemiss (1947), by Nathan and Helmer (1947), by Randolph and Kac (1946), by Sisam (1926), by Nowlan (1932), and by Mason and Hazzard (1935).

are algebraically equivalent because

$$(-1)^n r - a \cos(\theta + n\pi) - b \sin(\theta + n\pi) = \begin{cases} r - a \cos\theta - b \sin\theta & (n \text{ even}) \\ -r + a \cos\theta + b \sin\theta & (n \text{ odd}). \end{cases}$$

For another example, there are four algebraically non-equivalent forms for the equation of the locus defined by

(3) $$r - \cos(\theta/2) = 0.$$

The other forms are

$$-r + \sin(\theta/2) = 0, \quad r + \cos(\theta/2) = 0, \quad \text{and} \quad -r - \sin(\theta/2) = 0.$$

The student should be made aware of the lack of uniqueness of the polar equation of a locus from the start. The first step in the discussion of the locus of an equation should be to find the alternate forms of the equation. The students are delighted to learn that they can graph several equations at once!

Plotting. By using the alternate equations one can frequently shorten the table values or the work required to prepare it by listing a value of r for each of the alternate equations and restricting θ to the interval $0 \leq \theta \leq \pi$. For example, to plot the graph of equation (3), the table of values would have values of θ in one column and values of r in four parallel columns. Each of these latter columns would show the values of r obtained from one of the alternate equations.

Symmetry. The usual tests for symmetry of a polar locus are based on conditions which are sufficient to show symmetry. The failure of the test to show symmetry does not show the lack of symmetry. By utilizing the alternate equations of the locus one can state a necessary and sufficient test for symmetry which requires a single substitution for each type of symmetry.

The locus of $F(r, \theta)$ is symmetrical with respect to (i) *the pole,* (ii) *the polar axis or* (iii) *the perpendicular to polar axis at the pole if, and only if,* (i) $F(-r, \theta) = 0$, (ii) $F(r, -\theta) = 0$, *or* (iii) $F(-r, -\theta) = 0$ *is algebraically equivalent to* $F(r, \theta) = 0$ *or one of its alternate forms.*

After a type of symmetry has been found, some of the r-columns of the table of values may be omitted and the symmetry used to complete the curve.

Intersections. The points of intersection are found by solving simultaneously all the pairs of equations made up of an equation for the first locus and an equation for the second locus. This can be systematically accomplished by first introducing an integer valued parameter into each equation, by then restricting the values of the parameters to sets of values which yield algebraically non-equivalent equations, and finally solving the two equations simultaneously in terms of the two parameters. If there are but few alternate forms for each equation, it may be simpler to solve the various pairs of equations than to solve the equations involving the parameters.

This procedure does not include the pole among the points of intersection unless the two curves have a common tangent at the pole. In general, an intersection at the pole is most readily found by noting that both curves pass through the pole.

Example. Find the intersections of the cardioid and the limaçon whose equations are

$$r = 2 + 2 \cos \theta, \qquad r = 10 \cos \theta - 2.$$

Each of these equations has an alternate form, namely,

$$r = 2 \cos \theta - 2, \qquad r = 10 \cos \theta + 2.$$

If one solves the two given equations simultaneously, he finds the points on the inner loop of the limaçon $(3, \pm\pi/3)$. The simultaneous solutions of the two alternate equations are the same points but the coördinates appear in the form $(-3, \pm 2\pi/3)$. The simultaneous solutions of a pair of equations made up of one given equation and the alternate of the other equation are the points on the outer loop of the limaçon $(2, \pm\pi/2)$, which may appear in the form $(-2, \pm\pi/2)$. The pole is also an intersection.

Conics. If the equation of a conic

(4) $$Ax^2 + 2Bxy + Cy^2 + 2Dx + 2Ey + F = 0$$

is changed to polar coördinates, the polar coördinate equation is factorable. If the conic passes through the pole, one factor leads to the equation $r = 0$ and the other factor leads to the unique equation of the conic. If the conic does not pass through the pole, it has two equations and these equations are obtained by equating the factors to zero individually.

The result of substituting $x = r \cos \theta$ and $y = r \sin \theta$ into equation (4) is

(5) $$(A \cos^2 \theta + 2B \cos \theta \sin \theta + C \sin^2 \theta)r^2 + 2(D \cos \theta + E \sin \theta)r + F = 0,$$

which can be abbreviated to

(6) $$Hr^2 + 2Gr + F = 0.$$

When θ is increased by π, H is not changed and the sign of G is changed. If the conic passes through the pole, $F = 0$ and equation (6) may be factored into

$$r = 0 \quad \text{and} \quad Hr + 2G = 0.$$

This second equation has no alternate forms because the change which is made to find an alternate equation merely changes the sign of both terms. If the conic does not pass through the pole, $F \neq 0$ and, after multiplication by H, (6) may be factored into

$$Hr + G + \sqrt{G^2 - HF} = 0 \quad \text{and} \quad Hr + G - \sqrt{G^2 - HF} = 0.$$

Each of these equations is an alternate form of the other because the substitution which is used to find alternate forms changes each one into the other. Also they are the only algebraically non-equivalent forms of the equation.

This illustrates a situation which should be true in general. If the rectangular equation of a locus be changed to polar coördinates, the polar equation is factorable and the factors are a complete set of alternate equations for the locus.

A LESSON IN GRAPHING*

F. MAX STEIN, Colorado State University

The place was my office in the mathematics building. I answered a knock at my door.

"Good morning, Sir or Madam, as the case may be." (It was obvious that he had learned his salesmanship according to the book.) "I am Mr. Cross Hatch, your friendly graph paper salesman. I sell graph paper."

"No, thanks," said I. "I already have a sheet of graph paper."

"But, just a minute," said Cross, putting his foot in the door, "I am well stocked with a variety seldom found elsewhere."

"What kinds do you have?" I asked, playing the part of the perfect straight man.

"I have the usual rectangular and polar coordinate paper, with which everyone is acquainted, but I also have log and log-log." (He appeared to stutter.) "I have power paper of all orders, exponential paper, sine paper, and cosine paper as well as sine-sine and cosine-cosine paper." (He started his stuttering again.)

"What can I do with these various kinds of paper that I can't do with the sheet I already have?"

"Oh, this paper isn't for you," said C. H., now appealing to my vanity. "This paper is for your students. It will make graphing much simpler for them, albeit perhaps somewhat monotonous. Your students will have to know only how to draw a straight line. Also, think how much simpler your work will be when you check graphs."

Now C. H. was hitting home to me, since I appreciated lack of work with the best. I could see how $y=x$ or $y=2x$, or even $y=mx$, could be graphed as a straight line, but how could the graph of the parabola $y=x^2$, for instance, be represented by a straight line? The answer to my unasked question was not long in forthcoming.

"For $y=x$," said Mr. Hatch, "you have a line through the origin inclined at 45° to the x-axis on your graph paper. Here the intervals along each axis are taken to be equal, Fig. 1a. Suppose you let each interval along the x-axis be taken as two units but keep the units the same on the y-axis. The graph of $y=x$ is still a line through the origin, but it is now inclined at 63°26′ to the x-axis, Fig. 1b. I think that even you can see that $y=2x$, or $y=mx$, m finite, can be represented by a 45° line on the proper paper if you turn the argument around. You see that the problem is not one so much of graphing now as in choosing the proper graph paper."

Without giving me time for a retort to his obvious dig at my mentality, he hurried on.

"Now to graph $y=x^2$, say, the student only has to pick the proper sheet of graph paper, draw a 45° line, and hand in his work, Fig. 2a. The paper he would

---

* From MATHEMATICS MAGAZINE, vol. 36 (1963), pp. 249–252.

A LESSON IN GRAPHING

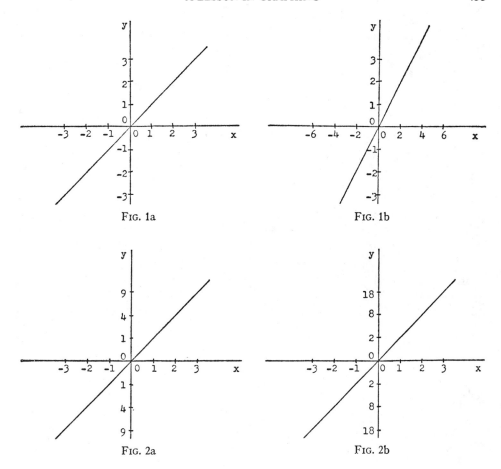

FIG. 1a

FIG. 1b

FIG. 2a

FIG. 2b

have to choose would have the intervals marked off as 1, 4, 9, $\cdots$ on the positive y-axis. Since all of our paper is square, the student would only need to turn his paper through 90° and make a few obvious changes and he could graph $y = \sqrt{x}$. I'm assuming your students are intelligent enough to handle the graph in other quadrants correctly in each case."

"Ah, ha!" thought I. "I can cut down on the number of kinds of graph paper my students will have to stock. All I need to do is to have them mark the intervals as 1^n, 2^n, 3^n, $\cdots$ along the positive y-axis for any positive integer n, and then they can graph $y = x^n$ on their usual paper as a straight line inclined at 45° to the positive x-axis."

"We also carry a little more expensive line of paper—multipurpose paper, we call it—so that students won't need to stock so many different kinds," said C. H., not bothering to comment on my unspoken thoughts. "This paper has a semi-elastic base so that it can be stretched without returning to its original size or it can be shrunk by dipping it in water for a short time. With this paper students can graph $y = 2x^2$, or $y = mx^2$, in the same manner as they would graph

$y=x^2$ by merely shrinking the paper and then halving the length of each subinterval along the y-axis, Fig. 2b. The intervals along the y-axis will then read 2, 8, 18, ⋯."

"Note," he continued, "that the paper for $y=x^2$, say, has two positive y-axes but no negative ones and that they both increase without bound both above and below $y=0$. However, our exponential paper, for graphing $y=e^x$, increases without bound above $y=1$ (no $y=0$ in this case), but it approaches zero as a limit below the x-axis, Fig. 3a. For our log paper this same phenomena is observed only with respect to the x-axis, Fig. 3b."

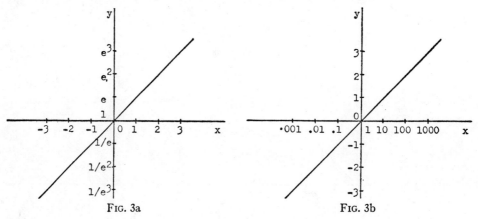

FIG. 3a   FIG. 3b

"This is a curious situation," I observed. "For all of your previous samples you had equal units marked off at equal subintervals on the x-axis, but now your equal units are on the y-axis. Elucidate," said I, displaying a multisyllable word.

"Actually," said C. H., "it is a matter of choice. My company has decided that for functions which are concave upward the equal units shall be on the x-axis and vice versa for functions concave downward."

"What advantage does this log-log paper have?" I asked after catching a glimpse of his next sample, Fig. 4. "It seems to me that, for log y = log x and

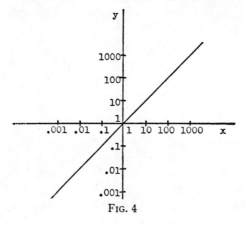

FIG. 4

as far as we are concerned, we have $y=x$ and thus have the same units along each axis as on my paper."

Patiently C. H. led me along. "Suppose you have a set of points on the same graph with both coordinates near 10, another set with coordinates near 100, and a final set with coordinates near 1000, and you wish to plot these all on the same graph. Could you tell much from your graph?"

I was beginning to see the light. To get all of these points on the same graph I would need to crowd the first set to such an extent that they would be meaningless. I started to fill out an order blank.

"Notice how points whose coordinates are very small positive values can also be put on this graph, and the entire gamut will have meaning. Also observe that the axes cross at the point $(1, 1)$ and not at $(0, 0)$." Hatch rambled on and on not appearing to notice that I was ripe for closing the sale.

Eventually, after the order for a generous supply of his entire stock had been signed and I was alone once more, I began to wonder—"Which kind of paper is best for graphing $y=\sqrt{25-x^2}$? Can I put 720° or only 180° on my polar graph paper and simplify my graphing in polar coordinates?" These questions led to others which in turn spawned others. I consoled myself; at least, I can now check a great portion of the homework papers involving graphing in much less time hereafter.

THE EQUATION OF A SPHERE*

MURRAY S. KLAMKIN, Ford Scientific Laboratory

A neat derivation for the equation of a circle passing through three given points was recently given by Bond (Math. Gazette, May 1966, Classroom Note 137). Here we extend the method to find the equation of a sphere passing through the four given points $P_r(x_r, y_r, z_r)$, $r=1, 2, 3, 4$.

It follows immediately that an equation of a two parameter family of spheres passing through the two points P_1 and P_2 is given by

(1) $$L + \lambda M + \mu N = 0,$$

where

$$L(x, y, z) \equiv (x - x_1)(x - x_2) + (y - y_1)(y - y_2) + (z - z_1)(z - z_2),$$

$$M(x, y, z) \equiv \begin{vmatrix} x & y & z & 1 \\ x_1 & y_1 & z_1 & 1 \\ x_2 & y_2 & z_2 & 1 \\ x_3 & y_3 & z_3 & 1 \end{vmatrix}, \quad N(x, y, z) \equiv \begin{vmatrix} x & y & z & 1 \\ x_1 & y_1 & z_1 & 1 \\ x_2 & y_2 & z_2 & 1 \\ x_4 & y_4 & z_4 & 1 \end{vmatrix}.$$

---

* From MATHEMATICS MAGAZINE, vol. 42 (1969), pp. 241–242.

Here, $L=0$, $M=0$, $N=0$, are the equations, respectively, for a sphere with $\overline{P_1P_2}$ as a diameter and two planes passing through the sets of three points (P_1, P_2, P_3) and (P_1, P_2, P_4). λ and μ are now determined such that (1) is satisfied also by points P_3 and P_4. Whence, $\lambda = -L(P_4)/M(P_4)$, $\mu = -L(P_3)/N(P_3)$. (There's only one simple determinant to evaluate since $M(P_4)+N(P_3)=0$.)

We could, of course, have written down the equation of the sphere (or circle) immediately in terms of the determinantal equation

$$\begin{vmatrix} x^2+y^2+z^2 & x & y & z & 1 \\ x_1^2+y_1^2+z_1^2 & x_1 & y_1 & z_1 & 1 \\ \cdot & \cdot & \cdot & \cdot & \cdot \\ \cdot & \cdot & \cdot & \cdot & \cdot \\ \cdot & \cdot & \cdot & \cdot & \cdot \\ x_4^2+y_4^2+z_4^2 & x_4 & y_4 & z_4 & 1 \end{vmatrix} = 0.$$

But then there would be considerable computation involved in expanding out the determinant.

Bond's method can also be extended to determine the equation of a conic passing through the four given points $P_r(x_r, y_r)$, $r=1, 2, 3, 4$ and whose axes are parallel to the coordinate axes.

We start with the equation

(2) $\qquad \lambda(x-x_1)(x-x_2) + \mu(y-y_1)(y-y_2) + \nu R = 0$

where

$$R \equiv \begin{vmatrix} x & y & 1 \\ x_1 & y_1 & 1 \\ x_2 & y_2 & 1 \end{vmatrix}.$$

Here, $\lambda(x-x_1)(x-x_2) + \mu(y-y_1)(y-y_2) = 0$ is the equation of a family of conics with axes parallel to the coordinate axes and for which $\overline{P_1P_2}$ is a diameter. The ratios $\lambda:\mu:\nu$ are now determined by requiring (2) to be also satisfied by points P_3 and P_4.

The conic could turn out to be a parabola and in that case its axis will only be parallel to one of the coordinate axes. It also may turn out that the last two equations for determining $\lambda:\mu:\nu$ are linearly dependent. In this case, there will be a family of conics with the desired property. This case will arise if we started out with four points which were given as the intersection of two conics of the appropriate type.

An alternative way of obtaining the desired conic is to start from the equation given by Loney (The Elements of Coordinate Geometry, Part I, Macmillan, London, 1954, pp. 378–379):

(3) $\qquad L_1(P_1, P_2)L_2(P_3, P_4) = \lambda L_3(P_1, P_4)L_4(P_2, P_3)$

where $L_i(P_r, P_s) = 0$ is an equation of the straight line containing points P_r and P_s. λ is now determined such that the xy term cancels out in (3).

BIBLIOGRAPHIC ENTRIES: CURVES AND SURFACES

Except for the entries labeled MATHEMATICS MAGAZINE, the references below are to the AMERICAN MATHEMATICAL MONTHLY.

1. R. M. Winger, Note on rational curves with trigonometric parameter, vol. 41, p. 368.
2. L. S. Johnston, A note on the conics, vol. 42, p. 501.

    Surprising relationship among various key points on standard conics.

3. W. T. Stratton, A study of general polar tangent curves, vol. 43, p. 398.
4. E. A. Whitman, Some historical notes on the cycloid, vol. 50, p. 309.
5. J. L. Coolidge, The story of tangents, vol. 58, p. 449.

    An in-depth discussion of what tangents are; includes calculus.

6. T. F. Mulcrone, The names of the curve of Agnesi, vol. 64, p. 359.

    Some of the history associated with the curve also known as "The Witch of Agnesi".

7. J. D. E. Konhauser, On a particular plane section of the torus, MATHEMATICS MAGAZINE, vol. 38, p. 161.

    A proof that the intersection of a plane through the center of a torus and tangent to the surface of the torus consists of two intersecting circles.

8. C. S. Ogilvy, Generating a hyperboloid, MATHEMATICS MAGAZINE, vol. 39, p. 276.

    Derivation similar to Wilson, AMERICAN MATHEMATICAL MONTHLY, vol. 18, p. 158.

10

AREA AND VOLUME

THE VOLUME OF THE SPHERE*

HENRY L. COAR, University of Illinois

The solids of revolution, that is, the solids generated by the revolution of a plane figure about an axis, offer many interesting problems to the student of synthetic geometry of three dimensions. The question of obtaining the volumes and total areas of solids, generated by the revolution of rectilinear figures, presents no particular difficulty, but when we attempt to find by purely synthetic means the volumes and superficial area of solids formed by curvilinear figures, the problem is not always simple without making some assumptions regarding limits. A well-known problem of this kind is to find the volume and area of an anchor-ring. A most interesting problem in this line is that of obtaining the volume of a sphere regarded as a solid of revolution. In the following proof, which I have not been able to find published anywhere, no assumptions of any kind regarding the existence of the limits are necessary.

We need consider only the hemisphere, which is generated by the revolution, through 360°, of a quadrant of a circle about a radius.

Let AOB be the quadrant of a circle and let us divide the radius OB into n equal parts. Then construct a set of inscribed and a set of circumscribed rectangles as indicated in the figure.

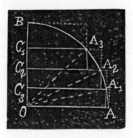

If now we rotate the complete figure through 360° about the radius OB, the quadrant will generate a hemisphere, while each of the inscribed as well as each of the circumscribed rectangles will generate a right circular cylinder. Let us designate

---

* From AMERICAN MATHEMATICAL MONTHLY, vol. 10 (1903), pp. 9–11.

the sum of the volumes of the cylinders generated by the inscribed rectangles by V_1, that of the cylinders generated by the circumscribed rectangles by V_2, and the volume of the hemisphere by V. We will first prove that both V_1 and V_2 approach V as their limit as n increases indefinitely, and will then find an expression for either of these and obtain its limiting value. To prove the first we have always $V_1 < V < V_2$, hence

$$V_2 - V < V_2 - V_1 \quad \text{and} \quad V - V_1 < V_2 < V_1.$$

But $V_2 - V_1$ is the volume of the lowest circumscribed cylinder, *i.e.*,

$$V_2 - V_1 = \frac{r}{n} \cdot \pi r^2 = \frac{\pi r^3}{n}$$

where r is the radius of the circle. Hence

$$V_2 - V < \frac{\pi r^3}{n} \quad \text{and} \quad V - V_1 < \frac{\pi r^3}{n}.$$

It follows therefore that each of these differences can be made less than any assignable positive quantity, and hence

$$\lim_{n \doteq \infty} V_1 = V \quad \text{and} \quad \lim_{n \doteq \infty} V_2 = V.$$

We thus prove that our hemisphere is actually the limit of the figures in question.

Let us now obtain an expression, say for V_2, and find its limiting value. We see that V_2 is the sum of n right circular cylinders. Let us find their volumes and add. For convenience denote the volumes of these n cylinders, beginning with the lowest one, by $v_1, v_2, \ldots v_n$. We shall then have

$$v_i = \frac{r}{n} \cdot AO^2 \cdot \pi = \frac{\pi r^3}{n},$$

$$v_2 = \frac{r}{n} \cdot A_1 C_1^2 \cdot \pi = \frac{\pi r}{n}\left(r^2 - \frac{r^2}{n^2}\right) = \frac{\pi r^3}{n}\left(1 - \frac{1}{n^2}\right),$$

$$v_3 = \frac{r}{n} \cdot A_2 C_2^2 \cdot \pi = \frac{\pi r}{n}\left[r^2 - \left(\frac{2r}{n}\right)^2\right] = \frac{\pi r^3}{n}\left(1 - \frac{2^2}{n^2}\right),$$

$$\cdots\cdots\cdots\cdots$$

$$v_n = \frac{r}{n}(A_{n-1}C_{n-1})^2 \cdot \pi = \frac{\pi r}{n}\left[r^2 - \left(\frac{\overline{n-1}\,r}{n}\right)^2\right] = \frac{\pi r^3}{n}\left(1 - \frac{(n-1)^2}{n^2}\right).$$

Hence

$$V_2 = v_1 + v_2 + \cdots + v_n = \frac{\pi r^3}{n}\left[1 + \left(1 - \frac{1}{n^2}\right) + \left(1 - \frac{2^2}{n^2}\right) + \left(1 - \frac{3^2}{n^2}\right) + \cdots \right.$$

$$\left. + \left(1 - \frac{(n-1)^2}{n^2}\right)\right]$$

$$= \frac{\pi r^3}{n}\left[n - \left(\frac{1}{n}\right)^2 - \left(\frac{2}{n}\right)^2 - \left(\frac{3}{n}\right)^2 - \cdots - \left(\frac{n-1}{n}\right)^2\right]$$

$$= \pi r^3 \left\{1 - \frac{1}{n^3}\left[1^2 + 2^2 + 3^2 + \cdots + (n-1)^2\right]\right\}.$$

Now the series $1^2 + 2^2 + 3^2 + \cdots + (n-1)^2$ is the sum of the squares of the first $n-1$ positive integers. The formula for the sum of the squares of the first n integers is given in any College Algebra under "piles of shot" and is

$$1^2 + 2^2 + 3^2 + \cdots + n^2 = \frac{n(n+1)(2n+1)}{6}.$$

Hence

$$1^2 + 2^2 + 3^2 + \cdots + (n-1)^2 = \frac{(n-1)n(2n-1)}{6}.$$

Substituting this in the expression for V_2 we have

$$V_2 = \pi r^3 \left[1 - \frac{1}{n^3} \cdot \frac{(n-1)n(2n-1)}{6}\right]$$

$$V_2 = \pi r^3 \left[1 - \tfrac{1}{6}\left(1 - \frac{1}{n}\right) \cdot 1 \cdot \left(2 - \frac{1}{n}\right)\right].$$

Now proceed to the limit and we have

$$V = \lim_{n \doteq \infty} V_2 = \pi r^3 \left[1 - \tfrac{1}{6}(1) \cdot 1 \cdot (2)\right] = \pi r^3 \left[1 - \tfrac{1}{3}\right] = \tfrac{2}{3}\pi r^3$$

which is the well-known result.

THE EXPRESSION OF THE AREAS OF POLYGONS IN DETERMINANT FORM*

R. P. BAKER, University of Iowa

The area of the triangle the rectangular coördinates of whose vertices are (x_1,y_1); (x_2,y_2); (x_3,y_3), is

$$\frac{1}{2}\begin{vmatrix} x_1 & y_1 & 1 \\ x_2 & y_2 & 1 \\ x_3 & y_3 & 1 \end{vmatrix}.$$

For the area of a quadrilateral whose vertices are 1,2,3,4, diagonals (13) and (24) and such that circuits (123), (134) have the area on the left, we have

$$\frac{1}{2}\begin{vmatrix} x_1, y_1, 1, 0 \\ x_2, y_2, 1, 1 \\ x_3, y_3, 1, 0 \\ x_4, y_4, 1, 1 \end{vmatrix} \equiv \frac{1}{2}\begin{vmatrix} x_1, y_1, 1, -k \\ x_2, y_2, 1, 1-k \\ x_3, y_3, 1, -k \\ x_4, y_4, 1, 1-k \end{vmatrix}.$$

The case of the pentagon or polygon of more sides than 5 is different.

Suppose that the area P of the pentagon can be expressed by

$$\lambda \begin{vmatrix} x_1, y_1, 1, a_1, b_1 \\ x_2, y_2, 1, a_2, b_2 \\ x_3, y_3, 1, a_3, b_3 \\ x_4, y_4, 1, a_4, b_4 \\ x_5, y_5, 1, a_5, b_5 \end{vmatrix} = \lambda \begin{vmatrix} x_1, y_1, 1, 0, 0 \\ x_2, y_2, 1, a'_2, b'_2 \\ x_3, y_3, 1, a'_3, b'_3 \\ x_4, y_4, 1, a'_4, b'_4 \\ x_5, y_5, 1, a'_5, b'_5 \end{vmatrix}$$

the latter being obtained from the former by subtracting multiples of columns.

Expanding by Laplace's method in minors of the first three columns we get the area as a sum of multiples of triangular areas all having the point 1 as vertex. The multipliers must obviously be equal. Hence all the determinants $\begin{vmatrix} a'_i & b'_i \\ a'_j & b'_j \end{vmatrix}$ must be equal. This is impossible, for

$$\frac{1}{2}\begin{vmatrix} a_2, b_2, a_2, b_2 \\ a_2, b_3, a_3, b_3 \\ a_4, b_4, a_4, b_4 \\ a_5, b_5, a_5, b_5 \end{vmatrix} \equiv (23)(45)-(24)(35)+(25)(34) \equiv 0,$$

---

* From AMERICAN MATHEMATICAL MONTHLY, vol. 11 (1904), pp. 227–228.

which cannot be satisfied by

$$(23) = (45) = (24) = (35) = (25) = (34).$$

The general case fails in consequence of a similar identical relation among the determinants of a matrix. This relation can be expressed as the expansion of a determinant of $2(n-3)$ rows which can be symbolized by $\dfrac{A\,|\,C}{B\,|\,C}$, where A, B, C denote, respectively,

$$\begin{vmatrix} 0,0,\ldots,a_{n-2} \\ 0,0,\ldots,b_{n-2} \\ \cdots\cdots\cdots \\ 0,0,\ldots,l_{n-2} \end{vmatrix}, \begin{vmatrix} a_2,a_3,a_4,\ldots,a_{n-2} \\ b_2,b_3,b_4,\ldots,b_{n-2} \\ \cdots\cdots\cdots \\ l_2,l_3,l_4,\ldots,l_{n-2} \end{vmatrix}, \begin{vmatrix} a_{n-1},a_n,a_2,a_3,\ldots,a_{n-4} \\ b_{n-1},b_n,b_2,b_3,\ldots,b_{n-4} \\ \cdots\cdots\cdots \\ l_{n-1},l_n,l_2,l_3,\ldots,l_{n-4} \end{vmatrix}.$$

When this is expanded according to Laplace's method in determinants of $(n-3)$ rows, we get

$$(2,3,\ldots,n-3,n-2)(n-1,n,2,\ldots,n-4)$$
$$-(2,3,\ldots,n-3,n-1)(n-2,n,2,\ldots,n-4)$$
$$+(2,3,\ldots,n-3,n)(n-3,n,2,\ldots,n-4)=0,$$

which cannot be satisfied if these determinants are all equal.

NOTE ON THE VOLUME OF A TETRAHEDRON IN TERMS OF THE COORDINATES OF THE VERTICES*

L.E. DICKSON, University of Chicago

1. Quite a variety of propositions of solid analytic geometry are needed for the usual derivation of the volume of a tetrahedron (cf. C. Smith, p. 24). If, as in the present note, we give an elementary proof making use merely of the concept of coordinates, we are in a position to apply the result to derive† easily several of the initial propositions in solid analytics, *e.g.*, that the equation of any plane is of the first degree, and conversely.

The plan of the proof (§ 3) is entirely obvious. The only novelty lies in a certain device which yields the result without computation. This device will first be illustrated in deriving the area of a triangle (§ 2).

2. Let the vertices of a triangle $\triangle$ taken in counter-clockwise order be (x_1,y_1), (x_2,y_2), (x_3,y_3). Then $\triangle$ can be expressed in terms of three right trapezoids with parallel sides y_i. The area of a right trapezoid with parallel sides y_1 and y_2, and base b, is $\tfrac{1}{2}b(y_1+y_2)$, being half of the rectangle of height y_1+y_2 and base b. Hence

$$2\triangle = (x_1-x_2)(y_1+y_2)+(x_2-x_3)(y_2+y_3)+(x_3-x_1)(y_3+y_1).$$

---

* From AMERICAN MATHEMATICAL MONTHLY, vol. 14 (1907), pp. 117–118.

† For plane analytics, this plan is followed in the chapter on graphic algebra in the writer's *College Algebra* (John Wiley and Sons).

The device consists in setting $s = y_1 + y_2 + y_3$. Then
$$2\triangle = (x_1 - x_2)(s - y_3) + (x_2 - x_3)(s - y_1) + (x_3 - x_1)(s - y_2).$$
Since each x occurs once positively and once negatively, the terms in s evidently cancel. The remaining terms give the expansion, according to the second column, of
$$\begin{vmatrix} x_1 & y_1 & 1 \\ x_2 & y_2 & 1 \\ x_3 & y_3 & 1 \end{vmatrix}.$$

3. Consider any tetrahedron $T = P_1 P_2 P_3 P_4$, the notation for the vertices being chosen so that P_1 is above the plane of P_2, P_3, P_4, while the latter lie in counter-clockwise order when viewed from P_1. Denote P_i by (x_i, y_i, z_i); its projection on the x, y-plane is $Q_i = (x_i, y_i, 0)$. Now T can be expressed in terms of four truncated right triangular prisms $P_i P_j P_k Q_i Q_j Q_k$, i, j, k denoting three of the four numbers 1, 2, 3, 4. The area of $Q_i Q_j Q_k$ is given by a determinant (§ 2). Applying the formula (§ 4) for the volume of a truncated right prism, we get

$$6T = D_4(z_1 + z_2 + z_3) + D_3(z_1 + z_2 + z_4) + D_2(z_1 + z_3 + z_4) - D_1(z_2 + z_3 + z_4),$$

where

$$D_1 = \begin{vmatrix} x_2 & y_2 & 1 \\ x_3 & y_3 & 1 \\ x_4 & y_4 & 1 \end{vmatrix}, D_2 = \begin{vmatrix} x_1 & y_1 & 1 \\ x_3 & y_3 & 1 \\ x_4 & y_4 & 1 \end{vmatrix}, D_3 = \begin{vmatrix} x_1 & y_1 & 1 \\ x_4 & y_4 & 1 \\ x_2 & y_2 & 1 \end{vmatrix}, D_4 = \begin{vmatrix} x_1 & y_1 & 1 \\ x_2 & y_2 & 1 \\ x_3 & y_3 & 1 \end{vmatrix}.$$

The device consists in setting $s = z_1 + z_2 + z_3 + z_4$. Then
$$6T = D_4(s - z_4) + D_3(s - z_3) + D_2(s - z_2) - D_1(s - z_1).$$
Here the terms free of s equal the expansion, according to the third column, of

$$D \equiv \begin{vmatrix} x_1 & y_1 & z_1 & 1 \\ x_2 & y_2 & z_2 & 1 \\ x_3 & y_3 & z_3 & 1 \\ x_4 & y_4 & z_4 & 1 \end{vmatrix}.$$

The terms multiplying s derived from the others by replacing each z by -1 and hence equal the expansion of a determinant derived from $-D$ by replacing each z by 1. But a determinant with two columns alike vanishes identically. Hence $T = \frac{1}{6} D$.

4. The volume of a truncated right prism P, whose base is a triangle $\triangle$, and lateral edges are a, b, c, is $\frac{1}{3}(a + b + c)\triangle$. This may be proved as in the geometries, or very simply as follows. Let $a \geq b \geq c$. Let the edge c be DE, d the side of triangle opposite D, h the perpendicular from D to d, so that $\triangle = \frac{1}{2} hd$. The plane through E parallel to $\triangle$ divides P into a right prism of volume $c\triangle$ and a pyramid with summit E, altitude h, and base a trapezoid with parallel sides $a-c$, $b-c$, and common perpendicular d. The area of the trapezoid is $\frac{1}{2}d(a+b-2c)$; the volume of the pyramid is therefore $\frac{1}{3}\triangle(a+b-2c)$. Adding $c\triangle$ to the latter, we get $P=\frac{1}{3}\triangle(a+b+c)$.

To give another proof, extend b (upwards) the length $a-b$. Thus to P we add a triangular pyramid with summit E, and base a right triangle of legs $a-b$ and d, and hence of volume $\frac{1}{2}d(a-b)\cdot\frac{1}{3}h = \frac{1}{3}\triangle(a-b)$. Next, extend c the length $a-c$. We thus add on a pyramid of summit E, and base equal to $\triangle$, and hence of volume $\frac{1}{3}\triangle(a-c)$. By these additions, P becomes a right prism of volume $\triangle a$. Hence

$$P + \tfrac{1}{3}\triangle[a-b] + \tfrac{1}{3}\triangle[a-c] = \triangle a, \quad P = \tfrac{1}{3}\triangle[a+b+c].$$

SOME LIMIT PROOFS IN SOLID GEOMETRY*

W. R. LONGLEY, Yale University

1. Introduction. The proofs of the formulas for the volume of a cone and the volume of a sphere involve limiting processes which have always been difficult for students of elementary geometry, and usually the teacher dismisses them with an informal explanation. The proofs given in this paper are new in the field of elementary geometry, although they appear to be more direct and natural than the traditional methods. They have been tried out a number of times with college freshmen and are well within the grasp of the majority. It is not unlikely that high school students, also, may find that this treatment is illuminating.

The method is that of the integral calculus and it possesses certain theoretical advantages which are important. In the first place the method is general. The same procedure is followed for any pyramid, for any cone, and for a sphere. It introduces at an early stage of mathematical training the fundamental summation idea of the integral calculus without first developing the machinery of integration. The student who will continue the study of mathematics cannot meet this idea too often or in too many different forms. The pupil who will never study the calculus gets a brief glimpse of a powerful general method which has practical applications in certain approximate formulas for computation.

The underlying concept of dividing a solid by parallel section planes is used in most of the current texts in connection with a triangular pyramid but it is quite worth while to develop this concept a little more because of its fundamental character. As applied to the sphere this concept seems much simpler than the usual one based upon spherical sectors. It is also advantageous to use the same concept in connection with the sphere that is used with the pyramid and cone.

The method involves an algebraic formula which may very properly be taken into geometry without proof. But the proof can be made by mathematical induc-

---

* From AMERICAN MATHEMATICAL MONTHLY, vol. 31 (1924), pp. 196–202.

tion and here again is an opportunity for the student to get a glimpse of a powerful general method which may never come to his notice again.

Assuming that the volume of a prism and of a cylinder are known, the only theorems to which direct reference must be made are the following:

THEOREM A. *If two straight lines are cut by three or more parallel planes, the corresponding segments are proportional.*

THEOREM B. *The area of a section of a pyramid, or cone, parallel to the base is to the area of the base as the square of its distance from the vertex is to the square of the altitude of the pyramid, or cone.*

THEOREM C. *The sum of the squares of the integers from 1 to n is given by*

$$1^2+2^2+3^2+\cdots+n^2 = \frac{n(n+1)(2n+1)}{6} = \frac{n^3}{3} + \frac{n^2}{2} + \frac{n}{6}.$$

2. The volume of a pyramid. The figures represent a triangular pyramid, but the discussion applies to a pyramid having any polygon for a base.

Suppose a lateral edge, AP, of a pyramid (Fig. *a*) is divided into a certain number (four in the figure) of equal parts, and that planes parallel to the base are passed through each point of division, including the vertex. These parallel planes cut out sections similar to the base and divide the altitude into the same number of equal parts as the edge AP. Suppose that a prism, with lateral edges parallel to AP and altitude equal to the distance between consecutive parallel planes, is constructed on each section as lower base. The prisms include the pyramid completely and lie partly outside of it. These prisms form a set of circumscribed prisms.

Suppose that a prism (Fig. *b*), with lateral edges parallel to AP and altitude equal to the distance between consecutive parallel planes, is constructed on each section as upper base. These prisms lie entirely within the pyramid and form a set of inscribed prisms.

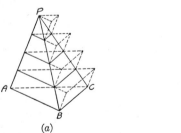
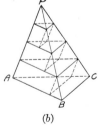

(a)   (b)

The total volume of a set of circumscribed prisms is the sum of the volumes of the separate prisms. It is obvious that the total volume of a set of circumscribed prisms is greater than the volume of the pyramid.

Suppose now that AP is divided into twice as many equal parts as before so that new section planes are passed midway between the original ones, and that a new set of circumscribed prisms is formed as before. It appears that the new set of prisms will be entirely within the old set. Hence the total volume of the new set will be less than the total volume of the old set, but will be greater than the volume of the pyramid.

It appears that, as the number of prisms is increased, the total volume of the set of circumscribed prisms will decrease, but will always be greater than the volume of the pyramid.

Similarly, it appears that, as the number of prisms is increased, the total volume of the set of inscribed prisms will increase, but will always be less than the volume of the pyramid.

Referring to the figures and counting down from the top, it is apparent that the first prism of the inscribed set is equal to the first prism of the circumscribed set. This is true also of the second prisms, third prisms, etc., until all the prisms of the inscribed set have been exhausted. There will be left only the lowest prism of the circumscribed set without any counterpart in the inscribed set. Hence for a given number of divisions of the edge AP the total volume of the set of circumscribed prisms exceeds the total volume of the set of inscribed prisms by the volume of the lowest prism of the circumscribed set. As the number of divisions of AP is increased, the altitude of each prism is decreased; hence the volume of the lowest prism of the circumscribed set is decreased, and can be made as small as we wish by making the number of divisions of AP sufficiently large. Hence the difference between the total volume of the set of circumscribed prisms and the total volume of the set of inscribed prisms can be made as small as we wish. In other words, as the number of divisions of AP is increased, the total volume of each set of prisms approaches the same limiting value.

The preceding discussion justifies the following definition:

Definition. The volume of any pyramid is the limit approached by the total volume of a set of circumscribed (or inscribed) prisms as the number of prisms is increased indefinitely.

THEOREM. *The volume of a pyramid is equal to one third the product of the area of the base and the altitude.*

Given the pyramid P-ABC (Fig. a). Denote the volume by V, the area of the base by b, and the altitude by h.

To prove $V = bh/3$.

Proof. 1. Suppose the edge AP to be divided into n equal parts by passing planes parallel to the base. These section planes divide the altitude into n equal parts. (Th. A.)

2. Suppose a set of circumscribed prisms to be constructed and let the prisms be numbered $1, 2, 3, \ldots, n$ from the vertex downwards. Then the altitude of each prism is h/n.

3. Let b_1 denote the area of the base of the first prism. The distance from the vertex to the base of the first prism is h/n. Hence

$$\frac{b_1}{b} = \frac{(h/n)^2}{h^2}, \quad \text{or} \quad b_1 = \frac{b}{n^2}. \qquad \text{(Th. } B.\text{)}$$

4. The volume V_1 of the first prism is

$$V_1 = b_1 \times \frac{h}{n} = \frac{bh}{n^3}.$$

5. The distance from the vertex to the base of the second prism is $2h/n$. Hence, with similar notation,

$$\frac{b_2}{b} = \frac{(2h/n)^2}{h^2}, \quad \text{or} \quad b_2 = \frac{4b}{n^2}, \quad \text{and} \quad V_2 = \frac{4bh}{n^3}.$$

6. Proceeding in this way,

$$V_3 = \frac{9bh}{n^3}, \quad V_4 = \frac{16bh}{n^3}, \quad \cdots \quad V_n = \frac{n^2 bh}{n^3}.$$

7. Let V_c denote the total volume of the set of circumscribed prisms. Then

$$V_c = \frac{bh}{n^3} + \frac{4bh}{n^3} + \frac{9bh}{n^3} + \cdots + \frac{n^2 bh}{n^3} = \frac{bh}{n^3}[1^2 + 2^2 + 3^2 + \cdots + n^2].$$

Hence

$$V_c = bh\left[\frac{1}{3} + \frac{1}{2n} + \frac{1}{6n^2}\right]. \qquad \text{(Th. C.)}$$

8. As n is increased, the terms $1/2n$ and $1/6n^2$ are decreased and approach zero as a limit, and V_c approaches the value $bh/3$ as a limit.

9. Hence $V = bh/3$. \hfill (Def.)

The preceding result can be obtained also by using inscribed instead of circumscribed prisms. If V_i denotes the total volume of a set of inscribed prisms obtained by dividing the altitude into n equal parts, it can be shown that

$$V_i = bh\left[\frac{1}{3} - \frac{1}{2n} + \frac{1}{6n^2}\right].$$

3. The volume of a sphere. To obtain a formula for the volume of a sphere, an argument analogous to that given for a pyramid may be used.

Consider a hemisphere and suppose the radius OA, perpendicular to the base, is divided into a certain number (four in the figure) of equal parts and that planes parallel to the base are passed through each point of division, including the point A. These parallel planes cut out circular sections. Suppose a right circular cylinder with altitude equal to the distance between consecutive parallel planes is constructed on each section as lower base. These cylinders, which include the hemisphere completely and lie partly outside of it, will be called a set of circumscribed

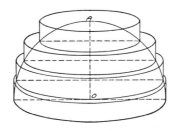

cylinders. The total volume of a set of circumscribed cylinders is the sum of the volumes of the separate cylinders. It is obvious that the total volume of a set of circumscribed cylinders is greater than the volume of the hemisphere. Suppose now that OA is divided into twice as many equal parts as before so that new section planes are passed midway between the original ones, and that a set of circumscribed cylinders is formed as before. It appears that the new set of cylinders will lie entirely within the old set. Hence the volume of the new set will be less than the volume of the old set, but will be greater than the volume of the hemisphere. It appears that as the number of cylinders is increased, the volume of the set of circumscribed cylinders will decrease and become more and more nearly equal to the volume of the hemisphere.

As in the discussion of the volume of a pyramid, a set of inscribed cylinders may be formed by constructing cylinders on each section as upper base. The total volume of a set of inscribed cylinders is always less than the volume of the hemisphere and increases as the number of cylinders is increased. The difference between the total volumes of the set of circumscribed cylinders and the set of inscribed cylinders approaches zero as the number of cylinders is increased indefinitely, and the limiting value of the total volume of either set is the volume of the hemisphere.

The preceding discussion justifies the following definition:

Definition. The volume of a hemisphere is the limit approached by the total volume of a set of circumscribed cylinders as the number of cylinders is increased indefinitely.

THEOREM. *The volume of a hemisphere of radius R is $\frac{2}{3}\pi R^3$.*

Given a hemisphere of radius R and volume V. To prove $V = \frac{2}{3}\pi R^3$.

Proof. 1. Suppose the radius OA to be divided into n equal parts by passing planes parallel to the base.

2. Suppose a set of circumscribed cylinders to be constructed and let the cylinders be numbered $1, 2, 3, \ldots, n$ from the base upwards. The altitude of each cylinder is R/n.

3. The radius of the first cylinder is R and its volume V_1 is

$$V_1 = \pi R^2 \times \frac{R}{n} = \pi \frac{R^3}{n}.$$

4. Let $r_2 \, (= C_2 B_2)$ denote the radius of the second cylinder. From the figure

$$r_2^2 = R^2 - (OC_2)^2 = R^2 - \left(\frac{R}{n}\right)^2 = R^2\left(1 - \frac{1}{n^2}\right).$$

Hence the volume V_2 of the second cylinder is

$$V_2 = \pi r_2^2 \times \frac{R}{n} = \pi \frac{R^3}{n}\left(1 - \frac{1}{n^2}\right).$$

SOME LIMIT PROOFS IN SOLID GEOMETRY

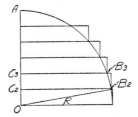

5. The radius $r_3(=C_3B_3)$ of the third cylinder is given by

$$r_3^2 = R^2 - (OC_3)^2 = R^2 - \left(\frac{2R}{n}\right)^2 = R^2\left(1 - \frac{4}{n^2}\right).$$

Hence the volume V_3 of the third cylinder is

$$V_3 = \pi r_3^2 \times \frac{R}{n} = \pi \frac{R^3}{n}\left(1 - \frac{4}{n^2}\right).$$

6. In a similar way the volume of each of the n circumscribed cylinders may be calculated.

7. Let V_c denote the total volume of the set of n circumscribed cylinders. Then

$$V_c = \pi \frac{R^3}{n} + \pi \frac{R^3}{n}\left(1 - \frac{1}{n^2}\right) + \pi \frac{R^3}{n}\left(1 - \frac{4}{n^2}\right) + \cdots + \pi \frac{R^3}{n}\left(1 - \frac{(n-1)^2}{n^2}\right).$$

Hence, by collecting terms,

$$V_c = \pi \frac{R^3}{n}\left[n - \frac{1^2 + 2^2 + \cdots + (n-1)^2}{n^2}\right].$$

8. The sum of the squares of the integers from 1 to $n-1$ is

$$1^2 + 2^2 + \cdots + (n-1)^2 = \frac{(n-1)(n)(2n-1)}{6} = \frac{n^3}{3} - \frac{n^2}{2} + \frac{n}{6}. \quad \text{(Th. C.)}$$

Hence

$$V_c = \pi R^3 \left[1 - \frac{1}{3} + \frac{1}{2n} - \frac{1}{6n^2}\right] = \pi R^3 \left[\frac{2}{3} + \frac{1}{2n} - \frac{1}{6n^2}\right].$$

9. As n is increased, the terms $1/2n$ and $1/6n^2$ are decreased and approach zero as a limit, and V_c approaches $\frac{2}{3}\pi R^3$ as a limit. Hence the volume of the hemisphere is $\frac{2}{3}\pi R^3$. (Def.)

COROLLARY. The volume of a sphere of radius R is given by

$$V = \frac{4}{3}\pi R^3.$$

THE ANALYTIC DETERMINATION OF THE AREA OF A TRIANGLE IN TERMS OF ITS SIDES*

K. P. WILLIAMS, Indiana University

It is an interesting problem in the theory of functions to find an explicit expression for a function from its general properties. Thus we obtain the factor form for sin x from a knowledge of its zeros, and a series of fractions for cot x from a knowledge of its poles. In each case a delicacy of analysis is required in order to determine completely the desired expression, and it is necessary to know something about the function in the entire complex plane.

One can adopt a similar point of view towards a formula that gives the area or volume of a geometric figure, or one side in terms of other sides and angles. A difficulty now arises. The geometric configuration exists only for certain values of the variables, but the formula has an analytic sense for unrestricted ranges. There is consequently a question of uniqueness if a geometric formula is thought of in the broad unrestricted light of function theory. In this note we shall examine the area of a triangle as a function of its sides from the point of view of analysis.

Let the sides of the triangle be x, y, and z, and denote the area by A. We wish an analytic expression for A in terms of x, y, and z. The variables x, y, z are restricted to positive or zero values. But they are not independent, since the relation $|y-z| < x < |y+z|$ is to be satisfied. For a given y and z the excepted values of x are then of two types: (1) negative and complex values, (2) certain positive values. There is a distinctly different origin for these two exclusions. The first arises from the fact that we are concerned with lines; the second from the fact that the lines must form a triangle. It can be expected that the formula that gives A will not react in the same manner if numbers of the first and second excluded classes, respectively, be substituted. One does not care what the result of substituting a number of the first class may be. But one does expect that the value of A for an excluded positive value of x will itself indicate that no triangle exists for such a value. This leads to the postulate:

If positive values for x,y,z, that are inconsistent with x,y,z being the sides of a triangle, be substituted in the formula that gives the area A, the value found for A must be either negative or complex.

Other properties of A are obvious geometrically.

(1) A must be symmetric in x,y, and z.

(2) If y and z are regarded as fixed, A must have real positive values for $|y-z| < x < y+z$.

(3) A must vanish for $x=|y-z|$ and $x=y+z$.

(4) If x,y,z are all multiplied by k, A must be multiplied by k^2.

By considering a right triangle we have the boundary condition,

(5) If $x^2 = y^2 + z^2$ we must have $A = \frac{1}{2}yz$.

We construct a function $F(x,y,z)$ that satisfies the postulate and the five

---

* From AMERICAN MATHEMATICAL MONTHLY, vol. 34 (1927), pp. 360–362.

conditions. Write
$$F(x,y,z) = (y+z-x)(x-y+z)f(x,y,z).$$
Changing the variables cyclically, we obtain
$$F(y,z,x) = (z+x-y)(y-z+x)f(y,z,x).$$
Since we must have $F(x,y,z) = F(y,z,x)$, it is seen that $y-z+x$ must be a factor of $f(x,y,z)$ and $y+z-x$) a factor of $f(y,z,x)$. But $y+z-x$ is obtained from $y-z+x$ by the cyclic change.

Write then
$$f(x,y,z) = (y-z+x)\phi(x,y,z),$$
so that
$$F(x,y,z) = (y+z-x)(x-y+z)(y-z+x)\phi(x,y,z).$$

Since the first three factors on the right form a symmetric quantity, it follows that $\phi(x,y,z)$ is symmetric. If x,y,z be replaced by kx,ky,kz, the first three factors on the right are multiplied by k^3. This suggests that it may be possible so to determine $\phi(x,y,z)$ that $F(x,y,z)$ gives A^2 and not A. We can write
$$F(x,y,z) = (x^2-y^2-z^2+2yz)(y+z-x)\phi(x,y,z).$$

Substituting $x^2 = y^2 + z^2$ we see that $(y+z-x)\phi(x,y,z)$ must $=(yz)/8$ when $x^2 = y^2 + z^2$.

The simplest choice is obviously $\phi(x,y,z) = (y+z+x)/16$. It follows that the function $\frac{1}{4}[(y+z-x)(x-y+z)(y-z+x)(x+y+z)]^{1/2}$ satisfies conditions (1)...(5). It also satisfies the postulate, since for $x < |y-z|$ or $x > y+z$ the function has an imaginary value. It is to be noted however that for $-|y+z| < x < -|y-z|$ the function has a real value, and vanishes for $x = -|y-z|$, and $x = -[y+z]$. Since, however these values of x are themselves totally inapplicable to any triangle, the fact does not, in accordance with the remark made before, arouse suspicion against the formula.

It remains to investigate the uniqueness of the function. For this purpose we consider
$$\tfrac{1}{4}[(y+z-x)(x-y+z)(y-z+x)(x+y+z)]^{1/2}\psi(x,y,z).$$

That some of the conditions would be satisfied by various choices of ψ is obvious. For instance if we put
$$\psi(x,y,z) = e^\rho, \quad \text{where } \rho = (x^2-y^2-z^2)(y^2-z^2-x^2)(z^2-x^2-y^2),$$
we shall have a function that satisfies (1),(2),(3),(5), and which furthermore has no positive zeros other than those given in (3).

In order to be consistent with (4) it is necessary that $\psi(kx,ky,kz) = \psi(x,y,z)$. Consider x,y,z as the rectangular coordinates of a point. It is evident that there is a bundle of rays lying in the positive octant forming a cone shaped solid, having the vertex at the origin, such that the coordinates of every point on one of the rays, and only the coordinates of such a point, can be the sides of a triangle. The last equation written shows that the function $\psi(x,y,z)$ is constant along any one ray. Assuming that $\psi(x,y,z)$ is continuous for $x=0, y=0, z=0$, we see that $\psi(x,y,z)=$ constant. By considering a right triangle, it is found that the constant is 1. We therefore conclude that the area of the triangle is

$$A = \tfrac{1}{4}\left[(y+z-x)(x-y+z)(y-z+x)(x+y+z)\right]^{1/2}.$$

It could have been expected that in addition to the conditions (1),...,(5) some other condition would need to be added. The restriction imposed is very mild and suffices to remove ambiguity.

A HISTORICALLY INTERESTING FORMULA FOR THE AREA OF A QUADRILATERAL*

J. L. COOLIDGE, Harvard University

The amount of information available in the literature of mathematics bearing on the quadrilateral, the general quadrilateral, the cyclic quadrilateral, the complete quadrilateral, *etc.*, is discouragingly large. Yet there seems to be no part of the science so far from exhaustion as elementary geometry. It has seemed to me that there must be connecting links between different known formulas which were worth investigating, not only for their own sakes but also for historical reasons. Here is one which, so far as I can find out, is new, and which seems to me to come into that category.

Let the pairs of opposite sides of a quadrilateral be aa', bb' while the diagonals are cc'. The first diagonal shall be divided by the intersection into the parts

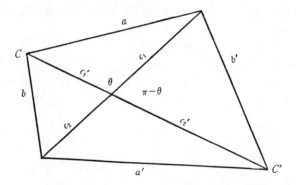

$c_1 c_2$ while the second is divided into the parts $c_1' c_2'$. Let one angle of the diagonals be θ, let the area be A, and finally let

$$2s = a + a' + b + b'.$$

---

* From AMERICAN MATHEMATICAL MONTHLY, vol. 46 (1939), pp. 345–347.

Then
$$4A^2 = c^2 c'^2 \sin^2 \theta,$$
and
$$a^2 = c_1^2 + c_1'^2 - 2c_1 c_1' \cos \theta,$$
$$\frac{c_1 c_1'}{2} = \frac{c_1^2 + c_1'^2 - a^2}{4 \cos \theta}.$$

Similarly,
$$\frac{c_2 c_1'}{2} = -\frac{c_2^2 + c_1'^2 - b^2}{4 \cos \theta},$$
$$\frac{c_2 c_2'}{2} = \frac{c_2^2 + c_2'^2 - a'^2}{4 \cos \theta},$$
$$\frac{c_1 c_2'}{2} = -\frac{c_1^2 + c_2'^2 - b'^2}{4 \cos \theta}.$$

Adding, we obtain
$$\frac{cc'}{2} = \frac{(b^2 + b'^2) - (a^2 + a'^2)}{4 \cos \theta}.$$

Hence
$$A^2 = \frac{c^2 c'^2}{4} [1 - \cos^2 \theta]$$
$$= \frac{c^2 c'^2}{4} - \frac{[(b^2 + b'^2) - (a^2 + a'^2)]^2}{16}$$
$$= -\frac{(a^4 + a'^4 + b^4 + b'^4) + 2(a^2 + a'^2)(b^2 + b'^2) - 2(a^2 a'^2 + b^2 b'^2)}{16} + \frac{c^2 c'^2}{4}$$
$$= (s-a)(s-a')(s-b)(s-b') - \tfrac{1}{4}[a^2 a'^2 + b^2 b'^2 + 2aa'bb' - c^2 c'^2] \cdot$$
$$A^2 = (s-a)(s-a')(s-b)(s-b') - \tfrac{1}{4}[aa' + bb' + cc'][aa' + bb' - cc'].$$

Now a very famous theorem due to Ptolemy (*circa* 139 A.D.) tells us that a necessary and sufficient condition that the vertices of a quadrilateral should lie on a circle is that
$$aa' + bb' = cc'.$$

But in that case we have for the area of a cyclic quadrilateral
$$A = \sqrt{(s-a)(s-a')(s-b)(s-b')},$$
an almost equally famous formula due to Brahmagupta (*circa* 728 A.D.)

I should, perhaps, mention in this connection that in 1842 there appeared in Grunert, *Beiträge zur reinen und angewandten Mathematik*, vol. 2, two proofs by Bretschneider and Strehlke, both rather clumsy I think, of the formula
$$A^2 = (s-a)(s-a')(s-b)(s-b') - aa'bb' \cos^2 \frac{C + C'}{2}.$$

SOLID ANGLES*

J. W. CELL, North Carolina State College

The concept of a solid angle is seldom introduced in the modern course in trigonometry for engineering students. Yet, to name two applications, this geometrical idea is convenient in studying light and in studying solenoid electrical coils. This topic can be introduced naturally in spherical trigonometry or at the same time that the term radian is defined in plane trigonometry. Even a brief discussion of the concept is preferable to its complete omission and the basic definition is easy. In the following paragraphs we outline some of the facts about solid angles.

DEFINITION. *Let C be a curve bounding an area A on a unit sphere. By the solid angle subtended by A we mean the figure composed of all the rays issuing from the center of the sphere and passing through points of C. The solid radian measure of this angle is defined to be the area of A.*

A polyhedral angle is a well known example of a solid angle. From the definition it follows that the maximum value for a solid angle is 4π solid radians, since the surface area of the unit sphere is 4π square units.

THEOREM 1. *The volume of a spherical sector is given by*

$$V = R^3\theta/3,$$

where θ is the solid angle at the vertex of the spherical sector.

For a unit sphere we see that $\theta = 3V$. This fact can be compared with the corresponding property for a plane angle, namely that $\phi = 2A$, where A is the area of the plane sector with vertex angle ϕ.

THEOREM 2. *The surface area of a zone on the surface of a sphere is given by $S = R^2\theta$, where θ is the solid angle which the zone subtends at the center of the sphere.*

If we recall that the surface area of a zone is given by $S = 2\pi R H$, where H is the altitude of the zone and R is the radius of the sphere, then another theorem follows from the preceding one.

THEOREM 3. *The solid angle at the vertex of a right circular cone with altitude h, radius r, and slant height s, is*

$$\theta = 2\pi(1 - h/s) = 2\pi\left(1 - \frac{h}{(h^2 + r^2)^{1/2}}\right) = 2\pi(1 - \cos\alpha),$$

where α is the plane angle between the axis of the cone and an element on the surface of the cone.

---

* From AMERICAN MATHEMATICAL MONTHLY, vol. 48 (1941), pp. 136–138.

From the formula for the surface area of a spherical triangle, we obtain the following:

THEOREM 4. *The solid angle which is subtended by a spherical triangle with angles A, B, and C (each measured in radian measure) is given by*

$$\theta = (A + B + C) - \pi,$$

the spherical excess in radian measure.

THEOREM 5. *The solid angle subtended at the vertex of a pyramid, with rectangular base (sides a and b) and with its vertex h units above one corner of the base, is given by*

$$\theta = \tan^{-1} \frac{ab}{h(a^2 + b^2 + h^2)^{1/2}}.$$

One could prove this theorem by considering the surface area intercepted, between the sides of the pyramid, on a unit sphere with center at the vertex of the pyramid. One could then divide the surface area into two right spherical triangles and apply Napier's rules to determine the angles. From this theorem we obtain immediately the following:

THEOREM 6. *The solid angle at the vertex of a right pyramid with square base (altitude h and side of base 2a) is given by*

$$\theta = 4 \tan^{-1} \frac{a^2}{h(h^2 + 2a^2)^{1/2}} = 4 \sin^{-1} \frac{a^2}{a^2 + h^2}.$$

Problem 1. Given that the number of lumens illumination on a plane area, caused by a point source of light, is equal to the candle power of the light multiplied by the solid angle which the plane area subtends at the light. Determine the number of lumens illumination:

a. On a circular area of radius 6 feet if the light is 60 candle power and is 8 feet directly above the center of the area. (Ans. 48π lumens).

b. On a square area of side 6 feet if a light of 100 candle power is placed 10 feet above the center of the area. (Ans. $400 \sin^{-1} (9/109)$ lumens).

Problem 2. Compare the total amounts of illumination on the floor of a room which measures 20 feet by 20 feet and has a ceiling height of 20 feet if:

a. One light of 1000 candle power is placed directly over the center of the floor and in the ceiling. (Ans. $4000 \sin^{-1} (1/5)$ lumens).

b. One light of 250 candle power is placed at each upper corner of the room. (Ans. $500 \pi/3$ lumens).

c. One light of 250 candle power is placed at the middle of each side of the ceiling. (Ans. $2000 \tan^{-1} (1/3)$ lumens).

Problem 3. Timbie and Bush in *Principles of Electrical Engineering*, third edition, page 368, show that for any point on the axis of a circular solenoid,

$$H_z = \frac{2\pi NI}{L}\left\{\frac{L/2 + y}{\{r^2 + (L/2 + y)^2\}^{1/2}} + \frac{L/2 - y}{\{r^2 + (L/2 - y)^2\}^{1/2}}\right\},$$

where N is the number of turns of wire in the length of the coil L, r is the radius of the solenoid, I is the current which the coil carries, H_z is the magnetic flux in the direction of the axis of the solenoid, and y is the distance from the center of the axis of the solenoid to the point in question.

A solenoid, for the purpose of this problem, can be thought of as a coil of wire wound uniformly around an oatmeal box, with but a single layer of wire.

a. Show that this formula can be stated in the simpler form,

$$H_z = (NI/L)(\theta),$$

where θ is the solid angle which the solenoid subtends at the point in question.

b. Give the special values for the point at the middle and at one end of the solenoid. What do these become if the solenoid is long compared to its diameter?

Remark. In problems such as the preceding one it is sometimes more convenient to determine the solid angle subtended by the opening and then to subtract this from the proper value. Thus, to determine the solid angle subtended by a circular cylinder at the middle of one end, it is easier to compute the solid angle subtended by the opening (the solid angle at the vertex of a right circular cone) and subtract this value from 2π.

THE AREA OF A TRIANGLE AS A FUNCTION OF ITS SIDES*†

VICTOR THÉBAULT, Tennie, Sarthe, France

1. Historical remarks. The first mention of the rule giving the area of a triangle as a function of the three sides is found in the works of Heron of Alexandria (1st century). Although it is now believed that this rule pre-dates Heron, demonstrations of it are in his two works, *Metrics* and *Treatise on the Diopter*.

In the book of the three Arabian brothers, Mohammed, Ahmed, and Alhasan, (9th century) we encounter a new demonstration, the first which came to us in Europe. It was reproduced by Leonardo of Pisa in his *Practical Geometry* (1220) and then by Jordanus Nemorarius (13th century), and by most of the geometers of the Renaissance. It is curious that Heron, the Hindus, as well as all the authors we have cited, made an application of this rule to the triangle of sides 13, 14, and 15, whose area is 84. One is led to ask if these three numbers have a common origin, but, as Chasles had observed, the Greeks, the Hindus, and the Arabs may very well have separately become aware of the fact that 13, 14, 15 are the smallest integers which give a rational area for an acute angled triangle.

---

* From AMERICAN MATHEMATICAL MONTHLY, vol. 52 (1945), pp. 508–509.

† Translated from the French by Howard Eves.

THE AREA OF A TRIANGLE AS A FUNCTION OF ITS SIDES

One finds, still later, other new proofs of the rule by Newton in his *Universal Arithmetic* (1707); by Euler in the *Recent Commentaries of Petersburg* (v. I, 1747, p. 48); by Boscovich in volume V of his *Works* concerning optics and astronomy (1785). This last demonstration is obtained by trigonometric considerations.

2. New demonstration. The author has previously given a very short geometrical demonstration of the formula under consideration (*Mathesis*, 1931, p. 27), and here is another equally simple.

Being given a triangle ABC, ($BC=a$, $CA=b$, $AB=c$, $a+b+c=2p$), let (see figure) B' and B'', C' and C'' be the orthogonal projections of the vertices B and C on the bisectors AD and AD' of angle A. Rectangle $AB''MC'$ has dimensions equal to BB' and CC'', and is the sum of two rectangles, $AB''BB'$ and $B'BMC'$. The first of these rectangles is equal in area to triangle BAE, which has AB' for altitude and $BE = 2\,BB'$ for base. The second is equal in area to triangle BEC, which has $NC = B'C'$ for altitude and BE for base. Thus rectangle $AB''MC'$ is equal in area to triangle ABC. Similarly, rectangle $AC''NB'$ is equal in area to triangle ABC, for this rectangle is the difference of rectangles $AC''CC'$ and $C'CNB'$, respectively equivalent to triangles CAF and CBF, the difference of which is triangle ABC. We thus have

(1) $$BB' \cdot CC'' \cdot CC' \cdot BB'' = (\text{area } ABC)^2.$$

Now, the hyperbola ($\mathcal{H}$) which has foci at B and C and which passes through A, has for tangent at A the interior bisector AD of the angle formed by the radii vectors AB and AC, and the ellipse ($\mathcal{E}$) having foci at B and C and passing

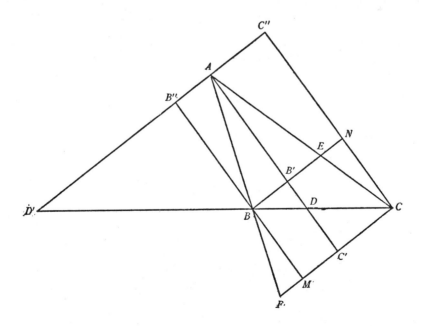

through A has for tangent at A the exterior bisector of angle A. Therefore*

(2) $$BB' \cdot CC' = \tfrac{1}{4}[a^2 - (b - c)^2]$$
$$= \tfrac{1}{4}(a + b - c)(a - b + c) = (p - c)(p - b),$$

(3) $$BB'' \cdot CC'' = \tfrac{1}{4}[(b + c)^2 - a^2]$$
$$= \tfrac{1}{4}(b + c + a)(b + c - a) = p(p - a),$$

whence, by virtue of (1),

$$(\text{area } ABC)^2 = p(p - a)(p - b)(p - c).$$

The consideration of the conics $(\mathcal{K})$ and $(\mathcal{E})$ remarkably simplifies the calculation of the products (2) and (3), which can also be obtained directly by the evaluation of the segments BB', BB'', CC', CC'' as functions of the sides a, b, c of triangle ABC.

FORMULA FOR THE AREA OF A TRIANGLE[†]

M. K. FORT, Jr., University of Virginia

We prove in Theorem 1 that a certain determinant is an invariant. Theorem 1 is then used to prove the well known formula

$$A = 1/2 \begin{vmatrix} x_1 & y_1 & 1 \\ x_2 & y_2 & 1 \\ x_3 & y_3 & 1 \end{vmatrix}$$

for the area of a triangle.

THEOREM 1. *If P_1, P_2, P_3 are points in a plane, and these points have coördinates (x_1, y_1), (x_2, y_2), (x_3, y_3) and (x_1', y_1'), (x_2', y_2'), (x_3', y_3') respectively in rectangular coördinate systems C and C' (which we shall assume to have the orientation commonly used in analytic geometry texts); then*

$$\begin{vmatrix} x_1 & y_1 & 1 \\ x_2 & y_2 & 1 \\ x_3 & y_3 & 1 \end{vmatrix} = \begin{vmatrix} x_1' & y_1' & 1 \\ x_2' & y_2' & 1 \\ x_3' & y_3' & 1 \end{vmatrix}.$$

---

* *The product of the focal perpendiculars on any tangent to a central conic, $x^2/\alpha^2 \pm y^2/\beta^2 = 1$, is constant and equal to β^2.* See almost any analytical geometry text, *e.g.*, Fine and Thompson, p. 85. A synthetic demonstration may be found in Macaulay's Geometrical Conics (2nd ed.), p. 112. (H. Eves)

[†] From AMERICAN MATHEMATICAL MONTHLY, vol. 54 (1947), pp. 337–339.

The transformation from the C' system to the C system is given by equations of the form

$$x = mx' - ny' + h$$
$$y = nx' + my' + k$$

where $m^2 + n^2 = 1$. Therefore

$$\begin{vmatrix} x_1 & y_1 & 1 \\ x_2 & y_2 & 1 \\ x_3 & y_3 & 1 \end{vmatrix} = \begin{vmatrix} mx_1' - ny_1' + h & nx_1' + my_1' + k & 1 \\ mx_2' - ny_2' + h & nx_2' + my_2' + k & 1 \\ mx_3' - ny_3' + h & nx_3' + my_3' + k & 1 \end{vmatrix}.$$

The determinant on the right side of the above equation can be simplified by subtracting the proper multiples of the last column from the first two columns. If we do this we get

$$\begin{vmatrix} x_1 & y_1 & 1 \\ x_2 & y_2 & 1 \\ x_3 & y_3 & 1 \end{vmatrix} = \begin{vmatrix} mx_1' - ny_1' & nx_1' + my_1' & 1 \\ mx_2' - ny_2' & nx_2' + my_2' & 1 \\ mx_3' - ny_3' & nx_3' + my_3' & 1 \end{vmatrix}.$$

The determinant on the right is equal to

$$(m^2 + n^2) \begin{vmatrix} x_1' & y_1' & 1 \\ x_2' & y_2' & 1 \\ x_3' & y_3' & 1 \end{vmatrix}.$$

Since $m^2 + n^2 = 1$, we see that

$$\begin{vmatrix} x_1 & y_1 & 1 \\ x_2 & y_2 & 1 \\ x_3 & y_3 & 1 \end{vmatrix} = \begin{vmatrix} x_1' & y_1' & 1 \\ x_2' & y_2' & 1 \\ x_3' & y_3' & 1 \end{vmatrix}.$$

THEOREM 2. *If P_1, P_2, P_3 are the vertices of a triangle and the cyclic order $P_1P_2P_3P_1$ induces a counter-clockwise orientation on the boundary of the triangle, then the area A of the triangle is given by*

$$A = 1/2 \begin{vmatrix} x_1 & y_1 & 1 \\ x_2 & y_2 & 1 \\ x_3 & y_3 & 1 \end{vmatrix}.$$

Choose a coördinate system C' so that P_1 is at the origin and P_2 is on the positive x'-axis. It follows from the fact that $P_1P_2P_3P_1$ induces a counter-clockwise orientation, that P_3 must be in either the first or second quadrant. For this choice of C' we now see that $x_1' = y_1' = y_2' = 0$, that x_2' is the length of the side P_1P_2, and that y_3' is the length of the altitude perpendicular to this side. Thus the area of the triangle satisfies $2A = x_2' y_3'$. We now apply Theorem 1 and obtain

$$\begin{vmatrix} x_1 & y_1 & 1 \\ x_2 & y_2 & 1 \\ x_3 & y_3 & 1 \end{vmatrix} = \begin{vmatrix} 0 & 0 & 1 \\ x_2' & 0 & 1 \\ x_3' & y_3' & 1 \end{vmatrix} = x_2' y_3' = 2A.$$

Therefore

$$A = 1/2 \begin{vmatrix} x_1 & y_1 & 1 \\ x_2 & y_2 & 1 \\ x_3 & y_3 & 1 \end{vmatrix}.$$

In a similar fashion we can prove:

THEOREM 3. *If P_1, P_2, P_3 are the vertices of a triangle and the cyclic order $P_1P_2P_3P_1$ induces a clockwise orientation on the boundary of the triangle, then the area A of the triangle is given by*

$$A = -1/2 \begin{vmatrix} x_1 & y_1 & 1 \\ x_2 & y_2 & 1 \\ x_3 & y_3 & 1 \end{vmatrix}.$$

THE METHOD OF ARCHIMEDES*†

S. H. GOULD, Williams College

The works of Archimedes have come down to us in two streams of tradition, one of them continuous, the other broken by a gap of a thousand years between the tenth century and the year 1906, when the discovery of a manuscript in Constantinople brought to light an important work called the *Method*, on the subject of integration.

Newton and his contemporaries in the seventeenth century were much puzzled by one aspect of the integrations to be found in the continuous tradition. In the books on the *Sphere and Cylinder*, for example, it is clear that the somewhat complicated method employed there for finding the volume of a sphere represents merely a rigorous proof of the correctness of the result and gives no indication how Archimedes was led to it originally. The discovery of 1906 removes the veil, at least to some extent.

The newly discovered *Method* consists of imagining the desired volume as cut up into a very large number of thin parallel slices or discs, which are then suspended at one end of an imaginary lever in such a way that they are in equilibrium with a solid whose volume and center of gravity are known. Thus, in

---

* From AMERICAN MATHEMATICAL MONTHLY, vol. 62 (1955), pp. 473–476.

† An address to the Mathematical Association of America at the 1953 Summer Meeting in Kingston, Ontario, Canada.

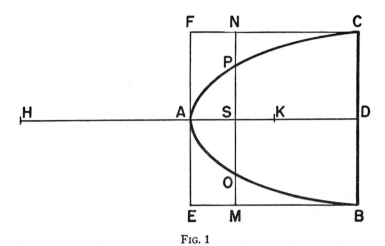

Fig. 1

Proposition 4 of the *Method*, Archimedes shows that the volume of a paraboloid of revolution is one-half of the volume of the circumscribing cylinder by slicing the two solids (see Figure 1 which represents a plane section through their common axis AD) at right angles to AD. For let us take HAD to be the bar of a balance with $HA = AD$ and with the fulcrum at A, and imagine the circle PO to be removed from the paraboloid and suspended at H. Since $AD/AS = DB^2/SO^2$ in the parabola BAC, we have

$$\frac{HA}{AS} = \frac{AD}{AS} = \frac{MS^2}{SO^2} = \frac{\text{(circle in cylinder)}}{\text{(circle in paraboloid)}},$$

so that, by the law of the lever, the circle in the cylinder, remaining where it is, is in equilibrium with the circle from the paraboloid resting in its new position. If we deal in the same way with all the circles making up the paraboloid, we find that the cylinder, resting where it is with its center of gravity at the midpoint K of AD, is in equilibrium about A with the paraboloid placed with its center of gravity at H. Since $HA = AD = 2AK$, the volume of the paraboloid is therefore one-half of that of the cylinder, as desired.

Many accounts of the *Method* have been given since its discovery in 1906; for example, by T. L. Heath in his *Supplement to the Works of Archimedes*, Cambridge, 1912. In all of them, as in the original work of Archimedes himself, we are invited to *imagine* the lever and the objects suspended from it. But if we construct an *actual* lever and *actual* discs, the various figures, which may be spheres, cones, *etc.*, see below, will be observed to balance, slice by slice, as successive slices are added. The whole procedure then becomes a picturesque and effective illustration of the concept of an integral as the limit of a sum.

To find the volume of a sphere, a problem which Archimedes considered so important that he asked to have the result engraved on his tombstone, a cone and a sphere are together weighed against a cylinder (see Figure 2 and the accompanying sketch). Here the circle NM, resting where it is in the large

462 AREA AND VOLUME

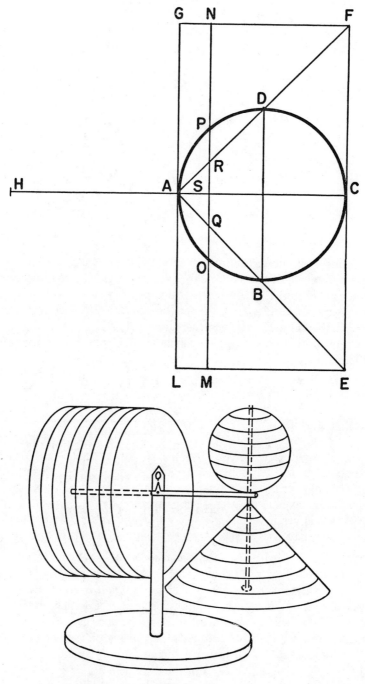

Fig. 2

cylinder $GLEF$ is in equilibrium about A with two circles placed at H, the one circle PO being taken from the given sphere and the other RQ from the cone FAE. For we have

$$OS^2 + QS^2 = OS^2 + AS^2 = AO^2 = CA \cdot AS = MS \cdot SQ$$

and therefore

$$\frac{HA}{AS} = \frac{MS}{SQ} = \frac{MS^2}{MS \cdot SQ} = \frac{MS^2}{OS^2 + QS^2}.$$

Thus, by the law of the lever as before,

one-half of cylinder equals cone plus sphere

from which, since the cone is one-third of the cylinder,

sphere equals one-sixth cylinder.

Thus the cylinder circumscribed about the sphere, being one-quarter as great as the large cylinder $GLEF$, is three-halves as great as the sphere, which is the result stated on the tombstone of Archimedes.

If squares are substituted for the circles of cross-section in these figures, the argument remains unchanged and we have the solution of another famous problem (Proposition 15 in the *Method*), namely to find the volume common to two right circular cylinders intersecting at right angles.

The actual models were constructed by D. A. Eberle of the Psychology Workshop at Purdue University. The various slices were cut from a piece of white pine 1/2″ thick and 7″ wide. Thus the cylinder $GLEF$ is composed of seven slices, each with a diameter of 7″. The seven slices for the cone, being first cut as stepwise increasing cylindrical discs with easily calculated radii, were placed all together on a mandrel passing through a 3/16″ hole through their centers and were then shaped down on a lathe, a procedure found to be especially necessary for the square cross-sections in the problem of the intersecting cylinders. The lever itself is a piece of steel 9″ by 1/2″ by 1/32″, placed so that its 1/2″ face is vertical. In each disc a thin slit was cut with a fine hacksaw from edge to center so that the disc could be slipped onto the lever.

A SIMPLER PROOF OF HERON'S FORMULA*

CLAUDE H. RAIFAIZEN, M.I.T.

A few years ago while attempting a proof of Heron's formula for my own amusement, I discovered one which, compared to Heron's complicated geometrical proof and to the trigonometric proofs for it, is quite simple.

In any triangle, for at least one of the vertices, the perpendicular from that vertex to the line containing the opposite side intersects that side. In what follows we will assume that in triangle ABC one such vertex is C. (See Figure 1.)

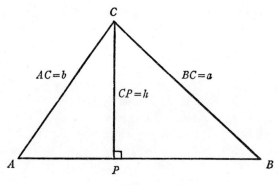

FIG. 1.

(1) Letting $CP=h$ we see that by the Pythagorean formula $AP=\sqrt{b^2-h^2}$, $PB=\sqrt{a^2-h^2}$, and hence

$$c = AB = AP + PB = \sqrt{b^2 - h^2} + \sqrt{a^2 - h^2}$$

(2) $\sqrt{a^2-h^2} = c - \sqrt{b^2-h^2}$; $\quad a^2-h^2 = c^2 - 2c\sqrt{b^2-h^2} + b^2 - h^2$;

$a^2 - b^2 - c^2 = -2c\sqrt{b^2-h^2}$; $(a^2-b^2-c^2)^2 = 4c^2(b^2-h^2) = 4b^2c^2 - 4c^2h^2$.

(3) Observing that $ch=2A$, where A is the area, we have $(a^2-b^2-c^2)^2 = 4b^2c^2 - 16A^2$.

(4) Then

$$\begin{aligned}
16A^2 &= 4b^2c^2 - (a^2 - b^2 - c^2)^2 \\
&= (2bc - a^2 + b^2 + c^2)(2bc + a^2 - b^2 - c^2) \\
&= [(b^2 + 2bc + c^2) - a^2][a^2 - (b^2 - 2bc + c^2)] \\
&= [(b + c)^2 - a^2][a^2 - (b - c)^2] \\
&= (b + c + a)(b + c - a)(a - b + c)(a + b - c) \\
&= (a + b + c)(a + b + c - 2a)(a + b + c - 2b)(a + b + c - 2c).
\end{aligned}$$

---

* From Mathematics Magazine, vol. 44 (1971), pp. 27–28

(5) Letting $a+b+c=2s$ we get $16A^2=(2s)(2s-2a)(2s-2b)(2s-2c)$ and
$$A^2 = s(s-a)(s-b)(s-c).$$

BIBLIOGRAPHIC ENTRY: AREA AND VOLUME

1. R. E. Mortiz, On the generalization of a theorem in solid geometry, AMERICAN MATHEMATICAL MONTHLY, vol. 15, p. 95.

>The volume of a truncated right triangular prism is the product of the area of its base by one-third the sum of the lateral edges. This is generalized to regular prisms with n faces using trigonometric identities.

AUTHOR INDEX

Numbers in italic type refer to bibliographic entries.

Abeles, F., *299*, 339
Abeles, Francine, 339
Adler, Claire, 276
Akerberg, Bengt, 179
Albert, A. A., 275
Albert, R. G., 164
Allen, E. F., *385*
Allen, E. S., 357
Allison, David, 176
Amir-Moéz, A. R., 211
Andrews, G. E., 81
Anning, N., *282*
Anning, Norman, 371
Apostol, T. M., *341*

Bailey, H. W., *371*
Baker, R. P., 441
Ballantine, J. P., 265, 266, *282* (*twice*), 409
Bankoff, L., *299, 342*
Barrow, D. F., *282*
Bartlow, T. L., *260*
Beaman, E., *282*
Beckenbach, E. F., *105*
Bell, E. T., 159
Bellman, Richard, 74
Bender, H. A., *218*
Bettinger, A. K., 201
Bingley, G. A., *129*
Bird, M. T., 368
Bleick, W. E., 293
Boas, R. P., Jr., 65, *282*
Bourbaki, N., *71*
Bourne, Samuel, 227
Boyer, C. B., *105, 418*
Bradley, A. D., *218*
Bradley, H. C., 302
Brauer, Alfred, 60
Brickman, Louis, 222
Brink, R. W., 1
Brown, T. A., *260, 418*
Buck, R. C., 165
Burns, J. C., *299*
Burton, L. J., 198, *218*
Bussey, W. H., *174, 251*

Cairns, S. S., *341*
Cajori, Florian, 283
Carmichael, R. D., 184
Cell, J. W., 454
Chakerian, G. D., *341*
Chu, T. S., 199
Coar, H. L., 438

Coburn, R. K., *183*
Coe, C. J., 226
Coffman, R. T., 393, 395
Conrad, S. R., *342*
Cooke, W. P., 67, *183*
Coolidge, J. L., *437, 452*
Court, N. A., 303
Coxeter, H. S. M., *341*
Craig, H. V., 209
Cronwall, T. H., *282*

Dancer, Wayne, 190
Dean, G. R. 261, *371*
Dederick, L. S., 220
DeMar, R. F., *341*
Dickson, L. E., *218*, 247, 442
Dorwart, H. L., 195, *225*
Dowling, R. J., 114
Dubisch, R., *105*
Duncan, D. C., 361
Dunkel, Otto, 285
Dwyer, D. T., 232

Eagle, E. L., 200
Easton, Joy B., 382
Ehrmann, M. C., *353*
Eisenman, R. L., 417
Engelhardt, J. O., 289
Ercolano, J. L., 350
Ericksen, J. L., 388
Eves, H., *341*

Flanders, H., *178*
Ford, L. R., 273
Fort, M. K., Jr., 231, 458
Fort, Tomlinson, 138
Franklin, P., *418*
Fuchs, L., 123
Fulton, C. M., 203, *225*
Frame, J. S., *218*
Franklin, P., *105*

Gandz, S., *260*
Garfunkel, J., *342*
Garver, Raymond, 253
Gaughan, E. D., 127
Gauntt, Robert, 109
Gehman, H. M., 270
Gibbens, G., *353*
Gilbert, G., 294

AUTHOR INDEX

Goldberg, M., *341, 342*
Goldberg, Michael, 331
Golomb, Michael, 204
Golomb, S. W., 257
Goodman, A. W., *260*
Gould, S. H., 460
Graustein, W. C., *282*
Grecos, A. P., *260*
Green, T. M., 91
Greenstein, D. S., 174
Greenwood, G. W., 405
Greenwood, R. E., Jr., *218*
Grossman, H., *341*
Grossman, H. D., *298*

Halfar, Edwin, 108
Halmos, P. R., *72*
Hammer, P. C., 428
Hausmann, B. A., *225*
Hawthorne, F., 370
Hazard, W. J., *298*
Heaton, H., *282*
Hedberg, E. A., *218*
Hellman, M. J., 278, 281
Hemminger, R. L., 80
Hempel, C. G., 34
Henderson, A., *282*
Henriquez, Garcia, 269
Hildebrandt, T. H., 15
Hoffman, L., *341*
Hohn, F. E., *225*
Holladay, J. C., *105*
Hood, R. T., *114*
Householder, A. S., 194
Huff, G. B., *282*, 380
Hurwitz, W. A., 161

Ingraham, A., *298*
Isaacs, Rufus, 352

Jacobson, R. A., *341*
Jacobson, W. I., *341*
James, G., *282*
Jerbert, A. R., 409
Johns, A. E., 364
Johnson, P. B., *353*
Johnson, R. A., *341*
Johnsonbaugh, R. F., 107
Johnston, L. S., 255, 362, 372, 426, *437*
Jonah, H. F. S., 28
Jones, J. P., 110
Jones, R. T., *299*
Juredini, G. M., 424

Karpinski, L. C., *282*
Kasner, E., *341*
Kattsoff, L. O., *225*

Kazarinoff, N. D., 250, 313
Keller, M. W., 28
Kempner, Aubrey, 420
Kennedy, E. C., 267
Kershner, R. B., 214
Kirchner, R. B., 311
Klamkin, M. S., 233, *260*, 277, *282*, *341*, 435
Klee, V. L., Jr., 173
Knebelman, M. S., 307
Knuth, D. E., *105*
Konhauser, J. D. E., *437*
Kravitz, Sidney, 325

Lacey, O. L., *341*
Lambert, W. D., 386
Landis, W. W., 354
Lange, L. H., 180
Lange, L. J., *114*
Lariviere, R., *418*
Larsen, H. D., 69
Leader, S., 412
Lehmer, D. H., 218
LeVan, M. O., *105*
Levy, H., *72*
Lewin, M., *299*
Longley, W. R., 444
Lubin, C., *298*
Lunn, A. C., 356
Lyon, Richard, 106

MacDonald, I. D., 76
MacDonnell, D., 294
MacDuffee, C. C., *71*
Maier, E. A., 109, 111
May, K. O., 64, 89
May, Kenneth, 408, 412
McEwen, W. R., 385
McShane, E. J., 192
Milenkovic, V., *299*
Mills, C. N., *418*
Moise, E. E., *72*
Moore, D. H., *219*
Moran, D. A., 349
Morley, F. V., *299*
Mortiz, R. E., *465*
Moser, Leo, 173
Mott, T. E., 414
Mulcrone, T. F., *437*
Mullin, R. C., 175
Musselman, J. R., *385*

Neidhard, G. L., *299*
Newman, D. J., 178, *260*
Nielsen, K. L., *71*
Niven, I., *341*
Niven, Ivan, 111
Nowlan, F. S., 45
Nymann, J. E., 258

AUTHOR INDEX

Ogilvy, C. S., *225, 353,* 393, *437*
Oglesby, E. J., 263
Ore, O., *72*

Paradiso, L. J., *218,* 251
Parker, E. T., *72*
Pascual, Michael, 381
Pavlick, F., *342*
Pedoe, D., *385*
Pedoe, Daniel, 206, 295
Pennisi, L. L., *114*
Peterson, B. B., *299*
Piranian, G., *72*
Polya, G., *71, 174*
Polya, George, 290
Porges, A., *385, 418*
Putnam, T. M., 391

Rabson, Gustave, 109
Raifaizen, C. H., 464
Rainich, G. Y., 226
Ransom, W. R., *114, 218,* 389, 413
Richardson, Moses, 20, 54
Richmond, D. E., 117
Robinson, L. V., 73
Robinson, R. M., *282*
Rosenbaum, R. A., 256
Ruddick, C. T., *404*

Sallee, G. T., *341*
Sandham, H. F., *341*
Sastry, K. R. S., *299*
Schaaf, W. L., 71
Schaumberger, Norman, 210
Scheid, F., *418*
Schelkunoff, S. A., 115
Schorling, R., *71*
Scott, Carol S., 408
Scudder, H. R., 342
Shaer, J., *342*
Shisha, O., *353*
Sholander, M., *105*
Silver, A. L. L., 83
Smail, L. L., *129*
Smiley, M. F., 128
Spickerman, W. R., *299*
Spiegel, M. R., *282*
Spitznagel, E. L., Jr., 100
Stark, M. E., *251*
Stein, F. M., 432
Stein, R. G., 177
Steinberg, Leon, 222

Stoll, R. R., 410
Stouffer, E. B., *260*
Stratton, W. T., *437*
Subbarao, M. V., 110
Szele, T., 123

Taussky, O., *260*
Thébault, Victor, 456
Thielman, H. P., 61, *71*
Thomas, J. M., *218*
Thompson, S. L., *218*
Thornton, H. B., 360
Toporowski, S., 110
Trigg, C. W., *299*

Usiskin, Z., *105, 299*
Uspensky, J. V., 185

Venkatachaliengar, K., *299*

Wagner, R. W., 196, 429
Wallace, A. D., 44
Ward, M., *174*
Ward, Morgan, 106
Wayment, S. G., *299*
Weaver, J. H., *385*
Webster, R. J., 297
Weil, A., *71*
Weitzenkamp, Roger, 313
Westlund, Jacob, 219
Weyers, D. V., *114*
Whitford, D. E., 277, *282*
Whitman, E. A., *437*
Wilansky, A., *174*
Williams, C. W., *341*
Williams, K. P., 450
Williams, W. L. G., 139
Williamson, J., *260*
Wilson, J. P., 418
Winger, R. M., 130, 146, *437*
Wood, F. E., 198
Woods, Roscoe, 189
Wrench, J. W., Jr., *218*
Wunderlich, M., 80
Wylie, C. R., Jr., 343

Yates, R. C., 373, 375, *385*
Yocom, K. L., *341*
Young, F. H., 208, 378, 387, 388
Young, J. W. A., 151

Zuckerman, H. S., *341*